W9-BGL-225

ANNUAL REVIEW
OF ENERGY

EDITORIAL COMMITTEE (1985)

HARVEY BROOKS
JOEL DARMSTADTER
JACK M. HOLLANDER
JAMES HOWE
S. S. PENNER
DAVID STERNLIGHT
IRVING WENDER

Responsible for the organization of Volume 10
(Editorial Committee, 1983)

HARVEY BROOKS
JOEL DARMSTADTER
JACK M. HOLLANDER
JAMES HOWE
DAVID STERNLIGHT
IRVING WENDER
MELVIN K. SIMMONS (Guest)

Production Editor   WENDY A. CAMPBELL
Indexing Coordinator   MARY A. GLASS
Subject Indexer   STEVEN SORENSEN

TJ
163.2
.A55
Vol. 10

ANNUAL REVIEW OF ENERGY

VOLUME 10, 1985

JACK M. HOLLANDER, *Editor*
The Ohio State University

HARVEY BROOKS, *Associate Editor*
Harvard University

DAVID STERNLIGHT, *Associate Editor*
Pasadena, California

ANNUAL REVIEWS INC.   4139 EL CAMINO WAY   PALO ALTO, CALIFORNIA 94306 USA

GOSHEN COLLEGE LIBRARY
GOSHEN, INDIANA

ANNUAL REVIEWS INC.
Palo Alto, California, USA

COPYRIGHT © 1985 BY ANNUAL REVIEWS INC., PALO ALTO, CALIFORNIA, USA. ALL RIGHTS RESERVED. The appearance of the code at the bottom of the first page of an article in this serial indicates the copyright owner's consent that copies of the article may be made for personal or internal use, or for the personal or internal use of specific clients. This consent is given on the condition, however, that the copier pay the stated per-copy fee of $2.00 per article through the Copyright Clearance Center, Inc. (21 Congress Street, Salem, MA 01970) for copying beyond that permitted by Sections 107 or 108 of the US Copyright Law. The per-copy fee of $2.00 per article also applies to the copying under the stated conditions, of articles published in any *Annual Review* serial before January 1, 1978. Individual readers, and nonprofit libraries acting for them, are permitted to make a single copy of an article without charge for use in research or teaching. This consent does not extend to other kinds of copying, such as copying for general distribution, for advertising or promotional purposes, for creating new collective works, or for resale. For such uses, written permission is required. Write to Permissions Dept., Annual Reviews Inc., 4139 El Camino Way, Palo Alto, CA 94306 USA.

International Standard Serial Number: 0362-1626
International Standard Book Number: 0-8243-2310-6

Annual Review and publication titles are registered trademarks of Annual Reviews Inc.

Annual Reviews Inc. and the Editors of its publications assume no responsibility for the statements expressed by the contributors to this *Review*.

TYPESET BY AUP TYPESETTERS (GLASGOW) LTD., SCOTLAND
PRINTED AND BOUND IN THE UNITED STATES OF AMERICA

PREFACE

We have come a long way in the twelve years since the first oil embargo brought to our attention the precariousness of the world's energy situation. Our data and our analytic tools have grown steadily more accurate and precise. We have gained a far better understanding of the economics of energy, thanks to the unplanned experiments prompted by the behavior of oil prices over this period. We have mounted and sustained a number of important long-range research and development programs.

All energy issues, of course, are not yet settled. Foreign exchange constraints on many countries are still severe, and they still limit energy options. The future of nuclear fission power remains clouded, in spite of the existence of numerous and well-defined technical solutions. The assessment of energy-related risks still provokes debate over fundamental questions of analysis.

Yet this 10th volume of the *Annual Review of Energy* reflects a certain amount of leisure for calm assessment of the possibilities. Developments in the oil markets, the advance of research and development programs, and more confident assessments of the technologies and economics of energy give us the time and the tools for strategic thought. We have the leisure to speculate on the origin of natural gas, the future shape of nuclear power systems, and the potential for long-range international cooperation in energy research and development. We have the leisure to review a decade of successes and failures in United States energy policy. We have the leisure to consider energy risks calmly. For the near future, at least, time may be on our side.

Our expanded knowledge of energy issues should help us use this relative leisure productively. Information gathered patiently over the past dozen years has brought to light many new patterns and possibilities. For example, we can assess with some confidence energy use in the Third World, a task that would have been impossible only a few years ago. Brazil's experience with ethanol fuels provides the benefit of a technical and economic experiment that could have been carried out perhaps nowhere else. The accident at the Three Mile Island nuclear facility has given us (along with a warning) a treasury of information on the accident behavior of reactor cores and power plants. We have the data and the analytic apparatus to make sense of patterns in industrial and residential energy use, of the performance of passive solar design in buildings, and of the potential of new photovoltaic technologies.

There is still far to go, but we have attained a striking maturity in our thinking about energy since the publication of Volume 1 of this series in 1976. The challenge now is to sustain this momentum, using our new tools and our current leisure wisely, without complacency, for the benefit of people everywhere.

CONTENTS

viii   CONTENTS *(continued)*

ANNUAL REVIEWS INC. is a nonprofit scientific publisher established to promote the advancement of the sciences. Beginning in 1932 with the *Annual Review of Biochemistry*, the Company has pursued as its principal function the publication of high quality, reasonably priced *Annual Review* volumes. The volumes are organized by Editors and Editorial Committees who invite qualified authors to contribute critical articles reviewing significant developments within each major discipline. The Editor-in-Chief invites those interested in serving as future Editorial Committee members to communicate directly with him. Annual Reviews Inc. is administered by a Board of Directors, whose members serve without compensation.

1985 Board of Directors, Annual Reviews Inc.

Dr. J. Murray Luck, Founder and Director Emeritus of Annual Reviews Inc.
   Professor Emeritus of Chemistry, Stanford University
Dr. Joshua Lederberg, President of Annual Reviews Inc.
   President, The Rockefeller University
Dr. James E. Howell, Vice President of Annual Reviews Inc.
   Professor of Economics, Stanford University
Dr. William O. Baker, *Retired Chairman of the Board, Bell Laboratories*
Dr. Winslow R. Briggs, *Director, Carnegie Institution of Washington, Stanford*
Dr. Sidney D. Drell, *Deputy Director, Stanford Linear Accelerator Center*
Dr. Eugene Garfield, *President, Institute for Scientific Information*
Dr. Conyers Herring, *Professor of Applied Physics, Stanford University*
Mr. William Kaufmann, *President, William Kaufmann, Inc.*
Dr. D. E. Koshland, Jr., *Professor of Biochemistry, University of California, Berkeley*
Dr. Gardner Lindzey, *Director, Center for Advanced Study in the Behavioral Sciences, Stanford*
Dr. William D. McElroy, *Professor of Biology, University of California, San Diego*
Dr. William F. Miller, *President, SRI International*
Dr. Esmond E. Snell, *Professor of Microbiology and Chemistry, University of Texas, Austin*
Dr. Harriet A. Zuckerman, *Professor of Sociology, Columbia University*

Management of Annual Reviews Inc.

John S. McNeil, Publisher and Secretary-Treasurer
William Kaufmann, Editor-in-Chief
Mickey G. Hamilton, Promotion Manager
Donald S. Svedeman, Business Manager
Richard L. Burke, Production Manager

ANNUAL REVIEWS OF
Anthropology
Astronomy and Astrophysics
Biochemistry
Biophysics and Biophysical Chemistry
Cell Biology
Earth and Planetary Sciences
Ecology and Systematics
Energy
Entomology
Fluid Mechanics
Genetics
Immunology
Materials Science

Medicine
Microbiology
Neuroscience
Nuclear and Particle Science
Nutrition
Pharmacology and Toxicology
Physical Chemistry
Physiology
Phytopathology
Plant Physiology
Psychology
Public Health
Sociology

SPECIAL PUBLICATIONS
Annual Reviews Reprints:
   Cell Membranes, 1975–1977
   Cell Membranes, 1978–1980
   Immunology, 1977–1979

Excitement and Fascination
   of Science, Vols. 1 and 2

History of Entomology

Intelligence and Affectivity,
   by Jean Piaget

Telescopes for the 1980s

A detachable order form/envelope is bound into the back of this volume.

Ann. Rev. Energy. 1985. 10 : 1–34
Copyright © 1985 by Annual Reviews Inc. All rights reserved

AMORPHOUS SILICON ALLOYS FOR SOLAR CELLS

K. W. Mitchell

ARCO Solar, Inc., P.O. Box 2105, Chatsworth, California 91313

K. J. Touryan

Mt. Carmel Trust/Flow Industries, 6200 Plateau Drive,
Englewood, Colorado 80111

1. INTRODUCTION

1.1 Scope of Article

Although photovoltaic conversion of sunlight into electricity is an old technology, only recently have intense efforts been made to reduce the cost of conversion and thereby create an economically competitive source of electricity. Interest in thin-film photovoltaic technologies has increased dramatically in industry and government, with special attention focused on amorphous silicon. One important reason is that hydrogenated amorphous silicon materials (expressed by the abbreviated notation a-Si : H) are stronger absorbers of solar rays than conventional crystalline silicon materials now available on the market. Thus, photovoltaic devices that use amorphous silicon require less material than devices using crystalline silicon.

This article presents a critical review of recent progress in developing materials and processes for amorphous silicon solar cells. The review attempts to balance technical and nontechnical information.

The review is divided into seven sections. Section 1 provides a general overview of the subject and puts it in historical perspective. Sections 2–5 (Mitchell) give technical details on the properties of amorphous silicon, common fabrication processes, device structures, and device characteristics. Sections 6 and 7 (Touryan) are more general and provide information on research and development efforts around the world. Section 7 concludes the review and makes some predictions for the future.

1

0362–1626/85/1022–0001$02.00

1.2   Perspectives on Amorphous Silicon

The investigation of the structural, chemical, electronic, and optical properties of amorphous Si alloys and their application as electro-optic elements has received much attention, especially since 1976. Mahan & Stone of the Solar Energy Research Institute (SERI) published an amorphous Si bibliography consisting of 2246 references covering the period through 1981 (1). Several conferences and reviews have also been devoted to the subject (2–10).

Important advantages of hydrogenated amorphous silicon photovoltaic technology include : (a) a well developed feedstock industry and technology base; (b) deposition techniques adaptable to the fabrication of large-area solar modules; (c) the opportunity for relatively low temperature (less than 300°C) fabrication processes; (d) reduced material requirements compared to those of crystalline Si, because layers 1 micrometer or less absorb sunlight strongly; and (e) the adaptability of the materials to the technique of stacking layers of solar cells, in which each layer is highly responsive to a different color in the solar spectrum, so that more electricity is produced per unit area.

Several major aspects of a-Si : H materials require further development and optimization: (a) a poor minority-carrier lifetime requires drift field collection, (b) cell performance can degrade with illumination, (c) materials cannot be heavily doped, and (d) poor-quality alloys presently inhibit tandem cell development (see Section 4.3).

1.3   Historical Review

Hydrogenated amorphous silicon was first investigated by Sterling et al (11) in 1955, but the role of hydrogen was not emphasized until about 1975 (12, 13). Nonhydrogenated amorphous Si contains many dangling or broken bonds that create a large density of localized states throughout the energy gap of the semiconductor. Hydrogen can terminate these dangling bonds and remove the localized states from the energy gap. Consequently, a-Si : H exhibits many properties not evident in nonhydrogenous amorphous Si.

Chittick et al (13a) observed a large photoconductive effect in their films, whereas the photoconductivity of sputtered or evaporated a-Si films is negligible. In 1972, Spear & LeComber (14) demonstrated that the Fermi level in a-Si : H could be moved by the electric field generated in a MOS transistorlike structure. In a-Si, the density of gap states is so large that the Fermi level is pinned near midgap. In 1974, Engemann & Fisher (15) observed relatively efficient photoluminescence in a-Si : H at a temperature of 77 K; photoluminescence is not measurable in a-Si because the nonradiative recombination lifetime is very short.

Electronic devices using a-Si : H were made in 1968 at the Standard

Telecommunications Laboratories in Harlow. In 1975, Spear & LeComber (16) published a detailed study of the substitutional doping of a-Si : H. Prior to that time, it was thought that one could not dope amorphous semiconductors because the amorphous material would locally arrange itself to satisfy the normal valence state of the dopant atom. The first published work on a-Si : H devices appeared in 1976 (17, 18), and since then interest in amorphous silicon alloys has increased. Considering the recent evolution of this technology, it is remarkable that in 1983 50,000 m² of amorphous silicon photovoltaic devices were sold, out of a total of 200,000 m² of photovoltaic modules produced in the world. Almost all amorphous silicon cells were fabricated in Japan for low-efficiency consumer electronics applications, such as pocket calculators.

The historical growth of solar cell efficiency for small and large areas is shown in Figure 1. The most rapid progress in cell efficiency has occurred since 1978. For small-area (~ 1 cm²) solar cells used in research laboratories, the cell efficiency was 4% in 1978, reached 10.1% in July 1982, and is presently reported to be 11.5%. These achievements are for glow-discharge-deposited amorphous Si p-i-n solar cells. For larger solar cells, the p-i-n Si : H alloy cell efficiency was 2% in 1979 for an area of 50 cm² and is now over 9% for a gross area of 790 cm² (see Section 4 below).

Strong private activity in the United States has resulted in commercially available amorphous silicon photovoltaic modules, and with advances in the technology the cost of electricity from amorphous silicon will continue to decline. The goal is to be competitive with traditional bulk power generators for electric power.

2. PROPERTIES OF AMORPHOUS SILICON

The design and performance of amorphous Si solar cells are determined by the underlying physiochemical and electro-optic nature of the materials. The following sections briefly describe the properties of amorphous Si that are particularly relevant to solar cells.

2.1 *Optical Absorption Properties*

An excellent review of the optical properties of amorphous silicon hydride is provided by Cody (20). The optical absorption coefficient α of a semiconductor determines the fraction of absorbed light, $A(E)$, where E is the photon energy. For a specular sample of thickness W, neglecting multiple reflection effects, $A(E)$ is given by:

$$A(E) = 1 - \exp(-\alpha(E)W).$$

As shown in Figure 2, the absorption coefficient of a-Si : H is greater than that of crystalline Si above about 1.7 eV photon energy, although less than

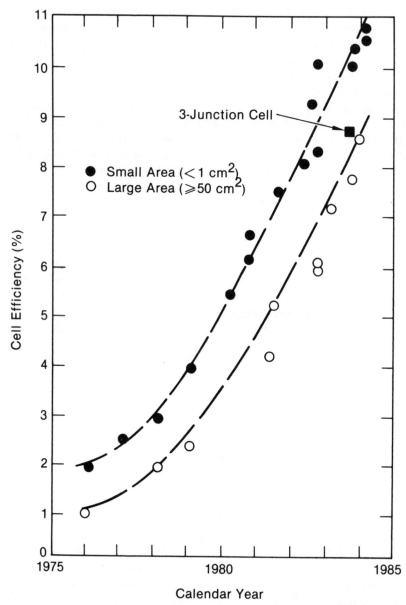

Figure 1   Progress in improving the efficiency of amorphous Si p-i-n solar cells prepared by glow discharge (19).

that of a direct gap semiconductor such as GaAs (10). The optical absorption coefficients of a-Si can be approximated by (21, 22):

$$\alpha(1/\mu m) = 47\,(E - E_g)^2/E$$

for absorption coefficients greater than 10^4 cm^{-1}, where E_g is the band gap

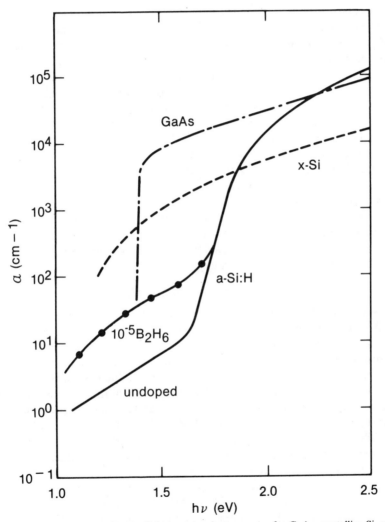

Figure 2 Optical absorption coefficient versus photon energy for GaAs, crystalline Si, and undoped and lightly boron doped (10^{-5} B$_2$H$_6$ concentration in the gas phase) amorphous Si (10, 20).

energy. Near the band edge, an additional term associated with the band-tailing states must be included:

$$\alpha(1/\mu m) = \alpha_0 \exp[(E - E_i)/E_0],$$

where E_0 is typically on the order of 50–79 meV and α_0 and E_i are empirical parameters.

The number of photons absorbed also depends on the nature of sunlight. The SERI AM 1.5 Global spectrum has achieved acceptance in the photovoltaic community (23). Based on this spectrum, assuming 100 mW/cm² intensity and the absorption coefficients given above, the maximum short-circuit currents for layers of 1.4 eV, 1.7 eV, and 2.0 eV band gap cannot exceed about 33 mA/cm², 23 mA/cm², and 15 mA/cm² respectively. The amount of current generated for each of these band gaps as a function of optical path length is given in Figure 3. As shown, most of the current is generated within one micron of thickness.

Several factors influence the shape of the a-Si:H absorption curve, such as deposition conditions and hydrogen content (20). In addition, introducing dopant gases during the a-Si:H film growth, such as phosphine for donor-doping or diborane for acceptor-doping, increases the optical absorption near the band edge, as shown in Figure 2. These relatively large infrared absorption tails are attributed primarily to creation of dangling bonds (24). Finally, alloys can be formed to change the band gap, such as a-SiGe (25–29) and a-SiSn (30–33) to lower the band gap or a-SiC (33–38) and a-SiN (33–39) to increase the band gap.

2.2   Transport of Photogenerated Carriers

In solar cells, the transport of photogenerated carriers occurs either by diffusion in the absence of an electric field or by drift where collection is assisted by an electric field. The principal parameter defining carrier transport is the product of the mobility and lifetime of the excess carriers.

The disordered nature of amorphous semiconductors leads to a high density of states within the forbidden energy gap, in the form of band tail states and dangling bonds. Both the mobility and lifetime of free carriers are substantially reduced compared with those in crystalline materials. Drift mobility measurements give values of about 1 and 10^{-3} cm²/V-sec at room temperature for electrons and holes with extended states values of 13 and 0.67 cm²/V-sec respectively (40). Using the junction recovery method, the lifetimes of excess carriers in a-Si were found to be 10–30 μsec for injection current densities of about 10 mA/cm² (41). The method of layer formation, the substrate material, and incorporated impurities all affect the resultant carrier transport properties. Development of new measurement techniques

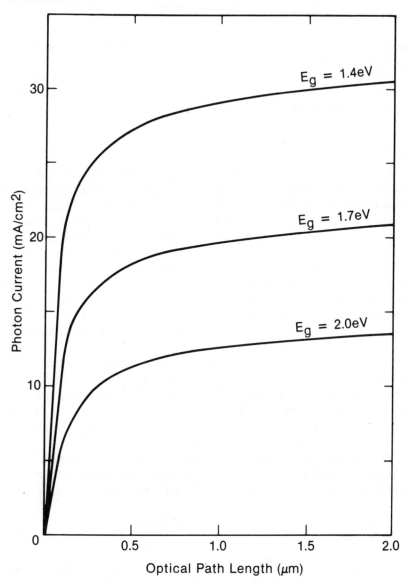

Figure 3   Total photon current versus optical path length for amorphous Si band gaps of 1.4, 1.7, and 2.0 eV.

and modeling of the carrier transport mechanisms in amorphous materials are receiving much attention in the literature (2–9).

For amorphous Si, the principal carrier transport mechanism is by drift collection under the influence of a large electric field. A significant consequence of drift collection is that collection of the photogenerated current becomes voltage-dependent, influencing the fill factor of the cell as discussed in Section 5.3. As a result also, most of the highest efficiency amorphous Si solar cells are based on p-i-n devices, in which most of the light is absorbed in the i-layer, where it can be collected under the influence of an electric field.

2.3 *Electronic Dopants*

Demonstration of substitutional doping of a-Si by Spear & LeComber (16) established that hydrogenation of amorphous silicon reduced the density of states in the energy gap, enabling the Fermi level, E_f, to be shifted by more than 1 eV by the addition of Column I, III, or V elements from the periodic table. Phosphine and diborane are the most commonly used gases for donor- and acceptor-doping respectively. From conductivity and thermo-power measurements, E_f in boron-doped a-Si films is about 0.5 eV above the valence band (42), while in phosphorus-doped a-Si films, E_f is about 0.2 eV below the conduction band (43). For a p-i-n solar cell, the open-circuit voltage V_{oc} is limited by the built-in potential V_{bi}, which is determined by the relative positions of the Fermi levels in the p and n layers with respect to the local vacuum level.

Doping a-Si layers appears to cause a substantial amount of defect formation, evidenced by the increase in the infrared absorption tail (see Figure 2) and by luminescence, electron spin resonance (ESR), and deep level transient spectroscopy (DLTS) data as shown in Figure 4. As a result, the fraction of electrically active dopant atoms, as shown in Figure 5, is small and decreases with increasing dopant concentration (24).

Alternative doping strategies include the use of a-Si alloys with larger band gaps, especially a-SiC : H (33–38), and the formation of microcrystal-line doped layers (44, 45) to enhance the electrical conductivity of the films.

3. AMORPHOUS SILICON DEPOSITION PROCESSES

A variety of techniques have been used to deposit amorphous Si alloys, including glow-discharge (7, 10, 19, 46), sputtering (47), chemical vapor deposition (48), and photochemical vapor deposition (49). The most prevalent deposition approach is RF glow-discharge using pure silane (SiH_4). An RF glow-discharge system is illustrated in Figure 6. The substrates are heated to the range of 200–400°C and can be placed on either or both electrodes. The systems usually operate at 13.56 MHz, with power densities between 0.1 and 2.0 W/cm^2 and system pressures between 5 and 250 mtorr. Silane flow rates of 30 standard cm^3 per min result in film growth rates of 1–15 Å/sec, depending on the RF power, background pressure, and substrate temperature. Other gas sources, such as SiF$_4$ (49a) or mixtures of SiH$_4$ in H$_2$, He, or Ar, have also been used.

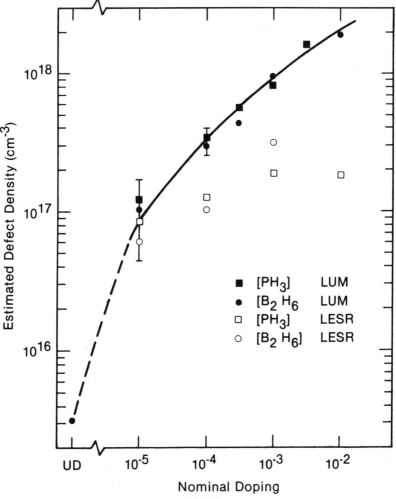

Figure 4   Defect densities versus doping concentration as estimated by light-induced electron spin resonance (LESR) and luminescence measurements (24).

Alloys can be formed by introducing other gases during the deposition. To increase the band gap, methane (CH_4) or ammonia (NH_3) is added to form a-SiC:H (33–38) or a-SiN:H (33–39). Lower band gaps are achieved using germane (GeH_4) or tetramethyltin to form a-SiGe:H (25–29) or a-SiSn:H (30–33). Microcrystalline Si:H (44, 45), used especially in the doped layers, cannot be formed using pure SiH_4 and requires high dilutions (1:9) with H_2 or inert gas. In addition, high RF power densities aid in formation.

Chemical vapor deposition (CVD) of a-Si:H [see review by Kaplan (48)] usually involves the thermally assisted decomposition of silane or higher silanes (Si_2H_6, Si_3H_8, etc). SiH_4 requires higher temperatures on the order of 600°C to achieve reasonable deposition rates. In contrast, the higher

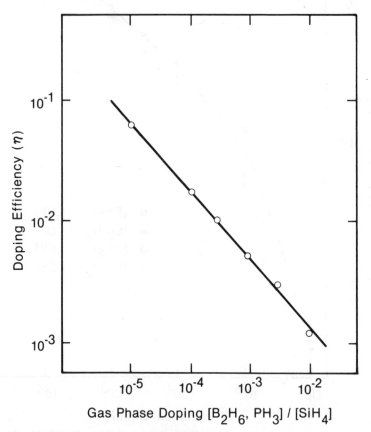

Figure 5   Doping efficiency for both boron and phosphorus obtained from the defect densities estimated in Figure 4 (24).

silanes are less stable chemically and can be deposited below 450°C. Figure 7 illustrates different CVD approaches. For the cold wall atmospheric pressure system, silane is decomposed on a substrate placed on the heated susceptor. In low-pressure CVD (LPCVD), layers can be deposited uniformly on a multiple number of substrates. In the HOMOCVD approach, silane is homogeneously predecomposed and is deposited on a cooled substrate. Recently photochemical vapor deposition, in which ultraviolet light is used to excite and dissociate the reactant gases, has shown promise (49), with reported solar cell efficiencies of 8.6%. Mercury vapor is added to enhance the deposition rate. In addition, acetylene (C_2H_2) and dimethylsilane may replace methane for the formation of the a-SiC:H layers.

Large-area deposition techniques are required for industrial processing. In 1984, the SERI Amorphous Silicon Research Project office awarded three-year cost-shared [US Department of Energy (DOE) 70%, private 30%] contracts to Chronar, Solarex, 3-M, and SPIRE to demonstrate stable p-i-n solar cells of at least 12% efficiency over 1 cm², to demonstrate

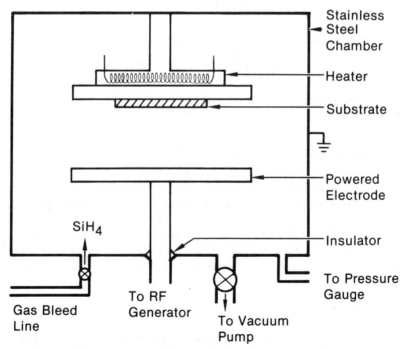

Figure 6   Cross-section of an RF capacitive glow-discharge deposition system.

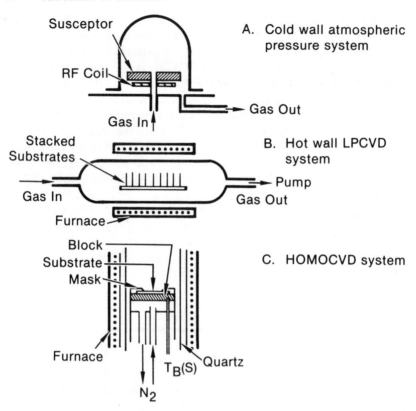

Figure 7 Schematics of CVD reactors: (*a*) cold wall atmospheric pressure system; (*b*) hot wall low pressure CVD system; (*c*) HOMOCVD system (48).

stable 8% efficient submodules of at least 1000 cm^2, and to establish a technology base in amorphous Si alloys leading to 20% efficient devices. The approaches, summarized in Figure 8, include multiple-chamber, two- or three-electrode glow-discharge processing on glass or flexible polymer substrates.

Many deposition alternatives exist for amorphous Si alloys. A successful a-Si photovoltaic industry will require selecting processes that allow uniform deposition of high-quality materials over large areas at sufficient rates.

4. AMORPHOUS SILICON DEVICE STRUCTURES

Historically, several a-Si device structures have been investigated, such as Schottky barrier (SB), metal-insulator-semiconductor (MIS), and p-i-n or

Company	Approach	Company	Approach
Chonar	• Single-junction p-i-n cells • RF glow discharge 2-electrode • 3 chambers • Glass substrate	3-M	• Single-junction p-i-n cells • RF glow discharge 2-electrode • 3 chambers • Flexible polymer substrate
Solarex	• Single-junction p-i-n cells • DC glow discharge 3-electrode • 3 chambers • Glass substrate	SPIRE	• Multi-junction a-Si alloy cells • RF glow descharge • 6 chamber • Glass substrate

a-Si Deposition Method

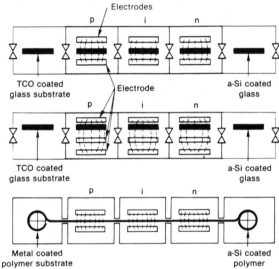

Three-Chamber Reactors for Depositing p-i-n Material

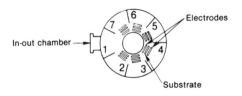

Six-Chamber Reactor for Depositing Material for
Multi-Junction Cells

Figure 8 Technical approaches of programs sponsored by SERI Amorphous Silicon Research Project, initiated in 1984.

n-i-p homojunction, heterojunction, and stacked junctions. The convention used in referring to p-i-n and n-i-p cells is that the light is incident on the first layer listed. Homojunction refers to use of a single material, whereas heterojunction refers to the use of a different material for the front window layer. Stacked junctions, described in more detail in Section 4.3, are those in which junctions are optically stacked to convert the sunlight incident on the cell to electricity.

4.1   Simple Devices: Schottky Barrier and MIS Cells

SB solar cells are formed by depositing a metal onto the semiconductor layer. In the case of a-Si, SB solar cells are fabricated on metallized substrates such as Mo, Cr, or Ni on glass, stainless steel, or plastic film by first depositing a thin (100–300 Å) n + layer to form the back contact, then an a-Si i-layer (nominally slightly n-type) about 5000 Å thick, and finally a metal such as Pt, Rh, Pd, Ni, etc. Although Carlson & Wronski demonstrated a 5.5% efficient a-Si/Pt device in 1977 (50), the V_{oc} is usually limited to about 600 mV and was surpassed by other device structures.

MIS cells are made by growing a thin native oxide by heating in air or by depositing an insulator (such as titanium dioxide, silicon nitride, or niobium pentoxide) between the a-Si and the metal. Again, although open-circuit voltage up to 875 mV has been observed, the performance of these devices is limited. Further reviews of SB and MIS devices can be found elsewhere (1, 17).

4.2   p-i-n and Heterojunction Cells

The predominant a-Si solar cells are p-i-n type devices and the related heterojunction structures with an a-SiC or other wide band gap window. The two broad categories of devices are superstrate cells, in which the light is incident through glass, and substrate cells, in which the active device layers are deposited on metal foils or metallized insulators and the light is incident through a top transparent conductor. Two typical p-i-n type devices are illustrated in Figure 9. A transparent conductor is required on the front of the cell to act as a partial antireflection coating and to reduce the series resistance losses of the thin front layer.

Heterojunction solar cells using wide band gap window(s) usually of a-SiC:H have achieved the highest efficiencies to date. By increasing the band gap of the front doped–amorphous layer, more light is allowed into the i-layer, where it can be collected. Normally the front doped–armorphous layer is inactive because of the large number of dopant-created defects, as discussed in Section 2.

The efficiency and photovoltaic parameters of a variety of a-Si solar cells are summarized in Table 1. As illustrated in Figure 1, not only has the efficiency increased in recent years, but the device areas have increased as well. Several groups have now reported efficiencies over 10%; the highest efficiency devices tend to be heterojunction cells. In addition, fabrication of modules is a primary concern of many groups. A large area (3716 cm²) monolithic series-connected module with a total area efficiency of 5% has been reported, as have other modules with total-area (900 cm²) efficiencies of 7.9% and active-area (790 cm²) efficiencies of 9% (61) using thin-film silicon-hydrogen alloys.

4.3   Improving Overall Efficiency: Stacked Junction Cells

Studies have shown the large leveraging effect of photovoltaic module efficiency on the economics of photovoltaic systems, especially for central utility applications (62). The maximum theoretical efficiency of single-junction solar cells is limited by the trade-off between increased current and lower voltage as the energy band gap is decreased. An alternative is to divide the solar spectrum into energy bands and optimize individual

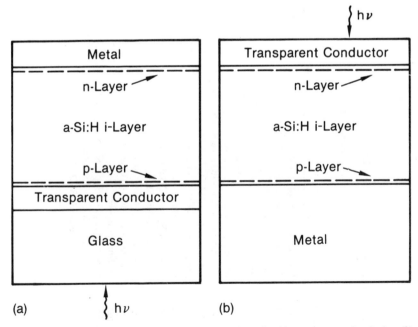

Figure 9   Cross sections of typical p-i-n a-Si:H solar cells: (*a*) superstrate p-i-n device; (*b*) substrate n-i-p device.

Table 1 Performance of a-Si:H solar cells

Type	Configuration	Area (cm²)	Eff (%)	V_{oc} (mV)	J_{sc} (mA/cm²)	FF	Ref.
SJ	ITO/nipni a-Si/p poly-Si/Al	0.44	12.5	1325	14.2	0.66	51
HJ	Ag/ni a-Si/p a-SiC/text. TCO/glass	1.0	11.5	869	18.9	0.70	52
HJ	Al/ni a-Si/p a-SiC/SnO₂/glass	4.15	10.2	865	16.1	0.73	53
HJ	Ag/ni a-Si/p a-SiC/SnO₂/glass	1.09	10.1	840	17.8	0.676	36
HJ	Me/mi a-Si/p a-SiC/TCO/glass	0.04	10.0	847	17.3	0.68	54
HJ	Me/mi a-Si/p a-SiC/TCO/glass	1.0	9.2	870	15.5	0.68	55
SJ	ITO/nipnipn a-Si/i a-SiGe/p a-Si/S.S.	0.09	8.5	2200	6.74	0.57	28
Mod	Me/mi a-Si/p a-Si/TCO/glass	100	8.1	11.96V	15.6	0.61	54
HJ	Me/mi a-Si/p a-SiC/TCO/glass	100	8.0	850	14.4	0.654	55
pin	ITO/n μc-Si/ip a-Si/S.S.	1.2	7.8	860	13.9	0.655	45
SJ	Al/nipni a-Si/pa-SiC/SnO₂/glass	4.15	7.7	1710	6.23	0.71	56
HJ	Al/ni a-Si/p a-SiC/SnO₂/glass	1.0	7.7	880	14.1	0.62	35
SJ	ITO/nipn a-Si/i a-SiGe/p a-Si/Steel	0.25	7.7	1410	9.6	0.57	28
pin	ITO/nip a-Si/SnO₂-ITO/glass	0.16	6.9	874	13.4	0.59	57
pin	ITO/nip a-Si/Steel	0.04	6.9	857	13.0	0.62	58
SJ	Al/ni a-SiGe/pni a-Si/p a-SiC/SnO₂/glass	4.15	6.8	1450	7.64	0.61	56
pin	ITO/n μc-Si/ip a-Si/S.S/polyimid	0.09	6.4	910	10.6	0.62	59
Mod	Al/ni a-Si/p a-SiC/SnO₂/glass	930	6.0	21.735V	0.376A	0.68	60

junctions for each band. Typically, two cell electrical connection schemes are considered: four-terminal connection, in which the maximum power points of each junction are tracked separately, and two-terminal connections, in which the two junctions are constrained to pass the same electrical current. This approach, commonly called stacked junction, tandem junction, or multiple junction, projects theoretical efficiencies of 20–60% (63–66). Early emphasis was placed on the demonstration of this concept using high-quality single-crystal device technologies, especially with Si and III–V compounds such as AlGaAs, GaInAs, and GaAsSb (67).

Applying the stacked cell concept to amorphous materials gives estimated efficiencies of 20% for two junctions and 24% for three junctions (68, 69). Figures 10 and 11 present plots of efficiency profiles in band gap space and of efficiency vs top cell band gap of the two-junction stacked device and the top junction for the amorphous thin-film cases under four-terminal and two-terminal operation. As shown in Figure 10a, efficiencies above 20% are predicted for top cell band gaps of 1.55–2.1 eV and bottom cell band gaps of 0.95–1.4 eV. About two-thirds of the efficiency is contributed by the top cell. Two-terminal operation restricts the domain of allowable band gaps to achieve 20% efficiency.

Efficiencies of 7.7% for a-Si:H iso-band gap cells (56), 10.2% for a-Si:H/a-Si:H/a-SiGe:H three-cell stacked cells (28a), and 12.5% for a-Si:H polycrystalline Si cells (51) have been reported. Further improvements require optimization and advances in the state of the art of the required material and device technologies.

4.4   Cost-Effective Amorphous Silicon Modules

Several advantages are evident for amorphous Si modules compared with the conventional single-crystal Si modules. The first is the substantial reduction in the semiconductor material required. The a-Si:H layers are on the order of 0.5–1.0 μm in thickness, compared with about 300 μm for crystalline Si. In addition, as discussed in Section 3, amorphous Si can be deposited over large areas, offering opportunity for large-area patterning techniques such as laser scribing (52), to define the cell areas, and cell interconnection, to form monolithic series-connected panels. Finally, processing of continuous films is a future possibility (70, 71).

5.   DEVICE CHARACTERISTICS OF AMORPHOUS SILICON SOLAR CELLS

Maximizing the efficient conversion of sunlight into electricity by amorphous Si solar cells requires an understanding of the underlying mechanisms that control the generation and collection of electrical carriers and the

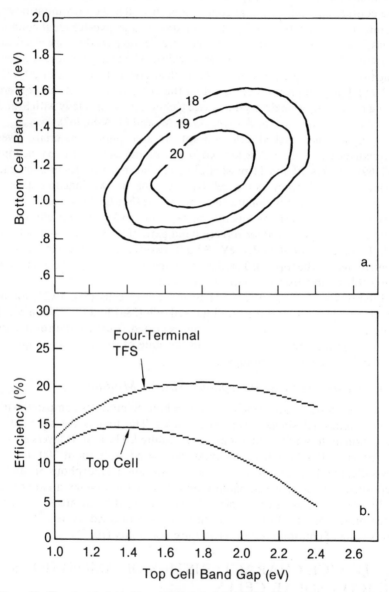

Figure 10   Four-terminal Air Mass 1.5 Global photovoltaic efficiency for amorphous thin film-devices: (*a*) multiple-junction efficiency versus top cell and bottom cell band gap; and (*b*) top cell and total efficiency versus top cell band gap (69).

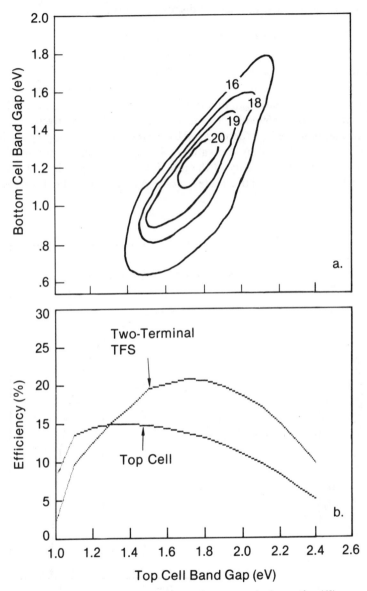

Figure 11   Same as Figure 10 except for two-terminal operation (69).

separation of these carriers by the solar cell junction to create a photovoltage. These processes are discussed below.

5.1   Current-Voltage Characteristics

The current-voltage curve of an 11.5% efficient Sanyo a-Si : H cell (52) under 100 mW/cm² Air Mass 1 illumination (Figure 12) illustrates the definition of the short-circuit current density J_{sc}, the open-circuit voltage V_{oc}, the maximum power point P_{max}, and the fill factor FF. The V_{oc}, J_{sc}, and FF for the 1 cm² cell are 0.869 volts, 18.9 mA/cm², and 0.70 respectively.

The magnitude of V_{oc} is determined by the band gaps and doping levels of the p-, i-, and n-layers, by the value of J_{sc}, by the temperature and light intensity, and by the quality of the junction characterized by the reverse saturation current J_0 and the diode factor n. As shown in Table 1, values for

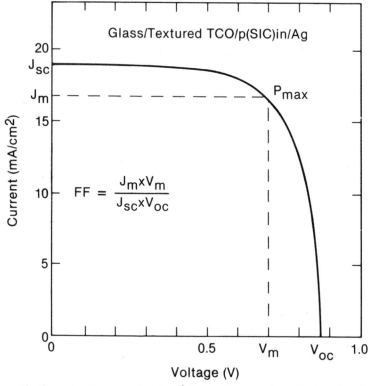

Figure 12   Current-voltage curve for a 1 cm² Sanyo p-i-n amorphous Si solar cell under Air Mass 1.5 direct illumination (52).

p-i-n cells range from 840 mV to 910 mV. The maximum values measured as a function of light intensity and temperature suggest attainable limits for the devices. Figure 13 shows a plot of short-circuit current versus open-circuit voltage for a 10.2% efficient, 4.15 cm^2 ARCO Solar thin-film silicon:hydrogen alloy cell with V_{oc}, J_{sc}, and FF values of 0.865 V, 16.1 mA/cm^2, and 0.73 respectively (53). The curve, indicative of a good solar cell, is extremely linear with J_{sc} proportional to light intensity, J_0 equal to 7.8×10^{-11} A/cm^2, and a diode factor n, of 1.76. Open-circuit voltages up to 1.038 V at about 25 suns (2.5 W/cm^2) intensity and up to 1.15 V at 193 K (1 sun) were measured on this device. Values above 1 V have been measured at 1 sun for similar devices in the laboratory, suggesting that substantial opportunity still exists.

5.2   *Spectral Response*

The spectral response of amorphous Si solar cells under reverse bias (typically between -0.5 V and -2 V) is a good indicator of the optical absorption characteristics of the cell. Under reverse bias, the electric fields in the i-layer are typically high enough to give unity collection efficiency for photons absorbed in this layer. Figure 14, for example, shows the spectral response at -2 V bias of a glass/transparent conductor/p-i-n Si:H cell (53).

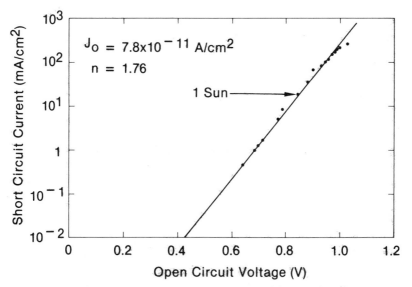

Figure 13   Short-circuit current versus open-circuit voltage plot for a 4.15 cm^2 ARCO Solar. Thin Film Si Transparent Conductor/p(SiC)-i-n/Aluminum solar cell (53).

Also plotted is the theoretical curve, given by:

$$QE = [(1-R)\exp(-\alpha_p W_p)][1-\exp(-\alpha_i W_i)]$$
$$\times [1+Rb\exp(-\alpha_i W_i - 2\alpha_n W_n)]$$

which contains a front reflection loss, R, an inactive p-layer, complete collection from the i-layer, and reflection from the back contact, Rb, where α_p and W_p, α_i and W_i, and α_n and W_n are the optical absorption coefficients and layer thicknesses of the p-, i-, and n-layer respectively. The absorption coefficients are of the form described in Section 2.1.

The shapes of the curves indicate that the quantum efficiency for wavelengths longer than 600 nm is determined by the soft absorption edge of the i-layer. For shorter wavelengths the spectral response is dominated by the photon absorption in the p-layer, from which no carriers are collected, and by the cutoff of the transparent conductor (TC), which accounts for the drop in quantum efficiency below 360 nm and for some transmission losses into the semiconductor layer at longer wavelengths. Additional optical effects include reflection from the back contacts, especially for aluminum or silver, which enhances the red response for thin

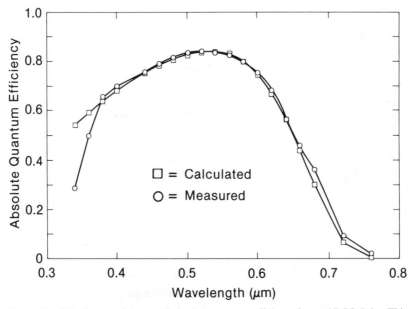

Figure 14 Calculated and measured absolute quantum efficiency for an ARCO Solar. Thin Film Silicon Transparent Conductor/p(SiC)-i-n/Aluminum solar cell at $-2V$ reverse bias (53).

devices (72), and texturing of the TC, which increases the optical path length in the amorphous Si : H.

Identification of the loss mechanisms in the spectral response helps focus research efforts to maximize the photogenerated current from the cells. Substantial gains in short-circuit current can be achieved through optical enhancement techniques (73–75), which significantly increase the collection efficiency for the longer wavelength edge of the spectral response curve, as shown in Figure 15. This is demonstrated in the use of the textured transparent conductor along with a silver back reflector to achieve the high

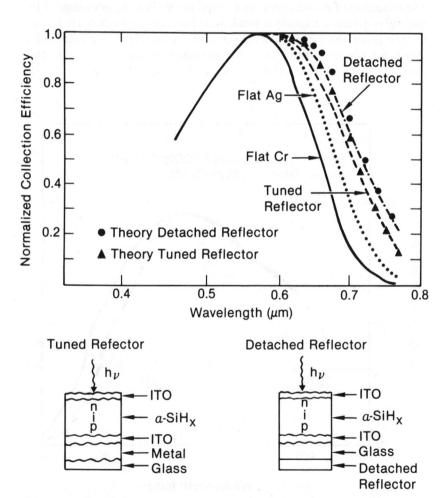

Figure 15 Normalized collection efficiencies for the unenhanced and enhanced a-Si solar cells indicated in the cross-section (73).

short-circuit current of 18.9 mA/cm^2 for the 11.5% Sanyo cell described in
Section 5.1. The spectral response curve in Figure 16 shows the collection
enhancement at the longer wavelengths over the conventional cell structure. Another loss of current arises from the absorption of light in an
inactive p-layer. Modifications to the p-layer deposition process have
reduced these losses, as demonstrated in Figure 17, so that most of the light
absorbed in the intrinsic layer is efficiently collected under reverse bias (53).

5.3 *Voltage-Dependent Current Collection*

The collection of photocurrent in amorphous Si : H cells, as evidenced by
spectral response, is strongly dependent on bias voltage, as shown in Figure
18. As a result, the shape of the light current-voltage curve, and thus the fill
factor, are affected by this voltage-dependence in addition to series and
shunt resistance and the quality of the diode. Several theories (53, 76–81)

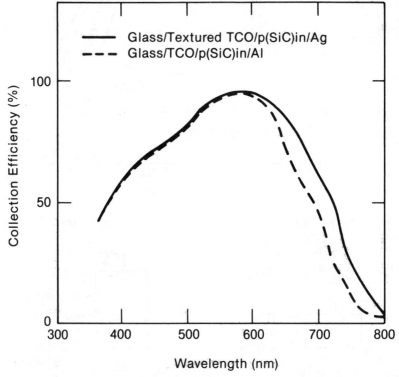

Figure 16   Absolute collection efficiency for the Sanyo cell indicated in Figure 12 compared to
a typical a-Si solar cell.

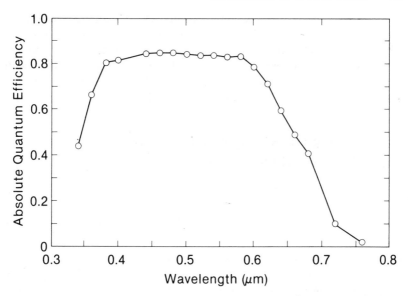

Figure 17   Absolute quantum efficiency for 10.2% efficient, 4.15 cm² ARCO Solar TFS solar cell (53).

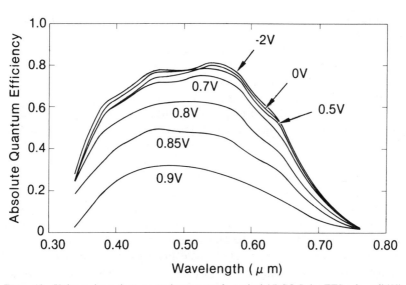

Figure 18   Voltage-dependent spectral response of a typical ARCO Solar TFS solar cell (53).

have been proposed to describe this behavior which relates to electric-field-assisted carrier collection. The profiles of the electric field and of the carrier lifetime and mobility in the i-layer are important. In addition, the front and back interface regions can also strongly influence the current collection and thus the fill factor. Improvements in current collection leading to fill factors above 0.80 are expected to come from improved material quality and device design.

6. STATUS OF AMORPHOUS SILICON SOLAR CELL TECHNOLOGY

6.1 *Economic Factors*

As alluded to in Section 1, high conversion efficiencies do not guarantee commercial success of a solar cell technology. The most important consideration is the cost of the photovoltaic energy compared with that available from other sources. There is a general expectation that once solar cell arrays are available for about $2 per annual kilowatt hour (AkWh), photovoltaics will penetrate the bulk power market (residential and commercial). Industry analysts forecast future sales of more than $10 billion for the year 2000 (81a). It should be noted that worldwide photovoltaic sales have risen from just over 5 MW in 1981 to an estimated 25 MW in 1984. For amorphous silicon, today's main market is in consumer products in which the cost of the cell is a relatively small fraction of the total cost of the product.

Utility-interconnected photovoltaic systems in either distributed or central station form must be considered in the context of long-range utility system planning. Utilities generally decide on the installation of a new generating plant on the basis of life-cycle costs, which include capital and operating costs over the life of the facility.

One way to combine capital and operating costs on a consistent basis is shown in Figure 19, where estimates of levelized electricity cost for a range of new conventional generation alternatives are plotted against capacity factor (82, 83). Capacity factor is the ratio of the average power output of a plant over a specified period of time to its nameplate rated peak power output. When the required capacity factor is above 15–20%, new coal plants are projected to provide the least expensive electricity. Oil-fired generators are projected to be cost-effective if they are required to operate for peak power needs. Clearly, if photovoltaic systems are to become a significant source of energy, they must be able to generate electricity at costs below those for oil. Whether they must compete directly with electricity from coal or other options will depend, in large part, on the influence of institutional and resource availability issues, which in turn will depend on the utility.

As a baseline for determining photovoltaic cost and performance targets, an allowable current-dollar levelized energy cost of $0.15 per kWh was chosen by the Electric Power Research Institute, corresponding to $0.065 per kWh in constant 1982 dollars. As seen in Figure 19, this cost is midway between projected energy costs with new oil plants and new coal generation and is comparable to the cost of energy from new coal plants operating at capacity factors that can be expected from photovoltaic generation.

The trade-offs between allowable module cost, module efficiency, and levelized electricity cost are illustrated in Figure 20. Key assumptions for this analysis included the algorithm relating sunlight availability to energy delivery, a 12.5% discount rate, and an 8.5% inflation rate (as Figure 19). The DOE cost (in 1982 dollars) and efficiency goals for a flat-plate photovoltaic system for the late 1990s are as follows: module efficiency of 13–17%, measured at 28°C and AM1.5; module cost of $40–75/m^2; area-related balance-of-system cost of $50/m^2; power-related balance-of-system cost of $150/kW; and system life expectancy of 30 years. The area-related costs include those for array structure and foundation, site preparation, land, and installation. Power-related costs include those for panel connections, power conditioning subsystem, tracking mechanisms, and the high-voltage transformer.

As one specific example for today's technology, the estimated cost of the ARCO Solar 6.5 MW Carrisa Plains project is indicated to be $3.51/AkWh (84a). This cost may be broken down as follows: modules in racks,

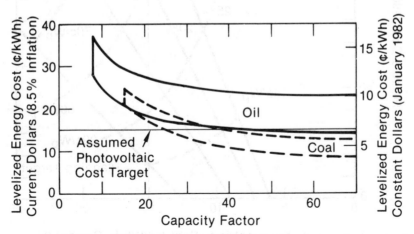

Figure 19 Levelized electricity cost versus capacity factor for a range of conventional generation options. Levelized electricity cost is presented in both current and constant dollar terms over the period 1982–2012 (84).

$1.92/AkWh; power conditioning, pedestals, drives, installation, and wiring, $1.59/AkWh. Increasing the module efficiency to 15% (midway in the EPRI analysis) would reduce the total to $2.65/AkWh. To achieve the $2.00/AkWh goal, absolute costs of both photovoltaic modules and the balance of systems must be reduced. The thin-film research activities currently under way are aimed at the dual targets of efficiency and cost reduction.

6.2 Efforts Around the World

Based on continuing improvements in a-Si:H conversion efficiency and scientific understanding during the past decade, amorphous silicon cell technology is considered a leading contender for economic power generation.

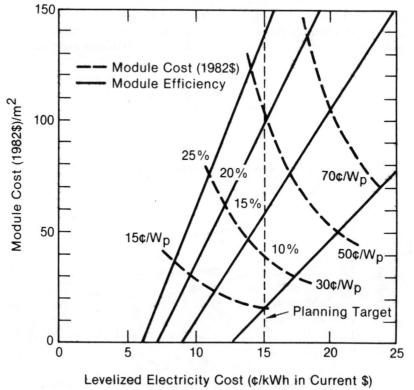

Figure 20   Module cost and efficiency versus 30-year levelized electricity costs for flat-plate PV systems ($50/m² area-related balance-of-system costs) (84).

The key competitors in a-Si:H cell technology development are in the United States, the European Economic Community, and Japan.

From 1975 through fiscal year 1983, the US government spent more than $700 million for photovoltaic research and development. Since 1977, $27 million has been allocated for amorphous silicon research. In Japan, cumulative government sponsorship totals about $24 million.

As noted in Section 3, in early 1984 four US companies were funded by the US government to conduct two research and development efforts: to improve cell efficiency and stability of single-junction a-Si:H cells, and to identify and develop new a-Si:H alloy materials and multijunction cells made from these materials with the long-term objective of achieving cell efficiencies greater than 18%.

In May 1984, the above joint efforts between federal government and industry were integrated into an expanded five-year research plan (84) by DOE, based on the need for strong cooperation among government, industry, and universities with research activities in two main areas: single-junction thin films and multijunction concepts. These concepts are organized under five major categories: (a) materials research; (b) cell research and development; (c) submodule research; (d) process development; and (e) module research and development.

Over and above joint government-industry ventures, a number of companies around the world have substantial research under way, funding for which exceeds government spending.

6.3   U.S. Policy

Photovoltaics as a source of electric energy will succeed, and the world leader in this success will be the United States. We have the ability not only to develop and produce solar modules, but also to use them. We have the technical skill, the land, the sunlight, and the demand. We owe our good position in part to nature for the resource, but in greater part to the quality of American science and engineering. Government support and tax incentives, along with the commitment of American industry, have bolstered photovoltaics.

The greatest single achievement of the US national photovoltaics research and development program is the high quality of US commercial solar cell module products, recognized worldwide. DOE program engineers worked closely with industry in quality assurance, field testing, failure analysis, and module engineering.

As thin-film modules appear in the marketplace, a repeat performance is appropriate. However, greater US program achievements lie ahead, in the products of long-term research and development. Federal research and development should continue to be concentrated in areas too costly or risky for private industry, such as new materials and fundamental modeling of device physics. DOE must plan to do a few things well. The program must be stable, with a long time horizon and multiyear commitments, especially to the schools, which provide the most valuable technical resource—people.

The US photovoltaic manufacturing industry has also been aided by solar residential and solar business tax credits. Such incentives promote response directly to the current marketplace. They enable the small manufacturer and all the distributors and dealers to benefit. In this fashion, the customer determines the winner and the government helps all.

7   CONCLUDING REMARKS

All new high technology industries are research and development intensive and experience continual price reductions and plateaus, entries and departures of participants, redeployments of assets and personnel, and successes and failures of technologies and products. Photovoltaics is no different. The business has expanded rapidly since the mid-1970s. The industry is dynamic; both refinement of existing technologies and development of new technologies are proceeding rapidly. The photovoltaics product is the ultimate capital good—pay all the money now and get free energy for the next 20 or 30 years. Selling any capital good, especially for export in today's environment of extremely high real interest rates, a soaring dollar, and expectation of flat energy prices, is difficult. However, despite these uncertainties, photovoltaics is a reliable, safe, economical, and durable source of electricity that is being deployed throughout the world for applications that include consumer products as well as utilities (84a–88).

Thin-film amorphous silicon alloys for solar cells represent a viable option currently under commercialization, and a strong consensus exists that multijunction or tandem amorphous silicon cells will be one of a number of technologies that will exceed the efficiency targets.

The achievable conversion efficiency for multijunction devices is at least 20%. Important technical problems that need to be solved include a basic understanding of the physics of amorphous silicon alloys; a resolution of durability concerns; the control of interfaces and surfaces; and the development of systems capable of uniformly depositing the material over large areas.

Technology is not the only governing factor for the large-scale market penetration of photovoltaic technologies. Economic, political, institutional, and environmental issues are crucial. We are all concerned about leadership these days. The United States must lead world trends and innovations in photovoltaic technology. We have a strong base of expertise in fundamental science, and we must apply it vigorously. The following points should be kept in mind in planning:

1. Federal incentives can play an important role in this infant industry. These incentives should make minimum calls on public funds, but give maximum support to legitimate governmental goals. A five-year extension of the tax credits would allow orderly planning rather than time-constrained decision making.
2. Concessionary financing for export sales would be helpful. Our prime international customers tend to be needier nations and concessionary financing is available from our competitor nations.
3. Publicity and public education on the advantages of photovoltaics are critical to acceptance of the quiet photovoltaic electric generator.
4. Utility demonstrations help the industry work with its real long-term customer. A building block approach to creating confidence will lead to long-term success. Demonstration answers, in an orderly fashion, many questions that utilities, investor-owned and municipal, have regarding interconnections, quality of power, and ease of operation and maintenance.

The marketplace will yield technology differentiation. Substantial private research and development money is being spent in the belief that success will be rewarded. The industry views the future with optimism and confidence—confidence in the people and confidence in photovoltaics technology.

ACKNOWLEDGMENTS

The coauthors gratefully acknowledge advice from the SERI photovoltaics staff, especially Don Ritchie, Jack Stone, Ed Sabisky, and Amir Mikhail; and from Charles Gay, Elliot Berman, and Don Morel of ARCO Solar, Inc. The support and assistance of the ARCO Solar Technical Information group is also gratefully appreciated.

Literature Cited

1. Mahan, A. H., Stone, J. L. 1981. Amorphous silicon bibliography. *Solar Cells* 4:205–448. 1983–84. *Solar Cells* 7:347–426
2. Tanka, K., Shimiza, T., eds. 1983. *Proc. 10th Int. Conf. on Amorphous and Liquid Semiconductors, Tokyo*, Aug. 22–26. *J. Non-Cryst. Solids* 59–60:1–1325
3. Fritzsche, H., Kastner, M. A., eds. 1984. *Proc. Int. Topical Conf. on Transport and Defects in Amorphous Semiconductors*,

Bloomfield Hills, Mich. March 22–24. *J. Non-Cryst. Solids* 66:1–392
4. Hamakawa, Y., ed. 1983. *Japan Annual Review in Electronics, Computers, and Telecommunications:* Vol. 6 *Amorphous Semiconductor Technologies and Devices.* New York: OHMSHA, Ltd./North Holland
5. Joannopoulos, J. D., Lucovsky, G., eds. 1984. *Topics in Applied Physics.* Vol. 55: *The Physics of Hydrogenated Amorphous*

Silicon I-Structure, Preparation, and Devices. Berlin: Springer-Verlag

6. Joannopoulos, J. D., Lucovsky, G., eds. 1984. Topics in Applied Physics. Vol. 56: The Physics of Hydrogenated Amorphous Silicon II-Electronic and Vibrational Properties. Berlin: Springer-Verlag

7. Madan, A. 1984. Amorphous silicon solar cells. Silicon Processing for Photovoltaics, ed. K. V. Ravi, C. P. Khaltak. New York: North Holland. In press

8. Madan, A., Von Roedern, B. 1984. Surfaces and contacts to amorphous semiconductors. Amorphous Semiconductor Handbook, ed. H. Henish. Inst. Amorphous Studies. In press

9. Pankove, J., ed. 1984. Semiconductors and Semimetals. Vol. 21B: Hydrogenated Amorphous Silicon. New York: Academic

10. Carlson, D. E. 1984. Solar Energy Conversion. See Ref. 5, pp. 203–43

11. Sterling, H. F. The Preparation of High Purity Silicon: Decomposition of Silane. Standard Telecommunication Labs. Limited, Harlow, Essex, England. Rep. 1403/1955/99

12. Lewis, A. J., Connell, G. A. N., Paul, W., Pawlick, J. R., Temkin, R. J. 1974. Hydrogen incorporation in amorphous germanium. Proc. Int. Conf. Tetrahedrally Bonded Amorphous Semiconductors, ed. M. H. Brodsky, S. Kirkpatrick, D. Weaire, pp. 27–32. New York: Am. Int. Phys.

13. Triska, A., Dennison, D., Fritzche, H. 1975. Hydrogen content in amorphous Ge and Si prepared by RF decomposition of GeH_4 and SiH_4. Bull. Am. Phys. Soc. 20:392

13a. Chittick, R. C., Alexander, J. H., Sterling, H. F. 1969. Preparation and properties of amorphous silicon. J. Electrochem. Soc. 116:77

14. Spear, W. E., LeComber, P. G. 1972. Investigation of the localized state distribution in amorphous Si films. J. Non-Cryst. Solids 8–10:727–38

15. Engemann, D., Fisher, R. 1974. Influence of preparation conditions on the radioactive recombination in amorphous silicon. Proc. 12th Int. Conf. Phys. Semicond., ed. M. H. Pilkuhn, pp. 1042–46. Stuttgart: Teubner

16. Spear, W. E., LeComber, P. G. 1975. Substitutional doping of amorphous silicon. Solid State Commun. 17:1193–96

17. Spear, W. E., LcComber, P. G., Kinmond, S., Brodsky, M. H. 1976. Amorphous silicon p-n junctions. Appl. Phys. Lett. 28:105–7

18. Carlson, D. E., Wronski, C. R. 1976. Amorphous silicon solar cells. Appl. Phys. Lett. 28:671–73

19. Sabisky, E., Wallace, W., Mikhail, A., Mahon, H., Tsuo, S. 1984. Advances and opportunities in the amorphous silicon research field. Proc. 17th IEEE Photovoltaic Specialists Conf., pp. 217–22. New York: IEEE

20. Cody, G. D. 1984. The optical absorption edge of amorphous silicon hydride. See Ref. 9. In press

21. Cody, G. D., Tiedje, T., Abeles, B., Brooks, B., Goldstein, Y. 1981. Disorder and the optical-absorption edge of hydrogenated amorphous silicon. Phys. Rev. Lett. 47:1480–83

22. Cody, G. D., Wronski, C. R., Abeles, B., Stephens, R. B., Brooks, B. 1980. Optical characterization of amorphous silicon hydride films. Solar Cells 2:227–43

23. Bird, R. E., Hulstrom, R. L., Lewis, L. J. 1983. Terrestrial solar spectral data sets. Solar Energy 30:563–73

24. Street, R. A., Biegelsen, D. K. 1984. The spectroscopy of localized states. See Ref. 6, pp. 195–260

25. Dong, N. V., Danh, T. H., Le Ny, J. L. 1980. Electrical properties of hydrogenated amorphous SiGe alloys. Proc. 3rd E.C. Photovoltaic Solar Energy Conf., pp. 830–34. Boston: Reidel

26. Nakamura, G., Sato, K., Yukimoto, Y., Shirahata, K. 1980. Amorphous $SiGe_x$ for high performance solar cell. See Ref. 25, pp. 835–39

27. Singh, P., Galley, D., Fagen, E. A., Dalal, V. L. 1981. Optical and electrical properties of amorphous silicon-germanium alloy films. Proc. 15th IEEE Photovoltaic Spec. Conf., pp. 912–16. New York: IEEE

28. Nakamura, G., Sato, K., Yukimoto, Y. 1982. High performance tandem type amorphous solar cells. Proc. 16th IEEE Photovoltaic Spec. Conf., pp. 1331–37. New York: IEEE

28a. Yang, J., Mohr, R., Ross, R. 1984. High Efficiency Amorphous Silicon and Amorphous Silicon-Germanium Tandem Solar Cells. Proc. 1st Int. Photovoltaics Science and Engineering Conf. (A-IIA-L6). Tokyo, Japan: The Japan Times, Ltd.

29. Yukimoto, Y. 1983. Hydrogenated a-SiGe alloy and its optoelectronic properties. See Ref. 4, pp. 136–47

30. Verie, C., Rochette, J. F., Rebouillat, J. P. 1981. Novel amorphous semiconductor for solar cells: the silicon-tin alloys system. See Ref. 27, pp. 251–52

31. Williamson, D. L., Deb, S. K. 1983. Mossbauer spectroscopy of amorphous silicon-tin-hydrogen alloys. J. Appl. Phys. 54:2588

32. Mahan, A. H., Williamson, D. L.,

Madan, A. 1984. Properties of amorphous silicon tin alloys produced using the radio frequency glow discharge technique. *Appl. Phys. Lett.* 44:220–22

33. Kuwano, Y., Ōbnishi, M., Nishiwaka, H., Tsuda, S., Fukatsu, T., et al. 1982. Multi-gap amorphous Si solar cells prepared by the consecutive, separated reaction chamber method. See Ref. 28, pp. 1338–43

34. Pankove, J. I. 1978. *US Patent No. 4,109,271*

35. Hamakawa, Y., Tawada, Y., Nishimura, K., Tsuge, K., Kondo, M., et al. 1982. Design parameters of high efficiency a-SiC:H/a-Si:H heterojunction solar cells. See Ref. 28, pp. 679–84

36. Catalano, A., D'Aiello, R. V., Dresner, J., Faughnan, B., Firester, A., et al. 1982. Attainment of 10% conversion efficiency in amorphous silicon solar cells. See Ref. 28, pp. 1421–22

37. Tawada, Y. 1983. Preparation and valency electron control of a-SiC:H. See Ref. 4, pp. 148–60

38. Evangelisti, F., Fiorini, P., Giovannella, C., Patella, F., Perfetti, P., et al. 1984. Photoemission studies of a-SiC:H/a-Si and a-SiC:H/hydrogenated amorphous silicon heterojunctions. *Appl. Phys. Lett.* 44:764–66

39. Hirose, M. 1983. Preparation and properties of a-SiN:H. See Ref. 4, pp. 173–80

40. Tiedje, T., Cebulka, J. M., Morel, D. L., Abeles, B. 1981. Evidence for expotential band tails in amorphous silicon hydrogen. *Phys. Rev. Lett.* 46:1425–28

41. Snell, A. J., Spear, W. E., LeComber, P. G. 1981. The lifetime of injected covers in amorphous silicon p-n junctions. *Phil. Mag.* 43:407–17

42. Jan, Z. I., Bube, R. H., Knights, J. C. 1980. Field effect and thermoelectric power on boron doped amorphous silicon. *J. Appl. Phys.* 51:3278–81

43. Spear, W. E. 1977. Doped amorphous semiconductors. *Adv. Phys.* 26:312

44. Matsuda, A., Yamasaki, S., Nakagawa, K., Okushi, H., Tanaka, K., et al. 1980. Electrical and structural properties of phosphorus-doped glow-discharge Si:F:H and Si:H films. *Jpn. J. Appl. Phys.* 19:L305

45. Tanaka, K., Matsuda, A. 1983. Basic properties of plasma-deposited microcrystalline Si. See Ref. 4, pp. 161–72

46. Spear, W. E., LeComber, P. G. 1984. Fundamental and applied work on glow discharge material. See Ref. 5, pp. 63–118

47. Thompson, M. J. 1984. Sputtered material. See Ref. 5, pp. 119–75

48. Kaplan, D. 1984. CVD material. See Ref. 5, pp. 177–201

49. Tanaka, T., Kim, W. Y., Konagi, M., Takahashi, K. 1984. Amorphous silicon solar cells prepared by photochemical vapor deposition. See Ref. 52, pp. 563–66

49a. Madan, A. 1984. Devices using fluorinated material. See Ref. 5, pp. 245–83

50. Carlson, D. E., Wronski, C. R. 1977. Solar cells using discharge-produced amorphous silicon. *J. Elec. Mat.* 6:95–106

51. Hamakawa, Y., Okuda, K., Takakura, H., Okamoto, H. 1984. a-Si:H/poly Si:H stacked solar cell having more than 12.5% efficiency. See Ref. 19, pp. 1386–87

52. Nakano, S., Kawada, H., Matsuoka, T., Kiyama, S., Sakai, S., et al. 1984. Laser patterned integrated type a-Si solar cell module. *Proc. 1st Int. Photovoltaic Science and Engineering Conf.*, pp. 583–86. Tokyo: Japan Convention Services, Inc.

53. Mitchell, K. W. 1984. Device characterization and analysis of thin-film silicon:hydrogen alloy solar cells. See Ref. 52, pp. 687–89

54. Ohnishi, M., Nishiwaki, H., Tsuda, S., Nakamura, N., Nakano, S., et al. 1984. New fabrication method for a-Si solar cells. See Ref. 19, pp. 70–75

55. Sakai, H., Maruyama, M., Yoshida, T., Ichikawa, Y., Kamiyama, M., et al. 1984. Preparation of a-Si:H films and devices in the interdigital-vertical-electrode deposition apparatus. See Ref. 19, pp. 76–81

56. Morel, D. L., Wieting, R., Mitchell, K. W., Rumburg, J. P. 1984. Design and performance of thin film silicon hydride tandem modules. See Ref. 52, pp. 567–70

57. Konagai, M., Su Lim, K., Sichanugrist, P., Komori, K., Takahashi, K. 1982. The effect of residual impurity B or P on photovoltaic properties of amorphous silicon solar cells. See Ref. 28, pp. 1321–26

58. Kuwano, Y., Ohnishi, M., Nishiwaki, H., Tsuda, S., Shibuya, H. 1981. Photovoltaic behavior of amorphous Si:H and Si:F:H solar cells. See Ref. 27, pp. 698–703

59. Okaniwa, H., Nakatani, K., Yano, M., Asano, M., Suzuki, K. 1982. Preparation and properties of a-Si:H solar cells on organic polymer film substrate. *Jpn. J. Appl. Phys.* 21 (Suppl. 21–2):239–44

60. Morel, D. L., Rumburg, J. P., Gay, R. R., Turner, G. B., Mitchell, K. W. 1984. The effect of module design on energy delivery performance of thin film silicon:hydrogen power modules. See Ref. 19, pp. 374–78

61. Curry, R., ed. 1984. *Photovoltaic Insider's Report*. Vol. III, No. 11 (Nov.) Suppl.

62. Taylor, R. W. 1982. *Photovoltaic Balance of System Assessment*, EPRI AT: 2474 (June)
63. Mitchell, K. W. 1978. High temperature operation of two-junction photovoltaic converters. *Proc. Intl. Electron Device Meet., Washington, DC*, Dec. 4–6, pp. 254–57
64. LaMorte, M. F., Abbott, D. H. 1980. Cascade solar cell design for high temperature operation. *Trans. Elect. Dev.* 27: 831–40
65. Fan, J. C. C., Tsaur, B. Y., Palm, B. J. 1983. High-efficiency crystalline tandem cells. *Proc. SPIE* Vol. 407, *Photovoltaics for Solar Energy Applications* II: 73–87
66. Wu, C. H., Williams, R. 1983. Limiting efficiencies for multiple energy-gap quantum devices. *J. Appl. Phys.* 54: 6721–24
67. Mitchell, K. W. 1981. High efficiency concentrator cells. See Ref. 27, pp. 142–46
68. Rothwarf, A. 1984. An assessment of tandem solar cell prospects using amorphous silicon alloys. See Ref. 19, pp. 712–14
69. Mitchell, K. W. 1984. Detailed modeling of multiple junction thin film solar cells. See Ref. 52, pp. 691–94
70. Ovshinsky, S. R. 1984. Roll-to-roll mass production process for a-Si solar cell fabrication. See Ref. 52, pp. 577–82
71. Okaniwa, H., Nakatoni, K. 1983. Flexible substrate solar cells. See Ref. 4, pp. 239–50
72. Den Boer, W., VanStrijp, R. M. 1983. Computer simulation of the optical behavior of amorphous silicon solar cells. *Solar Cells* 8: 169–78
73. Deckman, H. W., Roxlo, C. B., Wronski, C. R., Yablonovitch, E. 1984. Optical enhancement of solar cells. See Ref. 19, pp. 955–60
74. Deckman, H. W., Wronski, C. R., Witzke, H. 1983. Fabrication of optically enhanced thin film a-Si:H solar cells. *J. Vac. Sci. Tech.* A1: 578–87
75. Deckman, H. W., Dunsmuir, J. H. 1982. Natural lithography. *Appl. Phys. Lett.* 41: 377–79
76. Crandall, R. 1980. Field-dependent quantum efficiency in hydrogenated amorphous silicon. *Appl. Phys. Lett.* 36: 607–8

77. Plattner, R. D., Pfleidere, H., Rouscher, B., Kruhler, W., Moller, M. 1981. Spectral response and capacitance measurements of a-Si solar cells. See Ref. 27, pp. 917–21
78. Gutkowicz-Krusin, D., Wronski, C. R., Tiedje, T. 1981. Carrier collection efficiency of a-Si:H Schottky-barrier solar cells. *Appl. Phys. Lett.* 38: 87–89
79. Okamoto, H., Kida, H., Nonomura, S., Hamakawa, Y. 1983. Variable minority carrier transport model for amorphous silicon solar cells. *Solar Cells* 8: 317–36
80. Faughnan, B., Moore, A., Crandall, R. 1984. Relationship between collection length and diffusion length in amorphous silicon. *Appl. Phys. Lett.* 44: 613–15
81. Hamakawa, Y., Okamoto, H. 1984. Device physics and optimum design of the amorphous silicon solar cells. See Ref. 4, pp. 192–203
81a. US Dept. of Energy Publication. 1984. *Photovoltaics Electricity from Sunlight.* DOE/NBMCE-1075
82. Taylor, R. 1983. *Photovoltaic Systems Assessment: An Integrated Perspective.* EPRI Rep. AP-3176-SR
83. DeMeo, E. A., Taylor, R. W. 1984. Solar photovoltaic power systems: electric utility R&D perspective. *Science* 646: 245–51
84. *Amorphous Silicon Research Project, Five Year Research Plan.* 1984. SERI Rep. SP-211-2350
84a. Shushnar, G., Reinoehl, R. F., Arnett, J. C. 1984. Analysis of Costs for a 5MW Photovoltaic Power Plant. *Proc. 1st International Photovoltaic Science and Engineering Conf.* Tokyo, Japan: The Japan Times, Ltd.
85. Maycock, P. 1984. *Photovoltaic technology performance cost and market forecast to 1995: a strategic technology and market analysis.* Photovoltaic Energy Systems, Inc., Alexandria, Va, Sept.
86. Witwer, J. G. 1983. *Photovoltaic Power Systems Research Evaluation.* EPRI Rep. AP-3351
87. Marchetti, C., Nakienovic, N. 1979. *Market Penetration in Energy in a Finite World*, p. 253. Prog. Leader W. Hafele, Cambridge, Mass: Ballinger
88. Touryan, K. J. 1984. U.S. solar route. *Geopolitics of Energy* 6: 4–9

Ann. Rev. Energy. 1985. 10 : 35–52
Copyright © 1985 by Annual Reviews Inc. All rights reserved

THE THREE MILE ISLAND UNIT 2 CORE: A POST-MORTEM EXAMINATION

R. S. Denning

Nuclear Technology and Physical Sciences Department, Battelle's Columbus Laboratories, 505 King Avenue, Columbus, Ohio 43201

INTRODUCTION

On March 28, 1979, an accident occurred at the Three Mile Island Unit 2 (TMI-2) nuclear reactor that had a dramatic impact on public perceptions about the potential hazards of nuclear power. The direct consequences to the health of the neighboring public were extremely small, but the belief that the consequences could have been much more severe reduced confidence in the nuclear industry's ability to build and operate plants and in the federal government's ability to regulate them safely. Not surprisingly, the accident also had a major effect on the regulatory agency, the US Nuclear Regulatory Commission (NRC). Some aspects of the accident presented evidence of deficiencies in the existing regulatory process and in the regulations that control the design and operation of nuclear power plants. The group most affected by the accident, however, has been the nuclear industry. Utilities that own nuclear plants have been required to make costly modifications to existing units and those under construction. Furthermore, nuclear power plants, which at one time had appeared to be the most economic means of ensuring the availability of inexpensive power for projected future demands, were now recognized as potential economic liabilities. No new nuclear power plant has been ordered in the United States since 1978, and a number of plants under construction have been cancelled.

One purpose of this article is to review what actually happened in the accident, as revealed by instrumentation at the time of the accident, by post-accident analysis, and by data collected during the continuing recovery operation. The article provides a general description of the current physical

35

0362–1626/85/1022–0035$02.00

state of the TMI-2 core, recognizing that much more information will be obtained in the next year, during defueling operations. Another purpose of this article is to look beyond the details of the TMI-2 accident sequence to describe the implications of this one sequence to our understanding of severe accident behavior, of the potential consequences of reactor accidents, and of the safety of nuclear power plants in general.

DESIGN AND OPERATION OF A NUCLEAR POWER PLANT

The fault that initiated the TMI-2 accident was a minor system failure. In a large, complex system like a nuclear power plant, equipment failures will occasionally result in the need to shut down the reactor. In recognition of the possibility of these events, safety systems are provided to permit the plant to be brought to a safe shutdown condition. In particular, a number of key safety functions must be established and maintained.

To explain why these safety functions are important, it is necessary to review briefly how a nuclear power plant operates. Figure 1 is a schematic drawing of a nuclear plant of the Three Mile Island type. The uranium that

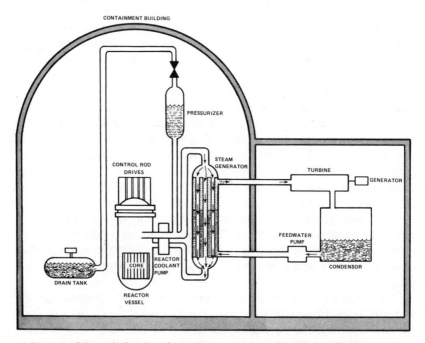

Figure 1   Schematic drawing of a nuclear power plant of the Three Mile Island type.

fissions to produce nuclear energy is in the form of uranium dioxide, a ceramic material with a high melting temperature. The uranium dioxide, in cylindrical pellets, is contained in long, thin tubes of Zircaloy, an alloy of the metal zirconium similar to steel but with better nuclear properties for use in a reactor. These fuel rods are grouped into bundles and stacked vertically to form the core region of the reactor. The core of the reactor is approximately a right circular cylinder, 12 feet in diameter and 12 feet high.

When control rods of neutron-absorbing material are withdrawn from the core, the reactor "goes critical" (a self-sustaining chain reaction is established), releasing energy in the fuel. This energy is transferred from the fuel to water that is pumped through the core. The heated water (primary system water) then flows through piping to the steam generator, where the heat is transferred through tubes to boil water (secondary system water) to produce steam. In the TMI-2 type of reactor the primary system is maintained at high pressure [2200 pounds per square inch absolute (psia)] so that it will not boil. The secondary system is at lower pressure (1000 psia), so that boiling will occur on the secondary side of the steam generator. The steam thus produced turns a turbogenerator that produces electricity. The steam is subsequently condensed and returned to the steam generator as feedwater.

Fission involves the interaction of a neutron and a uranium atom to produce two atoms of lighter weight, more neutrons (on the average approximately 2.5), and energy. The energy, of course, is eventually converted to electricity. The neutrons released by the reaction proceed to cause other fissions, in a chain reaction. The lighter atoms, called fission products, remain largely in the matrix of the fuel where they were produced.

Despite the popular misconception, a nuclear reactor cannot explode like a nuclear weapon. It is physically impossible. The concern about reactor safety is entirely associated with the possibility of the release of fission products from the plant. Fission products are radioactive (emit ionizing radiation, which is potentially harmful to living tissue). As long as the fission products are retained in the fuel rods, members of the public cannot be exposed to this radiation. If fission products are released from the fuel and carried by air or water outside the plant, then people can be exposed to radiation, with possible harmful effects depending on the level of exposure. A number of barriers in the plant prevent this from happening. The Zircaloy tubing that contains the fuel is sealed to prevent those fission products that migrate out of the fuel matrix from escaping. The reactor coolant system pressure boundary (reactor vessel and piping) is leak-tight, providing the second barrier to release. Finally, an additional leak-tight barrier is provided by the containment building, which is designed to withstand high pressure loads without leaking.

In the event of an accident or off-normal condition requiring shutdown, a few key functions must be established and maintained to place the reactor in a safe, stable condition and to protect the barriers to fission product release. These functions are illustrated in Figure 2. The first key function is to turn off the fission reaction by inserting the neutron-absorbing control rods. Although this greatly reduces the heat being produced in the fuel, it does not completely stop heat production, since the radioactive fission products release heat as well as radiation.

The second key function is to provide a pathway for transferring heat from the fuel to the environment. For heat to be removed from the fuel, the fuel must be covered with water. The emergency core cooling system is designed to supply water to the primary system in case it leaks from the system in a pipe break accident or is boiled away. During powered operation, heat is removed from the reactor coolant system through the steam generator. In an accident, heat may also be rejected through the steam generators or, when the system has been depressurized, through the residual heat removal system, which is the system used to remove heat when the reactor is shut down for refueling. Eventually the heat is released to the atmosphere, through a cooling tower, or to a body of water, such as an adjacent river.

The third key function is to remove radioactivity (e.g., fission products) from the containment atmosphere. The TMI-2 type of plant is equipped with a water spray system that can effectively sweep radioactive material from the containment atmosphere.

The final function is to ensure containment isolation. This system closes

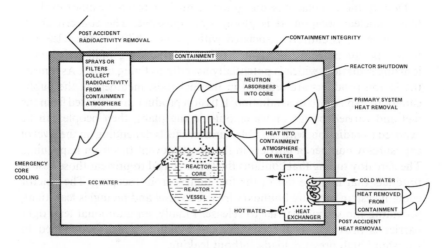

Figure 2   Safety functions of a nuclear power plant.

any valves or openings from the containment that could serve as a pathway for fission product release in an accident.

REVIEW OF THE ACCIDENT

The accident began at 4:00 AM[1] on the morning of March 28, 1979. This is time zero for the following description of the accident scenario. Monitoring systems, sensing a disturbance, automatically stopped the normal operating (main) feedwater system, which provides water to the secondary side of the steam generators, and activated the emergency feedwater system. However, the system did not operate properly. A testing error, days before the accident, had left block valves closed in the lines of the emergency feedwater system. Lights on the control panel should have indicated the improper positions of the valves, but operators did not recognize the misalignment, apparently because of the profusion of other indicators and lights. The emergency feedwater pumps were turned on, but no water flowed to the steam generators; thus, one of the essential functions, the ability to remove heat from the primary system, was lost early in the accident. Although subsequent analyses have indicated that malfunction of the emergency feedwater system did not have a major impact on the course of the accident (the closed valves were discovered eight minutes after the accident began), it is one of the first examples of the role of human error in the accident.

Within one second of shutdown of the main feedwater system, the turbine and generator were also tripped. Following turbine trip, energy removal through the steam generators was less than the reactor power, and the pressure in the primary system began to rise. At three seconds a pilot-operated relief valve opened at the top of the pressurizer to relieve pressure, and at eight seconds the reactor shutdown system was automatically activated on a high-pressure signal. The neutron-absorber rods were rapidly inserted into the core, and the chain-reaction terminated.

By 13 seconds after the start of the accident, the pressure had decreased below the set point of the relief valve, at which it should have closed. Instead it remained open. This failure was the direct cause of the TMI-2 accident. For a number of hours this valve remained stuck open, allowing reactor coolant to be lost from the system. High temperatures, measured by a thermocouple downstream from the valve, should have indicated to the operators of the plant that the valve had not closed. Because of a history of high leakage through the valve, however, the operators were accustomed to high temperature readings from this instrument, and they did not recognize the elevated temperature as unusual.

[1] The event times are based on values reported in References 2 and 9.

One feature of the accident that was particularly confusing to the operators was the behavior of the water level in the pressurizer (see Figure 1). In normal operation the pressurizer is approximately half full of water. The pressure in the reactor coolant system is maintained by controlling the temperature in the pressurizer using electrical heaters. With an open valve at the top of the pressurizer, the pressurizer filled with water and remained essentially filled throughout the accident. In the meantime, however, the mass of water in the balance of the primary system was decreasing. Instead of single-phase liquid water, a mixture of steam and water was being pumped around the reactor coolant system by the reactor coolant pumps. Operators did not recognize the possibility that the water inventory in the reactor coolant system could be decreasing when the pressurizer at the top of the system was full of water. Thus, when the emergency core cooling system was automatically activated to resupply water to the reactor coolant system, the operators reduced the flow of water makeup to prevent overfilling the pressurizer.

Within the first hour of the accident, the amount of steam in the reactor coolant system was high enough to cause vibrations in the reactor coolant pumps. At 1 hour and 13 minutes, two of the four reactor coolant pumps were turned off, and at 1 hour and 41 minutes the last two coolant pumps were turned off. Figures 3 and 4 illustrate the condition of the water in the reactor coolant system before and after the pumps were shut down. At this point, the liquid water and entrained vapor separated, and the water level fell to the top of the core.

As water continued to boil in the core, the water level dropped below the top of the core. The uncovered fuel began to heat above the boiling temperature of the water. Thermocouples above the core, showing steam temperatures higher than the boiling point, should have unambiguously indicated to the operators that the core was exposed. Unfortunately, the operators were unable to interpret the significance of these readings, and they remained unaware that water was being lost from the system and that the core was being threatened.

The next 80 minutes was the critical period for the core. The water level dropped to a low level in the core. The exposed fuel and control rod materials were severely damaged, fission products were released from the fuel, and hydrogen was produced by the reaction of steam with overheated Zircaloy.

At 2 hours and 18 minutes a block valve in the same line as the stuck-open pilot-operated relief valve was closed by the operators, stopping the loss of primary coolant from the reactor. After unsuccessful attempts to restart reactor coolant pumps, at 2 hours and 54 minutes one of the reactor coolant pumps was operated, pushing some water into the core, and quenching and fragmenting the overheated fuel. Twenty-six minutes later, the emergency core cooling system was finally started manually, refilling the core region with water. Portions of the core were so badly damaged, however, that they remained blanketed in steam for a number of days.

Hydrogen produced while the core was uncovered was released to the containment over the next few hours. At 9 hours and 50 minutes, rapid combustion of the hydrogen (deflagration) occurred in the containment building, rapidly increasing the containment pressure. Sensing high pressure, monitoring systems activated the containment spray system automatically. The sound of the explosion was heard in the control room, but not until a day later was the significance of the event fully recognized.

At 15 hours and 50 minutes a reactor coolant pump was operated, flow was established through the loops, heat was transferred to the steam

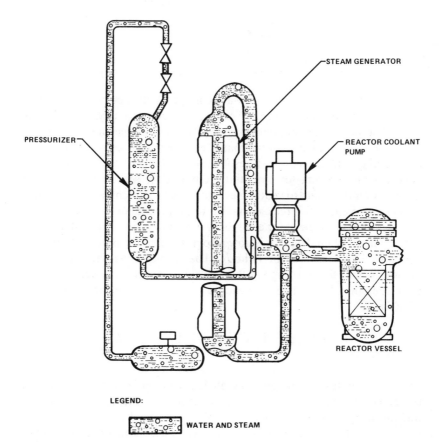

Figure 3   Reactor primary system conditions before pump trip.

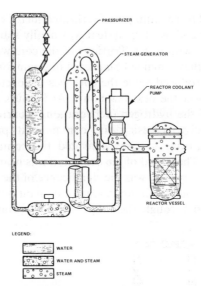

Figure 4   Reactor primary system conditions after pump trip.

generator, and a stable situation was achieved. At this point, the accident phase was essentially terminated. Yet it was not until two days later that the severity of the accident became apparent and concern was raised for public safety. Leakage of highly contaminated water from systems that penetrate the reactor containment had resulted in a difficult radioactive waste management problem for the utility. On Friday, March 30, it became necessary for the utility to vent some fission gases to the environment. It was the measurement of this release that led to the arousal of the NRC and the consideration of the need for emergency action to protect the public. With the advantage of hindsight, it is now recognized that the potential threat to the public had been over for 48 hours. The problems that faced the NRC, the utility, and the governor of the State of Pennsylvania were a lack of understanding of what had happened, as well as a severe communications problem.

RESPONSE OF THE NRC AND THE NUCLEAR INDUSTRY

The nuclear industry is closely regulated. The potential hazards of nuclear power have been recognized since the concept of producing electricity from nuclear energy was first proposed. How, then, could an accident of this severity have happened? Following the accident, a number of commissions were established by different groups to answer this question. The best known were the President's Commission on Three Mile Island (1) for the executive branch, headed by John Kemeny, and the TMI Special Inquiry Group (2) for the Nuclear Regulatory Commission, headed by Mitchell Rogovin. A combination of malfunctions and errors were found to be the direct cause of the accident. More important, these problems were found to have arisen from some basic deficiencies in the way plants are regulated and operated, and particularly in the training of operators.

Although equipment mulfunctions were involved in the accident (the most important being the failure of the relief valve to close), in general the design of equipment (including safety systems) appears to have been adequate. The design philosophy for nuclear power plants recognizes that equipment malfunctions will occur. Provisions are made for redundant and diverse safety systems to ensure that these malfunctions will not lead to conditions that can severely damage the plant or threaten the safety of the public. Damage occurred to the TMI-2 plant not because the equipment was inadequate, but because the operating staff misinterpreted the condition of the plant and either prevented safety systems from functioning or failed to use them effectively. Furthermore, the poor performance of the operating staff in controlling the accident did not result from the quality of the staff on duty. Rather, in general there had been inadequate training in severe accident behavior and inadequate consideration of human factors in the design of nuclear power plant control rooms. The design and licensing practices had emphasized the performance of hardware but had not recognized the role of the operator in severe accident response.

The various review committees made a number of recommendations for changes in the regulatory process, the operation of nuclear power plants, and the training of operators, and also recommended specific hardware modifications. A plan for implementing these changes was developed by the NRC (the TMI Action Plan) (3, 4); it has been generally carried out over the last several years.

In addition, the Nuclear Regulatory Commission undertook a major research effort to develop a better understanding of severe accident behavior (5). Phase I of this research effort is nearing completion. In addition to undertaking basic studies of severe accident processes, the NRC is using the result of this research to (6)

1. Reassess the risk to the public from severe accidents in nuclear power plants.
2. Evaluate the cost-effectiveness of additional modifications to plant hardware or operations beyond those in the TMI Action Plan.
3. Reassess the potential off-site consequences of severe accidents and modify existing regulations, if appropriate.

In 1975 the Nuclear Regulatory Commission had issued the results of a major study (WASH-1400) (7) of the risk to the public from nuclear power reactor accidents. This was the first serious attempt to predict the likelihoods and consequences of severe accidents. The results of WASH-1400 have been used extensively in characterizing the risk from reactors, in planning for emergency response in the event of an accident, and in preparing environmental impact statements for nuclear power plants. The risk predicted in WASH-1400 is very small in comparison to other risks to which the public is exposed (such as those from natural events such as tornadoes, lightning, and earthquakes, and from man-caused events like automobile accidents and dam failures). However, the potential consequences of some types of accidents (of very low probabilities) were predicted to be quite large, with the potential for lethal doses of radiation to people near the plant. Following the TMI-2 accident, however, questions arose about whether the severe consequences predicted for some accident types in WASH-1400 were realistic or whether a more detailed analysis of these accidents would indicate much smaller consequences.

Despite the severity of damage to the reactor core and the large quantity of fission products released from the fuel in the TMI-2 accident, the actual release of radioactivity from the plant was quite small. The NRC has undertaken a reassessment of the methods used to predict the release of radioactivity in severe accident sequences to determine if previous analyses have been too conservative. In general, the more detailed computer models that have been developed since the TMI-2 accident indicate that the off-site consequences of severe accidents would be substantially less than suggested by earlier analyses (8).

For example, in the station blackout sequence (loss of all off-site and emergency on-site power), which dominated the risk from the TMI-2 type of plant in WASH-1400, 70% of the core inventory of radioactive iodine was predicted to be released to the environment. In contrast, improved analyses indicate that only about 5% of the radioactive iodine would be released in this accident sequence if the containment were to fail early in the accident. If, as the analyses indicate, the containment remains intact for more than 12 hours, the estimated release would be less than 0.3%. A reduction in the estimated release of the volatile fission products (such as iodine) by one to two orders of magnitude would virtually eliminate the potential for lethal doses of radiation to the public in this accident. Since the effectiveness of fission product retention mechanisms depends on the accident sequence and the plant type, the possibility of public fatalities due to radiation sickness in a severe nuclear accident cannot be completely precluded. They appear to be even less likely than previously thought, however. The

American Physical Society is undertaking an independent review of the improved methods for analyzing the release of fission products in severe accidents. Its report is scheduled for release in early 1985.

The nuclear industry has also taken action to prevent the occurrence of another TMI-2 accident. An Institute for Nuclear Power Operations (INPO) was established to provide a mechanism for self-regulation by the industry. The institute focuses on the management of operations at nuclear power plants, evaluating the training of staff and the adequacy of procedures. INPO also undertakes periodic inspections of plants and recommends improvements to plant managers.

A Nuclear Safety Analysis Center (NSAC) was established by the nuclear industry to collect and analyze data from the TMI-2 accident (9). NSAC has subsequently been incorporated into the Electric Power Research Institute (EPRI), the centralized research organization of the electric utilities. The industry has also funded a major research effort through EPRI to develop a better understanding of severe accident behavior. The NRC and EPRI research efforts are largely complementary.

Additionally, the industry supports a program to evaluate the need for modifications in the design and operation of nuclear power plants to protect the public against the risk of severe accidents. The first phase of the Industry Degraded Core Rulemaking (IDCOR) (10) program was completed in 1984. The results of the program have been presented to the Nuclear Regulatory Commission through a series of technical information exchange meetings. Although final decisions on the need for additional hardware modifications (beyond the TMI Action Plan) for nuclear power plants must await the mid-1985 completion of some studies by the NRC, preliminary conclusions of the NRC evaluations (6) appear to be consistent with those of the IDCOR program. Major modifications to existing plants do not appear warranted.

Under the sponsorship of the US Nuclear Regulatory Commission and the Electric Power Research Institute, a number of large experimental programs have been undertaken in the United States to develop a better understanding of severe accident behavior. It is difficult and expensive to simulate the conditions typical of a severe accident. For this reason the TMI-2 accident in itself represents a unique research opportunity. Many severe accident processes that are difficult to reproduce in an experimental facility occurred in the TMI-2 accident. To ensure that potentially valuable research data are not lost in the process of recovery from the accident, a four-party agreement was arranged between General Public Utilities (the power plant's owner), the US Department of Energy, the Nuclear Regulatory Commission, and the Electric Power Research Institute. The

GOSHEN COLLEGE LIBRARY
GOSHEN, INDIANA

Department of Energy has since established a Technical Integration Office that has overseen the collection of data at the TMI-2 plant that have research significance (11).

POST ACCIDENT EXAMINATION OF THE PLANT

Recovery and cleanup of the TMI-2 plant have required a herculean effort. If the TMI-2 accident has indicated that the potential health effects of reactor accidents are less than previously thought, it has also indicated that the economic impact of an accident can be disastrous. Decontamination and recovery of the plant have proceeded through a number of stages. Before entering the containment, it was necessary to decontaminate the adjacent auxiliary and fuel handling buildings. Cleanup and chemistry control of reactor coolant system water are performed in the auxiliary building. Leaks from this system, which are a minor maintenance problem during normal operation of the plant, became a major pathway for the leakage of radioactivity and contamination of these outer buildings after core damage had occurred.

Another major obstacle to entry into the containment was the need to release the radioactive noble gases from the containment atmosphere to the environment. This purging of the containment atmosphere, although consistent with radiation protection standards for the release of radioactive materials, was not surprisingly met with public resistance. Purging of the containment was undertaken from June 28 to July 11, 1980.

The first human entry of the containment took place on July 23, 1980, and since that time a large number of entries have been made. As expected, all levels of the building were found to be highly contaminated. Very high radiation levels were found in the basement area, which was flooded with approximately 700,000 gallons of water during the accident. Samples of this water and the layer of sludge on the floor of the basement were found to contain large quantities of the more volatile radioactive fission products from the core. Signs of the hydrogen burning event were also evident, in the forms of charred paint and electrical cables and a partially melted telephone.

The next major step in plant recovery involved decontaminating the upper floors of the containment building, to reduce recovery crews' exposure to radiation, and then processing the contaminated water in the basement.

In June 1982 (12) a small TV camera was inserted through a penetration in the upper head of the reactor vessel down into the core region. Surprisingly, the TV pictures indicated a large voided region near the top of the core and evidence of significant core disruption. In September and

October 1983, small samples of fuel debris were remotely collected and removed from the core. The following picture of the current condition of the core has emerged. Significant portions of the reactor core have been turned into rubble. In general a few rows of intact rods stand at the periphery of the core region. The top $4\frac{1}{2}$ feet of the central portion of the core region has been voided (13). Below the void is a debris bed of fuel rubble of unknown depth (debris samples have been taken from as deep as 22 inches into the debris). The fuel debris is gravel-like in appearance and contains fragmented fuel and cladding material, as well as some porous material that was once molten and has resolidified. The top of the debris pile is very uneven, indicating that preferential regions of higher flow were established. Figure 5 shows a profile of the voided region of the core.

In early 1985, when this article was edited, defueling of the reactor was expected to begin in the spring of 1985. The loose debris was to be effectively vacuumed out of the core region until a more solid structure was reached. Whether the more resistant layer would be composed of stubs of the lower

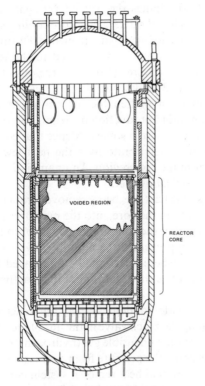

Figure 5   Profile of damaged core.

ends of fuel assemblies or a solidified layer of previously molten material, or whether debris would fill the entire core region, was still the subject of speculation.

The data collected from TMI-2 are consistent with the following scenario of core damage during the accident. The water level dropped to a minimum level of approximately three feet from the bottom of the core (based on an interpretation of the behavior of in-core neutron detectors). Within the next six months it should be possible to verify or alter the estimated minimum water level based on probing and removal of the debris bed. Core heat-up and some liquefaction of zirconium and uranium dioxide core material occurred; significant transport of molten core material also occurred from the core region to the lower plenum of the reactor vessel. Silver-indium-cadmium control rod material must also have melted in the accident. Droplets of control rod material have been found on upper plenum surfaces. When a reactor coolant pump was operated at 2 hours and 54 minutes, the introduction of reactor coolant into the region of overheated and partially liquefied fuel resulted in quenching and fragmentation of the fuel. Over the next 20 minutes the water level again decreased and fuel debris was reheated. When the emergency core cooling system pumps were actuated, the overheated fuel was again quenched. During the next 12 hours of the accident, makeup water was provided to the reactor in sufficient quantity to keep the core covered (14). Some regions of the core remained quite hot, however. At 3 hours and 45 minutes a change in geometry (perhaps the slumping of molten fuel or control material) apparently occurred, resulting in rapid production of steam. Based on the behavior of in-core instrumentation, there is some evidence that conditions in the core may have continued to deteriorate over the next few hours despite the availability of water in the core region. An unambiguous interpretation of these data is not possible, however.

Over most of this time heat was being removed from the core by makeup water that flowed through the core, into the pressurizer, and out the relief valve into the basement of the containment. Hydrogen in the piping of the reactor coolant system prevented the natural convective flow of steam to the steam generators, where heat could be removed from the primary system. Figure 6 illustrates the condition of the reactor coolant system in this period.

An attempt was made to depressurize the reactor coolant system completely by opening the block valve in the line of the pilot-operated relief valve. The amount of noncondensible gas (hydrogen) trapped in the reactor coolant system prevented the pressure from decreasing to a level (less than 300 psi) at which the residual heat removal system could be operated. It is unlikely that this system would have been effective anyway.

Attempts were made during this time to restart some of the reactor coolant pumps. Apparently hydrogen trapped in the pumps prevented their operation. After approximately 10 hours enough hydrogen had escaped from the loop containing the pressurizer that some convective flow to the steam generator was established. At 15 hours and 50 minutes one of the reactor coolant pumps in this loop was successfully operated. At this point

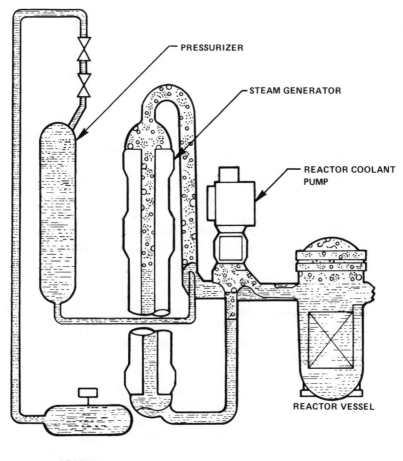

LEGEND:

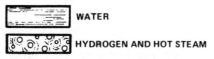

Figure 6   Reactor primary system conditions following core recovery.

the essential functions had been established. The block valve was closed, water was flowing through the core region, and heat was being rejected from the primary system through a steam generator. After 16 hours the plant was finally under control.

Following operation of the reactor coolant pump, the finer core debris was transported around the system, and some of the debris apparently was trapped in the lower plenum of the reactor vessel. Some of the lower regions of the core or lower plenum were apparently highly blocked, by either refrozen material or highly packed fine core debris, since thermocouple readings from these regions remained above the boiling temperature for a number of days after the reactor coolant pumps were restored to operating.

Approximately 50% of the core inventories of the fission products iodine, cesium, and noble gases (xenon and krypton) were released from the fuel (15). These are very large releases of radioactivity, similar (within a factor of 2) to the release from fuel that would have occurred in a complete core meltdown accident. Much of this release was probably due to the quenching and fragmentation of the fuel, however, rather than being the direct result of high temperatures.

Some of the fission products released from the fuel were deposited on structural surfaces in the reactor coolant system. Some of these structures have been removed from the upper plenum of the vessel and analyzed. The total quantities of fission products on these structures are small. However, it is likely that during the accident the quantities deposited were larger and have since dissolved in the flowing coolant water.

It is unlikely that a major source of fission products, other than noble gases, was ever airborne in the containment, since the principal pathway for release (through the pressurizer) was always blocked with water. Although the contamination level (primarily Cesium 137) was a cleanup problem, the quantity of fission products on the containment walls was a very small fraction of the core inventory.

Approximately 50% of the zirconium in the core was oxidized in the accident, resulting in the release of hydrogen to the containment atmosphere. It is estimated that approximately 700 pounds of hydrogen were consumed in the combustion event (16), introducing an overpressure of 28 pounds per square inch gage on the containment. Although the resulting pressure was close to the design pressure of the containment, it was well below the containment's expected failure pressure.

SUMMARY AND CONCLUSIONS

At this time there can be no question as to the severity of the TMI-2 accident. The fuel reached temperatures that resulted in liquefaction of core materials, major changes in core geometry, large releases of radioactive fission products from the fuel, a major release of hydrogen to the containment, and combustion of hydrogen in the containment. How close was the accident to complete core meltdown? The question is unanswerable. The length of time it took to quench the core completely (many days) certainly indicates severe blockage, and possibly the approach to a condition that could not have been reversed within the reactor vessel.

Would the accident have been significantly worse if it had gone to complete meltdown? Probably not. The release of fission products from fuel and the challenges to containment integrity were comparable to those predicted for complete core meltdown accidents. The cleanup problem, which is already a major economic strain on the utility's finances, would have been even greater, however, if complete core melting had occurred and the reactor vessel had been breached.

As a result of the accident, some deficiencies were identified in our understanding of severe accident behavior, in the manner in which reactors are regulated, and in the manner in which plants are operated. Improvements have been made in each of these areas. Although it is difficult to quantify how much these improvements have reduced the risk from severe accidents, there can be little question that the reactors are safer today as a result.

Finally, it is essential that the TMI-2 accident be placed in proper perspective. The three major consequences of the accident were: psychological damage to the neighboring population; economic loss to the utility, its stockholders, and its ratepayers; and increased radiation exposure to the cleanup crews. The most severe accident in the history of commercial nuclear power did not result in the loss of a single life.

Literature Cited

1. Kemeny, J. G. 1979. Report of the President's Commission on the Accident at Three Mile Island. Washington, DC: USGPO
2. Rogovin, M. 1980. Three Mile Island, Nuclear Regulatory Commission Special Inquiry Group. Washington, DC: USGPO
3. US Nuclear Regul. Commission. August, 1980. NRC Action Plan Developed as a Result of the TMI-2 Accident, NUREG-0660
4. US Nuclear Regul. Commission. October 31, 1980. Clarification of TMI Action Plan Requirements, NUREG-0737
5. Larkins, J. T., Cunningham, M. A. Jan. 1983. Nuclear Power Plant Severe Accident Research Plan, NUREG-0900
6. NRC Policy on Future Reactor Designs: Decisions on Severe Accident Issues in Nuclear Power Plant Regulations, NUREG-1070. To be published
7. Reactor Safety Study: An Assessment of Accident Risks in U.S. Commercial Nuclear Power Plants, WASH-1400. Oct. 1975
8. Gieseke, J. A., Cybulskis, P., Denning, R. S., Kuhlman, M. R., Lee, K. W., Chen, H. July, 1984. Radionuclide Release Under Specific LWR Accident Conditions, Vol. 5, BMI-2104
9. Analysis of Three Mile Island—Unit 2 Accident, NSAC-80-1. Revised March, 1980
10. Ground Rules for the Industry Degraded

Core Rulemaking Program, LR 81-017-01, Tech. Energy Corp., March 31, 1982

11. Carlson, J. O., ed. 1984. TMI-2 Core Examination Plan, EGG-TMI-6169

12. UPDATE—TMI Unit 2 Technical Information and Examination Program, Vol. 4, No. 1, Dec. 15, 1983

13. UPDATE—TMI Unit 2 Technical Information and Examination Program, Vol. 4, No. 2, June 25, 1984

14. Wooton, R. O., Denning, R. S., Cybulskis, P. Jan. 1980. Analysis of the Three Mile Island Accident and Alternative Sequences, NUREG/CR-1219

15. Davis, R. J., Tonkay, D. W., Vissing, E. A., Nguyen, T. D., Shawn, L. W., Goldman, M. I. Nov. 1984. Radionuclide Mass Balance for the TMI-2 Accident: Data Through 1979 and Preliminary Assessment of Uncertainties, GEND-INF-047

16. Henrie, J. O., Postma, A. K. March, 1983. Analysis of the Three Mile Island Unit 2 Hydrogen Burn, GEND-INF-023, Vol. 4

Ann. Rev. Energy. 1985. 10 : 53–77
Copyright © 1985 by Annual Reviews Inc. All rights reserved

THE ORIGIN OF NATURAL GAS AND PETROLEUM, AND THE PROGNOSIS FOR FUTURE SUPPLIES

Thomas Gold

Center for Radiophysics and Space Research, Space Sciences Building, Cornell University, Ithaca, New York 14853

INTRODUCTION

Any estimation of future supplies of hydrocarbons—oil and gas—is necessarily bound up with the question of the origin of these deposits. The theory of a biological origin, which has become accepted almost universally, would place fairly sharp limits on what the earth might provide. The total volume of sediments, the proportion of organic matter they may contain, and the fraction of this that may have been converted into accessible hydrocarbons have all been estimated. So long as many areas of sediments on the earth remain unexplored by the drill, a prognosis based on such evaluations can be a rough judgment at best. Still, it will give some indication of what can be expected, and it is not likely to be in error by orders of magnitude.

In the early 1970s, estimates based on these considerations had indicated that oil was running out—that the earth could provide for no more than 20 to 40 years' supply at the present rate of consumption. These estimates were widely accepted, resulting in drastic increases in the price of oil. It was the estimates, and not an actual shortage, that prompted these changes, with their profound effects on the world economy.

But what if the source material for oil and gas were not limited to organic deposits, but instead these substances derived wholly or in part from materials incorporated deep in the earth at the time of its formation? This

53

0362–1626/85/1022–0053$02.00

possibility has been considered in the past, and again in recent times with modern data (1–10). In that case one would make quite different and very much more optimistic estimates of future supplies. But is there any possibility of that? Is anything coming up from deep sources in the earth?

There is little doubt that large quantities of certain fluids—the so-called "excess volatiles"—have made their way up through the crust over geologic time (11). The water of the oceans, the nitrogen of the atmosphere, and the large quantities of carbon now resident in the carbonate rocks of the sediments must have been supplied in that way. The erosion of the basement rocks that produced the other components of the sediments would have provided only a very small fraction of these substances.

There is no agreement (and very little discussion) of the form in which the "excess" carbon came to be supplied to the surface, although it is clear that much of it was at some stage in the form of atmospheric CO_2, which then dissolved in the ocean water and finally precipitated to form carbonate rocks. The amount of CO_2 in the atmosphere or the oceans at present is a very small fraction of the quantity that has cycled through the system to produce the large deposits of carbonate rocks, which amount to as much as 20% of the mass of all the water in the oceans.

The usual assumption seems to have been that all this carbon came up from its original deep sources already in oxidized form, namely as CO_2. There is no compelling reason for such an assumption, and there are several good reasons for considering an alternative possibility, namely that methane or other hydrocarbon fluids were involved. In the oxidizing circumstances of the outermost crust and the atmosphere, most of the carbon would end up as atmospheric carbon dioxide in any case, and the subsequent deposition process of the carbonate rocks would be the same. What would be different is the identification of the unoxidized forms of carbon that we find in the outer crust and in the sediments.

When it was thought that the supply of carbon had all been in the form of CO_2, any unoxidized carbon had to be considered the result of photosynthesis in plants, in which sunlight had supplied the energy necessary for the dissociation of CO_2. The burning of such fuels, then, constituted the regaining of this "fossil solar energy." But if the primary carbon had come up from depth in the form of methane or other hydrocarbon fluids, the situation would be quite different. A small fraction of this, arrested on the way up and contained at accessible levels in the outer crust, could constitute an enormously larger energy resource than all the biological deposits. The methods of prospecting for oil and gas would then be quite different, and many new locations would come under consideration, which would have been excluded if one was looking for biological deposits only.

The Opposing Viewpoints

Why should we doubt the biological origin of the earth's hydrocarbons? Has not an immense amount of work demonstrated that the locations of hydrocarbon deposits, the chemical natures and contents of the oils, and the isotopic composition of the carbon, all fit the biogenic theory? Has the biogenic theory not provided the basis for the exploration which has been so successful in the past?

This is what is generally said, but when we look at the various points in detail the situation turns out to be not nearly so clear-cut. Several leading investigators, in both geology and chemistry, have expressed doubts. Hollis Hedberg, a well-known petroleum geologist, wrote, "It is remarkable that in spite of its widespread occurrence, its great economic importance, and the immense amount of fine research devoted to it, there perhaps still remain more uncertainties concerning the origin of petroleum than that of any other commonly occurring natural substance" (12). Sir Robert Robinson, the British organic chemist and Nobel laureate, who investigated the chemical nature of petroleum, wrote, "Actually it cannot be too strongly emphasized that petroleum does not present the composition picture expected from modified biogenic products, and all the arguments from the constituents of ancient oils fit equally well, or better, with the conception of a primordial hydrocarbon mixture to which bioproducts have been added" (13). And again, "It is believed that the arguments for a biological *and* an abiological origin of petroleum are alike incontrovertible. Hence a duplex origin of mineral oil is envisaged" (14). Today we have a great deal more information, which, although rarely examined from that viewpoint, greatly strengthens the duplex origin theory.

One way of discussing this complex subject, with its numerous, apparently conflicting clues, is to discuss the historical evolution of the ideas. The theory of a biological origin became quite firmly established in the second half of the last century, although at that stage no very detailed chemical arguments could be used. Coal, found mainly in sediments, and often with fossils, was firmly regarded as a biological product, and based on this, it was considered that immense amounts of plant material had been laid down over geologic time in circumstances where they escaped destruction by oxidation. It seemed reasonable to think that some fraction of this was converted into the hydrocarbon molecules of gas and oil.

Mendeleev was on the other side of the argument. His paper (in 1870) on the origin of petroleum ended with the words: "The capital fact to note is that petroleum was born in the depths of the earth, and that it is only there that we must seek its origin" (1). He provided many arguments for his conclusion, among them the regional patterns of petroleum occurrence and their relationship to underlying structures rather than to the nature of the sediments.

The Formation of the Earth

Until the middle of this century, it was thought that the earth had formed as a hot body, had been a ball of liquid rock, as such no doubt well mixed, and had then gradually cooled, producing a differentiated crust overlying a homogeneous mantle. In such an evolutionary history, there could not have been any hydrocarbons in the early earth, for even if they had been initially supplied in the formation process, they would surely have been destroyed. To account for the supply of excess carbon to the surface, one then thought in terms of oxidized carbon only, since that would be the stable form expected in such a case. Carbon dioxide was found to come out of volcanoes, apparently confirming this viewpoint.

The present understanding of the formation process of the earth is different. It has become quite clear that the earth, as well as the other terrestrial planets, accreted as solid bodies from solids that had condensed from a gaseous planetary disk. The primary condensates, ranging in size from small grains to asteroid-sized "planetesimals," all contributed to the formation of the final earth. An accretion from small grains only, could have led to a final body that was inhomogeneous with depth, since at successive epochs different materials might have been acquired. However, when such an accretion process was punctuated by major impacts of competing bodies that had formed on collision orbits, then an orderly, layered composition would be turned into an erratic patchwork. This appears to have been the case in the formation of the earth, just as it was in the formation of all other solid planetary bodies, demonstrated by the ubiquity of impact craters on them.

There is now a great deal of evidence from trace elements and their isotopes that have come up from depth, in various parts of the earth, that the mantle is indeed quite inhomogeneous (15, 16) and its chemical composition patchy, confirming that it has never been all molten. What melting has taken place seems to have been only a small proportion at any one time, but since this partial melting resulted in outpourings that formed the entire outer crust, it produced the impression of an earth cooled from liquid.

Abundance of Hydrocarbons in Solar System

Another part of the discussion, where the outlook in earlier times was quite different from that of today, concerned the derivation of hydrocarbons in the solar system. Hydrocarbons used to be thought of as substances that were specifically biological. Methane and other hydrocarbons were clearly the results of biological processes, and there was little indication that these molecules would be produced in other ways. In fact, when methane was first detected in the atmosphere of Jupiter, this resulted in publications suggesting that some form of life on Jupiter must be responsible.

Now we know not only that there is methane in the atmosphere of Jupiter, but that by far the largest proportion of the carbon in the entire planetary system is in the form of hydrocarbons. The greatest quantity is in the massive outer planets and their satellites. Jupiter, Saturn, Uranus, and Neptune have large admixtures of carbon in their extensive atmospheres, chiefly in the form of hydrocarbons—mainly methane. Titan, the satellite of Saturn, has methane and ethane (CH_4 and C_2H_6) in its atmosphere. Its clouds are composed of these substances, and it is likely that liquid methane-ethane mixes are on the surface below, and perhaps make up oceans, rivers, and frozen polar caps, more or less as water does on the earth. The asteroids—that swarm of minor planetary bodies between Mars and Jupiter—also seem to have hydrocarbons on their surfaces, and most probably in their interiors. These small bodies cannot retain any atmosphere, and therefore gaseous hydrocarbons cannot be present there. But their dark surfaces and details of their reflection spectra give the strong suggestion that tarlike substances are prominent on their surfaces.

The meteorites are considered to provide us samples of the materials from which the planets formed. One class—the carbonaceous chondrites—contain some volatile substances, and it is this class that is thought to have supplied the earth with most of its complement of volatiles. While carbon is a minor constituent of the other types of meteorites, it is present at a level of several percent in the carbonaceous chondrites, mostly in unoxidized form, with a certain fraction in the form of hydrocarbon compounds (17). The meteorites we now see would have lost any more volatile compounds in their $4\frac{1}{2}$-billion-year flight as small objects through the vacuum of space, but one would judge that at the time of their condensation in a dense nebula, other more volatile components, including the lighter hydrocarbons, would have been present also. If this type of material was incorporated in the earth, and later subjected to heat and pressure, it would separate out some of these volatile components. The mobile forms of carbon that would be expected would be chiefly methane, with an admixture of heavier hydrocarbon molecules. The pressure, temperature, and chemical surroundings in the earth would then determine what products of this would reach the near-surface domain.

There is no abundant meteoritic material that would outgas carbon dioxide, nor would such material be expected in the chemical circumstances that are believed to have existed at the time of planetary formation. Carbon dioxide may, of course, result from the oxidation of the hydrocarbons at depth, or on their way to the surface, but it is unlikely that oxidized carbon was the primary material that was incorporated at the formation, and provided all the surface carbon we now have.

The Stability of Hydrocarbons at Depth in the Earth

Another area where the arguments have changed greatly, especially in recent times, is the stability of hydrocarbons. It used to be thought that temperatures above about 600°C would dissociate methane, and that temperatures of 300°C were sufficient to destroy most of the components of natural petroleum. Since such temperatures are reached at depths of a few tens of kilometers in the crust, it seemed pointless to discuss an origin of hydrocarbons from nonbiological sources in the mantle. If the origin had to be found in the upper and cooler parts of the crust, then there was really no alternative to the biological theory.

This viewpoint about the thermal instability of hydrocarbons came about simply because the experiments had not been done at the appropriate pressures. Hydrocarbons appear to be greatly stabilized against thermal dissociation by high pressures, and we must therefore discuss the question of stability in the pressure and temperature regime that exists in most of the crust of the earth.

Figure 1 shows the approximate relationship between pressure (or depth) and temperature in continental and oceanic regions. Only intensely volcanic zones would fall outside the band so defined. The theoretical pressure-temperature line on which methane would begin to be destroyed (more accurately defined as the line where in equilibrium 95% would remain methane in the presence of hydrogen from the remaining 5%) is entirely above the pressure-temperature regime in the earth to a depth of 30 km. Down to that depth methane appears to be essentially stable against thermal dissociation. Even below that depth it would be only partially dissociated down to a depth of more than 300 km, where the fraction remaining as methane would have decreased to 10%. Perhaps somewhere around the 600-km depth one can place the lower limit of the possible existence of methane in the earth.

The question of the stability of hydrocarbons against oxidation, with oxygen that is available in the rocks, is bound up with the details of the outgassing process. If the gases ascend in regions of magma, then each bubble that forms has access to the oxygen available at the constantly changing surface of the bubble in the liquid rock. If the chemical equilibrium were then to favor the oxidation of the hydrocarbons and the partial reduction of the magma, this would indeed take place. For many types of magma, and for the high temperatures and low pressures that occur at shallow depths in a volcanic region, chemical equilibration would certainly go in that direction, and so it is no surprise that volcanoes generally emit carbon mainly in the form of CO_2, with only minor amounts as CH_4.

Where gases make their way through solid rock, as we shall discuss later, the process involves pore spaces and pathways held open by the gas pressure. In that case, no chemical equilibrium between the rock and the gas need be expected, since the gas comes into contact with only a very limited

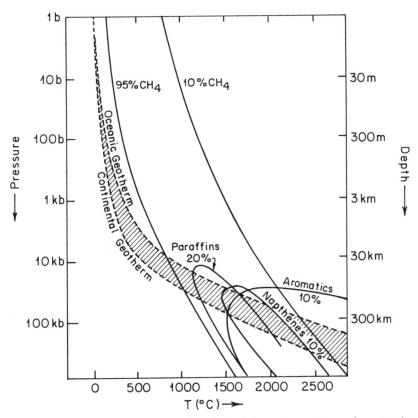

Figure 1 The shaded area represents the relation between temperature and pressure (or depth) in nonvolcanic areas. The stability limits of methane are indicated (known from experiments to approximately 1500°C). The beginning of the domains of equilibrium production of the major constituents of petroleum are shown, as deduced from thermodynamic calculations by Chekaliuk (22).

amount of rock on the surfaces of the pores. The amount of oxygen that can be made available by the rock surfaces bordering the gas-filled cracks is limited, and once this oxygen is used up no further oxidation will take place along any particular pathway. Methane has indeed been found deep in granitic rock in many locations (18–20), including the superdeep hole drilled in the Kola Peninsula (21).

What about the other hydrocarbon molecules that make up the bulk of petroleum? Thermodynamic calculations (22) have suggested not only that most of these molecules are stable in the pressure-temperature regime at a depth between 30 and 300 km, but indeed that they would form in equilibrium in a mix of carbon and hydrogen in that regime (Figure 1). If the carbonaceous chondrite type of material were packed in the earth, then, in that depth range, methane and a mix of heavier hydrocarbon molecules could be expected.

Transport of Heavier Molecules by Methane

At the high pressures ruling deep in the earth, methane, though technically a gas, behaves much like a liquid. It is a good solvent of other hydrocarbons (23, 24), and it will therefore mobilize heavier compounds, which on their own would have had viscosities too high to be mobile. It is such streams of methane that may have been making their way up towards the surface over most of geologic time and that may have supplied the bulk of the surface carbon. Heavier hydrocarbon molecules, as well as organometallic compounds and other trace substances, would be transported upward in that stream, and no doubt chemical processes would change the detailed molecular structure on the way up, as different pressures, temperatures, and catalytic actions of minerals were encountered. At shallow levels, as the pressure drops to values at which the solubility in methane becomes small, the heavier molecules will be deposited, resulting in the concentrated deposits of carbonaceous matter, including the deposits of petroleum.

Very high pressure experiments are required to define these data more precisely and to explain what chemical changes such a stream would suffer on its way to the surface.

Diamonds provide evidence that unoxidized carbon in very pure form does in fact exist at depths below 150 km. This is the depth required to produce this high-pressure form of carbon. A carbon-bearing fluid that could flow at such levels and deposit the very pure carbon must have been present to allow the mineralization to take place. Methane seems to be the best candidate for this. Fluids in high-pressure inclusions in natural diamonds have been analyzed and methane was, indeed, found to be one of the components (25, 26).

THE EVIDENCE OF "MOLECULAR FOSSILS"

The biological origin theories obtained their strongest support when molecules thought to be strictly biological were discovered to be present in most petroleum. Porphyrins and certain isoprenoids were the classes of complex molecules that seemed to be derived from molecules that are common constituents of plants and animals. It is not quite certain where we have to draw the line, especially in view of the findings in meteorites. The carbonaceous chondrites contain molecules built up almost certainly without the aid of biology, yet their virtual indistinguishability from biological products is striking. "If found in terrestrial objects, some substances in meteorites would be regarded as indisputably biological"— chemist Harold Urey concluded in 1966 (27). Quite probably some of these complex molecules form readily without the help of biology, and biology, when it developed, adopted them for its own purposes.

But even if we ignore the complex molecules in petroleum that we also find in meteorites, there still remain some specific biological markers. But do they point to the origin of the bulk substance, or merely to a contamination it has acquired? Most oils are found in porous sediments, which contain some biological materials. Oil is a good organic solvent, and, on migrating through the sediments, would necessarily pick up some of these materials. Recently, however, another important source of biological contamination has been identified; petroleum from many different sources, as well as coal, contain a class of molecules—hopanoids—derived from bacterial cell walls (28). It appears that none of these fuels were free from substantial alteration and contamination by bacteria. There is no suggestion that these bacteria provided the bulk of the unoxidized carbon compounds that led to the deposition of oil and coal; they merely invaded many, and perhaps all, existing deposits. While previously it was thought that one could distinguish biodegraded from nonbiodegraded petroleum, it now seems that a certain amount of degradation is essentially universal. All the specific biomolecules may be due to this bacterial processing, and therefore their presence does not give any evidence that the bulk source material itself was of biological origin.

Certain other properties of natural petroleum have also been identified as the results of biological contamination. Optical activity—that is a rotation of the plane of polarization of light—is rarely seen in nonbiological liquids. In petroleum, different boiling point fractions have shown various degrees of this effect, and of both signs (right-handed and left-handed). This implies that certain molecules that can form as either right- or left-handed show a numerical preference for one or other of the two types. This is common in biology, where the whole genetic system and the enzymes it creates are all asymmetric in this way. Without biology such asymmetries could also arise, but only in rather special circumstances.

What has now been seen is that these asymmetries are totally absent in oils found in reservoirs above a certain temperature (29). A sharp cutoff for the effect appears at about 66°C, and as this is much too low a temperature to result in the destruction of all the molecules concerned, this can only be interpreted as being a temperature at which a particular type of bacterial action (by no means all bacterial action) has ceased.

A similar situation exists with respect to another biological indicator frequently present in petroleum. This is the greater abundance of molecules with an odd carbon number than an even one in the paraffin component, for molecules with a carbon number in the range 20–40. It seemed improbable (though not impossible) that the frequent occurrence of this odd-even effect could be due to any nonbiological buildup or breakdown of such molecules, while biological processes were known to produce such effects. It has now been demonstrated that in the same circumstances in which the optical activity disappears in the oils of a region, as one goes from one level to a deeper and warmer one, the odd-even effect also disappears. Philippi (29) has demonstrated this for the oils of five geographically widely separated basins, and has concluded that this sharply temperature-dependent effect can be attributed only to a microbial action in which certain molecules are preferentially destroyed. He observes that this goes in step with other chemical transformations that must also be attributed to microbial action.

Biological markers are clearly abundant in petroleum. It is no longer clear, however, that any of them point to an origin of the bulk substance as a surface biological sediment produced with the aid of photosynthesis, as had been generally assumed. Microbiology in the ground is much more widespread than had earlier been recognized, and it has certainly introduced many, and quite possibly all, the biological evidence seen in petroleum.

THE CARBON ISOTOPE RATIO

The precise ratio of the two stable carbon isotopes, carbon-12 and carbon-13, is sometimes taken to identify carbonaceous material of biological origin. The organic carbon in plants is generally depleted in carbon-13 relative to the carbon-dioxide of the atmosphere, from which it was mostly derived. This is due to a selection favoring the light isotope in the chemical processes of photosynthesis. The resulting change in the abundance ratio of the two isotopes is rather large, and no comparably strong isotopic fractionation effect is noted in any single nonbiological chemical process in nature. As it was found that most petroleum and, even more, most natural methane, showed a marked depletion of the heavy isotope, it was natural to consider that this confirmed the theory of a biological origin for these substances.

It is beyond the scope of this report to go into the details of the numerous discussions in the literature of the carbon isotope evidence in relation to petroleum and natural gas (30). There is evidence, however, essential to the present discussion, for a progressive isotope fractionation process in the migration of gas and petroleum through the crust, leading to cumulative effects. As in industrial isotope fractionation processes, the final result can be a much larger fractionation than that produced in any single step. The simple argument—that the fractionation produced by biological processes is larger than any other, and that material showing such fractionation must be biological—is therefore invalidated (8). This fractionation process is discussed briefly in the appendix.

THE MECHANISM OF FLUID ASCENT THROUGH THE CRUST

If fluids bring up hydrocarbons from deep sources, it is important for several reasons to understand the details of the migration processes involved; chemical composition, trace element content, and especially the geographic and vertical distribution of deposits of oil and gas may depend on these details.

Diffusion of fluids from deep levels through the rocks of the crust is too slow to be important, even on the long time scale of geology. Only a bulk motion through cracks could provide a significant flow through the solid rocks. Transport from deep levels to the surface must therefore occur in two steps.

First, fluids must be assembled by diffusion over small distances into cracks, which are generated and held open by the fluid pressure. That, of course, is a process which must occur to initiate any form of outgassing. Volatile substances liberated under heat must separate out from the solids and build up pressure in the pores. If this process generated only small pores, diffusely distributed and separated from one another, no movement would result.

The second step in the upward transport process can only occur if a region contains a sufficiently concentrated source of fluids, so that the pores that are generated create an interconnected domain of fracture porosity (by a process of hydraulic fracturing, often called "hydrofracking" when done artificially). If such a fluid-filled domain grows to a sufficient size and volume, then buoyancy forces exerted on the light fluid in the denser rock will drive it upward. This would occur in the following way. The connected fluid-filled pores have a pressure gradient in them given by their density; the denser rock must have a higher pressure gradient in it over the same height interval. When such a porosity domain has a sufficiently large vertical extent, the pressure difference between fluid and rock will become so large that the strength of the rock is insufficient to withstand it. The fluid pressure will then become excessive at the top of the domain and force open new pore spaces, while at the same time it will be insufficient at the bottom of the domain to hold the pores open, against the greater rock pressure. In this way the fluids will force their way upward through the largely stationary rock (9, 10).

One can estimate the maximum vertical height that a domain can have before it must start this upward migration. This depends of course on the density of the fluid at that depth, on the density of the rock, and on the strength of the rock in tension (at the top of the domain) and in compression (at the bottom of the domain). Depending on these quantities, estimates would range mostly between 2 and 10 km.

This process will have many important consequences for oil and gas exploration. First it implies that only high concentrations of fluids can become mobile. Where there is an upward stream, it will soon use up all available oxygen on its pathways, and will then be largely protected from further oxidation.

At the shallow levels at which oil is found, it could have been carried only in small proportion by the stream of gas, which by then has reached a low pressure. Therefore a large amount of gas must have escaped in a region in order to have left behind a substantial deposit of oil. Chemical evidence of methane seepage at the surface overlying oil fields should then be the rule, and may make chemical prospecting much more helpful than had previously been thought. Oxidation products of methane, as well as helium and various organometallic complexes carried up in the stream, may all be useful for this purpose. Gas reservoirs, even of large size, can occur in regions of less prolific supply. They do not represent the residue of the stream, but the stream itself, and there is no need for a great oversupply to have been provided.

Regional and vertical patterns must be expected to result from such a mode of supply, in which oil and gas deposits are densely packed, and essentially fill all available zones of porosity that have adequate caprocks. Other local characteristics in the same region, such as the geologic age and setting of a formation, the type of sediment, and its present depth, should be of secondary importance. The significant questions to ask in any area would be whether it belongs to a generally petroliferous region, whether a rock there has sufficient porosity to become a reservoir, and whether there is an adequate caprock over it.

ACCESSIBLE DOMAINS OF HIGH FLUID PRESSURE

As we drill downward from the surface in most areas of the earth, we find some porosity and permeability in the rocks, and it is usual to have the pore spaces filled with water. The fluid pressure in the pores at any level is therefore determined by the height of the water column above it. This is generally referred to as "ordinary" or "hydrostatic" pressure. The pressure in the rock, determined roughly by the density of the rock, increases about $2\frac{1}{2}$ times as fast. With increased depth, the two pressures therefore get progressively more out of step, so long as there is sufficient permeability to maintain the pore-pressure at the hydrostatic value. Since the strength of the rock is limited, there must be a depth at which the rock fails and the pores are crushed shut. A layer of very low or zero permeability will result, which we have termed the "critical layer" (see Figure 2).

This situation is indeed often observed, and, depending on the type of rock and its strength, an effectively impermeable level is encountered in sedimentary rock at a depth that lies most commonly between 10,000 and 20,000 ft. Very often a continuous decline of porosity and permeability is

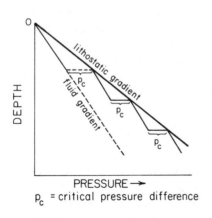

PRESSURE →

p_c = critical pressure difference

Figure 2 Schematic showing relation of pressure in rock and in fluid-filled pores. Fluids are lighter than rock, and the pressure gradient in a connected system of pores will therefore be smaller than the pressure gradient in the rock. When the compressive stress in the rock is not adequately compensated by the pore-fluid pressure, the rock collapses and closes the pores. Another fluid domain underneath the "critical layer" can then have a higher pressure in it, and thus hold the pores open again. In this way a stacked system of fluid-filled domains would set itself up, wherever there is a supply of fluids from below. The step-wise pressure system in the fluid would be an approximation to the pressure in the rock, with the difference between the two always limited by the maximum stress the rock can bear without collapsing.

observed as one drills to this effectively impermeable level, and this observation has been taken to suggest that no adequate porosity for production of oil or gas could be expected at still deeper levels. In several instances, however, where drilling was nevertheless continued to deeper levels, another zone of porosity and permeability was encountered, with a fluid pressure much in excess of the hydrostatic value, and sometimes quite close to the pressure given by the overburden weight of rock (the "lithostatic" value). It is clear that in these cases the fluid pressure, now disconnected by the impervious layer from the hydrostatic domain above, was at a pressure sufficiently high to balance the rock pressure and maintain porosity and permeability. This lower domain can exist only when it has an effective seal against the hydrostatic domain above. Otherwise the high-pressure fluids would, of course, simply be expelled upward, the pressure would drop, and the pores would collapse.

A layer of low permeability to shield and maintain a high-pressure domain below can arise in a variety of other ways as well, and it can then be at shallower levels than that at which it would be enforced by the ultimate crushing strength of the rock. Strata of very dense media, such as shale or salt, may serve this purpose, or dense layers may be generated by the cementing action of minerals deposited by the fluids. Whatever the origin, any such layer may serve to protect an underlying domain in which there is a pressure in excess of the hydrostatic one.

It used to be thought that the probability of encountering such "overpressured" domains with expanded porosity was very low, and that some special local circumstance would have been required to create it. Indeed, if the source materials for the fluids had all derived originally from surface deposits, this would be true. If, however, fluids have been penetrating from below, they will always arrive at each level at the pressure at which they will fracture the rock, a pressure which in practice is very close to the lithostatic one. In this case, no special circumstances are required to set up the high-pressure domain—it would be normal to expect such a domain under each petroliferous area that had obtained its supply from below.

The structure of permeable domains of limited vertical height, disconnected from each other by impermeable layers, is the only one in which fluids and rock can coexist in a static, or nearly static, condition. The finite strength of the rock, and the different densities of rock and fluid, allow no other configuration. The uppermost such domain is open to the surface, and therefore to atmospheric pressure. The first critical layer encountered on the way down, then, allows the transition to a higher pore pressure. In a strictly static case, one can expect such a series of domains to continue downward, making the fluid pressure approximate in a stepwise fashion to the lithostatic pressure. Where such a structure of vertically stacked domains has been established, a further supply of fluid from below must of course also allow an upward transportation to take place. It will not now be the process discussed earlier, which was that of a domain ascending through an initially tight and unfilled rock. In the present case the flow must take the form of a spillover from one domain to the next one above, either by a momentary rupture through the sealed layer between them, or by a gradual leakage through it, but in either case preserving the basic stepwise pressure profile.

High-pressure domains filled with methane or methane-saturated water, underlying oil and gas fields, have been seen in many instances. Paul Jones (31) has investigated this phenomenon and has concluded that "All Cenozoic oil-producing sedimentary basins of the world are geopressured at depth." The production of gas from such high-pressure domains has been very successful in Oklahoma, Texas, Louisiana and other locations, and many of the deep wells have flow rates and life expectancies rarely equalled by shallow wells.

If a region is petroliferous, due to its location overlying a productive area of the mantle, one would expect to see similarities in the chemical makeup of the fluids and in the trace elements that they carry, irrespective of the various traps in which the fluids may be found in the area. These chemical characteristics may be impressed not only on the oil found in a region, but also on the solid carbonaceous materials there, which in the conventional theories of petroleum origin have come to be regarded as source materials. Such patterns have indeed been identified, and appear to be the rule.

THE GEOGRAPHIC DISTRIBUTION OF OIL, GAS, AND ASSOCIATED TRACE ELEMENTS

Mendeleev, in 1870 (1), and many authors since (2–5, 8) have noted that the occurrence of petroleum seems to be more closely related to large-scale tectonic features of the crust than to the magnitude of organic sedimentary deposits. Within petroliferous provinces it is indeed common to find many different levels that are productive, and also to find that the detailed chemistry of the gas or petroleum, and the trace elements they carry, show striking similarities, even when quite different ages and geologic settings of individual oil and gas fields are involved.

The outstanding example of an oil province showing these features is the region of the Middle East, with the oil fields of Iraq, Iran, the Persian Gulf States, and Saudi Arabia. It is thought that 60% of the world's recoverable oil that has been discovered so far is in this region, whose area amounts to only 1/200 of the earth's land surface. Clearly some unifying feature has to be found that can explain this extraordinary concentration. One could not think that independent circumstances in different part of this region would have conspired to make it all so hydrocarbon-rich; that different epochs, different organic deposits in different geologic circumstances, and different caprocks could have been involved, and that it was merely a matter of chance that they had all, independently, been so favorable in this area.

Yet no common feature of all these fields has been noted. Some are in the folded mountains of Iran, some in the flat deposits of the Arabian desert. The oil and underlying gas fields span over quite different geologic epochs, have different reservoir rocks, and have quite different caprocks (32). The search for organic source rocks responsible for the world's largest oil fields has not led to any consensus. In different parts of this petroliferous province, sediments of quite different type and age have been suggested as the source rock, and evidently quite different materials serve as caprocks. The quantities of organic sediments have been regarded as inadequate for the production of all the oil and gas (32, 33), and would probably be seen to be much more inadequate still, if one allowed for the large natural seepage rate of the area, which has been known since ancient historical times. Thus Kent and Warman, who report on their detailed geologic studies of the area, write: "It is a remarkable circumstance that the world's richest oil-bearing region is deficient in conventional source rocks." And also: "The oil is distributed in reservoirs varying in age from middle Jurassic to Miocene, with maxima in the middle Cretaceous and Oligo-miocene. Despite this range of age and type of reservoir, there is a notable homogeneity in chemical composition of the oils, and there is a presumption that they have a common stratigraphic origin." It has also been noted that "most of the reservoirs are conspicuously full to the spill level" (34).

On the basis of a deep origin, this and many other regional concentrations of hydrocarbons would be ascribed to patches of a particularly hydrocarbon-rich mantle underlying the area. The mantle underneath the crust is in any case known to be of uneven composition, and the scale of this unevenness is recognized in other instances also to be much larger than the scale of crustal features and topography. A hydrocarbon-rich area of mantle will make the overlying crust oil- and gas-rich by filling and overfilling every available trap, no matter whether these traps are in flat plains or steep mountains, in old or young porous rock, or with deep or shallow caprocks.

Another petroliferous province that can be singled out as demonstrating a large-scale tectonic, rather than regional sedimentary relationship, is the island arc of Java and Sumatra and its tectonic continuation into Burma. The volcanic chain and the clearly associated earthquake belt begins in the East, at the western tip of New Guinea, and then runs through the small islands, to Java, Sumatra, and the Andaman Islands and then northward through Burma into China. The petroliferous province is mostly a belt to the inside of the active volcanic arc, almost over the entire length, and although the geologic setting varies greatly, as between the islands and the folded mountains of Burma, the oil fields seem to follow the arc with remarkable consistency.

The global map of seismically active areas, and of volcanic areas, shows an incomplete but significant relationship to the map of known hydrocarbon deposits (7). Different explanations for these relationships have been advanced, but it could be readily understood as due to an accelerated fluid transport in zones of active faults penetrating the crust.

Trace elements carried in oil and gas frequently show regional patterns that also seem to be unrelated to the sediments of the locality. The noble gas helium is clearly seen concentrated in petroliferous areas, and indeed all the world's commercial production of helium is obtained by separation from natural gas. Helium-rich regions are generally coincident with petroliferous regions. This clear relationship, which has been observed globally, will need to have an explanation in any theory of the origin of hydrocarbons.

Most of the helium that the earth possesses has resulted from the radioactive decay of uranium and thorium since the time of formation of the earth. A small proportion appears to have been incorporated as helium at the time of formation (35). The ratio of the two stable isotopes of helium, helium-3 and helium-4, is quite different in the two types. The helium produced by radioactivity is all helium-4, and only a very small admixture of helium-3 can result from a certain nuclear reaction with terrestrial lithium; the primordial helium, thought to be of an isotope ratio similar to that in the sun, has a much higher proportion of helium-3.

Chemical or biological processes cannot have concentrated the inert helium. However, uranium and thorium, the progenitors of helium-4, have undergone chemical concentrations, and helium-4 may show the resulting regional patterns. Where helium is found with an isotopic ratio higher in helium-3 than could have been produced on the earth, it has to be in part primordial. Any concentration of such helium is therefore indicative of a primary outgassing process that has channeled up helium from its diffuse distribution in the original rocks, from levels sufficiently deep not to have been outgassed before.

Methane containing such helium has been identified in numerous locations. The East Pacific Rise, the great rift of the Pacific Ocean, emits such gases along most or all of its length (36, 37). Methane in the volcanic rocks of Japan (38), methane in the rift of the Red Sea (39), and methane in the African Rift Valley (40) have all been found to have an admixture of this primordial helium, and this is suggestive that the methane, like the helium, is derived from deep sources, a suggestion strengthened by the absence of substantial organic deposits in some of the cases.

Nikonov (41), who has studied the global distribution of helium and the relationship to other gases, concluded that: "All the largest helium deposits are related to combustible gases and petroleum." Moreover, he showed that there is a clear relationship between the helium concentration and the distribution of the different types of hydrocarbon molecules; the highest helium concentrations occur typically in oil-gas fields and not in "dry" gas (42). In total quantity, one sees frequently more helium than could be accounted for by production from uranium and thorium within the sedimentary basin, and an inflow from elsewhere must have been responsible (43).

Helium occurrence shows geographical patterns on a much larger scale than those of individual oil and gas fields. Thus, almost all oil and gas fields of West Texas, New Mexico, Arizona, Kansas, Colorado, Utah, Wyoming, and Montana have higher helium contents than almost any oil or gas field east of that region (44, 45). These relationships cannot be accounted for in terms of radioactive minerals in each sedimentary domain, and it is difficult to see any way in which helium of an independent origin would have mixed with such consistency with organically produced materials, whose local density must surely span a wide range of values. It is also difficult to see on that basis why one does not find much helium unmixed with methane, or why the detailed mixture of the hydrocarbon molecules in a field should show any relationship to the proportion of helium present.

The noble gas argon, and in particular argon-40, the decay product of potassium-40, is present in some gas fields in extraordinarily high concentrations. A discussion of one instance by Pierce et al (43) contains the remark: "Explaining the radiogenic argon (0.1 percent by volume) in the Panhandle field presents a problem similar to that of helium. Calculation shows that the reservoir rock would have had to be about a hundred percent potassium to supply the argon present; the argon therefore also (like the helium) must have been derived from an external source."

Numerous other trace elements in oil and gas tell a similar story. Almost all oils of South America are peculiarly rich in vanadium (46). The vanadium-to-nickel ratio in oils tends to be characteristic of large geographic regions that span over many oil fields in a variety of geologic settings (47). Mercury is so abundant in some oil and gas regions that its detection at the surface is used for hydrocarbon prospecting (48). All these observations are not adequately explained by the individual properties of the sedimentary deposits, and one must assume that they reflect a larger scale pattern that exists at a deeper level.

CONCLUSIONS

If hydrocarbon fluids have been streaming up through the crust over geologic time, and have been responsible for supplying the great quantity of surface carbon of all the carbonate rocks and other carbon deposits, then the earth must have contained, and presumably still contains, quantities of these fluids that are many orders of magnitude larger than the quantities in all known oil and gas fields. Most of this is out of reach, at depths far too great for any foreseeable drilling technology. The practical importance lies in the quantities that have been arrested, temporarily on a geologic time scale, in the shallow depth range that can be economically exploited. Present drilling techniques seem to place the limit at a depth of approximately 30,000 ft.

Most of the oil and gas exploration has been in the shallow domain, whose pore spaces are generally open to the surface and where special caprocks of particularly low permeability were required to hold down an accumulation of oil or gas. It is this domain, of generally hydrostatic fluid pressures, that contains almost all the oil that has so far been found. Below the first critical layer, where the pressure is above the hydrostatic value, gas is common, but oil is rare. This distribution of oil and gas has usually been attributed to the inability of oil to survive for long periods at the higher temperatures at that depth. The alternative possibility is that oil is deposited from the stream at the levels where the sudden pressure drops occur and where the solubility of oil in gas is thereby diminished.

It is the shallow domain where oil is found that has formed the basis of conventional estimates of future supplies. If, however, gas has been streaming up from deep levels, then the type of reservoir where an expanded porosity has been built up underneath the first critical layer can be expected to constitute a major global resource. Instead of finding this an anomaly, in some special circumstances, it would now be the expectation that in all petroliferous areas, and in many such areas not yet identified, such reservoirs may exist. These reservoirs may have quantities of gas corresponding to the major carbon outgassing process of the earth, and not just to the small fraction that would be given by biological materials from the surface that have been buried in such a way as to avoid near-surface oxidation.

Gas exploration geologists who have opened up gas production from the deep domain foresee tapping enormous resources there. Robert A. Hefner III, a leader in this field, estimates that at least 140 trillion cubic feet of gas, and quite possibly as much as 300, can be produced from levels below 15,000 ft in the Anadarko Basin of Oklahoma alone. From extensive drilling experience there, he concludes that the entire deep high-pressure domain in this basin is one common reservoir. The productivity of the individual wells is of course dependent on the permeability of the local formation, but in no case is it limited by the availability of gas in the region. Will numerous other deep basins, still unexplored at depth, tell a similar story?

Deep sedimentary basins would continue to be the major target for exploration, simply because sedimentary rocks generally make the best reservoirs. The volumes of unexplored sediments in the world below 15,000 ft are still immense, and if a large fraction of these is as productive as the few areas that have been explored at that depth, the global resources of gas would be very many times larger than all the hydrocarbon resources that have been discovered to date. Not only large sedimentary basins, but many rift valleys that have been filled up with sediments, may turn out to be good reservoirs, as well as good conduits for the ascent of gas. Gas from this deep level may then be not only a much larger total resource, but also a resource much more widely distributed in the world than gas from the intensely petroliferous areas that produce oil; a globally much more even supply of fuel could result from this.

While sediments would still be the first choice, igneous and metamorphic rocks can also receive serious attention. Some types of volcanic rock have developed high porosities, and some faulting in them has produced media of good permeability. This is the case, for example, in the "Green Tuff" of Japan, where gas and oil production from volcanic rock is in progress (38). Metamorphic rocks may also have substantial porosity in zones of severe faulting. Also, ancient meteoritic impacts that have smashed crystalline rock can be expected to have produced large volumes of buried porosity (51). One such location in Sweden, the Siljan Ring, is at this time under intense investigation, and gives evidence of being a strongly petroliferous area.

What will be the effects of such a general outlook on the strategies and techniques of prospecting? Identifying areas of hydrocarbon-rich mantle, and of good conduits to the surface, will be a first priority. Extensions from known petroliferous areas, deep-seated tectonic patterns, and rifts and faultlines will assume greater significance. Source rocks and the biological contents of sediments will be unimportant, although the presence of such carbonaceous materials as have previously been regarded as source materials may be an indication of hydrocarbon flow in the area. They will then serve as indicators of hydrocarbon presence, but now just as likely below them as above.

Regions below existing oil and gas fields, including the exhausted ones, would now be promising. Locations that showed shallow oil and gas, in which deep drilling was abandoned because of declining porosity, may have to be re-explored to somewhat deeper levels to find regions of expanded porosity in the high-pressure domain below.

The extensive areas of tar sands may give evidence of massive outgassing, and may therefore point to underlying gas fields.

Porous volcanic or metamorphic rocks may be as promising as sedimentary ones. Very ancient rocks that are thought to predate the widespread occurrence of plant life, should equally come under consideration.

Chemical surface prospecting is the means of finding past or present surface seepage of methane and the substances with which it is frequently associated. This would assume a greater importance, first because it would be clearer why it should be successful, when one recognizes that massive seepage generally goes together with a large content. Second, the task is made easier than it was thought to be, when the estimated seepage rates may be so much larger than had seemed possible previously. Most oil in the world was discovered by following up surface oil seeps. It is likely that most gas will also be discovered by following up gas seeps. Whereas oil can be recognized by any person walking over the land, identifying gas seeps generally requires more sophisticated methods.

ACKNOWLEDGMENTS

Work at Cornell on the subject of the origin of natural gas is supported by the Gas Research Institute (Chicago), and experimental work has also received support from New York Gas Group (NYGAS) and the New York State Energy Research and Development Agency. The author also gratefully acknowledges the help of Dr. Steven Soter in many aspects of the work presented.

APPENDIX

Progressive Fractionation of Carbon Isotopes

It has long been noted that methane in the same vertical column in the ground tends to be progressively more depleted in carbon-13 as one goes from deep to shallow levels (30). To ascribe different origins to the methane at different levels seems improbable when it is so often found that the same region is methane-rich at all levels probed. It seems more likely that one is seeing methane at all stages of its expected upward migration. In that case, the isotopic change, so consistent in many instances, has to be attributed to an ongoing effect that continues to remove preferentially the heavy isotope during this migration.

A product of this process has now been discovered, namely a particular form of carbonate that lies characteristically over oil and gas fields (49). Methane and other hydrocarbons appear to have been oxidized, and the CO_2 so produced seems to have combined with oxides of calcium and other metals in the rocks to produce this "anomalous" carbonate. The carbonate is enriched in carbon-13 relative to the local methane and other hydro-carbons, showing that isotopic fractionation has progressively depleted the upward-moving hydrocarbon stream of carbon-13. Nevertheless, this carbonate usually has a lower carbon-13 content than "normal" carbonate derived from atmospheric CO_2, and for this reason is readily distinguishable from it.

Such a fractionation depends not only on the isotopic selection in the chemical processes that are involved, but also on differences in molecular diffusion speed, wherever diffusion in pore spaces, rock, or water is involved. For methane the latter effect is particularly large (3% slower speed for the heavy isotope, and therefore a longer residence and an enhanced chance of oxidation on a given path), and this probably accounts for the fact that methane shows a much larger fractionation effect than any other hydrocarbon.

At shallow levels, where this type of evidence has been obtained, the oxidation process appears to be mediated by bacteria. At deeper levels, where the temperature is too high for any of the methane-oxidizing bacteria to exist, this process, and the resulting fractionation, should be absent. Presumably at very much deeper levels, where the temperature is so high that oxidation can be thermally induced, there can again be such fractionation. The final carbon isotope ratio of upwelling hydrocarbons sampled at a particular level will then depend on (a) the original ratio of the deeply embedded carbonaceous source material (which may have different values in different regions); (b) the isotopic fractionation that occurs in the cracking process that splits off the more mobile components; and (c) the cumulative isotopic fractionation in the upward migration, as a proportion is lost to oxidation. The final product may then show a larger depletion in carbon-13 than any single-step biological process could have produced. Biology is involved in a sense, in that one fractionation process is due to selective bacterial destruction. The situation here is similar to that of the optical activity or the odd-even effect; in each case biology enters through the selective removal of some components, leaving the remainder with a biological imprint.

The quantities of anomalous carbonates that are seen overlying some oil and gas fields deserve comment; they often imply that the seepage of methane, and possibly of other hydrocarbons, has amounted to much larger quantities than those now thought to be in the underlying sediments.

The township of Cement in Oklahoma is one such example (50), where an area of approximately ten square miles has an abundance of such carbonate cement, sufficient to discolor and harden the surface soil, and to fill the porosity of the sedimentary rock beneath, down to a depth of at least 2000 feet. This patch overlies very neatly the most productive oil and gas fields of the region, and it appears that the cementing of the pore spaces actually helped to improve the quality of the caprock. An estimate of the mass of the isotopically anomalous carbon in place there, derived from an estimate of 15% porosity that is so filled, amounts to almost one billion tons. It would have taken more than 50 trillion cubic feet of methane to lay down this carbon. Presumably only a fraction of the seeping methane was converted into this carbonate and the quantity lost in the area must have been much larger still. The highest estimate of the entire gas content of the Anadarko Basin of Oklahoma, of which this area is a small part, amounts to 300 trillion cubic feet. While no claim for the accuracy of these figures can be made, they show that the production of this carbonate can be a major item in the discussion of both the quantitative aspects of the gas's source material and of the isotopic fractionation that has taken place.

Literature Cited

1. Mendeleev, D. 1877. L'origine du petrole. *Rev. Sci.* 2e Serie, 7e Annee, No. 18, pp. 409–16
2. Sokoloff, W. 1889. Kosmischer ursprung der bitumina. *Bull. Soc. Imp. Nat.* Moscou, Nouv. Ser. 3, pp. 720–39
3. Kudryavtsev, N. A. 1959. Geological proof of the deep origin of petroleum. Trudy Vsesoyuz. Neftyan. Nauch.-Issledovatel. Geologorazvedoch. Inst. No. 132, pp. 242–62
4. Porfir'ev, V. B. 1974. Inorganic origin of petroleum. *Am. Assoc. Petrol. Geol. Bull.* 58:3–33
5. Kropokin, P. N., Valiaev, B. M. 1976. Development of a theory of deep-seated (inorganic and mixed) origin of hydrocarbons. *Goryuchie Iskopaemye: Problemy Geologii i Geokhimii Naftidov i Bituminoznykh Porod*, ed. N. B. Vassoevich. Akad. Nauk SSSR, pp. 133–44
6. Gold, T. 1979. Terrestrial sources of carbon and earthquake outgassing. *J. Petrol. Geol.* 1(3):1–19
7. Gold, T., Soter, S. 1980. The deep-earth-gas hypothesis. *Sci. Am.* 242(6):154–61
8. Gold, T., Soter, S. 1982. Abiogenic methane and the origin of petroleum. *Energy Explor. Exploitat.* 1:89–104
9. Gold, T. 1984. The deep earth gas; pathways for its ascent and relation to other hydrocarbons. *1984 Int. Gas Res. Conf.* Rockville, MD: Government Institutes Inc.
10. Gold, T., Soter, S. 1985. Fluid ascent through the solid lithosphere and its relation to earthquakes. *Pure Appl. Geophys.* In press
11. Rubey, W. W. 1951. Geologic history of seawater—an attempt to state the problem. *Geol. Soc. Am. Bull.* 62:1111–47
12. Hedberg, H. D. 1964. Geologic aspects of origin of petroleum. *Am. Assoc. Petrol. Geol. Bull.* 48:1777–803
13. Robinson, R. 1963. Duplex origin of petroleum. *Nature* 199:113–14
14. Robinson, R. 1962. The duplex origins of petroleum. *The Scientist Speculates*, ed. I. J. Good, pp. 377–87. New York: Basic Books
15. Bailey, D. K., Tarney, J., Dunham, K. 1980. The evidence for chemical heterogeneity in the Earth's mantle: a discussion. *Phil. Trans. R. Soc.* 297:135–493
16. Hart, S. R. 1984. A large-scale isotope anomaly in the Southern Hemisphere mantle. *Nature* 309:753–57
17. Anders, E., Hayatsu, R., Studier, M. H. 1973. Organic compounds in meteorites. *Science* 182:781–90
18. Fritz, P., Frape, S. K. 1984. Methane in

the rocks of the Canadian shield. Univ. Waterloo. Private communication

19. Petersil'ye, I. A., Pripachkin, V. A. 1979. Hydrogen, carbon, nitrogen, and helium in gases from igneous rocks. *Geochem. Int.* 16(4): 50–55

20. Symposium on occurrence of petroleum in igneous and metamorphic rocks. 1932. *Am. Assoc. Petrol. Geol. Bull.* 16(8): 717–858

21. Kozlovsky, Ye. A. 1984. The world's deepest well. *Sci. Am.* 251(6): 98–104

22. Chekaliuk, E. B. 1980. The thermal stability of hydrocarbon systems in geothermodynamic conditions. *Degazatsiia Zemli i Geotektonica*, ed. P. N. Kropotkin, pp. 267–72. Moscow: Nauka

23. Zhuze, T. P., Iushkevich, G. I. 1957. Compressed carbohydrate gases—solvents of oil and oil residues. *Notices Acad. Sci.* SSSR No. 2, pp. 63–68

24. Price, L. C., Wenger, L. M., Ging, T., Blount, C. W. 1983. Solubility of crude oil in methane as a function of pressure and temperature. *Org. Geochem.* 4: 201–21

25. Deines, P. 1980. The carbon isotopic composition of diamonds: relationship to diamond shape, color, occurrence and vapor composition. *Geochim. Cosmochim. Acta* 44: 943–61

26. Melton, C. E., Giardini, A. A. 1975. Experimental results and a theoretical interpretation of gaseous inclusions found in Arkansas natural diamonds. *Am. Mineral.* 60: 413–17

27. Urey, H. C. 1966. Biological material in meteorites: a review. *Science* 151: 157–66

28. Ourisson, G., Albrecht, P., Rohmer, M. 1984. The microbial origin of fossil fuels. *Sci. Am.* 251(2): 44–51

29. Philippi, G. T. 1977. On the depth, time and mechanism of origin of the heavy to medium-gravity napthenic crude oils. *Geochim. Cosmochim. Acta* 41: 33–52

30. Galimov, E. M. 1969. Isotopic composition of carbon in gases of the crust. *Int. Geol. Rev.* 11: 1092–104

31. Jones, P. H. 1976. Gas in geopressured zones. *The future supply of nature-made petroleum and gas*, ed. R. F. Meyer, pp. 889–911. New York: Pergamon

32. Kent, P. E., Warman, H. R. 1972. An environmental review of the world's richest oil-bearing region—the Middle East. *24th Int. Geol. Congr.* (Sect. 5), pp. 142–52

33. Barker, C., Dickey, P. A. 1984. Hydrocarbon habitat in main producing areas, Saudi Arabia: discussion. *Am. Assoc. Petrol. Geol. Bull.* 68: 108–9

34. Tatsch, J. H. 1974. *Petroleum deposits: origin, evolution, and present characteristics*, p. 269. Tatsch Assoc. Mass: Sudbury

35. Craig, H., Lupton, J. E. 1981. Helium-3 and mantle volatiles in the ocean and the oceanic crust. *The Oceanic Lithosphere:* Vol. 7, *The Sea*, ed. C. Emililiani, pp. 391–428. New York: Wiley

36. Craig, H., Clark, W. B., Beg, M. A. 1975. Excess 3He in deep water on the East Pacific Rise. *Earth Planet. Sci. Lett.* 26: 125–32

37. Welhan, J., Craig, H. 1979. Methane and hydrogen in East Pacific Rise hydrothermal fluids. *Geophys. Res. Lett.* 6: 829–31

38. Wakita, H., Sano, Y. 1983. 3He/4He ratios in CH_4-rich natural gases suggest magmatic origin. *Nature* 305: 792–94

39. Burke, R. A. Jr., Brooks, J. M., Sackett, W. M. 1981. Light hydrocarbons in the Red Sea brines and sediments. *Geochim. Cosmochim. Acta* 45: 627–34

40. Deuser, W. G., Degens, E. T., Harvey, G. R., Rubin, M. 1973. Methane in Lake Kivu: new data bearing on its origin. *Science* 181: 51–54

41. Nikonov, V. F. 1973. Formation of helium-bearing gases and trends in prospecting for them. *Int. Geol. Rev.* 15: 534–41

42. Nikonov, V. F. 1969. Relation of helium to petroleum hydrocarbons. *Dokl. Akad, Nauk SSSR* 188: 534–41

43. Pierce, A. P., Gott, G. B., Mytton, J. W. 1964. Uranium and helium in the Panhandle gas field, Texas, and adjacent areas. *US Geol. Survey Prof. Pap.* 454-G

44. Tongish, C. A. 1980. Helium—its relationship to geologic systems and its occurrence with the natural gases, nitrogen, carbon dioxide, and argon. *US Bur. Mines Rep. Invest.* 8444. 176 pp.

45. Clark, M. 1981. Helium: a vital natural resource. *Geol. Survey Wyoming*, Public Inf. Circ. No. 16. 31 pp.

46. Kapo, G. 1978. Vanadium: key to Venezuelan fossil hydrocarbons. *Bitumens, Asphalts and Tar Sands*, ed. G. V. Chilingarian, T. F. Yen, pp. 213–41. Amsterdam: Elsevier

47. Tissot, B. P., Welte, D. H. 1978. *Petroleum formation and occurrence: a new approach to oil and gas exploration*, pp. 363–65. New York: Springer-Verlag

48. Rudakov, G. V. 1973. The relationship between incidences of mercury mineralization and the presence of oil and gas. *Geol. Zh.* 33(5): 125–26

49. Donovan, T. J., Friedman, I., Gleason, J. D. 1974. Recognition of petroleum bear-

ing traps by unusual isotopic compositions of carbonate-cemented surface rocks. *Geology* 2:351–54

50. Donovan, T. J. 1974. Petroleum microseepages at Cement, Oklahoma: evidence and mechanism. *Am. Assoc. Petrol. Geol. Bull.* 58:429–46

51. Donofrio, R. R. 1981. Impact craters: Implications for basement hydrocarbon production. *J. Petrol. Geol.* 3(3):279–302

Ann. Rev. Energy. 1985. 10 : 79–107

INTERNATIONAL COOPERATION IN MAGNETIC FUSION[1]

W. M. Stacey, Jr.

Nuclear Engineering, Georgia Institute of Technology, Atlanta, Georgia 30332

M. Roberts

Office of Fusion Energy, U.S. Department of Energy, Washington, DC 20545

INTRODUCTION

The development of magnetic fusion energy has been pursued for over three decades. During this time, the perceived motivation for and character of the magnetic fusion program has evolved considerably, and the nature of international cooperation in magnetic fusion has changed correspondingly.

Magnetic fusion research began with military interest and was classified until the late 1950s. With declassification, magnetic fusion entered an era of international cooperation, which has increased steadily to its present level.

During the 1960s, magnetic fusion research was essentially a basic science program. The fusion community, as individuals and as institutions, was drawn together by common scientific interests and by a growing recognition of the difficulty of the path that lay ahead. A number of informal working relationships established the basis for much of the more substantial cooperation that took place later.

An outstanding early example of international cooperation was the dissemination of the tokamak confinement concept. This concept was

---

[1] The US Government has the right to retain a nonexclusive, royalty-free license in and to any copyright covering this paper.

developed by Soviet scientists who, in the 1960s, announced achievement of then spectacular levels of confinement. The Soviet results were not fully accepted by the international community until a British team was invited to Moscow to confirm the measurements with superior diagnostics. European and American research emphasis then shifted quickly toward the tokamak; it remains the most advanced magnetic confinement concept and the major concept under development in the fusion programs of the world.

During the 1970s, magnetic fusion came to be widely considered as a possible future energy option and therefore a possible solution to the impending energy crisis which, stimulated by the OPEC oil policies, was then perceived. Plans were formulated to demonstrate the production of electricity from magnetic fusion early in the next century, and a fledgling fusion technology development program was initiated. As the magnitude of the task of developing fusion power became clearer, discussion of substantive international cooperation became more common among program leaders. The International Energy Agency (IEA), created by the Western nations in response to the OPEC oil embargo, included research in magnetic fusion among its cooperative activities, and a number of implementing agreements were concluded to effect cooperation in various areas of fusion technology. In 1978, the Soviet Union proposed to the Director General of the International Atomic Energy Agency (IAEA) that the principal magnetic fusion "countries" (the Soviet Union, United States, Europe, and Japan) join to design, construct, and operate internationally the next major tokamak experiment. This proposal led to the creation of the International Tokamak Reactor (INTOR) Workshop under IAEA auspices. The Magnetic Fusion Energy Engineering Act of 1980 (Public Law 96-386) reflected the optimistic spirit of the decade just ended by committing the US to demonstrate fusion power by the end of the century. The status of international cooperation in magnetic fusion at this time is well summarized in the proceedings of a National Science Foundation workshop (1).

Technical progress toward the goals defined in the 1970s has been impressive. Moreover, the substantive cooperative efforts begun in the 1970s are now bearing fruit. Nevertheless, the situation is changing once again during the 1980s. The energy crisis has disappeared in the public perception, and energy has become a less urgent concern to politicians and policy planners, to a great extent in the US and to a lesser extent in Europe and Japan. At the same time, the economic climate and mounting budget deficit (at least in the US) are causing reconsideration of all domestic government programs, including fusion. Simultaneously, the Heads of State and Government meeting at Versailles in 1982 identified fusion as an area in which increased international cooperation would be desirable (2), and the findings of a high-level fusion working panel toward that end have been endorsed at subsequent meetings. International cooperation in science, and thus also in magnetic fusion, is becoming a national policy of the US (3). The Secretary of Energy has expressed a personal commitment to international cooperation in fusion, particularly in the construction of major facilities.

International cooperation in magnetic fusion up to the present has been substantial and largely successful. However, this cooperation has, with few exceptions, been outside the mainline efforts of national programs. What is being considered now is a qualitatively, as well as quantititatively, different degree of international cooperation, which could lead to interdependent national programs.

Thus, this is a propitious time to evaluate past experience and examine future opportunities for international cooperation in magnetic fusion. That is the purpose of this paper; which draws heavily on a recently completed study (4) by the National Research Council, in which the first author participated, and on internal briefing documents (5) prepared by the second author in planning, negotiating, and implementing cooperative arrangements for the Department of Energy.

THE WORLD'S MAJOR MAGNETIC FUSION PROGRAMS

Major magnetic fusion programs are conducted in four areas of the world—the United States, the European Community (EC), Japan, and the Soviet Union. The four programs are of comparable magnitudes and are at comparable stages of development. In each of these programs a scientific feasibility experiment based on the most advanced magnetic confinement concept—the tokamak—either has recently started operation (in the US and the EC) or will start operation within the next one or two years (in Japan and the USSR, respectively). Smaller fusion programs are carried out in several other countries.

Broadly speaking, the near-term technical objectives perceived by leaders in the four programs are similar: 1. to maintain a vigorous scientific base program, 2. to address next-step tokamak issues with a major facility, 3. to develop improved magnetic confinement concepts, and 4. to address fusion technology issues. Pursuit of these objectives is financially constrained, to varying degrees, in each of the four programs.

Plasma physics has a strong experimental component. As commonly understood, world leadership in fusion generally resides in that country possessing the experimental facilities with the greatest ability to explore the myriad frontiers of plasma physics. For the past decade the United States has been the overall world leader in magnetic fusion, although other programs have on occasion led in certain areas.

The US (6–8) has a strong experimental tokamak program that has established many of the world record plasma physics parameters. Two of these experiments, the Tokamak Fusion Test Reactor (TFTR) and the Doublet III D (D-IIID), will continue to extend our knowledge of plasma physics for the next several years. While definition studies are in progress, plans for initiating a major next-step tokamak experiment in the US are indefinite at present. The US also has the leading experimental program in the tandem mirror confinement concept, which is the most advanced alternative concept. Smaller programs are going forward in other, less advanced alternative confinement concepts. The US has a strong base scientific program and has the broadest and earliest established fusion technology program. The budget for the magnetic fusion program in the US is somewhat less than $450 million in fiscal year 1985.

The EC program (9–11) is considered by its participants to be on the threshold of world leadership in fusion, on the basis of a new generation of tokamak experiments (JET, TORE SUPRA, ASDEX-U) that will be operating over the next decade. This view is shared by many in this country. The EC program managers believe that they can maintain this lead by constructing a major new tokamak experiment, Next European Torus (NET), to operate in the late 1990s. NET has physics objectives of achieving an ignited plasma and a long-burn pulse, as well as ambitious technological objectives. Planning and preconceptual design work for NET has been authorized by the Council of Ministers of the European Community and initiated at the technical level; decisions as to whether to proceed to engineering design and to construction are scheduled for 1988 and 1992, respectively. The EC has leading programs in the less advanced stellarator and reversed-field pinch (RFP) alternative concepts. Fusion technology programs are expanding in support of the NET activity. The EC fusion program is carried out in the various national fusion laboratories of member countries and is funded partly by each nation directly and partly by the EC. The budget for the European fusion program during the current five-year plan has been approximately two thirds of the US fusion budget.

The Japanese fusion program (12–14) is newer than the other three major programs, but it is moving rapidly toward full parity. The program of the Japan Atomic Energy Research Institute (JAERI), under the Science and Technology Agency, concentrates on the tokamak and on fusion technology. The JT-60 experiment, which will begin operation within the year, will have capabilities comparable to those of TFTR, but will not have the capability to actually produce energy from D-T fusion. Conceptual design studies are in progress for a major new tokamak experiment, the Fusion

Experimental Reactor (FER), to operate in the late 1990s. FER would have objectives similar to those of NET. The fusion technology program is approaching the US program in strength. The university fusion program, under the Ministry of Education, Science, and Culture, conducts basic scientific and technological research. This program is investigating three alternative confinement concepts, the tandem mirror, stellarator, and bumpy torus, as well as the tokamak. A fourth alternative confinement concept, the reversed-field pinch, is being developed under the small program in the Ministry of International Trade and Industry, as well as in the two previously mentioned programs. The role of industry in designing and supplying equipment for the Japanese fusion program is greater than it is in the other three major world programs. The budget for the Japanese fusion program is approximately two thirds that of the US fusion budget.

The USSR has a strong experimental tokamak program and renowned theoretical activity in support thereof. The T-15 experiment, which should be operational within the next couple of years, will have capabilities comparable to those of TFTR, but will also incorporate the more advanced superconducting technology. The Soviets have plans to build both an ignition experiment (T-14) and an experimental reactor (OTR) with objectives similar to those of NET and FER, but also incorporating a "hybrid" blanket element to test the production of fissionable material from fusion neutrons. The USSR has significant programs in the development of the tandem mirror and stellarator alternative confinement concepts and has a substantial fusion technology development program.

There is sufficient similarity in the status of development and near to intermediate-term objectives of the four major fusion programs to provide a technical basis for expanded international cooperation.

PAST, PRESENT AND PLANNED COOPERATION

Informal Information Exchange

Technical meetings, the technical literature, and individual visits have contributed to an open and relatively unrestricted exchange of information since magnetic fusion research was declassified in 1958. Basic concepts and experimental results have been effectively shared in this manner, to the benefit of all concerned. The adoption of the Soviet tokamak concept by all the major programs for their mainline research efforts, and the sharing of information developed in the US and USSR in the magnetic mirror concept area on "minimum magnetic field" and "tandem" configurations, are striking examples of informal cooperation greatly advancing the progress of fusion research.

The shared data bases that have gradually accumulated in the open literature in theoretical and experimental plasma physics and, more recently, in fusion technology, have been of similar importance to the advance of fusion research worldwide. The International Atomic Energy Agency (IAEA) is particularly active in fostering the exchange of information, through hosting a biennial conference on plasma physics and a series of workshops and specialist meetings on selected topics, and through publication of the international journal *Nuclear Fusion.*

Formal Information and Personnel Exchanges

Formal personnel exchange and scientific visits were initiated between the US and the USSR following the 1973 Nixon-Brezhnev agreement on cooperation in nuclear energy and have continued since, with only a few periods of inactivity occasioned by political discord. The present level of exchange consists of six or seven annual visits each way, with an average visit involving about 8–12 man-weeks. In general, these exchanges have been relatively open and productive in plasma physics areas, but have been much less productive in technology areas.

Following the agreement signed in 1979 on cooperation in energy research, cooperation between the US and Japan in magnetic fusion has flourished. This umbrella agreement covers personnel exchange and other activities (joint planning, joint workshops, joint experimental programs, and joint operation of facilities) to be discussed later. The present level of personnel exchange, exclusive of the other, specific activities, consists of about 12 visits each way annually, with an average visit involving three to five people.

The format of one to several weeks of attention to a limited range of topics provides for a more intensive information exchange than normally occurs in informal exchanges. In general, these formal exchanges have been useful, particularly in the early stages, when the senior people are involved and the amount of new information involved is large.

The US also has a bilateral agreement for the exchange of information with Australia, and agreements with the EC, Canada, and the People's Republic of China (PRC) are pending. A bilateral agreement between the EC and Japan is also being discussed.

Cooperative Planning

The fusion programs of the European Community countries are cooperatively planned under the auspices of the Commission of the European Communities. Each of the national programs is formally assigned responsibility for certain aspects of the overall EC program. Each national program is funded partly by the nation and partly by the EC.

Cooperative planning among national programs takes place in several specific scientific and technical areas. The US and Japan formally cooperate, under the 1979 Agreement on Cooperation in Energy Research, in planning the program of a Joint Institute of Fusion Theory based at two plasma theory institutes located in the US and Japan, and on the planning of research on various alternative (non-tokamak) confinement concepts. A Joint Planning Committee for Fusion Technology is charged with planning cooperative activities between the US and Japan in technology.

The US, the EC, Japan, Canada, and Switzerland formally cooperate under the auspices of the International Energy Agency (IEA) in planning neutron radiation damage programs.

A great deal of informal planning occurs among the US, the EC, and Japan. A sharing of information on the design, construction, and operation plans and results for the present generation of large tokamaks (TFTR-USA, JET-EC, JT60-Japan, T15-USSR) has occurred biennially through the IAEA Large Tokamak Workshop and through informal exchanges. An agreement is now being negotiated to formalize an ongoing exchange among the US, the EC, and Japan under the auspices of the IEA and to extend its scope to include technical preparations for the follow-on experiments.

Active informal cooperation in design, coordination of experimental programs, and sharing of results takes place among the US, the EC, and Japan in the heliotron-stellarator confinement research area and in the reversed-field pinch confinement research area. Negotiations are now in progress to formalize these two cooperative planning activities under the auspices of the IEA.

Personnel from Canada, Japan, and Europe participate, on an informal basis, in a US planning activity (FINESSE) to identify testing needs and requirements for fusion nuclear technologies (i.e. the technologies of those components used for heat removal and conversion and for radiation attenuation).

The Versailles Summit Process has led to a fusion working panel consisting of high-level policy-makers (at the assistant secretary level) and subpanels staffed by technical and programmatic leaders. The Subpanel for Planning and Collaboration on Major New Fusion Research Facilities is charged with identifying the nature and timing of major facilities that will be required. The Subpanel for Near-term Fusion Physics and Technology is charged with identifying areas in physics and technology where increased cooperation would be beneficial. There is also a Subpanel on Administrative Problems Affecting Fusion Cooperation. The participants in this planning activity are the US, Canada, Japan, the EC, the United Kingdom, France, Italy, and the Federal Republic of Germany.

The European cooperative planning under the EC has been quite successful in producing (and supporting) a European tokamak program that is on a par with and, in the judgment of many, will soon be ahead of any national program. This success is surely in part attributable to the formal requirement for cooperative planning in the EC charter and the associated EC budgetary control. The benefits derived to date from the other, less binding, cooperative planning activities discussed above have been much more modest, but they have helped to avoid unnecessary duplication of effort.

Cooperative Research and Development

Many cooperative research and development projects exist, and others are planned. These projects are characterized by the national facilities of one country being shared or utilized by other countries.

The US, the EC, Japan, and Switzerland cooperate through the IEA on superconducting magnet development in the Large Coil Task. The US planned and constructed, at a cost of about $40 million, the Large Coil Test Facility, with consultation with the other participants. Each of the participants designed and constructed one (the EC, Japan, Switzerland) or three (US) test coils, at costs ranging from $15–25 million, with construction and test results being shared. Failure to produce all of the coils on time and an attempt to modify the agreed-upon testing approach produced some discord, but these difficulties were overcome and coil testing has now begun successfully with three coils in place (in Japan, the EC, and the US).

The US, the EC, Japan, and Canada participate in an international research program on plasma-wall interactions, under the auspices of the IEA, which is carried out on the TEXTOR tokamak, which was constructed and is operated by the Federal Republic of Germany (FRG).

The ASDEX-U tokamak, which is being constructed in the FRG, will be the most advanced facility in the world for investigating the poloidal divertor impurity control scheme. The US and the EC are presently negotiating an agreement, under IEA auspices, for US scientists to participate in the ASDEX-U experimental program, and useful cooperative planning is taking place on an informal basis.

Many cooperative research activities have been initiated in the field of neutron radiation damage in fusion-relevant materials. One of the first attempts was associated with an intended new US facility for materials testing called the Intense Neutron Source (INS). When this project was unilaterally cancelled by the US for technical and budgetary reasons, efforts were made to find an appropriate replacement activity. The result was a focus on the Fusion Materials Irradiation Test (FMIT) in the IEA Agreement on Radiation Damage in Fusion Materials, involving the US, the EC, Japan, and Switzerland. Although the negative feelings resulting from the unilateral cancellation of INS have not disappeared, the new agreement has proceeded with two activities. The first annex to the agreement covers cooperative research and development associated with all aspects of the FMIT Facility. This cooperative R&D has been successfully carried out. Although construction of the facility was not included in the first annex, when the FMIT facility recently became the victim of budgetary constraints, the US proposed that the other participants contribute to the facility construction cost (over and above their costs associated with cooperative R&D tasks). Agreement has not yet been reached on this proposal. Even though the specific R&D activities in the first annex have been carried out to the extent possible, the indefinite deferral of the basic facility has led to disappointment. The second annex covers cooperative irradiations in existing test facilities in the participating countries.

Other neutron irradiation cooperative research activities have been more successful. The US and Japan are each contributing about $2 million annually for full operation and experimental utilization of the Rotating Target Neutron Source (RTNS) II facility in the US. Japan and the US are each spending about $1 million annually to conduct neutron irradiation tests in the High Flux Isotope Reactor (HFIR) and the Oak Ridge Reactor (ORR) in the US. The US supports the basic operating costs of both reactors. Japanese scientists participate in the experimental programs in all cases. The US is providing materials to be tested in the Japanese Fusion Neutron Source (FNS) project. These three cooperative activities, conducted under the US-Japan bilateral agreement, are beneficial to both parties.

Two small-scale tritium handling components, developed and tested in a hydrogen environment in Japan, are being tested in the Tritium Systems Test Facility (TSTA) in the US. This cooperation is carried out under the US-Japan bilateral agreement.

Collaborative Activities

We use the word "collaborative" to distinguish activities in which all participants have more-or-less equal voices in determining the objectives, design parameters, and experimental programs, and in which all participants share the costs proportionately.

Certainly the most outstanding example in fusion of a collaborative project is the Joint European Torus (JET). The member states of the EC, and Switzerland and Sweden, collaborated to design, construct, operate, and carry out the experimental program on the largest tokamak in the world. The collaboration is under the JET Joint Undertaking, an organizational entity provided for in the treaty that created the EC. The JET is generally acknowledged to be well conceived, well designed and well built, thus representing a very successful international collaboration. The collaboration provided the European nations with a world class experiment that no one of them, possibly excepting the Federal Republic of Germany, could have afforded alone.

The collaboration between the US and Japan (under the US-Japan bilateral agreement) on the Doublet III (D-III) tokamak experiment is another example of successful collaboration on a major facility. For five years (1979–1984), the US and Japan spent annually $20 million and $12 million, respectively, to upgrade and operate the US D-III facility. The collaboration enabled the heating capability of the machine to be upgraded and allowed full utilization of the facility, which resulted in the attainment of world record plasma parameters and allowed the Japanese to train the operations team for their subsequent JT60 experiment. Separate teams of US and Japanese scientists operated the facility in alternating shifts, an approach which led to a stimulating competition but to limited sharing of results. The agreement to collaborate on D-III has been amended to include a modification of the vacuum vessel, to achieve greater plasma performance capability, and extended to 1988, with the US and Japan contributing $16 million and $8.6 million, respectively.

The International Tokamak Reactor (INTOR) workshop was initiated collaboratively by the US, USSR, Japan, and the EC, under the IAEA, in 1979 to proceed in stages from feasibility assessment through concept definition and design to construction, and finally to operation, of the next major tokamak experiment beyond the present generation (TFTR, JET, JT60, T15). The first two phases of INTOR were successively completed in 1979 and 1981, but only the USSR was prepared in 1981 to make the type of commitment (nationally or internationally) that was implicit in going into the design phase. Subsequent phases of the INTOR Workshop were redefined to emphasize the focusing of worldwide effort on critical technical issues affecting such an experiment, rather than on detailed design and construction, and the workshop has subsequently advanced worldwide understanding of a number of critical issues.

The US is providing assistance to Spain, under a US-Spain Technical Commission, in the design of an advanced experimental stellarator to be built in Spain. An agreement for a collaborative program of construction and operation is being negotiated.

AREAS FOR INCREASED FUTURE COOPERATION

The areas in which increased international cooperation would be beneficial in the future follow from the general near-term objectives for fusion that were outlined in a previous section.

Plasma Physics Base Program

For confinement concepts that are at relatively early stages of development (the so-called alternative concepts, such as the stellarator and the reversed-field pinch), there are a large number of possible options to explore before determining which variant of the concept to pursue to the proof-of-principle stage and beyond. Cooperative planning of the design and experimental programs and sharing of access to the experimental results would benefit all parties by allowing a larger number of concepts and a wider range of options within a given concept to be investigated.

For confinement concepts that are at relatively advanced stages of development (such as the tokamak), there are a large number of specific questions that need to be intensively studied. Cooperative planning of the experimental programs on existing facilities and of additional required facilities would enable the physics data base for such concepts to be established for the benefit of all without requiring each nation to expand its program substantially. The pending ASDEX-U cooperation and the Large Tokamak agreement mentioned in the previous section are examples of this type of cooperation.

The requirements to establish the scientific data base are identified in detail in the Magnetic Fusion Advisory Committee Panel 1 and Panel 2 reports (7).

Fusion Technology

The development of fusion technology will require the acquisition of a basic properties data base for the various technologies, and the development and testing of components. Cooperative planning would greatly facilitate the development of this data base without requiring substantial expansion of any national program.

Component development and testing will require substantial resources. A variety of major and medium-scale test facilities will be required for accelerated materials testing, high-field magnet development, neutral-beam and radio frequency heating system testing, blanket technology testing, high-heat-flux component testing, and other tasks. Cooperation in the construction and utilization of these test facilities would certainly limit the increase in national resources required for fusion development. The Large Coil Task (LCT) is an example of this type of cooperation.

The requirements are identified in detail in references (15) and (16).

Major Fusion Experiments

The next-step experiment in the tokamak program is currently estimated to cost from somewhat less than one billion to several billion dollars, depending on the physics and technology objectives. The detailed characteristics of and requirements for such experiments have been developed in several national design efforts and in one representative international study (17). Future experiments at a similar stage for other confinement concepts will probably cost a similar amount. Such experiments are obvious candidates for international collaboration.

Specific topics for cooperation within these three areas are being explored by the Versailles Summit subpanels as well as in other ongoing discussions.

INCENTIVES AND DISINCENTIVES

There are a number of general incentives for increased levels of international cooperation in the future. Cooperation expedites progress through the sharing of information and facilities and makes possible a much broader and more diverse program. The most obvious incentive is a sharing of the increased costs associated with the further development of fusion and a reduction in duplication of effort. A collateral benefit is the reduction in risk that any one program would incur by narrowing its options or unilaterally constructing a major facility. Interaction among program leaders leads to better program direction. International cooperation in fusion also may serve broader political objectives.

There would seem to be three potential disincentives for international collaboration in magnetic fusion. International collaboration that leads to interdependence of the national fusion programs would inevitably result in a reduction in national control. Some technology transfer is inevitable. Collaboration could inhibit the United States in developing a competitive advantage in the future market place for fusion technology.

Incentives

The United States program has benefited from prior international cooperation in two quite different ways: resources were available to support efforts beyond what could be supported in the United States alone, and novel and unique foreign contributions have influenced the US program technically. One example of financial benefit is the Japanese contribution to the US D-III tokamak, which allowed installation of the additional heating equipment that led to the achievement of record plasma parameters. A prime example of technical benefit is the rapid worldwide adoption of the tokamak confinement concept, which was invented in the USSR. As a consequence of interaction among technical and program leaders, all four major programs have advanced more rapidly and with better direction than they would have without cooperation.

The scientific and technical strengths of the fusion programs of Europe,

Japan, and the Soviet Union are comparable overall to those of the United States' fusion program. In some few instances, the foreign programs are more advanced than the US program—e.g. the development of microwave plasma heating technology in the USSR, the development of the stellarator confinement concept in Germany and Japan, and the experience in operating tokamak experiments with superconducting magnets in the USSR and soon in France. The United States' fusion program could benefit from gaining access to foreign expertise, and foreign programs could benefit from access to US expertise; international cooperation could lead to a stronger and broader national program, in a technical sense, than would be possible with a purely national effort.

As the development of magnetic fusion moves forward, the financial incentives for international cooperation increase. The present level of funding is viewed as adequate for the near term to maintain a broad program of scientific research, including development of the necessary supporting technology. However, substantial additional resources are required to achieve the ultimate scientific goal of fusion research, a self-sustaining (ignited) fusion plasma, which now seems to be within reach, and beyond that to develop the component technologies required for the practical realization of fusion power. The required additional funding would be determined by the need for new major facilities and by the program schedule. Since the costs of major new facilities would be many billions of dollars, international cooperation could provide a means of limiting the increase in national resources that would be required for the timely development of fusion.

Direct investments in projects in other countries can sometimes yield necessary information and experience for far less money than would be required to produce that information and experience in a national program. The Japanese investment in the Doublet III experiment in the United States is a good example. The Japanese were able to obtain valuable experience in the operation of a large tokamak at far less cost than if they had had to build a tokamak experiment of comparable size.

Several large test facilities will be required for the development of fusion component technology. The total national outlay for such facilities could be limited if their support and use were shared internationally. Such international cooperation is already taking place in the Large Coil Test facility in the United States, in which American, European, and Japanese superconducting magnets are being tested.

The national outlay for major plasma experimental facilities could be limited by the formation of an international project in which the costs of design, construction, and operation are shared. The Joint European Torus is the prime example of this ultimate mode of international collaboration. A recent INTOR study concluded that an experimental reactor of the FER-NET genre could be constructed by four partners, on a basis such that each partner obtained the additional technical benefit associated with providing one fourth of each type of high-technology components, for an outlay by each partner of only 40% of the cost of building and operating a national project unilaterally.

International cooperation also provides a means of sharing risks. A number of different scientific or technical approaches to a given problem can be tried by partners in a cooperative effort in which the resulting information and experience are made available to all the partners, as in the LCT, where six alternative coil designs are being tested.

In some circumstances, international collaboration in magnetic fusion could also serve broader political objectives. Again, the Joint European Torus is a prime example; collaboration on a major project by the European countries serves the political objective of European unity. Fusion was identified at the Versailles Summit Meeting in 1982 as one of the areas in which greater international cooperation would be desirable. Collaboration on a major fusion project could be a major, visible facet of any future resumption of East-West detente, just as the Apollo-Soyuz docking was in the recent past.

Disincentives

An interdependency of the United States fusion program with foreign programs would mean that certain essential elements of the program were not under the direct control of the American program leaders. Such an interdependency would also inhibit flexibility in shifting national program priorities in response to technical developments and changes in national policy. Funds committed to international collaborative projects are, at least in part, funds that are not available for national projects, and in times of budget reductions the former commitment may be honored at the expense of commitments to national projects because of larger political considerations. Furthermore, interdependency is contrary to the competitive instincts of program leaders, scientists, and engineers, who may seek to achieve national prestige by maintaining world leadership.

Fusion technologies have both potential national security and long-term commercial implications. Thus, questions of technology transfer in the process of international cooperation and of the impact of such cooperation on the competitive position of American industry naturally arise. While some few of the technologies developed for magnetic fusion (e.g. tritium and high-power millimeter-wave generators) also have defense applications, the number of such technologies is small, and to date it has been possible, with appropriate preparation, to proceed with cooperative activities. Thus, the national security aspects of technology transfer, while of potential concern in a few specific areas, should not be a major consideration overall in international cooperation in fusion.

With regard to the commercial aspects of technology transfer, it is a matter of timing. There will not conceivably be a significant commercial market for fusion until well into the next century, although there are markets for other applications of the technology in the near term. The information developed over the next few decades will inevitably be widely disseminated well before the arrival of a commercial fusion industry. Thus, there is a "window of opportunity" for international cooperation in the development of fusion technology over the next couple of decades.

Considering the above factors, we agree with the National Research Council study (4) that, "on balance, there are substantial potential benefits of large-scale international cooperation in the development of fusion energy."

THE CONTEXT

A multitude of social, political, institutional, and other factors define the context within which increased international cooperation must be developed. Some of these factors are favorable, and some tend to constrain the allowable options.

Favorable Factors

Fusion is at a comparable stage of development in the United States, the EC, Japan, and the USSR; the broad near-term objectives are similar in the four programs. This circumstance creates many opportunities for cooperation that should make sense technically and programmatically to the potential partners.

Greatly increased resources would be required to maintain the breadth and depth of the national fusion programs while moving forward to explore a burning plasma in a major next-step experiment and to develop fusion technology. There seems to be an increasing body of opinion among responsible leaders in government and in the fusion programs of the United States, Japan, the USSR, and the EC that international cooperation in the use of national resources is desirable, and may be required, in the present economic environment (4).

Previous cooperative undertakings in fusion have been substantial and generally successful. The participants generally feel that they benefited from the cooperation.

The technical and program leaders in the US, the EC, Japanese, and Soviet fusion programs have come to know and respect each other through many years of open professional and social contact. This rapport provides a rather unusual basis on which to build in negotiating and carrying out cooperative activities.

Magnetic fusion is the subject of one of the working groups established by decision of the heads of state and government at the Versailles Summit Meeting in 1982. This working group has identified the importance and magnitude of the effort of developing fusion and has concluded that a substantial increase in the level of international cooperation is justified. The heads of state have endorsed the activities of the working groups. This type of high-level endorsement can potentially have a very positive effect in removing bureaucratic impediments that could otherwise hinder the working out of arrangements.

Potential Impediments

Foreign perceptions of the United States' commitment to fusion and to international projects will certainly affect the willingness of other countries to enter into collaborative projects. The US government is not perceived as having a firm commitment to develop fusion, despite the Magnetic Fusion Energy Engineering Act of 1980 (4). A clear policy statement on the goals of the US fusion program and a corresponding firm commitment to meet those goals may be prerequisites for establishing international collaborative projects on a major scale. It is noted that one of the principal reasons for the success of the French Super-Phenix project is a clear national policy that assigns the project high priority, strong technical and industrial support, and adequate financial support.

The United States is not perceived as having a realistic plan for the development of fusion (4). The goal of the US program, as stated in the 1983 Comprehensive Program Management plan, "... is to develop scientific and technological information required to design and construct magnetic fusion power systems." This statement does not contemplate the development of an industrial base for the fabrication of engineering components or the construction of either a demonstration or a prototype power reactor; rather, it leaves these latter concerns to industrial initiative. Since the other major fusion nations seem to consider the goal of their programs to be the development of fusion power through the demonstration reactor stage, including engineering component development, there is a possibility that policy incompatibility could inhibit the achievement of cooperative agreements.

Differing levels of definition and detail in national fusion research and development programs can complicate negotiations on specific international cooperative projects. The parties with the better defined programs start with the advantage of knowing more precisely where they want to go and what they need to obtain from the cooperative effort. The partners with the less well-defined programs are put at a disadvantage. Their choice is between accepting an agreement that may not be fully advantageous or delaying the negotiation until they can evolve suitable levels of detail in their own national program plans to match those of the other negotiators.

The stated goals and milestones differ somewhat among the programs of Japan, the EC, and the US, being more definite in the first two. Nevertheless, all these programs lack detail as to performance, schedules, and costs. This fact suggests the possibility, at least, of attaining a reasonable compatibility through program adjustments. It would be to the United States' advantage to develop a realistic plan for fusion development, with specific objectives and schedules, to guide negotiations on international cooperation (4).

Entering into arrangements for international fusion cooperation will require a certain degree of trust among the participants. In particular, once a medium to large-size collaborative project has been agreed upon, it is essential that the commitment to it continue during its life. This lifetime, of course, can cover a decade or two, a circumstance that presents a problem in light of the annual budget review process that governments traditionally use. From time to time it is suggested that a project be taken "off budget," so as not to be subject to annual budget changes. This move is never popular with legislators, who by such a process relinquish a certain degree of authority. This arrangement is possible, however, where some independent fee, collected from users, provides money for such a fund—like automobile and truck taxes for highway funds. It is possible that such a fund could be established by utility users. It is also possible that through rather formal legal instruments, such as treaties, a strong obligation can be created to support a project financially. The fact that international obligations exist will help to protect that funding from year to year. However, the risk continues that at a certain time in the project's life the budget resources needed will be terminated by one or more countries, leaving the remaining participants to complete the project on their own, a prospect that may not be acceptable or possible. Thus, improved international instruments should attempt to address this question and, to the extent possible, ensure a reliable supply of funds for the program.

There is fairly widespread criticism abroad of the United States as an unreliable partner in long-term research and development efforts (4). Almost all US commitments to projects in the past have been fulfilled, but a few have not, and those are remembered abroad.

Although government officials speak of the benefits and necessity of international cooperation, the technical leaders in the laboratories do not really want to undertake large-scale collaboration (4). This attitude results partly from the ingrained spirit of scientific competition that pervades fusion research today, and partly from an appreciation of the additional practical difficulties of collaborative projects. However, the attitude of these technical leaders is being tempered, at least in the United States and Japan, by sober appraisals of the economic and political situation.

The programmatic and technical decision-making process is quite different in the United States, the EC, and Japan. In the United States, major programmatic and technical decisions can be made by responsible individuals at senior levels, whereas in Japan such decisions are made only after lengthy review and discussion at lower echelons leading to a consensus. In Europe such decisions are made only after numerous committee reviews. These styles lead to quite flexible, if occasionally erratic, evolution in US programs and to slowly formulated and steady, if occasionally cumbersome, evolution in EC and Japanese programs (4). Accommodation of these different styles of decision-making is necessary for large-scale cooperation.

Technical opportunities for cooperation may at times be incompatible with broader US policy considerations (4). The USSR has proposed joint international construction of the next-step tokamak experiment, yet it is unlikely that US-USSR collaboration is possible at present. Japan is willing to discuss further major cooperation, but there exists a political sensitivity to Japan on economic grounds. On the other hand, the Europeans, with whom cooperation would be the least controversial, show the least interest at present.

The extent to which any national fusion program will be willing to rely on international cooperation is a policy issue the resolution of which may constrain international cooperation. The previous practice in the US appeared to be to carry out domestically any research and technology development considered vital to the mainline US programs, and to encourage international programs in other areas. This position, which was perhaps satisfactory earlier, is now changing toward support of significant international collaboration.

In recent years, there has been an increased emphasis on equity in international cooperative activities, in response to a widespread feeling that the US was giving away more than it gained in such activities. There is criticism abroad that the US has placed undue emphasis on quantifying equity in exchanges in too narrow a context. While the objective of equity is quite proper, it is important that it be viewed in a broad context and pursued with a balanced approach that supports the spirit of mutual trust and the sense of partnership in a great task that must underlie any successful international cooperative activity.

IMPLEMENTATION

Many factors must be taken into account if large-scale international cooperation in magnetic fusion is to be successful. These include instruments of agreement, joint planning prerequisites, institutional framework, timing, legal and administrative considerations, and personnel mobility.

Instruments of Agreement

In almost all countries, a treaty between nations is the most binding agreement that can be established. Under US law a treaty has the equivalent status of a law enacted by the federal government. Nations consider treaties as important national commitments. While a nation can abrogate its obligations under a treaty either by terms of the treaty itself or unilaterally, such an action is rare, affecting the credibility of a nation, and not lightly or often done. Because of the binding commitment contained in it, a treaty involves a greater degree of approval and, therefore, normally takes substantially longer for its development and approval. On the other hand, once established, a treaty constitutes a mechanism for maintaining a high degree of certainty concerning the agreed positions of the countries.

The heads of state and government of the seven major Western countries and the EC, starting with the Versailles Summit in 1982 and continuing through successive conferences at Williamsburg and London, have endorsed in principle the idea of international collaboration in magnetic fusion. These endorsements could be formally implemented through agreements by heads of state and government. However, the seven countries include Canada, which has only a minor fusion program. Also, the Versailles Summit process tends to emphasize separate countries in Europe, whereas in Europe the fusion program is unified within the EC.

Although it is possible that the leaders participating in the Heads of State and Government Summit Conferences could enter into an agreement, an alternative arrangement could be among Japan, the United States, and the EC, or between any two of the three. Such an agreement would carry the full weight of the governments in power.

Abrogation of the agreement by a signing head of state would be an unusual, but not impossible, act. On the other hand, succeeding heads of state could either confirm the previous agreement or disavow it. There tends to be a certain degree of continuity from one government to another, even if the political parties change, on matters that are more technical than political, as fusion would be. Thus, an abrogation of such agreement would not normally be expected, but the possibility would be greater than if a treaty were in force.

A great many of the agreements between governments are negotiated and signed by appropriate ministries. While these agreements carry the full weight of the governments' commitments, they are subject to changing governments and may be abrogated at the ministerial level.

Joint Planning

Resource allocations for the next few years have been planned in all the major fusion programs. Furthermore, any major cooperation must meet the requirements of the separate national programs and therefore must be preceded by substantial joint planning.

Joint planning can be initiated informally, reaching whatever consensus is possible and then relying in residual matters on decisions by individual nations or groups of nations. This approach would involve exchanges of information as to the plans of all parties, leaving participants to proceed according to their own particular goals. One ongoing example of such informal joint planning at the program management level is the annual meeting under IEA auspices of the fusion program leaders of the US, the EC, and Japan for the purpose of exchanging information about their programs. A similar exchange, which additionally involves the Soviet program leader, takes place annually under IAEA auspices during the meetings of the International Fusion Research Council (IFRC). On the technical level, the US, Japan, the EC, and the USSR have cooperated in the IAEA INTOR workshop in defining an international consensus on the objectives and characteristics of a next-step tokamak experiment and on the R&D required to support the initiation of such an experiment.

On the other hand, joint planning can be more formalized, either in an umbrella agreement or as a subsequent arrangement, with whatever greater degree of binding effect may be agreed. To be effective, a formal joint planning activity requires policy guidance from government program leaders and technical direction from leaders in the laboratories. The undertaking must also have continuity over many years. The bilateral planning activity between the US and Japan is a successful example of such a formal joint planning activity that covers a limited range of topics. The Versailles Summit subpanels are addressing a wide range of possible areas of cooperation in plasma physics, fusion technology, and major fusion facilities and could possibly evolve into joint planning activities.

While these existing joint planning activities have supported cooperation, they do not in their present forms provide an effective mechanism for coordinating international cooperation on a major scale. However, they do provide a basis on which to build such a mechanism, and they suggest the appropriate elements of an effective international mechanism that would integrate joint planning at the program management and technical levels.

At the program management level, the program leaders in the United States, the EC, and Japan (and the USSR if the political situation becomes more favorable) could meet periodically to discuss and reconcile their respective programs for the development of fusion and to review the recommendations developed by joint planning groups at the technical level in specific areas. The technical level joint planning groups could consist of small numbers of technical leaders from the laboratories in the respective areas. These groups could meet periodically to discuss material prepared at home by broader communities of experts, and should maintain continuity of participants.

At present, it would seem appropriate to establish technical level joint planning groups in three areas: fusion technology development, alternative confinement concept development, and tokamak development. (These areas are covered by the Versailles Summit subpanels.) The first two groups would plan, respectively, for collaboratively developing fusion technology and alternative confinement concepts. These tasks would include identification of the required information and facilities and recommendations for equitable sharing of costs, construction and operating responsibility, and results. The tokamak development group would coordinate the experimental programs on existing experiments and would plan the next-step tokamak experiment or experiments. This latter group would recommend objectives, design concept, schedule, and cost and would define the required supporting research and development for the next-step tokamak experiment. The INTOR workshop has shown that such tasks can be performed successfully by an international group.

Thus, we agree with the National Research Council recommendation (4) that "the United States should take the lead in consulting with prospective partners to initiate a joint planning effort aimed at large-scale collaboration."

Institutional Frameworks

There are existing international organizations under the auspices of which more extensive international cooperation could be carried out without the necessity of new implementing agreements. The International Atomic Energy Agency (IAEA) has organized cooperative fusion activities generally considered to have been useful. This vehicle for near-term international cooperation is currently limited because of overriding political difficulties between East and West. If the political situation should change so as to permit cooperation between the East and West on fusion, the IAEA could be an important organization bringing the parties together.

The International Energy Agency (IEA) is undertaking research and development projects in fusion, as evidenced by the following agreements:

"Implementing Agreement for a Programme of Research and Development on Plasma Wall Interaction in Textor," August 10, 1977; "Implementing Agreement for a Programme of Research and Development on Superconducting Magnets for Fusion Power," October 6, 1977; and "Implementing Agreement for a Programme of Research and Development on Radiation Damage in Fusion Materials," October 21, 1980. A number of other agreements are pending, as discussed previously. The IEA provides a ready international mechanism for cooperation in fusion among the OECD countries.

The IEA currently serves quite effectively as a mechanism for the participation of several nations in the utilization of a Large Coil Test Facility built by the United States, and of a plasma technology experiment (TEXTOR) built by the Federal Republic of Germany. Similarly, a continuing workshop (INTOR) to focus technical effort on the problems of the next major tokamak experiment and to exchange information on the supporting research and development is held by the United States, the USSR, Japan, and the EC under the auspices of the IAEA. The IAEA also sponsors a number of effective data assessment and technical information exchange activities. An expansion of such activities under these agencies is reasonable. Neither of these agencies or other existing international organizations would be suitable as sponsors for a major international project because they function primarily as coordinators and administrators, not as managers, and because they have their own priorities.

However, an existing international organization may provide a framework for initiating a project, as was the case with the Centre Européen pour Recherche Nucléaire (CERN). CERN was initiated by an organizing conference sponsored by the United Nations Educational, Scientific and Cultural Organization (UNESCO), in an action that was ratified three years later by enough countries to ensure 75% of the required funding. CERN went on to become a highly successful institution, with international participation in design and construction of facilities and in performance of experiments.

The United States could have a bilateral arrangement with the EC and one with Japan. In addition, Japan and the EC could have a bilateral arrangement. This form has the advantage of direct relations between two parties so that the cooperation and management might be somewhat less complex. On the other hand, a major participant would not be included; and, if additional bilateral arrangements were established, in the end it might be more, rather than less, complex than a multilateral arrangement. The United States and Japan have a bilateral "Agreement on Cooperation in Research and Development in Energy and Related Fields," dated May 2, 1979. In accordance with this agreement, the two countries exchanged notes dated August 24, 1979, establishing an agreement in fusion energy and have exchanged further notes establishing committees and providing for cooperation in the Doublet III project. The EC and the United States and the EC and Japan are currently discussing bilateral agreements.

The United States, the EC, and Japan could establish a multilateral arrangement that would involve all three groups. This form has the advantage of involving the principal participants in the West concerned with fusion, but it has the disadvantage of being more complex than a bilateral arrangement because of the number of participants.

Finally, an international project could be established as a legal entity empowered to enter into contracts and to employ personnel for major collaborative activities. Several successful examples are discussed in the next section. Such an international project would need to be structured to permit strong project management, while at the same time providing for the integration of national political and technical elements in the overview process.

Technical Definition

The initiation of a cooperative activity must be preceded by some process of arriving at an agreement on the technical objectives and characteristics of the object of the cooperation, and by a parallel process of arriving at an agreement on the financial, administrative, and legal aspects of the cooperation. These latter aspects are discussed elsewhere in this section.

Arriving at an agreement on the technical objectives and general characteristics of the object of the cooperation is at the heart of major cooperative activities upon which the involved parties depend for essential parts of their national programs. Joint planning to coordinate national programs can certainly help to establish a framework for arriving at such agreements, but in most instances a more intensive process specific to the cooperative activity is necessary before even a decision in principle to go forward with the cooperative activity can be made.

There is a broad spectrum of approaches that can be taken in attempting to arrive at an agreement on the technical objectives and general characteristics of a potential cooperative activity. At one end of this spectrum of processes is the process in which each potential partner proceeds to define the objectives and characteristics of its own desired national activity, with periodic meetings and perhaps exchanges of observers to ensure an exchange of information. Such a process might evolve toward a sufficiently common definition of objectives and charac- teristics for the separate national activities that agreement on the objectives and characteristics of an international activity results. This process was used during the definition studies for the present generation of large tokamaks (TFTR, JET, JT60, T15), with the result that four large experiments with different characteristics but similar capabilities were conducted; however, this example is not a fair test of the process because the external incentive to collaborate was minimal at this earlier time. This process seems more likely to lead to the establishment and intractable defense of national positions than to the development of international consensus, given human nature.

An intermediate process would be that in which one nation defines a set of objectives and characteristics (and costs and schedules) and invites other nations to join in the activity so defined, possibly even after the activity has been initiated on a national basis. This approach was used by the US in initiating the Large Coil Task (LCT) cooperation and the Fusion Materials Irradiation Test (FMIT) Facility cooperation. In the first instance it was successful, and in the second instance it caused resentment and has not succeeded. The US initiated the Large Coil Task and committed itself to building the test facility, but then invited the Europeans and Japanese to participate at an early stage when they could still provide technical input to the facility design and test program. When the Large Coil Task was initiated, the European and Japanese programs in superconducting magnet development were quite small, and participation in LCT allowed European and Japanese fusion program managers to increase the scopes of their programs without sacrificing any elements of existing programs. On the other hand, the US had designed and developed FMIT before budgetary constraints motivated them to invite European and Japanese participation in facility construction, as distinct from facility use. Moreover, contributing to completion of the FMIT facility, as proposed by the US, would have required reprogramming limited research funds for fusion technology into a project in which European and Japanese scientists had had no direct part or benefit from formulating. These two examples are probably indicative of when unilateral project definition will succeed in initiating a cooperative activity—when participation is initiated early enough that the potential partners can provide technical input to the project characteristics and objectives and when participation results in a net benefit to the national fusion program of the invited partner, without requiring any significant reprogramming of his national program resources. Such situations may continue to arise for cooperative activities that are not in the mainlines of national programs, or that involve relatively modest financial contributions, but it is unlikely that major collaborative activities involving subjects at the cores of national programs can be initiated by this unilateral project definition approach.

At the other end of the spectrum is the process in which the potential partners cooperate to determine jointly the objectives, cost, schedule, and general characteristics of the proposed collaborative facility prior to the formal agreement to proceed with detailed design and construction. JET is a good example of this type of process—the JET Working Group, consisting of members from the participating countries, was established prior to the formal establishment of the JET Joint Undertaking, and the international JET Design Team worked for four years prior to the formal decision on construction. The INTOR workshop, in which US, Soviet, Japanese, and EC teams meet periodically to review national contributions and define homework tasks in the process of evolving an international consensus on the objectives and design concept for a next-step experiment, represents another good example of this mode of arriving at agreement (even though in this case it does not appear that the political situation will allow the project to move forward to design and construction). This approach has the important advantage of building a consensus and support at the technical level, as well as defining a project that can satisfy the high-priority national interests of the potential partners. There is the further advantage that this process can serve to broaden perceptions of how national interests can be served, thus facilitating agreement.

Legal and Administrative Considerations

A collaborative international project is complicated, but it can work if it is carefully planned and skillfully executed. Mechanisms must be established for creating an organizational entity and management structure. Procedures must be adopted for procurement, quality assurance, audits, and inspections. The authority of the project director, technical and political oversight mechanisms, national funding contributions, national industrial involvement, and priorities for operation of the facility must be established. Legal instruments of agreement must define the roles of national and local governments; provide guarantees of long-term commitment and provisions for withdrawal; regulate data communications; cover ownership of the facility; make provisions for liabilities, indemnities, and insurance against risk; and make provisions for taxes and customs duties. A Versailles Summit subpanel is addressing the issues of data communications and customs duties.

Personnel Mobility

Personnel mobility arrangements are intrinsic to substantive cooperation and have been in use for at least twenty years. Even so, there is as yet no simple and straightforward, standard method or form for their implementation. Mobility is easily described in principle, so that government-to-government agreements are easily reached on this point. Mobility also requires a legal agreement between the involved agents or contractors and the person affected. Since the relationships of the agents or contractors to their governments differ in ways that restrict their ability to sign contracts or assign rights to discoveries, it has not yet been possible to develop a single uniform, and therefore convenient, form of agreement. In fact, the lack of such agreements has been one of the principal hindrances in implementing some of the personnel assignments associated with the most successful bilateral activities.

Another aspect of personnel mobility is the set of personal difficulties associated with mobility. These difficulties are related to family or social dislocations, cultural differences, financial burdens, and diversion from professional development in one's home institution. Various steps have been taken to reduce these real difficulties. One is to recognize that a long-term personnel involvement may well be accomplished by a series of people traveling in a coordinated way rather than strictly by one person for an entire period. Another is to provide a collection and distillation of relevant foreign assignment experiences for prospective assignees. A Versailles Summit subpanel is dealing with the issue of personnel mobility.

EXAMPLES OF SUCCESSFUL COLLABORATION

For fusion the most relevant example of a major international collaborative project is JET (18). An important element of the JET success is the underlying legal framework. The project was set up as a joint undertaking by the member states of the European Community in 1978 under provisions of the 1957 Treaty of Rome, which established the European Community. Establishment of the joint undertaking was preceded by the JET Working Group (1971) and the JET Design Team (1973). Failure in the initial agreement to establish a mechanism for a decision on the site almost caused cancellation of the project in 1977.

The International Telecommunications Satellite Organization (INTELSAT) provides an example of the principle of "phasing in" an international project (19). Rather than attempting to define a complete set of international agreements at the outset, INTELSAT was established on an interim basis by an agreement that specified a time period for a study of the permanent form of the organization but did not set a deadline for the end of the interim arrangements. The permanent INTELSAT agreement, which was concluded six years later, provided for a phased shift from management of its space operations by the United States, as agent, to truly international management.

The success of any international collaboration depends on the extent to which technical considerations and political requirements can be merged at a given time. Although previous experience can provide guidance, the appropriate implementation structure for collaborative magnetic fusion projects must be designed specifically for the project at hand.

INTELSAT and JET provide relevant examples of combining technical and political elements in the decision-making chain for an international project. The following features of INTELSAT are noteworthy:

1. The Assembly of Parties, which meets every two years, is composed of all nations party to the Agreement, and is primarily concerned with issues of concern to the parties as sovereign states. The principal representation is provided by foreign ministers.
2. The Meeting of Signatories, which meets annually, is primarily concerned with financial, technical, and program matters of a general nature. The principal representation for each country is provided by the appropriate technical ministry.
3. The Board of Governors, which meets at least four times a year, has the responsibility for decisions on the design, development, establishment, operation, and maintenance of the international aspects of the project. The principal representation is provided by the officials concerned in their home countries with the operation and management of the project. Several expert advisory boards in technical, financial, and planning matters assist the Board.

The following aspects of the JET overview mechanisms are noteworthy:

1. The JET Council, assisted by the JET Executive Committee and the JET Scientific Committee, is responsible for the management of the project. The Council meets at least twice a year.
2. The Commission of the European Communities is responsible for financial decisions to the extent of its 80% contribution to the project.
3. National research organizations provide guidance to the JET Council on technical issues.
4. The EC Council of Ministers, with the assistance of the Committee of Permanent Representatives, is responsible for political decisions.

The Large Coil Task provides an example of finance and control in a national project with international participation. The United States funded, constructed, and operates the facility and pays the costs of its own test coils. The other participants pay for their own test coils. An executive committee, with one representative of each participant, decides on the test program.

Another relevant example of a major international project is the Super-Phenix project, a 1300-megawatt (electric) fast breeder reactor. This is a large-scale project that probably could not have been conducted without international collaboration. Super-Phenix is the result of agreements between the French and Italian governments for breeder development signed in 1974 and agreements between the French and Germans in 1976 on three levels: an agreement on breeder development policy between the governments; an agreement on research and development and the harmonization of national efforts between the nuclear research agencies; and agreements on commercial development among French, German, and Italian companies.

Several factors seem to underlie the success of the Super-Phenix project:

1. The French have provided strong project management and systems engineering on an extensive base of technology.
2. The French have majority control, and the other participants are junior partners. Management decision-making was clearly drawn from the beginning, with lines of authority established from the utility customer to the reactor developer and designer and to the component manufacturers.
3. The commitment of the parties stems from their lack of indigenous fossil fuels and natural uranium, and the imperative, as they perceive it, to develop breeder technology, which can make far more efficient use of natural uranium than can light water reactors.
4. There was a need to pool resources in such a large undertaking.
5. The Super-Phenix project developed against a background of other major cooperative efforts in Western Europe—in science, in aerospace and other multinational business ventures, and in economic union—that serve as trail markers.

CONCLUSIONS

There has been significant international cooperation in magnetic fusion to date. This cooperation has by and large been successful and forms a substantial basis on which to build increased cooperation.

There is sufficient similarity in the states of development and the near to intermediate-term objectives of the four major fusion programs to provide a technical basis for expanded international cooperation. The technical areas in which increased international cooperation could take place are relatively well defined.

The incentives for increased international cooperation seem to outweigh the disincentives.

There are a number of positive factors that would facilitate increased international cooperation. On the other hand, a number of real and perceived impediments would have to be resolved before substantial increases in the level of cooperation that would make national programs interdependent to any appreciable extent could be achieved.

The actual implementation of major collaborative activities requires that a host of complex issues, only a few of them technical in nature, be addressed.

On the whole, we conclude that increased international cooperation would benefit all participants and is feasible.

Literature Cited

1. International Collaboration in Fusion Energy Development. 1982. Special sect. in *Nucl. Technol./Fusion* 2:467–554
2. Scientific Cooperation Endorsed at Summit. 1983. *Science* 220:1252
3. A Political Push for Scientific Cooperation. 1984. *Science* 224:1317
4. National Research Council. 1984. Cooperation and Competition on the Path to Fusion Energy. Washington DC: Natl. Acad. Press
5. Roberts, M. 1983–84. Internal briefing documents. US Dept. Energy
6. US Dept. Energy. 1983. Comprehensive Program Management Plan for Magnetic Fusion Energy
7. Magnetic Fusion Advisory Committee. 1982–1984. Reports to Dept. Energy, Panels 1–7
8. Technical Panel on Magnetic Fusion of the Energy Research Advisory Board. 1984. Magnetic Fusion Energy R&D. DOE/S-0026
9. Palumbo, D. 1983. The Fusion Programme of the European Community. *Nucl. Technol./Fusion* 4:13
10. Commission of the European Communities. 1984. Report of the European Fusion Review Panel II. EUR FU BRU X11-213-84, Brussels
11. Commission of the European Communities. 1984. Proposal for a Council Decision (adopting a research and training program 1985 to 1989 in the field of controlled thermonuclear fusion). EUR FU X11/200-1, Brussels
12. Iso, Y. 1983. The Fusion Technology Program in Japan. *Nucl. Technol./Fusion.* 4:6
13. Japan Atomic Industrial Forum. 1983. Present Status of Nuclear Fusion Development in Japan
14. Yamamoto, K. 1984. The Fusion Program in Japan, *Proc. AIF Conf. Ind. Role Dev. Fusion Power, Washington, DC:* Atomic Energy Forum
15. National Research Council. 1982. Future Engineering Needs of Magnetic Fusion. Washington, DC: Natl. Acad. Press
16. Stacey, W. M. Jr., Bakser, C., Conn, R., Krakowski, R., Steiner, D., Thomassen, K. 1984. Future Technology Requirements for Magnetic Fusion—An Evaluation Based on Conceptual Design Studies. *Nucl. Technol./Fusion* 5:266
17. INTOR Group. 1980. International Tokamak Reactor: Zero Phase, STI/PUB/556; 1982. Phase One, STI/PUB/619; 1983. Phase Two A Part 1, STI/PUB/638. Vienna: IAEA
18. Wilson, D. 1981. *A European Experiment.* Bristol: Hilger
19. Leive, D. N. 1981. Essential Features of INTELSAT: Application for the Future, *J. Space Law* 9:45

Ann. Rev. Energy. 1985. 10 : 109–33

ENERGY USE IN CITIES OF THE DEVELOPING COUNTRIES[1]

Jayant Sathaye and Stephen Meyers

Energy Analysis Program, Lawrence Berkeley Laboratory, Berkeley, California 94720

INTRODUCTION

The movement of people from the countryside to cities is a significant feature of the recent development process of many Asian, African, and Latin American countries. Insofar as development is measured by production of goods and services (GDP), there is a high correlation between level of development and degree of urbanization. The urban population makes up 78% of the total population in the world's industrial, market-economy countries, 34% in the lower-middle-income countries, and 21% in the low-income countries (1).

Most people in the Third World still live in rural areas. But the proportion living in urban areas has grown between 1960 and 1982 from 17% to 21% in low-income countries and from 24% to 34% in lower-middle-income countries. Urbanization accelerated in many countries in the 1970s, as greater numbers of people migrated to cities seeking work. In many of the poorest countries (mostly in Africa), the growth rate of the urban population averaged more than 6% per year during the 1970s. Together, the developing countries now contain more than 250 cities with population over 500,000.

Urbanization brings with it changes in the ways resources are collected, distributed, and used. In rural areas, people are more directly involved in these activities. In cities, there is greater dependence on a supply network to meet these needs. As the cities of the developing countries rapidly grow,

---

[1] The US Government has the right to retain a nonexclusive royalty-free license in and to any copyright covering this paper.

there is increasing pressure on these networks, and in many cases concern about whether they will be able to cope with the growing demand (2). Understanding the nature of urban resource consumption, and in particular, the changes that take place as a country's population becomes more urban, is an important element in planning urban supply networks that will be able to accommodate future demands.

Energy and Cities

The cities of the developing countries are the gateways to the outside world, the places where modern influences first take hold. In the sphere of energy, "modern" has come to mean the widespread use of fossil fuels and electricity. As these energy sources have become available to fuel economic growth, patterns of energy use have evolved that are vastly different from those in the countryside. In Papua New Guinea in 1975, for example, per capita daily consumption of petroleum fuels (direct and indirect) was 139 MJ in the urban/industrial sector, and only 7.4 MJ in the rural sector (6). Urbanization has brought with it a transition away from traditional energy sources (including human power). Often, this transition has involved a shift from indigenous energy forms gathered by the users themselves to fuels that are not locally produced. The growth in income that has taken place in urban areas has also facilitated growth in demand for new services such as refrigeration, air-conditioning, and personal vehicle transportation.

The patterns of energy use in the cities of developing countries exhibit great variety. Estimated total energy demand per household in Mexico City was over 15 times that of Nairobi, for example (17, 23). There are pockets in nearly all Third World cities where the energy use patterns are similar to those found in warm-climate European cities. There are also areas where the patterns resemble more closely those of the rural areas from which many urban dwellers have recently arrived. The important difference is that energy resources are rarely available for free in urban areas, and the need to purchase energy creates problems for the urban poor.

The much higher level of per capita energy use in cities is a consequence of the quantity and kind of energy-using activities in cities. These activities fall into four categories:

1. Home-based activities
2. Transportation of people and goods
3. Production of goods
4. Provision of private and public services.

The increasing use of modern fuels to support these activities is, as mentioned, a characteristic of today's Third World cities. Indeed, in most countries the use of modern fuels is highly concentrated in the cities, and urban growth and rising demand for modern fuels are highly correlated (3).

Therefore, better understanding of the relationship between urban development patterns and use of modern fuels will allow governments to manage better the demand for fuels that may be difficult or expensive to acquire.

While the rural energy situation in developing countries has been the object of much study, there has been comparatively little research on urban energy use. To our knowledge, the only studies that have collected data sufficient to present a portrait of the entire pattern of energy use in a Third World city are Newcombe's studies of Hong Kong (4, 5) and Lae, Papua New Guinea (6). Other work has been done on pieces of the urban energy use picture, however, and there are also a number of government surveys that shed light on urban energy demand.

In this review, we present a framework for looking at urban energy demand and discuss findings from a number of studies and surveys. The nature of the literature that we have assembled restricts the scope of our discussion. Most of what we present concerns energy use by households, either in the home itself or in personal transportation. Energy use in commerce and in the provision of public services needs further study. We leave energy use in the production of goods, an area about which much has been written (6a), for another review.

ENERGY USE IN THE HOME

Energy use in the home depends on the activities engaged in by households and the kinds of devices and fuels used to help accomplish those activities. The former is conditioned largely by the physical and cultural environment in which people live. The latter is determined by the economic resources accessible to households and the general availability of different fuels and energy-using devices.

As households take on a more modern way of life, demands for new services arise, and there are changes in the manner in which old services are performed. In some cases, a new demand arises which traditionally did not exist (such as for refrigeration of food or for water heating for personal hygiene). In others, a traditional service (such as lighting or space cooling) is performed more adequately by the modern device/fuel combination. Replacement of human labor by non-human-powered machines is also a characteristic of the transition (especially in clothes care).

Most Third World urban households make use of a combination of traditional and modern methods. The mix varies widely among cities, but in most cities there are probably more households living in a largely traditional manner than in a fully modern one.

Convenience is a major motivating force in the transition to modern methods. So is the desire for more service: more cooling than ventilation provides, more entertainment than radio provides. Of course, considerations of style and image are at work in the development of new demands. What Newcombe writes about Papua New Guinea applies to many developing countries: "The patterns of personal energy use adopted by the European remnants of the colonial era and the newly emergent national elite are powerful determinants of aspirations for the future held by the majority . . . ."

The above forces have led to growing use of appliances in urban homes. Along with this growth, changes are taking place with respect to the devices and fuels used for cooking and lighting. The move from traditional fuels and devices to kerosene, gas, or electricity depends on household income. These changes generally lead to more efficient use of energy at the site of use, although conversion losses may be transferred to the power station. They also increase demand for fuels that may require foreign exchange resources or capital investment in the supply network.

In considering the evolution of home energy use, it is important to investigate:

1. The demand for new services and for new, more energy-intensive methods of performing traditional services.
2. The choice of devices and fuels to meet both traditional and new demands.
3. The "quantity" of service demanded (how much lighting, space cooling, etc).

These distinctions are important, for it is not energy per se that households demand, but rather particular services that may be met with varying quantities of different fuels. Some traditional services are being met in new ways as urbanization develops; some new services are emerging. In the following sections, we discuss key patterns of change with respect to energy-using household activities.

Cooking

Energy demand for cooking is the oldest and most common of household energy needs. Cooking is the primary energy-consuming activity in the vast majority of Third World homes, in both rural and urban areas. Results from a survey of semi-urban households in Pondicherry, India show that over 90% of total energy use in both low- and middle-income homes went for cooking (7). A survey of households in three Chinese cities found an even higher share of energy use for cooking (8).

The predominance of cooking in the household energy budget is a result of both the absence of other energy demands and the low efficiency of energy conversion associated with the use of traditional cooking fuels. As household income rises, these conditions change. Households are able to purchase new energy-using devices to meet various demands, and also to acquire different, more energy-efficient devices for cooking.

CHOICE OF COOKING FUEL   The choice of cooking fuels depends on preference, ability to afford to use a particular fuel, and the availability of fuels and devices. Urban areas are a fertile setting for change in cooking fuels because of both their income-earning opportunities and the availability there of different fuels.

The type of fuel used for cooking is usually a good indicator of economic status. As income rises, the main cooking fuel tends to move from firewood or charcoal to kerosene to liquefied petroleum gas (LPG) or electricity. Data from surveys of urban households in several Asian cities (9–12a) illustrate this pattern (Table 1). One pattern that emerges is a rise in the use

Table 1   Cooking fuels used in urban households[a]

Income group[b]	Percent of households in group using				
	Firewood	Charcoal	Kerosene	LPG	Electricity
Kuala Lumpur (1980)					
Lower-income (24)	4	15	75	25	19
Middle-income (54)	7	23	57	52	35
Higher-income (22)	0	17	19	87	50
Manila (1979)					
Lower-income (39)	9	1	35	45	11
Middle-income (50)	2	1	5	73	19
Higher-income (11)	1	0	1	78	19
Hyderabad (1982)					
Lower-income (43)	41	c	70	19	—
Middle-income (42)	24	c	65	54	—
Higher-income (15)	13	c	57	71	—
Bombay (1972)					
Lower-income (51)	10–30	10–30	98	9	—
Middle-income (44)	3–20	3–20	98	53	—
Higher-income (5)	3–10	3–10	77	94	—

Sources: References 9–12a.

[a] Data for Kuala Lumpur (Malaysia) and Hyderabad (India) reflect use of more than one fuel by households. Manila (Philippines) data refer to "usual" source of energy. Bombay (India) data refer to ownership of devices. The percentages of Bombay households owning a hearth for burning firewood or a stove for burning coal were 40, 23, and 13 for the respective economic classes.

[b] The percentages of households in each income group are given in parentheses. The income levels are not comparable between countries.

[c] Small amounts of charcoal are used at all income levels.

of kerosene with income, within the lower-income groups. As income increases further, however, kerosene gives way to LPG and/or electricity.

The kinds of cooking fuels used in different cities also reflect the relative economic situation of households (Table 2). Households in Kuala Lumpur and Manila are wealthier than those in the Indian cities; not surprisingly, use of LPG and electricity for cooking is more common in the former. Firewood, most of it purchased, continues to be used by many households in Hyderabad.

Income elasticities of cooking fuel consumption, calculated with data from a recent survey of 1800 households in Hyderabad, India, provide a quantitative estimate of changes in usage patterns with income (11). As expected, firewood turned out to be an "inferior" good—its consumption fell as income rose. A 10% increase in income was associated with an 8% decline in firewood consumption. For LPG, the income coefficient was almost 1, implying that LPG consumption rises in close proportion with income. Kerosene use showed a less continuous pattern than wood and LPG. Its use rose very sharply from the lowest income group and then fell off as income increased further and its use gave way to LPG and electricity.

Availability of fuels plays a key role in shaping patterns of change. In Hyderabad, for example, consumption of LPG is constrained by shortages and lack of access to LPG supplies. If supplies were adequate to meet the demand, the income elasticity reported above would be higher.

Availability strongly influences use of firewood, which remains a common fuel for the urban poor in many places. Where it is available for free or at low cost, it tends to be used, for reasons of both cost and habit. In many cities there are large numbers of newly arrived migrants to urban

Table 2 Cooking fuels used in urban households[a]

	Percent of households in group using				
	Firewood	Charcoal	Kerosene	LPG	Electricity
Kuala Lumpur (1980)	5	20	55	33	34
Manila (1979)	5	1	16	63	16
Hyderabad (1982)	29	[b]	65	42	—
Bombay (1972)	10–20	10–20	97	33	—

Sources: References 9–12a.

[a] Kuala Lumpur (Malaysia) and Hyderabad (India) data reflect use of more than one fuel by households. Manila (Philippines) data refer to "usual" source of energy. Bombay (India) data refer to ownership of devices. Thirty-one percent of Bombay households owned a hearth for burning firewood or a stove for burning coal.

[b] Small amounts of charcoal are used at all income levels.

areas, who tend to use energy in patterns similar to their earlier practices. Recent surveys of energy use in Kenya found that fuelwood accounted for 60% of total energy consumption among the lowest-income group of urban households (13). Kerosene made up only 8% of total consumption. In the Philippines, reliance on wood is much higher among households in the poorer socio-economic class in provincial urban areas, where wood is cheaper and more available than kerosene and LPG, than it is in Manila, the primary city (Table 3).

Lack of wood resources around urban areas has become a serious problem in many places, particularly in Africa. In Upper Volta, people have to travel 70–100 km along the main roads from Ouagadougou, the capital, to find wood (14). Yet wood is still by far the main energy source for cooking. Lack of wood can bring about a shift to a processed traditional fuel, charcoal. Usually produced from distant forests, it has become a common cooking fuel for the poor in many African cities. For example, in Monrovia, the capital of Liberia, over 80% of surveyed low-income families used charcoal for all or part of their cooking (15).

Lack of wood resources can also accelerate the transition to modern fuels, especially where they are locally available at a low price. In Indonesia, where kerosene is both more available than fuelwood and has historically been relatively inexpensive, three-fourths of all urban households reported owning a kerosene stove in 1976 (although fewer than that used kerosene as their main cooking fuel) (16). In China, where most forested areas were cut down long ago, coal is used for cooking by over three-fourths of urban households (8).

Ownership and use of more than one cooking device is common in urban areas. This is partly because different kinds of foods are best prepared with different types of fuel. A survey conducted in Nairobi found that low-

Table 3   Cooking fuel of the urban poor in the Philippines

	Metro Manila	Luzon	Visayas	Mindanao
	(percent of households)			
Wood	9	48	75	73
Charcoal	1	7	9	3
Kerosene	35	24	11	18
LPG	45	16	4	5
Electricity	11	3	0	1

Source: Reference 10.

income households using both charcoal and kerosene tended to use kerosene for quick food preparation and charcoal for traditional dishes (17). Households generally do not discard their old cooking device when they acquire a new, more modern one.

Use of more than one cooking fuel is clear from the Kuala Lumpur (9), Hyderabad (11), and Bombay (12) survey data. Among upper-income households in Kuala Lumpur, 87% used LPG, 45% used electricity, 23% used kerosene, and 19% used charcoal. (Continued use of charcoal by upper-income households is not unusual in many places, as certain popular foods require charcoal to cook in the desired fashion.)

In the Hyderabad study, the proportion of households using both wood and kerosene ranged from 24% in the lowest income group to 7–9% in the upper-income groups. The proportion using both kerosene and LPG ranged from 1% in the lowest income group to 35–40% in the upper-middle-income groups. The percentage drops for the highest-income groups, as use of only LPG becomes very common. In Bombay, most households with LPG stoves also owned kerosene stoves, even in the higher-income groups, but usage patterns were not surveyed.

Use of more than one cooking device and fuel is not restricted to middle- and upper-income households. Among the Bombay lower-income households, nearly everyone owned a kerosene stove, but 40% of the households also had a coal stove or fireplace for wood. Among the Hyderabad lower-income (0–1000 rupees per month) households, 21% used wood and kerosene, and 10% used kerosene and LPG.

Fuel switching can be either temporary or permanent. Households that own more than one cooking device have flexibility in responding to changes in the costs or availabilities of particular fuels. This allows temporary fuel switching. The acquisition of new cooking devices when poorer households gain more income may lead to a more permanent change, although going back to the old fuel often remains a possibility.

A survey of household energy demand in urban areas of the Philippines, conducted in 1979, provides some insight into the nature of fuel switching (10). About one out of six households had changed their cooking fuels in the few years prior to 1979. Almost half of those switched from LPG, 28% switched from kerosene, 14% switched from wood, and 9% switched from electricity. The fuel that households switched to is unfortunately not given. Since the period in question was a time of rising energy prices, it may be that most switchers moved to cheaper fuels. It is interesting that the incidence of switching was about the same across income groups. Almost two-thirds of the surveyed households expressed unwillingness to change fuels. Economic reasons ("more economical to use present fuel," "shifting from one fuel to another is expensive," "expenses will be the same if we change") were most often cited for this attitude. Questions of convenience and resistance to change were also a concern.

The cost of cooking, a function of the price of cooking fuels and the cost of the devices to use them, influences fuel choice. The latter can be an important barrier to the use of a new fuel. In Hyderabad, the prices of both kerosene and LPG per unit of "useful" energy are well below that of firewood.[2] This is confirmed by the fact that average energy expenditures were higher for households using only wood than for those using only kerosene, even though both consumed the same amount of "useful" energy. Habit, no doubt, is a factor in continued use of wood, but the cost of a kerosene stove seems to play a role. It is revealing that the proportion of Hyderabad survey households using only wood drops off sharply from the lowest-income group (35%) to the next-higher-income group (19%).

ENERGY EFFICIENCY OF COOKING   The amount of energy households use for cooking depends very much on the efficiency of energy conversion, which can vary greatly depending on the kind of fuel and device used. The effect of device efficiency is borne out by results from the Hyderabad survey. Average monthly energy use per person for cooking fell from 0.239 million Btu among very-low-income households, who rely heavily on wood, to 0.165 in the highest-income households, who mainly use LPG and, to a lesser extent, kerosene. Part of the decline is probably due to economies of scale in cooking, since the upper-income households average considerably more members than the poor households. But the use of more energy-efficient devices by the upper-income households probably is the most important factor.

The effect of use of more efficient devices can also be seen in results from the survey of Chinese households. Average monthly cooking fuel consumption per person was lowest in the homes where town gas was used (180 MJ) and highest where bituminous coal was used (630 MJ). Average consumption by households using coal briquettes or "comb-shaped" coal fell between these extremes.

ENERGY CONSUMPTION FOR COOKING   It is possible to use measured or estimated conversion efficiencies to calculate the approximate amount of energy that is actually providing the cooking service. Given the same cooking habits, this amount will be needed regardless of the type of fuel used. (In practice, use of a new cooking device often brings some change in cooking patterns.)

---

[2] "Useful" energy refers to the amount consumed net of conversion losses. For average efficiency of fuel combustion, the researchers used 0.17 for firewood, 0.20 for charcoal, 0.48 for kerosene, and 0.60 for LPG.

The trend for the Hyderabad households shows that "useful" energy consumption per person for cooking increases slightly as income rises. The basic monthly energy demand per person—around 75 MJ—is, however, nearly the same across the income spectrum.

Average monthly useful energy consumption per person for cooking among the surveyed Chinese households was between 90 and 95 MJ for all of the groups except the one that used coal briquettes, who averaged 65 MJ (the efficiency of briquette burning may be underestimated).

Lighting

The degree of indoor lighting is one of the main differences between Third World urban and rural homes. Most rural households use kerosene lamps that do not provide a great deal of light. (In some places they may be valued more for mosquito control than for lighting.) In urban areas, electric lighting is now widespread, although it is by no means universal. Where electrification has not yet reached—in the peripheral areas of large cities, where the poor tend to live, and in small cities and towns—kerosene is still used for lighting in many homes.

The use of electric lighting brings with it an increase in the amount and quality of service provided. The quantity of lighting service demanded by households, however, varies considerably. Wealthier families, living in larger homes, demand much more lighting than low-income households, who often only have one or two bulbs.

Electricity-Intensive Activities

With cooking, modernization brings changes in the fuel-device combinations used to provide an already existing activity. With lighting, the move from kerosene to electricity brings a much greater degree of service. With other activities, changes with modernization may provide more service than the old method, or may provide a service that was previously met with human energy or was not in demand at all. These features apply to three activities that could have a major impact on residential electricity demand in Third World cities: refrigeration, water heating, and air-conditioning.

Refrigeration is a service that is generally absent from the traditional setting. The percentage of urban homes with refrigerators varies from almost zero in China to over 75% in the wealthier cities of Asia (Table 4). A refrigerator is usually the first electricity-intensive appliance acquired when urban families have sufficient income. (Black and white televisions, which are often the first major electric appliance purchased, are much smaller consumers of electricity.) Data from the Philippines and Malaysia show that ownership of refrigerators increases sharply from the lowest-income group to the next highest (Table 5). In the upper-income groups, refrigerator ownership is nearly universal.

Table 4 Electric appliance ownership in urban areas

	Refrigerator	Air conditioner	Water heater
	(percent of households)		
China (1981)[a]	<1	—	—
Guatemala (1975)[b]	12	—	—
Liberia (1980)[c]	52	7	—
Manila (1979)	63	5	—
Malaysia (1980)[d]	70	9	12
Hong Kong (1974)[e]	80	13	9
Bangkok (1979)	82	15	4
Taipei (1979)	85	12	6

Sources: References 5, 8–10, 12, 18.
[a] Survey of 279 households in 3 cities.
[b] National survey of 854 households.
[c] Monrovia: survey of 89 low-income households.
[d] Kuala Lumpur and Kajang: survey of 535 households.
[e] Survey of 3977 households.

Water heating for purposes of hygiene is similar to refrigeration in that it is usually not found or is seldom used in a traditional setting. Although a great deal of domestic water heating is accomplished by nonelectric means (such as with piped gas) in the industrialized countries, in the Third World most modern water heaters use electricity. Malaysian survey data, which show low levels of water heater ownership for most income groups, indicate

Table 5 Electric appliance ownership and income in urban areas

Income group[a]	Percent of households in group using		
	Refrigerator	Air conditioning	Water heater
Malaysia (1980)[b]			
150–299 (3)	13	0	0
300–599 (20)	50	0	6
600–999 (28)	65	1	6
1000–1999 (29)	79	10	14
2000–4999 (15)	87	23	31
5000+ (5)	96	79	50
Philippines (1979)			
Low (49)	15	0	—
Middle (42)	66	2	—
High (9)	93	20	—

Sources: References 9–10.
[a] The percentages of households in each income group are given in parentheses.
[b] Kuala Lumpur and Kajang; units are M$/month.

that it is a lower priority for households than refrigeration. Low levels of water heater ownership are in part a consequence of the warm, humid climate of most Third World cities, and in part a result of the practice of heating water on the stove.

The tropical climate works in the other direction with respect to home cooling, the demand for which could have a major effect on electricity use. At present, air-conditioning is uncommon except among the wealthiest households. Determining the income level at which households tend to acquire air-conditioning is an important task for electric utility planning. It remains to be seen whether middle-class households will be satisfied with fans, which are quite common (see below), or will move to air-conditioning as soon as income allows.

Other Home Activities

Changes in the manners of supporting other activities also involve the acquisition of electricity-using appliances. For the most part, these appliances are not heavy electricity users. In places where household electricity use is very low, however, increasing use of such appliances could have a significant proportional effect on electricity demand. Ownership levels in cities of various countries are shown in Table 6. Ownership of TV sets has become common in many cities, and is nearly universal in the more modernized cities of Asia. Clothes washers are still for the most part not part of household appliance holdings.

Household Energy Consumption and Socio-Economic Class

The information presented in the previous sections illustrates how growth in family income brings with it changes in the fuels used to support the most

Table 6   Electric appliance ownership in urban areas[a]

	Radio	TV	Fan	Washer	Iron	Cooker
	(percent of households)					
China	39	66	45	2	1	—
Liberia	76	4	56	1	74	—
Guatemala	78	25	—	—	—	—
Manila	80	78	82	—	94	16
Malaysia	70	79	75	16	77	44
Hong Kong	~90	91	96	34	87	91
Bangkok	—	96	—	~5	—	84
Taipei	—	92	94	53	—	89

Sources: References 5, 8–10, 12, 18.
[a] See Table 4 for years covered.

important household activity, cooking, and increased demand for new household services and for new, more energy-intensive methods of supporting traditional activities. One result of this is that as incomes rise the relative shares of fuels in total energy consumption change. Results from a survey conducted in India in 1963–1964 show that as urban households' expenditure levels (a proxy for income) rose, the share of traditional fuels like wood and dung declined while that of modern fuels like coal (in 1963–1964) and electricity increased (12).

The share of electricity among Indian urban households in 1963–1964 (only 9% even in the highest income group) is very low by the standards of most of today's Third World cities. In other places, growing appliance holdings and more common use of electricity for cooking have made electricity a major energy source for homes. A survey of Bombay households in 1971 found that electricity accounted for over 40% of total energy use for the highest-income group (19). In the lowest-income group, it was only 5%. A 1974 survey of Hong Kong (5), a more modernized city than Bombay, showed a much smaller difference in the electricity share between the richest (40%) and the poorest (24%) groups. The electricity share in the lowest income group in Hong Kong is much greater than that in the lowest Bombay group, indicating both the use of more electric appliances and less use of inefficient cooking devices by the Hong Kong families.

Since in many places electricity demand requires the use of three to four units of some energy form (such as oil) to produce one unit of electricity, the effect of growing electricity demand is actually greater than is evident from looking only at direct energy use. When indirect (or "primary") energy use is counted, the average share of electricity demand for all Hong Kong households doubles from 31% to 61%.

Increased use of electric appliances and use of more efficient devices for cooking as income rises have opposing effects on overall energy consumption. In urban areas where the modernization process is not so far along, the result may be only a small increase in overall household energy use as income grows. In urban India in 1963–1964, for example, average per capita monthly total energy consumption was between 22 and 23 kg of coal equivalent for all but one of the income groups, and rose significantly only in the upper income group. It seems that greater efficiency of energy use in cooking by middle-income families balanced their use of more electric appliances. This effect is also apparent in survey data on Lae, Papua New Guinea, households. Average daily energy consumption (including wood) fell from 11.2 MJ per capita in "settlement" houses (the poorest) to 8.3 MJ per capita in "low-covenant" houses, who are less dependent on wood for cooking.

As overall income levels increase, and as the middle and upper classes become wealthier, the situation changes. Use of very-low-efficiency device/fuel combinations for cooking becomes less common among the poor, and ownership of electric appliances by well-off households becomes significant. Results of a 1971 survey of relatively modern (for India) Bombay showed a much greater increase in household energy consumption with income than was the case for urban India in 1963–1964. Average monthly energy consumption nearly doubled from the lowest- to the highest-income group, and there were also steady increases in the upper-middle groups. The Hong Kong survey found an even greater increase in energy consumption from the lowest- to the highest-income group.

Growth in Residential Demand for Modern Fuels

The growth rate in residential demand for modern fuels will depend largely on the movement of families to cities, the rate at which poor households move from traditional methods of meeting services to modern ones, and the extent to which middle-class households expand their appliance holdings. All of these factors will be affected by overall and particular economic conditions.

Among the fuels in use in Third World urban homes, the fastest growth is taking place in use of electricity. Room for growth in its use exists at nearly all income levels. The level of electricity consumption depends in large part on ownership of appliances, which is related to income level. Data from the Philippines suggest that the location of the household affects appliance ownership and electricity use. Among households in the same general socio-economic class, those living in the primary city, Manila, had greater appliance holdings and higher electricity consumption than did households living in more provincial urban areas (Table 7). This may be a function of

Table 7   Electricity use and the urban poor in the Philippines (1979)

	Manila	Luzon	Mindanao	Visayas
Avg. monthly consumption (kWh)	130	47	40	26
Electricity used (percent of respondents)				
Lighting	99	90	63	64
Ironing	87	48	14	4
Fan	64	32	9	10
TV	55	32	13	11
Refrigerator	27	16	12	2
Cooking	11	3	1	0

Source: Reference 10.

greater availability of appliances, or of the more modern life-style of Manila.

Rapid growth in appliance holdings can occur if economic conditions allow. Between 1962 and 1972 in Taipei, Taiwan, the average number of devices per 100 households increased from around zero to 77 for TV sets, from 5 to 99 for rice cookers, and from 2 to 48 for refrigerators (20). Growth between 1972 and 1982 was slower, though the use of air conditioners rose considerably.

Taiwan's rate of economic growth in the 1960s was exceptional. In many Third World cities, changes are now occurring more slowly. Lower-income households are somewhat stuck in existing energy use patterns due to lack of resources to change. On the other hand, there are a large number of households on the threshold of much higher electricity use. While demand models for household energy consumption have been developed in Taiwan (21) and elsewhere, further understanding of the dynamics of change is necessary to permit better prediction of future demand for electricity and other fuels.

ENERGY USE IN TRANSPORTATION

Energy use in transportation is determined by the demand for movement of people and goods. It is influenced both by the spatial environment in which people live and by the activities that entail movement, such as work, shopping, social interaction, and distribution of products.

The traditional human and animal-powered forms of transportation predominate in rural areas of developing countries. In the cities, walking is still a common form of transportation, but the use of motor-powered vehicles—motorcycles, cars, buses—has become an important feature of urban life. The annual rate of growth of the automobile stock between 1960 and 1970 was between 7% and 15% in most of the largest Third World cities (21a). Motorized vehicles increase speed of movement, but they also require the use of fuels that in many cases must be imported at high cost.

In considering energy use in urban transportation, it is important to be aware that, as with energy use in the home, it is particular services, not energy per se, that are desired. First, there are activities for which movement of people and/or goods is wanted, and a certain distance to be covered. Second, there is a choice of how the movement is to be accomplished: the mode of transport. The distance to be covered, time constraints, personal preference and economic resources, and the availability of different transport modes shape this choice. The amount of energy used to accomplish the movement is very much dependent on the transport mode chosen.

Mode of Travel

A wide range of transport options is often available in cities. These options, or transport modes, provide different types of service, measured in terms of speed and comfort of movement. They also require differing amounts of economic resources to make use of them.

Walking is of course the least expensive mode of travel, and it remains an important one in cities of developing countries. In Lae, Papua New Guinea, it was estimated that walking was the major transport mode for 20% of the people, compared to 15% for cars (6). Where longer distances are involved, buses are the primary transport mode for the urban poor and middle class.

The mode most commonly used by individuals and families varies with household income. As average income has grown, the use of personal motorized vehicles has risen. Mopeds and motorcycles are the lowest-cost mode of motorized personal transport. As Table 8 shows for Malaysia, they are used most frequently in the middle-income groups. Automobile ownership and use has become common in the more modern cities. Members of higher-income households are more likely to own or have access to and travel by car than are members of lower-income households. In Kuala Lumpur and Kajang in Malaysia, 40% of the surveyed households reported the car as their mode of travel to work. Among the poor households, very few used a car, but in the wealthiest group over 90% did so (Table 8).

Data on Philippine urban households also show that automobile ownership is very dependent on household income (10). In the upper 9% of households, 44% owned a car, 19% owned a jeep, and 13% owned utility

Table 8   Mode of travel to work in Kuala Lumpur and Kajang (1980)

Monthly income[a] (M$/month)	Percent of households in group using				
	Car	Taxi	Bus	Motorcycle	Others
150–299 (3%)	—	—	27	13	60
300–599 (20%)	8	2	30	22	39
600–999 (28%)	3	2	27	19	18
1000–1999 (29%)	48	4	11	18	18
2000–4999 (15%)	73	1	3	15	9
5000+ (5%)	92	—	—	—	8
Total	40	2	18	18	22

Source: Reference 9.
[a] Values in parentheses refer to percentage of sample population.

vehicles. In the middle-income group, only 5% owned a car and 6% owned a jeep. In the two lower-income groups, making up half the sampled households, there was no car ownership at all.

Car ownership also varies between large and smaller cities. In Manila, 11% of households owned a car. In provincial urban areas of the Philippines, the number was 3–4%. Higher incomes, a well-developed road system, and the greater distances to be traveled for daily activities probably account for the higher share in Manila.

Between countries, as within them, ownership of cars is higher where household incomes are higher. In Kuala Lumpur and Kajang, which are relatively affluent communities with a well-developed system of roads, over 40% of the households owned cars. In less affluent Manila, the share was only 11%.

Once a car is owned, the household tends to use it for much of its transport needs. In Nairobi, the share of total travel mileage accounted for by automobile travel increased from 3% in the lowest income group to 90% in the highest group (17). Automobile ownership also brings new demand for travel by facilitating trips that could not easily be made without a car.

Other Factors Affecting Transportation Energy Use

Growing use of cars has an important effect on energy use because cars tend to be an inefficient mode of transport relative to others available in the urban setting. In three Latin American countries, for example, automobiles are responsible for 65–75% of energy use in passenger transportation, but carry only 25–35% of the total volume of traffic (in terms of passenger-miles) (22). Energy efficiencies calculated for Hong Kong show the automobile being some 10–15 times more energy-intensive than buses (Table 9). For

Table 9   Energy efficiency in urban passenger transport

Mode	Hong Kong	Colombia	Brazil	Argentina
	(MJ per passenger-km)			
Automobile	1.5–3.9	3.17	3.24	3.99
Commuter rail	—	0.36	0.98	
Bus	0.18–0.24	0.59	0.33	0.59
Light bus	1.04	—	—	—
Taxi	3.34	—	—	5.69
Motorcycle	0.80–1.45	—	—	—

Sources: References 4, 22.

several Latin American countries, buses are estimated to be 5–10 times more energy-efficient than the automobile.

Energy efficiency depends in part on the load factor of the vehicle, i.e., how well its capacity is utilized. Traffic conditions also affect energy efficiency; congestion, which has become severe in many Third World cities, leads to less efficient operation of motorized vehicles.

The amount of energy used in a particular mode depends largely on the number of trips taken and their lengths. Studies of energy use in Nairobi and Mexico City show that the number of trips per day increases rapidly with rising incomes (17, 23). Average trip length is affected by particular features of the urban landscape, such as the degree of urban sprawl. Average trip lengths in Nairobi were reported to be 1.5–2.8 miles. Comparable figures for Mexico City, which is larger than Nairobi, were 3.5–6.0 miles. Automobile ownership encourages the undertaking of longer trips; the average car trip in Nairobi was almost twice as long as the average foot or bicycle trip.

Transportation Energy Use and Income

Because of the differences in modal choice, trip length, and number of trips, energy use in transportation varies enormously among income groups. Newcombe's 1974 survey of household energy use in Hong Kong found a 40-fold increase in per capita energy use for transportation (public and private) between the lowest- and highest-income groups (Table 10). Most of this increase was accounted for by purchases of gasoline and diesel fuel for

Table 10   Transport energy use and income in Hong Kong (1974)

Monthly income[a] (HK$/capita-month)	Public	Private	Total
		(MJ/capita-day)	
< 100 (3%)	0.13	0.13	0.26
101–150 (11%)	0.07	0.25	0.32
151–200 (18%)	0.21	0.05	0.26
201–250 (16%)	0.34	0.60	0.94
251–300 (10%)	0.49	1.38	1.87
301–350 (7%)	0.67	1.11	1.78
351–400 (4%)	0.49	1.58	2.07
401–550 (9%)	0.92	3.52	4.44
550+ (18%)	1.78	9.26	11.04

Source: Reference 5.
[a] Values in parentheses references to percentage of sample population.

private use. Consumption in this category showed a big jump from the second-highest to the highest-income group, reflecting frequent car use by the wealthiest families.

Estimates of transport energy use by households in Nairobi and Mexico City show a similar trend. In Nairobi, research indicated a 60-fold increase in energy use per household from the lowest to highest of the five income groups. As in Hong Kong, most of the increase occurred in energy used for car travel, which accounts for 30% of the energy used in the lowest-income group, compared to 95% in the highest-income group. For Mexico City, there was a 40-fold increase in the energy use per household. The automobile, again, is the main contributor to this increase.

Monthly expenditures on gasoline and diesel fuel reported by Kuala Lumpur and Kajang households also increased dramatically with income. Out of six income groups, the top group had an average expenditure of M$ 210, far above the average of only M$ 2 for the lowest income group. About 40% of the respondents in Kuala Lumpur reported having no expenditure for gasoline and diesel fuel.

Growth in Transport Energy Demand

There is a strong correlation between per capita income and the number of automobiles per capita (21a). The acquisition and use of private automobiles brings with it in turn a very large increase in per capita energy use for transport. If households are at a point where automobile ownership is a possibility, moderate increases in household income may lead to very substantial growth in the consumption of transport fuels. The price and availability of automobiles, the ability of city dwellers to purchase them, and the existence and adequacy of alternatives to private automobile use will be major determining factors of energy use in urban passenger transportation.

ENERGY USE IN THE COMMERCIAL/ INSTITUTIONAL SECTOR

The general category of commercial and institutional facilities includes a variety of energy users ranging from high-rise office buildings to neighborhood restaurants. There has been little study of energy use in this sector, in part perhaps because it is a relatively small user of petroleum fuels. In the more modern cities, however, growth in the sizes and numbers of buildings in this sector have become a major force pushing electricity demand upward.

The share of the sector's energy use in total urban energy use varies greatly among cities. In Hong Kong, with its many high-rise office buildings, hotels, and restaurants, it accounted for an estimated 22% of total primary energy use in 1971 (4). In less modern Lae in Papua New Guinea, the commercial/government sector's share of secondary energy use was only 5.4% in 1977 (6).

The importance of different activities within the commercial sector also varies among cities. In Hong Kong, with a population that traditionally eats out often, food and tourist-related services accounted for two-thirds of total sector energy use. In Lae, these activities accounted for a much smaller share.

The basic services needed for the operation of commercial/institutional establishments are lighting, ventilation and cooling, and in some cases, cooking and refrigeration. A survey of commercial establishments in Manila (24) showed that the most common applications of energy were for lighting (100% of respondents), ventilation using electric fans (71%), air-conditioning (68%), cooking (57%), and refrigeration (56%).

Energy costs can be a significant part of the total cost of operation. A survey of commercial establishments in Manila and other cities in the Philippines showed that, on the average, energy accounted for 9% of the total cost of operation. Restaurants, lodging places, and recreational facilities had higher (14–15%) energy costs than other establishments.

The Philippines survey also showed a keen awareness of the costs of energy. An average of 90% of the establishments surveyed claimed to practice energy conservation. Sixty percent of the respondents had turned off some lights, 31% had reduced their air-conditioning use, 24% had reduced their use of appliances, and 10% were driving less. Among restaurants and lodging places, 91% of the respondents practiced conservation and 87% had reduced their energy use. In the recreational facilities category, only 72% of respondents, the lowest percentage among all categories, practiced energy conservation. The 28% of respondents who did not practice conservation were aware that energy use was not minimal, but did not implement conservation measures because, in many cases, they deemed these "impractical or not applicable to their business." About 48% of the respondents were willing to switch fuels. Among these, 35% were willing to switch to wood and charcoal for cooking, and 22% were willing to switch to kerosene for lighting.

Growth in Energy Demand in the Commercial/ Institutional Sector

Because of the demand for lighting, ventilation and cooling, and refrigeration, a high share of the energy used in the commercial sector is in the form of electricity. (The share was 38% in Hong Kong, 82% in Lae.) Since most of the cities in the developing countries are located in the tropics, there is considerable potential in particular for increasing use of air-conditioning in new and old buildings. The extent to which building design makes use of natural cooling techniques will have an important effect on urban electricity demand.

URBAN ENERGY USE AND GOVERNMENT POLICY

There are many ways in which government actions affect energy use in the cities of the developing countries. These include policies that affect the prices of fuels, that restrict the use of certain fuels, that promote the use of new fuels, and that encourage the use of devices with greater energy efficiency. In this section, we do not aim to review the wide range of developments in this area, but merely give a flavor of the kinds of activities that are being undertaken.

Energy Use in Transportation

Transportation, so heavily dependent on oil, has been the main object of government attention. Government policies can affect choice of transport modes, the energy efficiencies of the various modes, and the overall demand for transportation. Policies concerning the price of transport fuels influence all of the above. Governments have not hesitated in passing along increases in the price of oil to consumers of gasoline, since it is regarded as a luxury fuel. Gasoline also usually carries high taxes. In general, the price of diesel fuel has not been increased as much as that of gasoline. This has prompted switching to diesel vehicles, which has resulted in more emissions to the already heavy pollution in urban areas.

Import tariffs on automobiles are one way the choice of mode in passenger transport is influenced. High tariffs in many countries discourage car purchase.

Another way is through subsidization of mass transit. Governments in many countries are making serious efforts to improve mass transportation by introducing newer, bigger, more efficient diesel buses. Actions designed to allow priority treatment of buses, to facilitate pedestrian and cyclist traffic, and to promote alternative public transportation modes also affect people's transportation choices. Such measures have the potential for improving urban traffic conditions as well as helping to conserve fuel.

Government policy can also influence the energy efficiency of particular transport modes. Policies that motivate auto manufacturers and assemblers to import fuel-efficient models have been used in Colombia and elsewhere. Many governments have increased the price of large, fuel-inefficient cars by raising taxes or import tariffs on them. The Philippine government has gone so far as to ban their importation, manufacture, and assembly.

Governments have also introduced programs to bring about the use of fuels produced from indigenous resources to substitute for gasoline and diesel. Use of alcohol fuels, to blend with gasoline or to be used alone has occurred on a large scale in Brazil, which has a unique combination of land for expanding sugar cane plantations, an integrated automobile industry, marketing outlets for gasohol, and a high dependence on imported petroleum. Other countries, such as the Philippines, have adopted programs to blend vegetable oils with diesel fuel. In another kind of fuel switching, electric vehicles are being considered, in countries with potential for hydro or geothermal development, as substitutes for gasoline and diesel-powered buses and trains. A study carried out for Bogota claims that the energy costs of electric trolley buses compare very favorably with available urban public transportation (24). Trolley buses have the additional advantage of being less polluting in the urban environment, as well as quieter.

The general demand for transportation is also subject to government influence. Measures such as weekend closings or shorter hours for gas stations have been used in many places. In the long run, of course, design of urban areas has a major impact on transportation needs.

Energy Use in the Home

Patterns of energy use in homes have been influenced by long-standing government subsidies of various fuels. Since the increase in oil prices, most governments have used some kind of exhortation to encourage energy conservation in the home, but welfare and political concerns have generally kept subsidies in effect.

Governments have also instituted programs to encourage, or at least maintain, the use of indigenous fuels by households. These include programs to encourage charcoal use, as in Papua New Guinea, and to allow wood use to continue, as with the tree plantations near Addis Ababa, Ethiopia.

One government policy that can affect energy use is tariffs on home appliances. With a large potential for increase in appliance ownership in many places, tariffs could be used to discourage purchase of particularly energy-intensive appliances or to encourage purchase of energy-efficient models.

Energy Use in the Commercial/Institutional Sector

Governments have paid relatively little attention to commercial and institutional energy use. In some places, government efforts have been directed at information dissemination, energy audits, and training courses. The Philippines has had an extensive program providing audits, free of charge, to commercial and industrial customers (25). Training courses for building supervisors or energy managers have also been effective in the

Philippines. Other more restrictive policies have included limiting lighting for advertisements to three hours in the evening (the Philippines) and restricting operation of elevators to above the fifth floor (South Korea).

In some places, governments are considering building energy standards that would require building owners to install and practice conservation measures. Officials in Singapore, for example, incorporated a set of building design energy conservation standards into the regulations for all new buildings in 1979. These were later extended to existing buildings. A recent analysis of possible modifications in Singapore's existing standards suggests that a 15–20% reduction in energy use is cost-effective and possible in the near term, and a reduction of up to 40% may be possible in the longer term (26). Other countries in the ASEAN region are also exploring energy-efficiency standards for reducing energy use in commercial buildings.

CONCLUSIONS AND AREAS FOR FURTHER RESEARCH

Energy use patterns in the cities of the developing countries have undergone considerable change during the past two decades. These changes are closely linked to growth in household income, which has allowed the acquisition of devices that use modern fuels. A major area of change has been in cooking, where use of traditional fuels like wood has given way to the use of kerosene, LPG, and electricity. The other key development in homes has been the growth in ownership of electric appliances. These include some devices, like refrigerators, water heaters, and air conditioners, that substantially increase household electricity demand.

A striking feature of energy use both in the home and in transportation is the large difference in the amount of energy used by households at different socio-economic levels. Major segments of the urban population in many cities own few appliances, and use most of their energy for cooking. For transport they either walk or use mass transit. In many cities, a small proportion of the households account for a large share of use of modern fuels, especially electricity. The extent to which lower-income households move into more electricity-intensive life-styles will depend very much upon both general income growth and income distribution.

Understanding the likely range of future energy requirements of the cities of the developing countries is an important task for urban planners. One of the main themes of this paper is that energy demand must be looked at in connection with the activities for which energy is used. Thus, information is needed on two fronts:

1. Changes in the number and kind of activities that involve energy use
2. The energy requirements associated with those activities.

One way of exploring these areas is to investigate the changes that occur with respect to energy-using activities as the income of urban residents grows. This approach was used by McGranahan and colleagues to study energy use in Mexico City and Nairobi. Their work, which involved estimating patterns of energy use by households in different income groups, provides a useful model for studies in other developing cities.

Research could be fruitfully directed toward key factors in each type of energy end-use. In the home, the main issue is the income levels at which changes in cooking fuels and appliance acquisitions occur. In transportation, we need to know the income level at which certain vehicles—especially the car—are purchased, and the ways in which the car is used. The factors influencing the choice of personal transportation or mass transit also bear investigation.

Along with better understanding of the acquisition of energy-using things, more information is needed on the real-life energy demands of particular household activities. For example, what does the use of air-conditioning or refrigeration mean in terms of energy requirements, and how does this vary with household income? How much energy does a typical middle-income family use in its car?

Cross-country research is useful in that it may reveal patterns that are similar in different places. This could allow the use of "rules of thumb" in cities where certain areas have not been studied. Combined with local information, such rules of thumb would allow a better assessment of future urban energy needs.

ACKNOWLEDGMENTS

This work was supported in part by the US Department of Energy, Office of Policy, Planning, and Analysis, under contract No. DE-AC03-76SF00098, and by the International Development Research Centre, Ottawa.

Literature Cited

1. World Bank. 1984. *World Development Report.* Washington, DC: World Bank
2. Meier, R., Berman, S., Campbell, T., Fitzgerald, C. 1981. *The Urban Ecosystem and Resource-Conserving Urbanism in the Third World Cities.* Rep. LBL-12604. Berkeley, Calif: Lawrence Berkeley Natl. Lab.
3. LBL International Studies Group. 1984. *Summary of Results and Progress Report for the Study of Energy Use in Developing Countries.* Berkeley, Calif: Lawrence Berkeley Natl. Lab.
4. Newcombe, K. 1975. Energy use in Hong Kong: Part II. Sector end-use analysis. *Urban Ecol.* 1:285–309
5. Newcombe, K. 1979. Energy use in Hong Kong: Part IV. Socioeconomic distribution, patterns of personal energy use, and the energy slave syndrome. *Urban Ecol.* 4:179–205
6. Newcombe, K. 1980. *Energy for Development: the energy policy papers of the Lae Project.* Canberra: Australian National Univ.
6a. Jankowski, J. 1981. Industrial Energy

Demand and Conservation in Developing Countries. Unpublished discussion paper. Washington, DC: Resources for the Future

7. Gupta, C. L., Rao, K., Vasudevaraju, V. 1980. Domestic Energy Consumption in India (Pondicherry Region). *Energy* 5: 1213–22

8. Zhu, H., Brambley, M., Morgan, R. 1983. Household Energy Consumption in the People's Republic of China. *Energy* 8: 763–74

9. Socio-Economic Research Unit. 1981. *The Use of Energy and Attitudes towards Energy Conservation among Urban Households.* Kuala Lumpur: Prime Minister's Dept.

10. Philippines Ministry of Energy. 1982. *Energy Sectoral Survey Series: 1979 Urban Household Energy Demand.* Manila: Ministry of Energy

11. Alam, M., Dunkerley, J., Gopi, K., Ramsay, W. 1983. *Fuelwood Survey of Hyderabad.* Washington, DC: Resources for the Future

12. Fernandez, J. C. 1980. *Household Energy Use in Non-OPEC Developing Countries.* Santa Monica, Calif: Rand Corp.

12a. Desai, A. V. 1981. Interfuel Substitution in the Indian Economy. Unpublished discussion paper. Washington, DC: Resources for the Future

13. Beijer Institute. 1984. *Energy and Development in Kenya: Opportunities and Constraints.* Stockholm: Beijer Inst.

14. Chauvin, H. 1981. When an African city runs out of fuel. *Unasylva* 33: 133

15. World Bank Urban Projects Dept. 1981. *Liberia: Monrovia Water, Power and Urban Projects.* Washington, DC: World Bank

16. Strout, A. 1978. *The Demand for Kerosene in Indonesia.* Cambridge, Mass: MIT

17. McGranahan, G., Chubb, S., Nathanns,

R., Mbeche, O. 1979. *Patterns of Urban Household Energy Use in Developing Countries: The Case of Nairobi.* Stony Brook, NY: Inst. Energy Res.

18. Ang, B. 1983. *Taiwan: Energy Outlook.* Rep. EDP 26. Cambridge: Energy Res. Group

19. Dutt, G. S. 1978. *Energy Use in Rural and Urban India.* Princeton, NJ: Cent. Environ. Stud.

20. Chang, F. 1983. *The Roles of Electric Power and Advanced Electrification in Taiwan.* Taipei: Taiwan Power Co.

21. Tzeng, G. H. 1984. *Characteristics and Demand Model for Household Energy Consumption in Taiwan.* Taiwan: National Chiao Tung Univ.

21a. World Bank. 1975. *Urban Transport.* Washington, DC

22. Inter-American Development Bank. 1982. *The Impact of Energy Costs on the Transport Sector in Latin America.* Washington, DC: Inter-American Development Bank

23. McGranahan, G., Taylor, M. 1977. *Urban Energy Use Patterns in Developing Countries: A Preliminary Study of Mexico City.* Stony Brook, NY: Inst. Energy Res.

24. Philippines Ministry of Energy. 1982. *Energy Sectoral Survey Series: Energy Consumption and Conservation Practices of Commercial Establishments in 1979.* Manila: Ministry of Energy

25. Ocampo, M. T. 1984. *Regulations in Energy Conservation and Energy Conservation in the Philippines.* Presented at ASEAN Conf. Energy Conservation in Buildings, May, Singapore

26. Levine, M., Curtis, R., Turiel, I. 1984. *Policies to Achieve Cost-Effective Energy Reductions in ASEAN Buildings.* Presented at ASEAN Conf. Energy Conservation in Buildings, May, Singapore

Ann. Rev. Energy. 1985. 10:135–64
Copyright © 1985 by Annual Reviews Inc. All rights reserved

ETHANOL FUEL FROM SUGAR CANE IN BRAZIL

Howard S. Geller[1]

American Council for an Energy-Efficient Economy, Washington, DC, 20036

INTRODUCTION

Production of sugar cane and ethanol are by no means new activities in Brazil. The nation was colonized to a large extent to produce sugar, and it was the world's leading sugar producer in the late 16th and 17th centuries. Legislation enacted in 1931 required the addition of ethanol to gasoline, and gasoline-ethanol blends of up to 40% ethanol have been used as an automotive fuel since then (1). From 1960–1975, ethanol production generally remained in the range of 400–700 million liters per year.

In November, 1975, Brazil initiated a program to raise the production of ethanol from sugar cane and increase the use of ethanol as a substitute for gasoline. The program was started partly because of the quadrupling of world oil prices in 1973, and also as a means of assisting the sugar industry in times of low international sugar prices (1). World sugar prices fluctuate considerably, and they were dropping rapidly in 1975. The practice of increasing ethanol production when low sugar prices prevail was already well established in Brazil.

The first phase of the alcohol program (1975–1979) involved mainly the expansion of distillery capacity at existing sugar mills. The blending of anhydrous ethanol with gasoline was continued, with the goal of reaching 20% ethanol in the blend by 1980 (2).

After the oil price shock in 1979, Brazil found itself spending nearly 50% of its export earnings on imported petroleum. At this time, a more ambitious second phase of the alcohol program was initiated, relying

---

[1] This study was completed while the author was working for the Companhia Energetica de Sao Paulo (CESP), Sao Paulo, Brazil.

135

0362–1626/85/1022–0135$02.00

largely on new, independent (also known as autonomous) distilleries. By 1981, independent distilleries accounted for over 70% of the new distilleries approved by the government, up from less than 25% during the first phase of the program.

The additional fuel was to be used in vehicles that ran on pure hydrated ethanol. Research and development at a government-supported laboratory proved that alcohol-fueled automobiles were feasible, and the technology was adopted by the Brazilian automobile industry.

National ethanol production, at 2.4 billion liters in 1978, was targeted at 10.7 billion liters by 1985 and 14.0 billion liters in 1987. Production of 14 billion liters would provide enough fuel alcohol to replace about 75% of the gasoline consumed as of 1978–1979. The second phase of the national alcohol program required concerted efforts by the petroleum industry, the automobile industry, sugar cane and alcohol producers, financing institutions, and policy-makers.

In terms of its production objectives, Brazil's ethanol program is highly successful. Table 1 shows the trends in alcohol production and use for 1976–1983. The production goal set for the first phase of the program was met. Production increased rapidly during the second phase of the program, with growth rates of 33% in 1982 and 41% in 1983. In early 1984, there were 304 distilleries in operation, with an additional 195 projects approved by the government. Based on the trends and current initiatives, it is likely that the 1985 production goal will be met. The fuel alcohol available at this production level is sufficient to replace about 60% of the gasoline that would be required in the absence of the alcohol program.

On the demand side, Table 2 shows the recent trends in the production of

Table 1 Trends in alcohol production and use[a] (billion liters)

Year	Total production	Fuel use	Nonfuel use	Exports	Additions to stock[b]
1976	0.64	0.17	0.32	0.03	0.12
1977	1.39	0.64	0.41	0.00	0.33
1978	2.36	1.51	0.37	0.02	0.46
1979	3.45	2.24	0.34	0.12	0.75
1980	3.68	2.68	0.49	0.38	0.12
1981	4.21	2.54	0.41	0.16	1.10
1982	5.62	3.70	0.39	0.34	1.20
1983	7.95	5.15	0.40	0.54	1.87

[a] Sources: (26, 28) and the Instituto do Acucar e Alcool, Sao Paulo.
[b] Includes losses and adjustments.

alcohol and gasohol vehicles. Sales of pure alcohol cars gained dramatically in 1983, representing approximately 75% of the light vehicles produced that year. In the domestic market, pure alcohol cars have been well accepted and accounted for nearly 90% of sales in 1983. There were approximately 1.5 million pure alcohol-fueled automobiles on the road in Brazil as of mid-1984.

Among efforts to develop and implement renewable energy sources worldwide, the Brazilian alcohol program stands out as one of the most significant accomplishments. However, important aspects of the ethanol program are controversial in both Brazil and other countries. Commentators in Brazil (3) vehemently argue about the "real cost" of ethanol production from sugar cane. Overseas, critics (4) attack the alcohol program for competing with food production, showing a negative energy balance, and being uneconomical.

The primary objective of this article is to analyze some of the controversial issues by reviewing up-to-date primary data and secondary studies available in Brazil. In particular, the issues of "food vs fuel," economics, employment, and energetics will be addressed. In addition, the article discusses some important trends, innovative developments, and policy issues that are occurring as part of the Brazilian experience with alcohol fuel.

FOOD VS FUEL PRODUCTION

National Perspective

The data on harvested crop area in Brazil (Table 3) shows that while there was a 12% increase in agricultural land area from 1976 to 1982, land devoted to basic food crops increased only 6.9%. This compares to a

Table 2   Trends in the sales of alcohol and gasohol vehicles[a]

Year	Alcohol vehicles (10^3)	Gasohol Vehicles (10^3)		
		Total	Domestic	Export
1980	241	775	—	—
1981	137	496	—	—
1982	234	452	352	100
1983[b]	583	212	92	120

[a] Sources: Comissao Executiva Nacional do Alcool, Brasilia and Associacao Nacional dos Fabricantes de Veiculos Automotores, Sao Paulo.
[b] Preliminary estimate.

Table 3 Trends in the land area used for crops[a]

Product	Area cultivated (10^6 hectares)			
	1976	1979	1982	1982/1976
Basic foods				
Corn	11.1	11.3	12.6	1.135
Rice	6.7	5.4	6.0	0.90
Beans	4.1	4.2	5.9	1.44
Wheat	3.5	3.8	2.8	0.80
Cassava	2.1	2.1	2.1	1.00
Subtotal	27.5	26.8	29.4	1.069
Export crops				
Soybeans	6.4	8.3	8.2	1.28
Cotton	3.4	3.6	3.6	1.06
Coffee	2.2	2.4	1.9[b]	0.86
Cocoa	0.4	0.4	0.5	1.25
Oranges	0.4	0.5	0.6	1.50
Castor oil	0.3	0.4	0.5	1.67
Tobacco	0.3	0.3	0.3	1.10
Subtotal	13.4	15.9	15.6	1.16
Sugar cane	2.6	3.1	3.9	1.5
Other	1.6	2.0	1.6	1.0
Total	45.1	47.8	50.5	1.12

[a] Source: (24).
[b] The land area devoted to coffee was abnormally low in 1982, most likely because of the low international coffee prices in 1981.

population increase of 16% during the same period. If the land area devoted to basic food crops kept up with population growth, then an additional 2.5 million hectares of food crops would have been needed in 1982.[2]

Table 3 shows that the cultivated area for sugar cane increased by about 1.3 million hectares from 1976 to 1982. Although this increase is sizable, it is much less than the increase in the land area used for the heavily exported commodity crops (2.2 million hectares). Indeed, the growth in the land area devoted to soybeans alone was about 50% greater than that for sugar cane. Soybeans have become the second largest food crop in Brazil (in terms

[2] Crop area is considered rather than crop production, because of the influence of climatic conditions on production and the difficulties of summing the production of different crop types.

of land area) in a period of about 10 years. Soy products, of which about 65% are exported, accounted for approximately 13% of Brazil's total export earnings as of 1981.

The growth in export commodity crops is not surprising, since the government has strongly encouraged and subsidized agricultural export commodities such as soy products. Government support included low-interest agricultural credit and other incentives similar to those provided for sugar cane and ethanol production in the early stages of the alcohol program.

Agricultural land allocation and crop production is also affected by trends in the prices farmers receive for different crops. In 1980–1982, there was a 38% decline in the real prices farmers received for domestic food crops on the average (5). Heavily exported crops (other than coffee) showed a 17% real price drop on the average during this period, while the price reduction for sugar cane was only 9%.

With these price trends, the devaluation of the cruzeiro in 1983 (which stimulates exports), and the domestic recession, the prospects for expanding per capita food production in Brazil are not encouraging. Indeed, the preliminary estimates of cultivated areas for 1983–1984 show a continuation of the trends described above. Partly because of drought in the Northeast and floods in the South, the land area planted with basic food crops is expected to decline from the 1982 level (6).

Contrary to the situation of basic food crops, the official forecast shows that the land area used for export crops and sugar cane will increase in 1983–1984. The land area devoted to soybeans, after remaining relatively constant for a number of years, is expected to grow by about 0.8 million hectares. Furthermore, the amount of new land devoted to soybeans is expected to be about twice as great as that devoted to sugar cane, once again demonstrating the importance of soybeans and other export crops.

The impact of the alcohol program on land utilization is by no means trivial. To meet the 1985 ethanol production goal of 10.7 billion liters, Brazil will have to plant approximately 1.6 million hectares of sugar cane in addition to the land area already used in 1982. Analysts (7) point out that the long-term objectives of the government to increase the production of basic food items, export crops, and ethanol would require unprecedented growth rates in the overall cultivated land area. However, the role of ethanol must be viewed in relation to export crops such as soybeans in what should be dubbed the "food vs fuel vs exports" issue.

Sao Paulo State

Sao Paulo is by far the leading alcohol producing state, accounting for approximately two-thirds of national production in recent years. On the local level, some substitution of sugar cane for other crops is certainly occurring in Sao Paulo. Data from the Institute of Agricultural Economics (8) pertaining to four major agricultural districts showed that as of 1981, about 36% of the new land area devoted to sugar cane since 1975 displaced other crops.

Aggregate data, however, presents a somewhat different picture. Table 4 shows the recent trends in agricultural land use in Sao Paulo state. It should be noted that in spite of the historical importance of coffee in Sao Paulo, the state is not currently among Brazil's major producers of either basic food crops or export commodity crops. Sao Paulo contained 8% of the land area devoted to basic food crops and about 13% of the land used for export commodity crops in 1982.

The state data indicates that in 1976 the total land area devoted to sugar cane was about 17.5% of the total crop area and 6% of crop and pasture area. Therefore, while the land area devoted to sugar cane increased 78% in 1976–1982, it started from a relatively low base.

Table 4 also shows that in 1976–1980, the total land area used for crops and pastures remained relatively constant in Sao Paulo. At the aggregate level, the increases in land area devoted to sugar cane and export crops came largely at the expense of plentiful pasture land. On the other hand, in 1980–1982, there was a sizable increase in the overall land area used for crops and pastures. This was more than enough to provide for continued growth in sugar cane production as a result of the alcohol program. Competition with food production declined as the program shifted its emphasis from distilleries attached to existing sugar mills to independent distilleries.

Table 4   Trends in the land area used for crops and pastures in Sao Paulo state[a]

	Land area utilized (10^6 hectares)			
Land type	1976	1978	1980	1982
Basic foods[b]	2.13	1.84	1.80	2.25
Export crops	1.84	2.21	2.29	2.04
Sugar cane	0.90	1.09	1.32	1.60
Other crops	0.22	0.27	0.29	0.30
Pasture	10.24	10.09	9.55	10.25
Total	15.34	15.50	15.25	16.44

[a] Sources: (7) and the Instituto de Economia Agricola, Sao Paulo.
[b] Basic food crops in this breakdown include corn, rice, beans, and cassava (wheat is included with "other crops").

THE COST OF ETHANOL PRODUCTION

Estimates (3) of the real economic cost of ethanol production vary: $0.18–0.48/liter or $35–90/bbl of gasoline replaced.[3] The large variation is due to conflicting primary data and differences in a number of major assumptions that go into any cost accounting. Of critical importance is the assumed exchange rate and the treatment of government subsidies.

A number of studies, as well as actual government payments during 1983, point to a real cost in the range of $45–60/bbl of gasoline replaced when the subsidies are taken into account and a parity exchange rate is used (a rate based on the parity value for exported goods). This cost range applies to Sao Paulo state and surrounding areas where alcohol production is both most intensive and most efficient.

Producer Cost Data

The major ethanol-producing association Copersucar released production cost data from the 1982–1983 harvest (9). The costs are based on a large number of production units located primarily in Sao Paulo state. However, Copersucar admits that the average production cost it presents does not reflect the overall mean, but rather is a "superior average" based on production units with relatively high costs. Copersucar takes this approach because it is arguing for higher prices from the government to ensure economic viability for a large majority of production units and to maximize its profits.

Copersucar uses a 12% profit margin on capital investments in its production cost estimates. However, the financial subsidies provided by the government are not included. The nonrepresentative sample and the profit margin factor largely offset the failure to include subsidies. Thus, the Copersucar data can be viewed as an approximation of the true (unsubsidized) production cost without profit.

Table 5 shows the cost breakdown according to Copersucar. In converting to US dollars, the official exchange rate was increased by a factor of 1.25–1.5 to compensate for the overvaluation of the cruzeiro at the time of analysis (10). The overall production cost with these assumptions is $0.28–0.33/liter of anhydrous ethanol or $49–59/bbl of gasoline replaced when pure ethanol is used in automobiles. It is seen that agricultural operations (sugar cane production) account for nearly two thirds and industrial

---

[3] Because of the increase in automobile efficiency when operating on alcohol rather than gasoline, one bbl of gasoline is replaced by approximately 1.2 bbl of hydrated ethanol (equivalent to about 1.12 bbl of anhydrous ethanol) when full replacement occurs. The replacement ratio is about 1:1 when blending 20% anhydrous ethanol and 80% gasoline.

Table 5 Estimated alcohol production costs in Sao Paulo state[a]

Component	Production cost		
	Cr$/liter[b]	US$/liter[c]	Percent
Agricultural			
Operating	36.3	0.118–0.142	42.5
Capital	8.9	0.029–0.035	10.5
Land	9.7	0.032–0.038	11.4
Subtotal	55.0	0.179–0.215	64.4
Industrial			
Operating	18.8	0.061–0.073	22.0
Capital	11.6	0.038–0.046	13.6
Subtotal	30.4	0.099–0.119	35.6
Total	85.4	0.278–0.334	100.0

[a] Costs apply to 1982–1983 as estimated by Copersucar (9). These production costs are somewhat greater than the true average since they were developed for the purpose of obtaining a higher purchase price from the government.
[b] Cruzeiro costs apply to Sept., 1982.
[c] The official exchange rate in Sept., 1982, is increased by a factor of 1.25–1.5 to account for the overvaluation of the cruzeiro at the time (10).

operations (distillery activities) about one third of the overall production cost. Furthermore, operating expenses are more than 2.5 times greater than capital-related expenses.

Social Cost

A major study was conducted to determine the "social cost" of ethanol production in 1980–1981 (10). The principal assumptions in this analysis include interest rates on capital investments increased by factors of 1.67–2.33 to remove government subsidies (equivalent to real interest rates of 6–14%); exchange rates increased by factors of 1.25–1.5 to compensate for the overvaluation of the cruzeiro; and direct labor valued at 0.5–0.65 times its actual cost to allow for the social benefit of job creation.

Using these assumptions and primary data from other studies and organizations yields a social cost for ethanol production in Sao Paulo state of $36–47/bbl of gasoline replaced in December 1980. The cost would be $41–54/bbl of gasoline replaced in mid-1983 dollars using US inflation rates as the basis for monetary correction. Even if labor is valued at its actual price, the social production cost in Sao Paulo is still only \$39–49/bbl of gasoline replaced (original dollars).

It should be noted that the cost for ethanol production is significantly higher in northeastern Brazil than in and around Sao Paulo state. For example, it is estimated that the social cost of ethanol production in Alagoas state, the largest alcohol producing state in the Northeast, is approximately twice that in Sao Paulo (10).

Government Payments

With the major devaluation of the cruzeiro in February 1983, and the monetary policy that followed, the official exchange rate was brought close if not equal to the parity level. In the period March–December 1983, the government, which guarantees the purchase of all ethanol that it has authorized, was typically paying alcohol producers about \$0.25/liter of anhydrous ethanol in Sao Paulo and other Central-South states (Instituto do Acucar e Alcool, personal communication). This corresponds to \$44.5/bbl of gasoline replaced.

However, it must be remembered that alcohol producers are benefiting from the subsidized loans extended by the government in past years. Producers who received loans in the initial phase of the program (1975–1979) are still paying nominal interest rates of 13–17% while the inflation rate is in excess of 200%. Beginning in 1980, new loans were indexed to inflation, although at a fraction that still resulted in a large subsidy, due to the high rate of inflation.

For loans issued in the initial phase of the program, the government is receiving essentially none of its money back, because of the large negative interest rates in real terms. In this case, the government is providing approximately \$0.06/liter in subsidy to producers who received loans in the initial phase of the program. Even with the decrease in subsidies that went into effect in 1981, the government still ends up paying about two thirds of the capital costs for the portion that it finances (11).

Hence, the overall production cost based on all government payments in the "full subsidy" case is about \$0.31/liter of anhydrous ethanol. The overall payment is about \$0.29/liter for producers that received loans as of 1981. These costs are equivalent to \$52–55/bbl of gasoline replaced. The upper level in this range is approximately what producers asked the government to pay for ethanol in 1983, assuming all subsidies are removed (12).

Higher Cost Estimates

As mentioned previously, some analysts (13) have estimated that ethanol production costs are as high as \$90/bbl of gasoline replaced. In making these estimates, the overvaluation of the cruzeiro in the past is not taken into account. Second, sugar cane production costs considerably in excess of official prices are assumed. Finally, ethanol is not given credit for its full performance advantage when replacing gasoline. If these factors were corrected, the ethanol cost would be in line with those presented above.

Comparison with Gasoline Costs

As of 1983, imported petroleum accounted for about 50% of total oil consumption in Brazil, in spite of a major (and successful) effort to increase domestic oil production by a factor of two or more during the early 1980s. Therefore, the cost of ethanol fuel is compared to the cost of gasoline derived from imported oil in this review.

Gasoline derived from $29/bbl imported petroleum is estimated to cost about $41/bbl ex-refinery in Brazil. In making this estimate, a $2/bbl shipping cost and $10/bbl refining cost are assumed (14). Thus, from a strict economic perspective, ethanol produced from sugar cane does not appear to be competitive with gasoline derived from imported petroleum at present.

However, the direct cost comparison neglects a number of important factors. Throughout the middle and late 1970s, Brazil had a net trade deficit. Marginal oil imports were debt-financed, and the interest charges on this debt were then capitalized (i.e. refinanced through new loans). Even with the large trade surplus that occurred in 1984, Brazil still had a negative balance of payments, due to the massive debt service requirement. Thus, marginal oil imports still contributed to the foreign debt. Alcohol production, on the other hand, is carried out primarily using local currency (for both capital and operating expenditures).

Marginal oil (and other) imports place a number of stresses on the Brazilian economy. First, because imports at the margin contribute to the foreign debt, a financing cost is involved. Second, because of the balance of payments problem, economic policy as of 1984 was strongly oriented toward encouraging exports rather than providing for domestic needs. The Brazilian government recognizes these problems and has implemented tight import restrictions. Also, the government applied a 20% surcharge to companies requiring hard currency to purchase foreign goods as of 1984.

Applying this 20% surcharge to the world petroleum price as of mid-1984 brings the total cost for gasoline derived from imported oil to $47/bbl. The overall cost comparison presented in Table 6 shows that ethanol, when blended into gasohol, is competitive with $47/bbl gasoline. However, for full gasoline replacement, ethanol still appears to be 6–19% more costly than gasoline. If world oil prices had stabilized at the 1980–1981 price (or if

Table 6 Overall cost comparison between ethanol and gasoline[a]

Parameter	Value
Ethanol	
Sugar cane cost	$10–12/ton
Ethanol yield	65 liters/ton
Distillation cost	$0.09–0.11/liter
Ethanol production cost	$0.264–0.295/liter
Replacement ratio for full gasoline replacement	1.2 liters of ethanol per liter of gasoline
Replacement ratio for 20% ethanol, 80% gasoline blend	1.0 liters of ethanol per liter of gasoline
Ethanol cost, full gasoline replacement	$50–56/bbl gasoline replaced
Ethanol cost, 20% ethanol, 80% gasoline blend	$42–47/bbl gasoline replaced
Gasoline	
Imported petroleum, including transport	$31/bbl
Import surcharge (20%)	$6/bbl
Refining gasoline	$10/bbl
Total gasoline cost	$47/bbl

[a] Costs in 1983 US$ assuming gasoline is derived from imported petroleum.

they rise to this level again), ethanol would be cost competitive for full gasoline replacement.

To justify full gasoline replacement at the mid-1984 world oil price, policy-makers can look to other benefits of ethanol production such as employment creation, rural development, increased self-reliance, and reduced vulnerability to crises in the world oil market.

Investment Costs for the Ethanol Program

Various sources (9, 15) show that the overall capital investment associated with a standard-size 120,000 liters per day ethanol facility was approximately $14 million at the official exchange rate in 1981–1982. This investment cost includes agricultural equipment and the distillery but excludes land. Correcting the exchange rate for the overvalued cruzeiro leads to an adjusted investment cost of $9.3–11.2 million. The industrial (distillery) phase of ethanol production typically represents about 60–70% of the capital investment per unit of production. If land is purchased for the provision of the sugar cane, the investment cost increases as much as 50–60% in the Central-South region and about 20–25% in the Northeast (16).

Based on these unit investment costs, meeting the 1985 alcohol program goal of 10.7 billion liters will require an overall investment of $4.7–5.7

billion excluding land (at a corrected exchange rate assuming production of 0.6 billion liters per year prior to the start of the program). This is equivalent to an investment cost of $0.47–0.57 per annual liter of capacity. Other studies confirm a capital requirement of this magnitude (1). If all land needed for sugar cane production is purchased, then the capital investment required by 1985 is in the range of $6.9–8.3 billion, or $0.69–0.83 per annual liter.

How significant is this investment? As of 1982, petroleum exploration and production in Brazil had a cumulative investment cost of $0.029/liter added to reserves (17). However, in 1982 alone, the capital investment in domestic petroleum exploration and production was about $2.8 billion, which added to oil reserves at a cost of $0.052/liter (in corrected dollars). Assuming an average production period of 15 years, the investment requirement for oil production capacity is $0.43 and $0.78 per annual liter in the cumulative and marginal cases, respectively.

In addition to the capital requirements for oil exploration and extraction, the investment cost for refining must be taken into account. Based on the most recent refinery constructed in Brazil, refining capacity costs $0.07 per annual liter. (O. Ivo, Petrobras, personal communication.) Thus, overall investment costs for domestic oil production are $0.50 and $0.85 per annual liter in the two cases.

To compare the investment requirements for petroleum and ethanol, it must be remembered that 1.2 liters of hydrated ethanol replace 1 liter of gasoline in pure alcohol vehicles. Thus, on a gasoline replacement basis, the investment requirement for alcohol production becomes $0.56–0.68 per annual liter excluding the purchase of land, and $0.83–1.00 per annual liter including land. Since some of the sugar cane used for alcohol production is bought or grown on rented land, ethanol production overall appears to be about as capital intensive as marginal petroleum production on a replacement basis.

EMPLOYMENT CONSIDERATIONS

Overall Job Creation

Although there are some discrepancies among studies of the number of jobs created per unit of alcohol production from sugar cane, it is clear that alcohol production in Brazil is highly labor intensive. Determining the number of jobs created is complicated by differences in labor productivity between and within regions, the high degree of temporary and in some cases unsalaried labor, and the dynamic nature of employment for alcohol production.

Table 7 presents estimates of the direct employment created by a typical

120,000 liters per day distillery in the Central-South and Northeast regions, based on a major study of this topic (16). It is seen that alcohol production is more than three times as labor intensive in the Northeast as in the Central-South region. This is a result of lower agricultural and industrial yields and less mechanization in the former area. The higher labor requirement in the Northeast is a major factor contributing to the greater production cost there.

It is seen from Table 7 that 75–90% of the total labor required for alcohol production is in the agricultural area. Agricultural labor is used mainly for the planting and harvesting of sugar cane.

In the standard method currently used for sugar cane production, nearly three times as much labor is needed in the six month harvesting season as in the six month land preparation and planting season (11). Consequently, there are about twice as many temporary jobs as permanent jobs in sugar cane production. Likewise, industrial labor utilization is highly seasonal, since distilleries typically operate 150–180 days per year during the harvesting season.

Based on the values shown in Table 7, it is possible to estimate the total

Table 7 Employment associated with a standard-size ethanol distillery[a]

	Total jobs		Total labor (p-yrs)	
	Region[b]		Region[b]	
Type	C-S	NE	C-S	NE
Agricultural				
Permanent	230	890	230	890
Temporary	450	1770	225	885
Subtotal	680	2660	455	1775
Industrial				
Permanent	85	85	85	85
Temporary	125	125	65	65
Subtotal	215	215	150	150
Total	895	2875	605	1925

[a] Based on a 120,000 liters per day distillery producing approximately 20 million liters of ethanol per year. Source: (16).
[b] C-S: Central-South region; NE: Northeast region.

number of jobs that will be directly created when the 1985 production goal is reached. For the incremental 10 billion liters that will be produced as a result of the alcohol program, about 620,000 workers will be engaged in the equivalent of about 420,000 full-time jobs.[4] Even though only about 20% of the alcohol will be produced in the Northeast, about 44% of the labor will be employed there, because of the large difference in labor intensity between regions.

To put these numbers in perspective, there were about 13 million persons in the agricultural work force in Brazil as of 1980 (18). However, these persons were not necessarily fully engaged. Furthermore, official statistics show that the size of the agricultural work force was virtually the same in 1980 as in 1970, even though there was a 50% increase in cultivated land area during this period.

The lack of growth in agricultural employment during the 1970s was due to increasing mechanization in general and to the low labor utilization in soybean production in particular. Data from three major agricultural districts in Sao Paulo state (16) show that soybean production requires only 25% as much labor per hectare cultivated as sugar cane, and 39–46% as much labor as corn. Sugar cane, on the other hand, is 49–84% more labor intensive than corn in these areas and is also more labor intensive than other major crops such as rice or beans.

Investment per Worker

Using the capital costs for ethanol production presented previously, the investment requirement per direct job can be determined. Using a parity exchange rate and including land costs as well as the investments in the distillery and farm equipment, the investment requirements are estimated at $23,000–28,000/person-year in the Central-South region and $6000–7000/person-year in the Northeast.

For comparison, government figures (19) show that direct job creation in the industrial sector requires an investment of $42,000/person-year on the average (1983 dollars). Especially capital-intensive activities, like the mineral products or paper and pulp industries, typically require about $70,000 per job position. Furthermore, it is reported (20) that the $5 billion oil refining-petrochemical complex at Camacari which began operating in 1978 required an investment of $200,0000 per worker. With an overall

---

[4] The estimate of total employment creation does not take into account any jobs lost due to substitution of sugar cane on existing crop and pasture land. This effect should be small since, as Table 3 shows, net food crop area has been maintained during the growth of the alcohol program. Furthermore, sugar cane cultivation is much more labor intensive than use of land as pasture (16).

investment cost nearly equal to that for the entire alcohol program up to 1985 (excluding land purchase), the Camacari complex produces only 5% as many jobs.

Social Factors Related to Employment

The high level of seasonal labor associated with ethanol production from sugar cane can result in low salaries, poor working and living conditions, and a lack of benefits for workers. Also, seasonality tends to promote undesirable practices, such as work-force migration and the use of labor contractors. This has adverse impacts on workers as well as communities near the production centers.

The severity of these problems varies substantially between distilleries and plantations. In some cases, workers are officially salaried and receive legally stipulated wages and benefits. A study of the social impacts of the alcohol program in different communities throughout Brazil (21) found that, as expected, labor conditions tend to be better in the more developed South and Southeast regions than in the Northeast and Central-West regions. In fact, sugar cane workers in one part of Sao Paulo state went on strike and won major concessions at the beginning of the 1984 harvest season.

Because of the problems associated with seasonal employment, a strategy has been developed for stabilizing the labor required for sugar cane production throughout the year (11). A relatively constant labor requirement can be achieved by extending the harvest period from six months to eight or nine months, which is made possible by planting cane varieties that mature at different times.

Extending the harvest also extends the working period for the distillery, thereby providing a 25% or greater increase in annual distillery output. This leads to more effective use of investment capital from the point of view of production as well as that of employment generation. In addition, it has been shown that ethanol producers would increase their profit margins by extending the harvest (11). A few producers in Sao Paulo are moving in this direction (W. Belik, Secretaria de Agricultura, Estado de Sao Paulo, personal communication).

ENERGY BALANCE

The energy balance for ethanol production (i.e. the amount of energy required to produce a unit relative to its energy content) appears to be quite favorable in Brazil. A comprehensive assessment (2) of direct and indirect energy consumption in the agricultural and industrial phases shows that

Table 8 Energy balance for the production of ethanol from sugar cane[a]

Area	Energy requirement	
	Mcal/ha/year	Kcal/liter
Agricultural		
Fuel	2240	620
Other	1560	430
Industrial[b]	—	470
Total	—	1520

[a] Based on an ethanol yield of 3600 liters per hectare cultivated. Energy requirements apply to the Southeast region. Source: (2).
[b] Energy inputs other than sugar cane.

external energy inputs are equivalent to only about 25% of the energy content of the alcohol produced. This result applies to the Southeast region and assumes a relatively high annual ethanol yield of 4700 liters per hectare. No credit is taken for bagasse residues not consumed in the distillery.

Using a more typical annual yield of 3600 liters per hectare cultivated, the energy balance shown in Table 8 is obtained. In this case, the external energy input of 1520 kcal/liter still represents only 30% of the energy content of the ethanol produced.[5] About 70% of the external energy inputs occur in the agricultural phase of production.

The energy balance in the Northeast region is expected to be even more favorable, since reduced mechanization and lower energy intensity should at least compensate for the lower alcohol yield. Ethanol production in Brazil from other feedstocks, such as cassava, sorghum, or corn, is estimated to have a somewhat poorer energy balance than the use of sugar cane (2). However, in all cases the energy content of the product is at least twice that of the inputs.

The energy balance for ethanol production from sugar cane or other agricultural feedstocks is much more favorable in Brazil than in the United States. This is due to the greater energy intensity of farming and the lower sugar cane yields in the United States. In fact, sugar cane production in the

[5] Anhydrous ethanol has a lower heating value, approximately 5100 kcal/liter.

United States is estimated to require more than five times as much energy per unit of output as is required in the Southeast region of Brazil (2). Consequently, ethanol production from sugar cane would have at best a slightly positive energy balance and possibly a negative net energy balance in the United States. Likewise, ethanol production from corn in the United States can show an unfavorable energy balance (22).

It is worth noting that the energy balance for the extraction and refining of gasoline in the United States is approximately the same as that for ethanol production in Brazil (23).

TECHNOLOGICAL TRENDS AND DEVELOPMENTS

A wide range of technological developments have accompanied the twelvefold increase in Brazilian ethanol production from 1976 to 1983. These developments are occurring in all phases of alcohol production and use—sugar cane cultivation, conversion to alcohol, and ethanol-fueled engines. Some of the more significant innovations and trends are described below.

Sugar Cane Production

Agricultural productivity data (24) shows that the average sugar cane yield in Brazil increased 31% in the decade beginning in 1974. The largest percentage increase occurred in the more advanced Southeast region. The improvement in sugar cane yield is attributed to increasing use of improved seed varieties, fertilizers, pesticides, and irrigation (25).

Further R&D on improved seed varieties and better management techniques are under way and continued increases in average cane yield are expected (26). Also, an effort is being made to increase the average sugar content of harvested cane by 20–30% (W. Belik, Secretaria de Agricultura, Estado de Sao Paulo, personal communication). Finally, new seed varieties and cultivation techniques are being developed for use with poorer quality soils.

Industrial Developments

Ethanol production from sugar cane involves three major stages: sucrose extraction, fermentation, and distillation. An ordinary distillery also contains boilers for steam production, and often steam turbines to generate motive power for internal consumption. Bagasse (the cane residue left after sugar has been extracted) is used as the boiler fuel.

Table 9 shows some of the technological options that are available and their characteristics. In particular, there are different extraction methods and steam production and utilization options. A conventional distillery is generally very inefficient in its use of bagasse and steam, employing relatively inefficient boilers, a single stage turbine, and low steam pressures. Thus, about 90% of the bagasse is consumed within the distillery. The remainder is discarded or returned to the soil.

There is growing interest in using bagasse and steam more efficiently within the distillery, to generate large amounts of excess bagasse. Excess bagasse can be sold as a by-product and ultimately used as a solid fuel. Another possibility would be to cogenerate electricity and steam efficiently (e.g. using a gas turbine) with any excess electricity supplied to the power grid.

Studies (27) have shown that simply using more efficient boilers and reducing steam losses in the plant can result in approximately 23% excess bagasse. Switching to diffusers for sucrose extraction requires less electricity and can yield about 38% excess bagasse. Diffusers, a recently commercialized extraction technique in Brazil, are beginning to be implemented in new distilleries.

Finally, increasing the steam pressure in the distillation phase along with other modifications, can lead to as much as 47% excess bagasse. It is estimated that this amount of excess bagasse can be produced in new

Table 9 Technologies available for use in ethanol distilleries[a]

Area	Characteristics[b]
Boilers	Low efficiency (eff = 0.65–0.75) (*)
	High efficiency (eff = 0.80–0.85)
Sucrose extraction	Mills (*)
	Diffusers
Turbines	Single stage (*)
	Multiple stages with backpressure or condensation
Steam pressure at boiler outlet	Low (22 bar) (*)
	Higher (43–61 bar)
Steam pressure in distillation phase	Low (2.5 bar) (*)
	Higher (6.5 bar)
Drivers of mills, choppers, etc.	Steam turbines (*)
	Electric motors

[a] Source: (27).
[b] A (*) indicates the most commonly used process in conventional distilleries.

distilleries at the attractive cost of about $0.34/GJ, assuming normal financing for the extra capital investment.[6] A number of other energy-conserving fermentation and distillation processes are under development in various research centers throughout the world (1).

Generating and using the maximum amount of excess bagasse would significantly raise the overall efficiency in the distillery, since only about 35% of the energy content of sugar cane is converted to ethanol (28). Equipment is now available in Brazil for drying, briquetting, or pelletting bagasse to facilitate its distribution and use.

Microdistilleries

For a typical 120,000 liter per day distillery, about 5500 hectares of sugar cane are required in Sao Paulo state. Due to this large sugar cane requirement and the relatively modern methods used for sugar cane production, the alcohol program is leading to a concentration in land holdings. Data from Sao Paulo (16) shows that more than 85% of the sugar cane is produced on plantations larger than 100 hectares, whereas this fraction is much lower for other crops (except soybeans). The formation of large sugar cane plantations has negative social impacts due to the loss of some small farms and the concentration of land holdings (29).

Microdistilleries producing less than 5000 liters of ethanol per day have been developed in Brazil. Since they require up to about 200 hectares of sugar cane, the large-scale deployment of microdistilleries could help reverse the trend toward increasing land concentration in conjunction with the alcohol program. Microdistilleries can have a number of other advantages as well.

Technically, microdistilleries are much simpler than conventional distilleries. The extraction and distillation phases in a microdistillery can be as efficient as in a normal distillery, although the fermentation effciency may be somewhat lower. Overall, microdistillery yields are already in the range of 55–60 liters per ton of cane, and yields as high as the 65 liters per ton typically obtained in regular-size distilleries in Sao Paulo should be possible with newer technologies and good management (30).

Regarding economics and labor, one study (30) concludes that a microdistillery can have about one third the capital cost and five times as many workers per unit of alcohol output as a regular-size distillery. The same study shows that microdistilleries can be competitive in terms of

[6] Reducing thermal losses, using diffusers for sucrose extraction, and shifting to elevated steam and distillation pressures are estimated to increase the capital cost for a normal-sized distillery by 7%, or $530,000 (27). Assuming a 15% annual capital recovery factor and increased operation and maintenance costs equal to 2% of the increased capital cost, approximately 35,000 tons per year of excess bagasse would be produced at a cost of $2.60/ton or $0.34/GJ.

production cost as long as the annual output is at least 400,000 liters, yields are in excess of 52 liters per ton, three shifts are used, and real interest rates are under 14%.

At the end of 1983, it appeared that there were about 7–10 micro-distilleries in Sao Paulo and nearby states (Instituto do Acucar e do Alcool, personal communication). Greater implementation had not yet occurred because the technical developments were recent, and because the reliability, performance, and ethanol quality from these units were somewhat uncertain (B. Johnson, University of Sao Paulo, personal communication). In particular, the combination of low capital cost and high efficiency has not been widely demonstrated.

Alcohol Use in Otto Cycle Engines

Ethanol (and methanol) have a number of advantages over gasoline in terms of engine and vehicle efficiency (1, 2). First, compression ratios as high as 12:1 can be used, compared to 6–8:1 for gasoline-fueled engines in Brazil. Second, the heat of vaporization of alcohol per unit of fuel energy is 6–8 times that of gasoline. Evaporation of alcohol can be carried out using the heat in the exhaust gases, through adiabatic evaporation within the engine, or via other methods. Third, it is possible to use leaner air-fuel mixtures with alcohol fuels. Finally, greater power is obtained per unit of engine volume with alcohol fuels; consequently, engine size and weight can be reduced.

The first generation of ethanol-fueled automobiles in Brazil (circa 1980) did not take full advantage of these opportunities. Brazilian automobile manufacturers, however, have modified and improved their alcohol engines and vehicles. New alcohol-fueled automobiles are now nearly 25% more energy efficient than their gasohol-fueled counterparts.[7]

Among the efforts to upgrade alcohol vehicles, Volkswagen of Brazil (32) has developed a new 1.6 liter engine which includes an increased compression ratio and a 12% lower specific fuel consumption compared to the engine it replaces. Ford of Brazil (33) also has developed alcohol engines with lower fuel consumption. In addition, Brazilian automobile manufacturers are continuing to improve the transmission ratios, air-fuel mixtures, and ignition timing in their alcohol vehicles (F. Nigro, Instituto de Pesquisas Tecnologicas, personal communication).

Due to the corrosiveness of alcohol, there were some problems with the

[7] The 1983 fuel economy ratings (31) show that counterpart models typically have a 20% lower fuel economy (km/l) when fueled with hydrated ethanol rather than gasohol. This corresponds to a 24% increase in efficiency with ethanol when the difference in calorific values is taken into account.

durability of carburetors and other vehicle components in the first generation of alcohol vehicles. Consequently, manufacturers changed materials or used protective coatings, and the quality of the ethanol was improved (1). As a result, vehicle durability is no longer considered a serious problem.

Because of the difficulties of operating diesel engines with alcohol, manufacturers are producing spark-ignition engines for use in ethanol-fueled buses, tractors, and smaller trucks. So far, these vehicles are used primarily within the sugar cane and alcohol industries. Conversion from a diesel to a spark-ignition engine does not lead to a significant increase in energy consumption, because of the greater efficiency when operating the latter on alcohol.

Diesel Engines

There is great interest in Brazil and other countries in developing alcohol-fueled diesel engines. Using alcohol fuels (ethanol or methanol) is a problem, because of their resistance to ignition at normal compression levels (characterized by a low fuel cetane number). Commercializing alcohol-fueled diesels is particularly important for broadening the "substitution base" as ethanol production expands, thereby ensuring further reductions in petroleum imports.

A number of options are under development for diesel engine conversion. These include use of dual injection, conversion to spark ignition, use of a glow plug for ignition, and use of fuel additives. Many of the researchers working on diesel engine conversion are also attempting to maximize fuel economy.

Dual injection involves use of a small amount of diesel fuel to initiate combustion. MWM of Brazil (34) has already introduced a series of dual fuel ethanol-diesel engines. Diesel replacement is 84% at full load and 72% at 50% load. With a slightly higher compression ratio compared to the original diesel engine, an increase in efficiency of about 7% results.

Researchers at the University of Sao Paulo in Sao Carlos (35) have experimented with 50–60-hp diesels, which they converted to run on ethanol by adding spark plugs and changing from direct injection to use of a carburetor. Increasing the compression ratio and leaning the air-fuel mixture compensate for the performance loss due to use of a carburetor. Results with a Perkins engine show efficiencies equal to or better than that with the original diesel engine.

An innovative direct injection diesel engine produced by the German company MAN was also converted to alcohol by adding spark plugs. This engine has been tested in Brazil as well as Germany, and efficiencies equal to or better than those with diesel fuel resulted (36). Unfortunately, MAN is not making engines in Brazil, and other manufacturers do not appear to be interested in this engine design.

The glow plug, an electronic resistance element used for ignition, is another option under development in Brazil for diesel engine conversion. A glow plug provides better ignition stability than spark plugs when used with alcohol fuels. So far, diesel engines have been operated successfully with glow plugs in the laboratory and efficiency improvements of up to 22% (using methanol) resulted (37). Research is beginning to address the questions of field performance and long-term durability.

Other Uses for Ethanol

It does not appear that ethanol is being used for domestic applications in Brazil. However, it is reported that ethanol can be used without difficulty as a cooking fuel or for lighting as long as a small amount of an illuminant is added (38). If used for cooking, ethanol could substitute for LPG, a petroleum product that is the predominant cooking fuel in Brazil. Ethanol could also replace wood and charcoal, which were the primary cooking fuels in 38% of Brazilian households in 1980 (18). Such a development would extend the benefits of the alcohol program to a greater portion of the population, since automobiles are still luxuries in Brazil.

Another promising area of application is the increased use of ethanol as a chemical feedstock. Ethanol is readily converted to ethylene and acetaldehyde which are major chemical building blocks. These and other ethanol derivatives can be used to produce plastics, solvents, lubricants, synthetic fibers, pesticides, and other products (38). In fact, ethanol was a major chemical feedstock in Brazil until the 1960s, when petroleum-based processes began to dominate (1). With incentives provided by the government, the use of ethanol as a raw material for the chemical industry recovered and advanced rapidly in the early 1980s (26).

Environmental Impacts

Distilleries produce about 13 liters of liquid effluent (stillage) per liter of ethanol. This effluent creates a serious pollution problem if disposed of in rivers and streams. Stillage, which is rich in minerals such as in potassium and organic materials, is increasingly being used as a fertilizer for sugar cane cultivation (40). Thus a potential contaminant has been converted into a useful product. Also, techniques have been developed for reducing the volume of stillage by a factor of up to 10 (41).

The substitution of alcohol for gasoline generally has very favorable air pollution impacts. First, there are no lead or sulfur emissions with alcohol-fueled vehicles. Furthermore, emissions of carbon monoxide and hydrocarbons are greatly reduced (26). Aldehyde emissions are much greater with ethanol, but this pollutant is still produced in relatively small quantities.

The main air pollutants of concern with ethanol fuel are some organic compounds, particularly unburned fuel and aldehydes. Emissions of both unburned fuel and aldehydes can be kept down if the fuel intake system is carefully controlled (2). While it is possible to use catalytic converters for greatly reducing aldehyde emissions, they have not been adopted in Brazil so far.

Little information is available on the long-term ecological consequences of sugar cane production in Brazil. Although sugar cane has been continuously produced for over 400 years, this has taken place on high-quality agricultural land. The ecological effects of recent developments such as increasing use of pesticides and chemical fertilizers and cultivation in poorer quality soils are uncertain.

POLICY ISSUES

Government Management

The role of the federal government in the development of ethanol fuel as a major energy source in Brazil cannot be overstated. The government has underwritten the program by providing highly subsidized financing for producers and a guaranteed market for the ethanol produced. The government also sets the prices at which it purchases alcohol from producers (with the intention of providing a fair profit margin) and for the sale of alcohol to consumers. While this strong government role has been vital to the success of the program, it has led to some problems.

By providing highly subsidized capital and a guaranteed market, the government naturally discourages competition and economic efficiency among producers. Indeed, close observers of the alcohol program claim many production units are poorly managed and inefficiently operated (B. Johnson, J. Wright, University of Sao Paulo, personal communication). The nearly complete removal of financing subsidies as of 1984 should help stimulate more efficient operation, at least among newer producers.

The rapid expansion of the alcohol program in the 1980s has created some problems. In particular, the demand for hydrated ethanol (in 100% ethanol-fueled cars) did not match the supply that the government was required to buy. In 1981, for example, hydrated ethanol consumption was only half of the 2.8 billion liters that was produced. Thus, the government ended up adding about 20% of total alcohol production to its stocks in both 1981 and 1982.

The oversupply was due largely to the plunge in alcohol vehicle sales during 1981. Production of alcohol vehicles, which reached 50,000 a month by the end of 1980, dropped to less than 5000 a month by mid-1981 (39). Also, the alcohol fraction in gasohol was kept at a relatively low level in the early 1980s.

Subsequently, the government took a number of steps to increase alcohol demand and improve the supply-demand balance. In early 1982, the government raised the alcohol fraction in gasohol and, through fuel pricing and vehicle tax policies, made it more attractive to purchase and operate alcohol vehicles. As a result, the sales and market share of alcohol vehicles increased dramatically throughout 1982 and 1983 (see Table 2). Second, the number of new distillery projects approved by the government dropped from 100 in 1980 to about 14 in 1982. In addition, the government is requiring producers to hold stocks longer in certain instances, it is authorizing some production without guaranteed purchase, and it is encouraging alcohol export.

Pricing Policy

Ironically, the large decrease or even removal of the financial subsidies for new alcohol producers could have some adverse economic impacts. To make new distilleries financed through conventional channels viable, the government will have to increase the price it pays for alcohol. As of late 1983, producers were asking for a 30% price hike, primarily because of decreasing subsidies (12). However, producers who entered the program at an earlier stage still benefit from the large financing subsidies they received. They will reap windfall profits if the purchase price is raised to accommodate new producers.

One solution would be to eliminate the financing subsidies for all producers (i.e. update the financing terms for old as well as new contracts). However, the government is legally bound to contracts previously made, and it is unlikely that the producers will accept such a change. Another solution would be for the government to set its purchase price on the basis of the amount of subsidy provided. The purchase price already varies according to the location and the type of alcohol produced. It would be relatively straightforward to base prices on the time of project financing as well.

Another pricing problem concerns government revenues from fuel sales. For many years, the Brazilian government set the sales price for gasoline (now gasohol) at levels at least twice the production, distribution, and marketing cost. Thus, gasoline sales have generated large revenues for the government. In 1983, the government was receiving approximately $0.25/liter above actual costs (based on imported petroleum), which led to a net profit of at least $2.5 billion.

As of late 1983, the price consumers paid for hydrated ethanol was 25% greater than the government purchase price. While this is enough to cover distribution and marketing, there is little or no excess revenue for the government. Furthermore, the ethanol sales price at the pump (Cr$/liter) was only 59% of that for gasohol. This provided an operating cost advantage of about 29% to owners of ethanol vehicles.

To reduce the loss of revenue as a result of the ethanol program, the government could raise the ethanol-to-gasohol price ratio. For example, raising the pump price ratio to 75% would still result in about a 10% operating cost advantage for ethanol-fueled cars. The market for ethanol vehicles should be well enough established to tolerate this narrowing in cost advantage. In mid-1984, the government took a step in this direction by raising the ethanol-to-gasohol price ratio to 65%.

Fuel pricing policy is also an important consideration for the conversion of diesel-fueled vehicles to alcohol. As of late 1983, the operating cost for normal diesel-powered vehicles was 25–35% lower than for the same vehicles running on ethanol (with either spark-ignition or diesel cycle engines). Thus, the selling price of diesel fuel will have to be increased substantially to make the shift to ethanol worthwhile.[8]

Siting and Production Policy

The government, through its authority for approving projects, can have a strong impact on the locations of distilleries and sugar cane plantations. The government can also influence the nature of ethanol and sugar cane production by favoring certain feedstocks, institutional arrangements, scales of production, and other factors. In the first phase of the program, the government routinely approved proposed projects. Consequently, alcohol production is now concentrated in Sao Paulo state, where there is a strong tradition of sugar cane and ethanol production and a high level of infrastructure.

As ethanol production moves toward and beyond the 1987 annual production goal of 14 billion liters, the government is exerting more influence on the siting and nature of production facilities. Because of the high fraction of total production in Sao Paulo and the possibility of a conflict with food crops there at the local level, the state government has been urging restrictions on the further expansion of alcohol production in Sao Paulo (40).

Alcohol producers, however, are requesting greater production quotas in

---

[8] The exception is in the sugar cane-alcohol industry where lower-cost, internally produced ethanol is available.

Sao Paulo state, in the prime agricultural zones as well as in more remote areas (W. Belik, personal communication). In some cases, they are even willing to support this expansion with private (as opposed to government) financing. If permission for increased alcohol production is granted, care should be taken to avoid a net loss of basic food crops.

The federal government has made an effort to expand alcohol production outside of Sao Paulo in areas such as the Central-West region where there are large amounts of available and suitable land, and where experience in sugar cane and ethanol production is rising. For example, the Central-Western states of Mato Grosso, Mato Grosso do Sul, and Goias, which accounted for only 2% of national sugar cane production in 1980, represented 13% of approved distillery capacity as of 1984 (26). Expansion in these areas helps to balance alcohol supply and demand geographically within the country, and to avoid conflicts with food production.

The World Bank, which began financing the Brazilian alcohol program in 1981, has tried to influence the siting and nature of projects. For example, projects in traditional sugar cane areas are given low priority for World Bank funds while small entrepreneurs, cooperatives, and projects in more remote regions are given high priority (26). Cooperative projects involve joint ownership and operation of a distillery by a number of sugar cane growers. If the implementation of cooperative projects is successful, it should help stem the trend toward increasing concentration of land ownership as a result of the alcohol program. However, cooperative projects are having a difficult time getting established because of institutional problems such as banks' requiring large amounts of collateral (B. Johnson, personal communication).

The government, through its authority for approving and financing distilleries, could stimulate other socially beneficial developments such as extension of the harvest period or use of manioc (cassava) as a feedstock. Ethanol production from manioc appears to be more labor intensive than that from sugar cane, has lower investment requirements, can be carried out using more marginal land, and would rely more on small and medium-size farms (1).

SUMMARY AND CONCLUSION

Competition with food production is one controversial aspect of Brazil's alcohol from sugar cane program. To meet the 1985 alcohol production goal, nearly 3.0 million hectares of sugar cane will be required. However, this is only 6% of the land area now being used for crops in Brazil. Planting of sugar cane for ethanol has been accompanied by a growth in the use of plentiful pasture and unutilized land for crop production. Furthermore, the new land area used for heavily exported commodity crops (particularly soybeans) has been considerably greater than the new land area devoted to sugar cane in recent years.

Another important consideration in biomass fuels programs is the net energy balance. With Brazil's highly labor-intensive methods of sugar cane cultivation, the direct and indirect energy consumption in the agricultural and industrial phases of ethanol production equals only about 30% of the energy contained in the alcohol produced. The energy balance for ethanol in Brazil is apparently no worse than the energy balance for the extraction and refining of petroleum in the US.

Perhaps the most controversial aspect of Brazil's alcohol program is cost. Data provided by alcohol producers as well as government payments indicate a real production cost of $45–60/bbl of gasoline replaced when pure ethanol substitutes for gasoline. For comparison, gasoline derived from $29/bbl imported petroleum is estimated to cost about $41/bbl ex-refinery in Brazil. However, this cost neglects the stress on the economy and society caused by importing oil. Assuming a 20% surcharge for imported oil, ethanol becomes competitive when it is used for blending with gasoline, and it is close to being competitive when used for full gasoline replacement. In addition, ethanol production appears to be about as capital intensive as domestic petroleum production in Brazil.

By 1985, the equivalent of about 420,000 full-time jobs will be created in rural areas as a result of alcohol fuel production. To put this figure in perspective, there were approximately 13 million persons in the agricultural work force in Brazil as of 1980, when agricultural employment was stagnating. Furthermore, alcohol production is a very effective use of scarce investment capital from the viewpoint of direct employment creation.

Substantial technological innovation is occurring as Brazil proceeds in its alcohol fuels program. These developments are in all relevant areas: sugar cane production, distilleries, and vehicles. First, the average sugar cane yield increased 31% in 1974–1983. Improved sugar cane varieties, new techniques for sucrose extraction, and microdistilleries have been developed and are emerging on a commercial basis. Methods for increasing the bagasse residue are available, and this material is starting to be sold and used as a solid fuel outside of distilleries. Spark-ignition alcohol engines have been perfected, and significant progress is occurring in the development of alcohol-fueled diesel engines. Finally, stillage, which poses a pollution problem if disposed of in rivers, is being productively used as a fertilizer on a wide scale.

There were bound to be some problems as ethanol production rapidly expanded and a new vehicle fuel was introduced on a massive scale. The ability to confront these problems and adopt new policies has been demonstrated. The government has started to exert greater influence over the nature of the program through its authority for approving and financing projects. Initiatives have been taken to increase sugar cane and ethanol production outside of Sao Paulo state, and the government is attempting to foster cooperative projects. In addition, steps have been taken to ensure a better match between ethanol supply and demand.

For these reasons, the first decade (1976–1985) of the Brazilian national alcohol fuel program should be regarded as a success.

The questions of how far Brazil can expand alcohol fuel production and whether Brazil's experience can be replicated in other countries remain. In 1984, the Brazilian government announced a goal of achieving energy self-sufficiency by 1993. As part of this plan, alcohol production is targeted at approximately 28 billion liters by 1993, about 3.5 times the 1983 production level (40). If ethanol derived from sugar cane continues to be the only source of alcohol fuel, then the planting of an additional 5–6 million hectares of sugar cane and the investment of approximately $10 billion will be required in the decade beginning in 1984. This represents a very ambitious undertaking, and it will have to be carefully planned if it is to be achieved without harmful effects on food production and land distribution. However, given the experience to date, it may indeed be a reasonable undertaking.

Some other developing countries are already moving toward producing ethanol fuel from sugar cane in large quantities. For example, Argentina, Kenya, the Philippines, and Zimbabwe have established programs to produce ethanol from sugar cane, for blending with gasoline. Zimbabwe's program appears to be relatively successful, while Kenya's effort is not (42). Of course, the viability and success of a program depends on a number of local factors.

In general, producing ethanol fuel is most likely to be feasible in countries that have traditions of producing sugar cane and ethanol from sugar cane, have excess arable land, and have a pressing need to reduce oil imports. In countries that are not as fortunate as Brazil in terms of arable land resources, it may be attractive to implement an alcohol fuels program in conjunction with efforts to increase agricultural productivities. Part of the alcohol produced could be used to operate irrigation pumps, which in turn could raise agricultural yields in order to provide sufficient food crops and sugar cane from a limited amount of arable land.

ACKNOWLEDGMENTS

The author would like to thank Jose Goldemberg and Jose Roberto Moreira for the opportunity to carry out this study.

Literature Cited

1. Rothman, H., Greenshields, R. N., Calle, F. R. 1983. *Energy from Alcohol the Brazilian Experience.* Lexington: Univ. Press Kentucky. 181 pp.
2. Moreira, J. R., Goldemberg, J. 1981. Alcohols—Its use, energy and economics—A Brazilian Outlook. *Resource Management and Optimization* 1(3):231–79
3. Rotstein, J., Homem de Melo, F. 1983. Os desenfoques do papel do alcool na politica energetica Brasileira. *O. Estado de Sao Paulo,* Oct. 13, p. 28; Nov. 29, p. 39
4. O'Keefe, P. 1983. Fuel ethanol: unrepeatable Brazil. *New Sci.* 99 (July 28): p. 295
5. Homem de Melo, F., Pelin, E. R. 1983. *Politica Agricola e Composicao da Producao.* Fundacao Instituto de Pesquisas Economicas. Univ. Sao Paulo
6. Homem de Melo, F. 1983. O que se pode esperar da agricultura. *Gazeta Mercantil,* Sao Paulo. Dec. 16, p. 4
7. Homem de Melo, F., da Fonseca, E. G. 1981. *Proalcool, Energia e Transportes.* Sao Paulo: Fundacao Instituto de Pesquisas Economicas. Livraria Pioneira Editora. 163 pp.
8. Pereira, A. 1983. Employment implications of ethanol production in Brazil. *Int. Labour Rev.* 122(1):111–27
9. Copersucar. 1983. *Atualizacao dos custos de producao de cana, acucar e alcool—safra 1982/83.* Sao Paulo
10. Comissao Executiva Nacional do Alcool. 1983. *Avaliacao Social do Programa Nacional do Alcool.* Brasilia, Brazil: Ministry of Industry and Commerce
11. Johnson, B., Fenerich, C. A., Fischman, R. 1983. *A. Extensao da Safra Canavieira Analise e Recomendacoes de Politicas.* Instituto de Administracao. Univ. Sao Paulo. 61 pp.
12. A Sopral preve prequizos. 1983. *Gazeta Mercantil,* Sao Paulo. Oct. 19
13. Homem de Melo, F., Pelin, E. R. 1984. *As Solucoes Energeticas e a Economia Brasileira.* Hucitec. Sao Paulo. 146 pp.
14. World Bank. 1981. *Staff Appraisal Report—Brazil Alcohol and Biomass Energy Project.* Washington, DC: World Bank
15. Renaul, D. 1983. Custo faz alcool depender da cana. *O Estado de Sao Paulo,* May 1. Sao Paulo
16. Coque e Alcool da Madeira S/A. 1983. *Alcool e Emprego: o Impacto da Producao de Alcool de Cana-de-Acucar e de Madeira no Geracao de Empregos.*

Brasilia, Brazil: Cadernos Coalbra. 170 pp.
17. Petrobras. 1983. *Petroleo Brasileiro—Preconceito e Realidade.* Rio de Janeiro
18. Fundacao Instituto Brasileiro de Geografia e Estatistica. 1983. *Anuario Estatistico—1982,* p. 134. Rio de Janeiro
19. Fundacao Instituto Brasileiro de Geografia e Estatistica. 1983. See Ref. 18, p. 433
20. Poole, A. 1983. *A Crise e a Oportunidade: Propostas para uma Politica de Combustiveis a Medio Prazo.* Latin American Institute for Economic and Social Planning, Santiago, Chile.
21. Johnson, B. 1983. *Community Impacts of Brazil's National Alcohol Program.* Instituto de Administracao. Univ. Sao Paulo
22. Chambers, R. S., Herendeen, R. A., Joyce, J. J., Penner, P. S. 1979. Gasohol, does it or does it not produce positive net energy? *Science* 206:789–95
23. Bullard, C. W., Penner, P. S., Pilati, D. A. 1978. *Resources and Energy* 1:267–313
24. Fundacao Instituto Brasileiro de Geografia e Estatistica. 1983. *Levantamento Sistematico de Producao Agricola.* Rio de Janeiro
25. Rieznik, P. 1983. *Tecnologia e Custos na Lavoura Canavieira.* Centro Brasileiro da Analise e Planejamento. Sao Paulo. 22 pp.
26. Pamplona, C. 1984. *Proalcool: Technical-Economic and Social Impact of the Program in Brazil,* The Sugar and Alcohol Institute, Belo Horizonte, 91 pp.
27. Ministerio das Minas e Energia e Electrobras. 1983. *Aproveitamento Energetico dos Residuos da Agroindustria da Cana-de-Acucar.* Rio de Janeiro: Livros Tecnicos e Cientificos Editora
28. Ministerio das Minas e Energia. 1983. *Balanco Energetico Nacional,* p. 81. Brasilia, Brazil
29. Goldemberg, J. 1982. Energy Issues and Policies in Brazil. *Ann. Rev. Energy* 7:139–74
30. de Sousa Dias, J. M. C., Novaes, F. V., da Cruz, E. R., Soares, R. P. 1983. *Avaliacao Tecnica e Economica do Funcionamento de Microdestilarias.* Empresa Brasileira de Pesquisa Agropecuaria. Ministerio da Agricultura. Brasilia, Brazil. 39 pp.
31. Secretaria de Tecnologia Industrial. 1983. *Guia de Consumo do seu Carro.* Brasilia, Brazil: Ministerio da Industria e do Comercio. 14 pp.
32. Schmidt, P. 1983. A Tecnologia Veicular da Volkswagen do Brasil: O Estado Atual e um Estudo para o Futuro Prox-

imo. *Anais do I Simposia de Engenharia Automotiva.* Secretaria de Tecnologia Industrial. Ministerio da Industria e do Comercio. Brasilia

33. Pinto, F. B. P. 1983. Otimizacao do Conjunto Motriz de Veiculos a Alcool. See Ref. 32

34. Dietrich, W., Bindel, H. W. H. 1983. O Desenvolvimento da Injecao Piloto para Uso de Alcoois em Motores Ciclo Diesel. See Ref. 32

35. Corsini, R. 1983. Conversao de Motores de Ciclo Diesel para Ciclo Otto a Alcool. See Ref. 32

36. Neitz, A., Chmela, F. 1979. MAN-FM Process to Enable Diesel Engines to Burn Methanol. *1st Int. Automot. Fuel Econ. Conf.,* Arlington, Va.

37. Nanni, N. 1980. Use of Glow-plugs in Order to Obtain Multifuel Capability of Diesel Engines. *4th Int. Symp. Alcohol Fuels Technol.,* Guaruja, SP. Brazil

38. National Research Council. 1983. *Alcohol Fuels Options for Developing Countries.* Board on Science and Technology for International Development. Washington, DC: Natl. Acad. Press. 101 pp.

39. Secretaria de Tecnologia Industrial. 1983. *Avaliacao do Programa Nacional do Alcool 1981/82—Aspectos Gerais.* Ministerio da Industria e do Comercio, Brasilia, Brazil. 28 pp.

40. O Futuro do Alcool em Sao Paulo. 1984. *Sao Paulo Energia* 1(4):24–25

41. Rosillo-Calle, F. 1984. A Reassessment of the Brazilian National Alcohol Programme (PNA). *Energy Digest* 3(4):29–32

42. Juma, C. 1984. *The Use of Power Alcohol in Kenya and Zimbabwe: An Overview.* Science Policy Research Unit. Univ. Sussex, England. 26 pp.

Ann. Rev. Energy. 1985. 10 : 165–99

INDUSTRIAL FUEL USE: STRUCTURE AND TRENDS[1]

Richard Thoreson and Richard E. Rowberg

Energy and Materials Program, Office of Technology Assessment, US Congress, Washington, DC 20510

James F. Ryan

Perkin-Elmer Corporation, Norwalk, Connecticut 06859

INTRODUCTION AND OVERVIEW

Between 1950 and 1973, US industrial output grew at an average annual rate of 3.7% (1). Over the same period, industrial energy consumption grew at an average rate of 3% (2), and energy inputs were commonly considered closely coupled with, if not proportional to, industrial output. This relationship changed dramatically during the last decade, as total energy use declined by 18% while total output (in constant 1972 dollars) rose by over 5% (Tables 1, 3). Over the same period, the use of premium fuel products from crude oil and natural gas declined by 24%, although electricity use increased by about 13% as it became the main source of energy for industry (if generation and transmission losses are included). Oil and gas are second and third, respectively, although their relative positions can shift quickly depending on boiler fuel switching in response to short-term price fluctuations.

The customary definition of the industrial sector in the US economy

---

[1] The US Government has the right to retain a nonexclusive, royalty-free license in and to any copyright covering this paper.

The information contained in this paper is based largely on analyses carried out by the Energy and Materials Program of the Congressional Office of Technology Assessment. The views expressed in the paper, however, are those of the authors and do not necessarily represent those of the Office of Technology Assessment or the Technology Assessment Board.

165

Table 1   Industrial fuel mix

Year		Oil	Gas	Electricity	Coal	Total
1973	Energy (quads)	9	10.2	8	4.3	31.5
	Share (%)	28.6	32.4	25.4	13.7	
1983	Energy (quads)	8	6.6	9	2.3	25.9
	Share (%)	31	25	35	9	

Source: (2).

includes manufacturing, mining, agriculture, and construction. By far the largest share of industrial energy use, about 82%, occurs in manufacturing, and therefore these activities are the primary subject of this article (Table 2). In comparison to other energy end-use sectors (e.g. residential, commercial, and transportation), the industrial sector has demonstrated exceptional flexibility in response to higher energy prices, energy supply uncertainties, and environmental restraints. In the last decade, the industrial share in total energy use declined from about 42% to about 37%. The industrial share of total oil use remained roughly constant, but industry's share of natural gas use declined from 46% to 39% and its share of electricity use declined from 40% to 36% over the same period (2).

Since the most recent peak in 1979, the decline in industrial fuel use has been even greater, including an absolute reduction of 8% in the consumption of electricity as well as declines in the use of every other major fuel. While total US fuel consumption declined by about 11% between 1979 and 1983, industrial fuel consumption declined by just over 20%. This occurred as a delayed reaction to earlier fuel price inflation, despite the fact that from 1979 to 1983 fuel prices were essentially flat or falling. A substantial part of the savings occurred because of a 4.7% decline in industrial product over the last four years. Most of the savings, however, occurred as energy input per unit of output declined by over 16%.

What do these recent trends imply for the future? Will rapid efficiency improvements and lagging output growth continue to reduce industrial energy needs? Or are the most recent savings entirely a delayed reaction to

Table 2   1982 fuel consumption: four divisions of industry[a]

Manufacturing	82%	Agriculture	4%
Mining	9%	Construction	5%

Source: Table 8.
[a] Electricity valued at 10,000 Btu/kWh.

the energy price inflation of the mid-1970s, a reaction that by now may have spent itself? A large delayed reaction between price change and market response can indeed be expected in traditional heavy industries that depend on long-lived capital. [See *Pulp and Paper (SIC 26), Chemicals (SIC 28)*, and *Petroleum and Coal Products (SIC 29)*.]

This paper describes recent industrial patterns and trends related to fuel consumption, and comments on future prospects. Our analysis suggests that there remains substantial delayed reaction among manufactures to the new energy price regime established in the 1970s, and, consequently, that substantial fuel savings per dollar of output can be expected in the future. The bulk of savings will occur as the mix of manufactured goods and services continues to shift toward less energy-intensive, higher value-added products, and as major installations for processing basic industrial materials become obsolete and are replaced (3). Imports of basic industrial commodities such as steel and petrochemical feedstocks may also restrict growth of domestic industry and thus industrial fuel consumption.

The first section below explores fuel-related industrial technology and economics by disaggregating total industrial energy use into first product mix, then generic fuel services, and finally fuel mix. The entire industrial sector is considered, to permit discussion of broad patterns and trends. The next section examines more detailed technical and economic relationships in three basic materials manufactures. These industries together used 57% of total manufacturing fuel in 1982 and 71% of premium fossil fuels. The final section summarizes past experience and extrapolates industrial energy use into the future.

INDUSTRIAL FUEL USE PATTERNS

As major factor inputs, fuel supplies determine industrial technology and output, and vice versa. There can be little doubt about the complexity and importance of this relationship after a decade of related macroeconomic uncertainties. The complexity is managed here by classifying industrial fuel use by product, by generic end-use service, and by fuel mix. Although they are closely related by technology and economics, each of these classifications is discussed separately below because each reveals distinctive patterns and trends.

Industrial Product Demand and Product Mix

A major factor limiting recent growth in industrial fuel use has been the relatively slow growth of industrial output. Between 1973 and 1983, total output grew on the average 0.5% per year, compared to 2% growth for the entire economy (Table 3). These two growth rates were much closer

Table 3   Output growth rates: four industrial divisions

Unit	1950–73	1973–83
GNP	3.7%	2%
Total industry	3.5%	0.5%
Manufacturing	4%	1.1%
Agriculture	1%	1.8%
Mining	2.4%	1%
Construction	3.2%	−1.6%

Source: (1).

between 1950 and 1973, when average industrial growth was 3.5% while the entire economy grew at 3.7%. This change in relative growth rates has lowered the growth rate of both industrial fuel consumption and total US fuel consumption, since the industrial sector uses much more fuel per dollar of output than other producing sectors in the economy.[2]

The relationship between output growth and fuel consumption is complicated, however, by the mix of industrial products. Product mix changes over time because different products grow at different rates (Table 3 and Table 5). That is important for fuel consumption because different products also use fuel inputs differently, and have a wide range of fuel inputs (measured in Btu) per dollar of value added (Table 4 and Table 6). In the last decade and before, manufacturing has been growing a bit faster than average for all industrial activities. Since manufacturing consumes 1.7–3.1 times as much energy per dollar as the other divisions of industry, its growth advantage tends to increase overall industrial energy-intensity.[3]

Manufacturing can also be disaggregated by two-digit Standard Industrial Classification (SIC) product categories (Tables 5 and 6). Both in growth rate and fuel intensity, the chemicals industry (SIC 28) stands out. Pulp and paper (SIC 26) is also notable for high energy-intensity and

[2] This difference in fuel intensity cannot be calculated exactly, but the numbers clearly suggest that it is large. In 1983, even after the relatively large reductions in industrial energy intensity, industry consumed about 25,000 Btu per dollar of output. If we disregard the fact that the rest of energy consumption goes primarily for residential and other personal uses (e.g. auto transportation), and that unlike commercial activities the value of this personal consumption is not included in gross domestic product, then the quotient of the remainder of economic output divided by the remainder of energy consumption—about 20,000 Btu per dollar of output—is still well below the industrial ratio. In other words, the industrial sector energy-intensity is several times larger than that of the commercial sector (1, 2).

[3] Construction is highly volatile, and average growth rate calculations are thus sensitive to starting and ending dates.

Table 4   1982 Industrial fuel intensities : four industrial divisions

	Gross product originating (billions of $)	Energy use (quads)	Btu's ÷ GPO × 10^4
Total industry	953	26.6 × 10^{15}	2.8 × 10^4
Agriculture, etc	86	1.1 × 10^{15}	1.3 × 10^4
Mining	125	2.5 × 10^{15}	2.0 × 10^4
Construction	124	1.4 × 10^{15}	1.1 × 10^4
Manufacturing	631	21.6 × 10^{15}	3.4 × 10^4

Source: (1, Table 8).

average growth over the last decade. Other relatively high growth two-digit SIC industries tend toward lower energy-intensity, and low growth industries tend toward higher energy-intensity, at least in the last decade. Fuel use patterns for SIC 26, 28, and 29 are considered in greater detail below.

As a preview of that discussion, and anticipating fuel demand projections

Table 5   Output growth rates : two-digit SIC manufacturers (in percent)

Industry (SIC #)	1950–1973	1973–1983
Food and kindred products (20)	3.3	0.6
Tobacco products (21)	2.6	−2.3
Textile mill products (22)	4.0	1.1
Apparel and other textile products (23)	3.1	1.3
Lumber and wood products (24)	3.5	0.1
Furniture and fixtures (25)	2.9	1.3
Paper and allied products (26)	4.1	1.2
Printing and publishing (27)	3.5	1.6
Chemicals and allied products (28)	6.1	2.2
Petroleum and coal products (29)	4.0	−0.9
Rubber and misc. plastic products (30)	6.3	1.6
Leather and leather products (31)	0.6	−1.8
Stone, clay, and glass products (32)	3.1	−0.5
Primary metal industries (33)	1.7	−2.7
Fabricated metal products (34)	3.8	0.1
Machinery, except electrical (35)	4.3	1.7
Electric and electronic equipment (36)	7.3	4.2
Transportation equipment (37)	4.9	−0.4
Instruments and related products (38)	6.2	4.2
Misc. manufacturing industries (39)	3.5	0.4
Total manufacturing	4	1.1

Source: (7).

Table 6   Fuel-intensity: two-digit SIC manufacturers

Industry (SIC #)	1982 value added (billions of 1982 $)	1982 energy (quads)	Btu/$ $\times 10^4$
Food and kindred products (20)	59.0	1.1	1.9
Tobacco products (21)	8.4	0.03	0.2
Textile mill products (22) Apparel and other textile prod. (23)	33.6	0.6	1.8
Lumber and wood products (24)	15.2	0.2	1.3
Furniture and fixtures (25)	9.1	0.1	1.1
Paper and allied products (26)	25.4	2.7	10.7
Printing and publishing (27)	37.3	0.2	0.5
Chemicals and allied products (28)	55.5	7.1	12.8
Petroleum and coal products (29)	24.2	2.7	11.2
Rubber and misc. plastic prod. (30)	19	0.4	2.1
Leather and leather products (31)	4.2	0.02	0.5
Stone, clay, and glass products (32)	17.4	1.2	6.9
Primary metal industries (33)	35.7	3.4	9.5
Fabricated metal products (34)	46.6	0.5	1.1
Machinery, except electrical (35)	79.3	0.5	0.6
Electric and electronic equip. (36)	64	0.4	0.6
Transportation equipment (37)	63.4	0.5	0.8
Instruments and related products (38)	22.7	0.1	0.4
Misc. manufacturing industries (39)	10.5	0.1	1.0
Total manufacturing	630.6	21.5	3.4

Source: (5, 7).

in the final section, we can briefly examine domestic production growth patterns for these three fuel-intensive manufactures. Chemicals has expanded rapidly because many of its major products were developed and first entered the marketplace in the last three decades. Consequently, they have been operating in the early or immature stages of their life cycles, when product demand can grow at very high rates. On the other hand, other basic materials-processing industries (including pulp and paper and petroleum refining) involve primarily mature products, which were developed and marketed well before World War II. In recent times, these mature industries have done well to keep pace with GNP.

Chemicals, along with pulp and paper, have had an additional advantage of remaining largely free of the import competition that has plagued the primary metals by crowding out domestic production. In fact, chemicals has been a major source of industrial exports, with net exports around $10 billion annually during the last decade (13). Petroleum refining has, of course, also suffered from foreign trade, not so much in terms of product imports, but rather in terms of cartel pricing imposed by exporting nations on imported crude oil used as refinery feedstock.

Generic Fuel Services

Fuels are consumed by industry to obtain end-use energy services and raw material feedstocks.[4] Total fuel consumption can be allocated to the following seven generic energy services (5, 8):

1. Space conditioning and lighting   5%
2. Direct process heat   29%
3. Steam   21%
4. Machine drive   21%
5. Mobile engines   5%
6. Electrolysis   4%
7. Raw materials   14%

Each of these end uses can be associated with conventional technology, preferred fuels, and general technical and economic constraints on fuel switching and efficiency improvements. Furthermore, technical and economic trends can be identified that help explain movements in total fuel consumption and fuel mix.

SPACE CONDITIONING AND LIGHTING   The same technologies are used by industry for space conditioning and lighting as in the rest of the economy; and over the last decade many of the same technological improvements have been made to control energy costs. However, since this generic end use constitutes a relatively small share of total industrial fuel use, compared to the residential and commercial use, these improvements have resulted in relatively small fuel savings for industry.

Fuel price inflation of the last decade has not changed the comparative advantages of different fuels. Electricity remains the obvious choice for lighting and cooling. Natural gas has maintained its superiority for space heating, on the basis of price and combustion properties, although in local fuel markets #6 residual oil may be less expensive.

DIRECT PROCESS HEAT   As a major generic end-use service, direct process heat is unique to industrial energy consumption. It involves the highest temperatures and pressures, with technology designed to precise specifications. Major process heating applications include chemical separation

---

[4] Energy services can be described at various levels of detail, from simple materials/energy balance equations to engineering process models, and ultimately to specific plant operations. To build national accounts of energy services, all of these levels must be used. Engineering process models are particularly helpful to make up for deficiencies in actual fuel use data.

and synthesis (see AMMONIA and ETHYLENE below), material drying, and metal heat treating. For many moderate- to low-temperature heating applications, steam from boilers is also used.

Fuel price inflation in the last decade has led to major improvements in fuel efficiency, based on substitution of capital and improved technology. Major gains have been achieved using computer controls to coordinate fuel consumption better with process heat requirements. Improved insulation around heaters and preheating of reactants with burner exhaust gases have also led to significant and widespread efficiency improvements. Other technical improvements are industry-specific, including improved catalysts, which reduce the necessary pressures, temperatures, and residence times for chemical reactions.

Besides these efficiency improvements, which can often be accomplished by retrofitting existing plants and equipment, larger improvements have been achieved with capacity replacement or new capacity. The most prominent example is in the steel industry, which can save 15–20% of total energy per ton of product by shifting from batch to continuous casting. Energy is saved because ingot stripping, reheating, and primary rolling are eliminated, and because less steel must be returned to the steel-making process in the form of unfilled ingot molds. Other examples include use of higher-temperature furnaces for hydrocarbon processing in the chemicals industry, which can significantly reduce residence time, with substantial savings of energy per unit of product.

Natural gas provides about 40% of direct process heat (8) because it is clean burning, fast, and generally lower priced than oil liquids offering comparable heating qualities. Large amounts of gas are used, especially on the Gulf Coast, where chemical and petroleum refining plants have been built to take advantage of local gas supplies. Oil accounts for the next largest fuel share, about 20%. The shares of both gas and oil would be substantially larger, 60% and 30% respectively, if coke for steel smelting and "still gas" in oil refining were excluded from total process heat as special heating applications.

The steel industry offers a prominent example of fuel switching to electricity. Thirty to forty percent of energy and 25% of total costs can be saved by producing steel from scrap, which is melted by electric-arc furnace, rather than from iron ore, which is processed using coke in a blast furnace (3). Because old integrated mills (which start with iron ore) are being replaced by new mini-mills (which start with scrap), fuel-efficiency improvements of all kinds can be accomplished in a cost-effective manner at the same time.

Outside of the steel industry, the fuel mix for process heating has not changed substantially, although electricity has been making significant gains based on infrared, induction, microwave, and laser heating processes. The general advantage of electricity lies in precise targeting and control of heat in relatively small-scale operations where convenience is at a premium. Such requirements are growing as the product mix shifts toward higher valued-added goods, produced by skilled workers and complex machinery operating in controlled environments.

MACHINE DRIVE   Machine drive includes all materials handling and processing by stationary equipment, except direct process heat. (See below for mobile equipment.) It includes pumping, rolling, grinding, mixing, compressing, refrigeration, blowing, and solids conveyance. Most machines are driven by electric motors, but larger units, in factories using large amounts of process steam, may be driven by steam turbines. Understandably, machine drive services are concentrated in industries producing basic industrial materials, such as chemicals, steel, oil refining, pulp and paper, and stone, clay, and glass.

Efficiency gains will result from possible improvements in the efficiency of electric motors (see below, ELECTRICITY). Some gains are likely in new capacity, where improved plant layouts and process sequencing can reduce the number of times a product is handled. Continuous casting in steel is a good example again, since it eliminates ingot handling.

Future demand for machine drive and the mix of electricity and steam will undoubtedly be affected by the relative growth of basic materials processing. To the extent that the basic industrial materials grow more slowly than the industrial average (Table 5), the share of machine drive in industrial energy services will decline.

MOBILE ENGINES   Mobile engines are used primarily by the nonmanufacturing activities: agriculture, mining, and construction. As the name implies, these engines are used to move and process materials that are spread around large areas, as in corn cultivation, or that are found in remote locations, as in coal mining or in oil drilling and production. The primary fuels are diesel and gasoline. About 75% of distillate fuel oil used by industry falls in this category (Ref. 28).

Efficiency improvements in diesel engines over the past decade have been small, since industrial diesels have always been relatively efficient and sized to match loads. Gasoline fuel-efficiency improvements have been more substantial, since gasoline-powered trucks can take advantage of engine improvements that have increased automobile fuel-efficiency. However, efficiency improvements for industrial vehicles have generally not been as large as those for personal vehicles, because industrial vehicles are already designed primarily for function and thus cannot obtain the efficiency gains from the performance, comfort, and style adjustments that substantially improve efficiency for personal transportation.

PROCESS STEAM   Steam from boilers is used throughout manufacturing for heat and power. Boilers, of all the devices for converting fuel into useful energy, offer the greatest opportunity to switch away from premium fuels. This opportunity exists because, in boilers, water is circulated through an integral system that is completely separate from the combustion gases. Replacing #6 residual oil or natural gas, which still account for over 65% of industrial steam (8), with coal and biomass can reduce fuel costs per Btu substantially, although the necessary boilers and associated solid fuels handling facilities are more expensive. (See INDUSTRIAL COAL USE below.)

In the last decade, boilers have increased their fuel-efficiency with computer controls to optimize air-fuel mixtures and with heat exchangers that use exhaust gases to preheat combustion gas. Both of these changes can often be retrofitted in existing boilers. In new installations, where a plant uses both steam and electricity and/or machine drive in large quantities, cogeneration can achieve very high fuel-efficiency, with a generator or power shaft driven by a steam turbine and the turbine exhaust used for process heat. Economic opportunities for cogeneration have significantly increased with the Public Utilities Regulatory Power Act (PURPA), which requires local public utilities to buy power from cogenerators at relatively high rates (3, pp. 55–56).

ELECTROLYSIS   Electrolysis is the use of electricity, instead of heat and pressure, to induce and promote molecular change. Its use is concentrated in the production of chlorine, alkalies, and aluminum. Over the last several decades, major efficiency improvements have been achieved in the production of chlorine by increasing the sizes of reactors (24, vol. 5).

MATERIAL FEEDSTOCKS   Hydrocarbon fuels, mainly oil and natural gas, are also used as chemical feedstocks (see *Chemicals*). By and large, this is the highest valued use for oil-based products because of the high market value of derivative chemicals and because substitute feedstocks such as coal or biomass involve much greater total costs or unproven process technology. As discussed above, the relatively recent development of the petrochemicals industry has led to rapid expansion in this use of oil. Past improvements in feedstock efficiency and those expected in the future are generally much less than for energy uses, at least in petrochemicals, because production processes have already been optimized to minimize feedstock inputs per unit of output.

Table 7 Industrial fuel mix (in percent)

	Oil	Natural gas	Premium fossil fuels	Coal	Electricity
1950	23	21	44	36	20
1953	23	23	46	34	20
1955	25	24	49	30	21
1960	27	29	56	22	22
1965	27	29	56	21	23
1970	27	33	60	16	24
1973	29	32	61	13	26
1975	29	30	59	13	28
1980	32	27	59	10	31
1983	31	25	56	10	34

Source: (2).

Industrial Fuel Mix

As a result of the interaction of product markets, industrial technology, and fuel prices, the industrial fuel mix changes over time. As shown in Table 7, the share of premium fossil fuels (oil and natural gas) peaked in 1973 and has declined steadily over the last decade. Although oil alone did not peak until 1980, its rise and the rapid decline of natural gas after 1973 are due to boiler fuel switching from gas to #6 residual fuel oil, a switch that can go either way on short notice because of dual firing capacity. Shares of the other two major energy sources have maintained steady trends over more than three decades, with electricity rising and coal declining (see above).

The turnaround in premium fossil fuels after 1973 can be explained generally by the sharp relative price growth after that year. The next section will review in considerable detail how this turnaround was accomplished in three energy-intensive industries. In the remainder of this section, trends in the use of electricity and coal are discussed, since they suggest opportunities and constraints for fuel switching away from premium fuels.

ELECTRICITY   For at least three decades, the use of electricity by industry has grown faster than total fuel consumption. Because of its high thermodynamic quality, electricity has a wide variety of applications for which it is uniquely suited. About 50% of industrial electricity use is for electric motors, including compressors, fans and blowers, and pumps. Next in size is electrolysis of aluminum and certain chemicals, which uses about 15–20% of electricity supplied to industry. After that comes electric process heating which accounts for 10% of current use. The latter is concentrated in those industrial processes that require high-temperature heat, such as cement, iron and steel, and glassmaking. The remainder of the electricity is used for lighting, electric controls, and heating and cooling systems.

From 1958 to 1973, electricity demand in manufacturing grew faster than total manufacturing output. The expansion of electrochemical processes in the chemical industry and the increased use of the basic oxygen furnace process in the steel industry were the major reasons for this. After 1973, growth of electricity demand dropped dramatically, and fell below that of manufacturing output. The decline in electricity demand growth was a reaction to the increase in electricity prices over that period. That it fell below growth of manufacturing output was due principally to a steady decrease in the proportion of that output from electricity-intensive industries.

Three main factors control the future of industrial electricity demand. The first is the potential for increased efficiency in current uses of electricity. The second is the growth potential in industries that are heavy users of electricity. The final factor is the potential for new uses of electricity that would replace existing processes that now use fossil fuels, or serve processes not now in existence.

A major contribution to increased efficiency will come from improvements in the efficiency of electric motors. Electricity use per unit of motor output, as a result of new semiconductor and control technology, could decrease by anywhere from 5 to 20% over the next 10–15 years depending on the price of electricity. Efficiency improvements of 20–30% also are technically possible for electrolysis of aluminum. Increased efficiencies are also possible in other electrical processes, such as metal cutting and welding, by use of new electric technologies.

The three major users of electricity—bulk chemicals, aluminum, and steel—account for more than a third of electricity use in industry. The chemical industry as a whole is expected to continue growing faster than total manufacturing output, but with a shift toward specialty chemicals that use relatively little electricity, and away from such commodities as chlorine, which are electricity-intensive. Steel makers are switching toward electric arc furnaces. Finally, the future of the aluminum industry will likely depend heavily on the potential for increased efficiency of electrolysis, but production growth is not expected to exceed total industrial output.

The biggest gains for electricity demand will come in the process heat category. Technologies such as plasma reduction and melting for primary metals production, and induction heating for shaping and forging, could have large effects on electricity demand. Other technologies, such as lasers and robotics, are likely to make much smaller contributions to electricity demand, because only small amounts of electricity are used for each application. The most immediate, potentially large source of new demand for industrial electricity is electric arc melting for steel.

A consequence of these competing factors is considerable uncertainty. If new technologies penetrate slowly, electricity demand in industry, for the next several years, will be dominated by the increases in efficiency and slowly growing electricity-intensive sectors. As a result the growth rate for electricity demand will likely trail that of total manufacturing output for that period. In the 1990s, however, the penetration of new electro-technologies should accelerate, particularly if the unique properties of electricity—e.g. the precision with which it can be controlled—can be fully used for new industrial processes. If this happens, electricity demand in the late 1990s could pick up dramatically, and once again outpace total output (9).

INDUSTRIAL COAL USE   The use of coal by industry is principally in two areas. The first is metallurgical coal, or coke plants, for the production of steel. The second is coal used in large boilers for the production of process steam. From 1958 to 1970, coal consumption by industry held fairly steady at around 175 to 200 million tons per year, about equally divided between metallurgical and steam coal. Since 1970, consumption has dropped steadily to the 1983 value of about 100 million tons. The use of metallurgical coal dropped even faster, so that it now makes up only about one third of industrial coal consumption. Steam coal use has remained fairly constant since 1974 at about 64 million tons per year.

The principal reason for the decline in the use of coke has been the drop in steel production by integrated steel makers. The decline of steam coal consumption before 1974 was a result of conversion to natural gas or oil, which were less costly when all the costs of burning coal were considered. The rapid rises in the prices of the former two fuels after 1974 halted that trend. Despite these price increases, however, the difference between the price of coal and those of oil and natural gas did not become large enough to overcome the total cost of using coal. These costs included those needed for coal transportation, storage, handling, and environmental controls, along with uncertainty about future environmental regulations. As a result, significant conversion to coal from existing oil and natural gas boilers has not occurred.

Projections of coal demand in industry indicate that metallurgical coal use will not grow significantly in the foreseeable future (10). Nearly all the growth in steel production in this country will likely be accounted for by electric arc processing of steel scrap. Steam coal demand is expected to grow, although current projections are well below those of a few years ago. The Energy Information Administration now projects that total industrial coal demand in 1995 will be about 170 million tons.

The stabilization of oil and natural gas prices has dampened even further the incentives to convert to coal. Only the largest boilers will even be considered as candidates for conversion to coal with current technology. While this could involve a substantial quantity of coal—about 45 million tons per year—many of these boilers will not be converted unless there is a very large increase in oil prices relative to coal, or new technology development significantly lowers the cost of conversion. In particular, development of pressurized fluidized bed combustion, coal-water mixtures, and small-scale coal gasification units could ease many of the storage, handling, and environmental control uncertainties. All of these technologies are well along in development and are being tested on a major scale, although primarily for electric untility applications. Continued development to a point at which industry will feel confident about the costs and reliability of conversion and operation will be needed, however, before significant commitment is made.

FUEL MIX DISAGGREGATED BY DIVISION AND MANUFACTURER   As a final, detailed summary of industrial fuel consumption, total fuel use can be disaggregated both by fuel and by consuming division and two-digit manufacturer (Table 8). The data show concentration among nine basic industries each with over a quadrillion Btu (a quad) of total consumption and more than half a quad of premium fuels. This concentration allows one analysis of industrial fuel consumption to focus on these most energy-intensive activities which are considered in some detail in the next section.

THREE BASIC MATERIAL PROCESSING MANUFACTURES

Compared to the rest of the US fuel economy, manufacturing employs by far the most diverse and complex set of fuel-using technologies. Furthermore, long-term trends in manufacturing fuel use are most strongly influenced by the progress of technology and product life cycles. The preceding discussion has identified key technological and economic factors and explored their relative importance for industrial fuel consumption in the past. However, there remain important details that can be understood only in terms of specific industries and their particular combinations of fuels with all other productive factors. Three two-digit SICs are considered here: pulp and paper (SIC 26), chemicals (SIC 28), and petroleum refining

Table 8 1982 Industrial fuel use disaggregated (in quads)

Sector (SIC #)	Petroleum products	Natural gas	Coal	Purchased[a] electricity	Total
Agriculture	0.62	0.11	—	0.33	1.06
Mining	0.45	1.39[f]	0.06	0.6	2.50
Construction	1.05[e]	0.14	—	0.2	1.39
Total nonmanufacturing	2.12	1.64	0.06	1.13	4.95
Food (20)	0.10	0.50	0.10	0.4	1.1
Tobacco (21)	0.00	0.00	0.01	0.02	0.03
Textiles (22)	0.06	0.13	0.04	0.32	0.55
Apparel (23)					
Lumber (24)	0.05	0.04	0.00	0.14	0.23
Furniture (25)	0.01	0.02	0.00	0.04	0.07
Paper (26)	0.3	0.43	0.26	0.6	2.7[b]
Publishing (27)	0.01	0.04	—	0.11	0.16
Chemicals (28)	3.10	1.80	0.44	1.8	7.14
Petroleum refining (29)	1.74[c]	0.65	0.01	0.33	2.73
Rubber (30)	0.03	0.09	0.02	0.21	0.35
Leather (31)	0.00	0.01	—	0.01	0.02
Stone, clay, and glass (32)	0.07	0.43	0.37	0.3	1.17
Primary metals (33)	0.1	0.74	1.2	1.34	3.38
Fabricated metal products (34)	0.02	0.18	0.01	0.26	0.47
Nonelectrical machinery (35)	0.03	0.13	0.02	0.31	0.49
Electrical equipment (36)	0.02	0.09	0.01	0.29	0.41
Transportation equipment (37)	0.03	0.11	0.04	0.3	0.48
Instruments (38)	0.01	0.02	—	0.07	0.1
Misc. manufactures (39)	0.00	0.02	0.00	0.03	0.06
Total manufacturing	5.68	5.43	2.53	6.88	21.64[d]

[a] kWh electricity = 10,000 Btu.
[b] Includes 1.1 Q of self-generated fuel.
[c] Includes still gas.
[d] Total is more than the sum of four columns due to inclusion of wood fuel in SIC 26.
[e] Includes asphalt.
[f] Includes lease and plant fuel.
Source: (2, 4, 5, 6, 8).

(SIC 29). They deserve specific attention because together they account for about 58% of manufacturing and about 48% of total industrial fuel consumption.

As indicated above, all three industries produce basic industrial materials. Consequently, they all use large amounts of direct process heat and steam. These two generic energy services account for the preponderance

(70%) of their total fuel use (6, 8). However, the full range of services will be described below, to integrate fuels into real production processes.

Pulp and Paper (SIC 26)

The pulp and paper industry makes pulp from wood feedstocks and then converts the pulp into paper and paperboard.[5] Of the four industries discussed here, the domestic pulp and paper industry is least affected by international trade. In fact, exports have been growing more rapidly than imports.[6] Since the sharp energy price growth started in 1973, domestic pulp and paper production has grown at about the same rate as total manufacturing (around 1%). Since 1950, pulp and paper has grown at an average annual rate of about 3.2% compared with about 3.1% for total manufacturing (Table 5).

As illustrated in Figure 1, logs are harvested (mainly in the Southeast, Northwest, and Northeast) and brought to large, centrally located mills where they are cut into short pieces, debarked, and then chipped. The strongest paper products are produced using full chemical pulping, in which wood chips are cooked with chemicals in an aqueous solution, at elevated temperatures (170°C or 350°F) and pressures, to dissolve lignin and other materials, leaving cellulose fiber. Pulp is also produced via various combinations of mechanical grinding, heat, and chemicals. Averaged over the entire industry, pulp production accounts for about 20% of energy use (24, Vol. 2).

For most writing paper (20% of total production), pulp must be bleached white without degrading the cellulose. Newsprint (16%) is only lightly bleached. Almost all bleaching is carried out with chlorine or chlorine compounds, leaving an effluent that must be biologically degraded to control environmental damage. Bleaching consumes about 5 million Btu per ton of white paper product, and overall it accounts for about 5% of total energy (3, 24).

The final, major stage of production is papermaking. Paper is formed by spraying a (99% water) pulp mixture on to a moving wire screen or filters, which catch and hold cellulose fibers in a thin sheet while letting water drain away. As the sheet moves along, water is removed first by suction, then by press rolls, and finally by evaporation as the sheet passes over steam-heated cylindrical dryers. By the end of the drying cycle, 95% of the water has been

---

[5] SIC 26 includes the manufacture of pulps from wood and other cellulose fibers and from rags, as well as the manufacture of paper and paperboard and the consumer and industrial products made from both materials.

[6] As a percentage of total domestic production, imports have declined from about 15% to about 13% over the last 20 years, while exports have increased their share over the same period from 3% to 7% (11, p. 2).

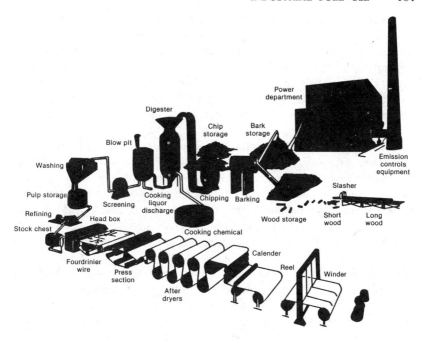

Figure 1   Pulp and paper production processes. Source (3).

removed. This final stage in the paper production accounts for about 40% of total energy (3, 24). Downstream fabrication and packaging of paper products for sale account for very little energy.

All pulp and papermaking process energy uses can be summarized in terms of three generic energy end-use services. Over 80% of the total fuel used is for steam and electric power. Steam and electric power are combined here because large cogenerators often produce both, and thus fuel use for one service cannot be separated from the other. Another 11% of total fuel is used for direct process heat.

FUEL USE PATTERNS AND TRENDS   Among industrial energy users, the pulp and paper industry is unique in its self-generation of energy and thus in its opportunity to replace premium fuels. In 1972, self-generated energy accounted for 40.5% of total energy; in 1983 this share had risen to 54.4%. This occurred at the same time that energy-intensity (per ton of product) declined by 14%. As a result of both adjustments, purchased fuel per ton of product declined by 33%; the use of liquid petroleum fuels declined by 55%; and the use of natural gas declined by 29%. Electricity use, however, increased by 48% (11).

Most self-generated energy (about 70% in 1983) is obtained by recycling the chemicals used for chemical pulping. After pulp digestion, a "black liquor" consisting of lignin, spent chemicals, and water is drained off. In a recovery boiler, the lignin is burned to generate steam at the same time that the sodium compounds used to separate lignin from cellulose are recovered for reuse. Between 1972 and 1983, energy from chemical recovery increased substantially, but the main new supply of energy was obtained from "hogged fuel" or timber removed from the forest at the time of primary pulp logging, which in the past had been left as trash.

The 14% improvement in overall energy-efficiency cannot be attributed to particular stages of production or to highly visible investments. Unlike in the steel industry, it did not occur because of increased product recycling (which eliminates energy for pulping) because the share of recycled paper remained roughly constant. Perhaps the only identifiable new technology in the last decade directly related to energy costs is computer-based process and boiler controls, which increased fuel-efficiency options across the board.

Recent data from the American Paper Institute show that the rising trend of energy's share in pulp and paper industry costs has been reversed as a result of fuel switching investments and the recent decline in premium fuel prices.[7] As a result, and as a result of previous exploitation of least-cost options, investment in fuel-efficiency improvements and fuel switching is likely to decline. Current industry literature suggests that new investment will not affect energy use significantly except for electricity.

The share of electricity in total energy use is likely to continue its long-term rise, due to demand growth for coated paper and the industry's desire, given high interest rates and large investments in plant and equipment, to reduce capital costs. The market for coated paper, which has a much shorter lifetime than fine writing paper, is growing rapidly because it costs less and because it can be used for color reproductions in news media. It costs less because it can be made from the whole tree, not just cellulose, and because, compared with conventional chemical pulping, it cuts capital costs per ton of product in half.

In addition, federal regulations concerning cogeneration have increased the attractiveness of that option for the industry. Although the paper industry has been the main supplier of cogenerated electricity in the past, in 1982–1983 total cogenerated electricity by SIC 26 increased by 9%, and current industry literature suggests it will expand rapidly in the future. Since fuel consumed by the cogeneration plant will now be charged to

[7] Personal communication, Ron Slinn, Energy Specialist, American Paper Institute. Energy costs as a percentage of sales were 6.9% in 1972, and 13.8% in 1982. In 1984, preliminary data indicated that this share would fall to about 12%.

Table 9 Chemicals: three-digit SIC groups

	Value added (% share)	Fuel use (% share)
Inorganic (281)	12	11
Plastics and synthetics (282)	13	8
Drugs (283)	19	a
Soaps, cleaners, toilet goods (284)	17	a
Paints and allied products (285)	5	a
Organic chemical (286)	20	62
Agricultural chemical (287)	8	15
Miscellaneous (289)	7	a

[a] Aggregated fuel use for the four groups together equals 5%.
Source: (12).

paper, and not to public utilities, total fuel-intensity for paper may increase as a result of this new economic trend.

Chemicals (SIC 28)

Among two-digit manufactures, chemicals (SIC 28) consumes by far the largest amount of fuel, with 27% of total industrial use and about 33% of industrial energy from premium fossil fuels (Table 8). Among energy-intensive basic materials industries, chemicals has by far the largest value added, and it was fifth among all two-digit manufactures in 1982.[8] In 1983, value-added SIC 28 was just under $60 billion, which is 60% larger than primary metals (SIC 33) and more than twice the output of pulp and paper (SIC 26) and petroleum refining (SIC 29).

In this very large sector, energy-intensity is uneven. In 1981, four of the eight three-digit subsectors [industrial organic chemicals (SIC 286), inorganic chemicals (SIC 281), plastic materials (SIC 282), and agricultural fertilizers (SIC 287)] used 96% of fuel (energy and feedstock) in 1981, while contributing only 54% of total value added (Table 9). Industrial organic chemicals (SIC 286) alone consumed over 60% of total fuel while producing only 20% of valued added.[9]

[8] SIC 28 includes all establishments that manufacture products predominantly by chemical processes. Three general product classes are included: 1. basic chemicals, such as acids, alkalies, salts, and organic chemical feedstocks; 2. synthetic fibers, plastics, and pigments used as industrial inputs; 3. finished chemicals for final sale, such as paints, fertilizer, and explosives (4).

[9] Total fuel consumption for SIC 286 is also underestimated in the census data because of the classification of petrochemicals processing in integrated oil refineries as refining (SIC 29) rather than SIC 286. Hence, the actual share of SIC 286 may be several points higher and total fuel consumption by SIC 28 somewhat higher than census estimates.

SIC 28 has also grown more rapidly in the last three decades than the other energy-intensive manufactures, and if that growth should continue it would substantially influence future growth in industrial energy use. Between 1950 and 1973, output (value added) of chemicals grew at average rate of 5.2%, compared to 3.2% for paper (SIC 26), 2.5% for refining (SIC 29), and −0.3% for primary metals (SIC 33). Since 1973, these growth rates have been generally lower, but chemicals still lead with 2.2% average annual growth, followed by paper at 1.2%, refining at −0.9%, and primary metals at −4.6% (Table 5).

This relatively rapid growth may be explained by the recent development of many chemicals and chemically derived materials (e.g. plastics and pesticides). Most paper, steel, and oil-based fuel products were far along in their commercial life cycles before most synthetic chemicals and fibers were discovered; and many of the latter found market niches where rapid growth and large profits could be sustained over several decades.

Besides contributing to industrial energy use, dynamic chemicals markets have other implications for fuel consumption and related analysis. Rapid growth and high profits support rapid capital turnover. Compared to the other three industries, chemical processing plant and equipment are typically more up to date, including the latest options for using expensive premium fuels and feedstocks efficiently, and chemical firms also have large budgets for research and development. Furthermore, compared to steel and paper, in the chemical industry oil and gas fuels and feedstocks account for much higher shares in total cost, and therefore this industry (along with petroleum refining) has for a long time made fuel-efficiency a major business objective. (See AMMONIA and ETHYLENE, below.)

On the other hand, these dynamic market conditions make it difficult to collect accurate historical data, because complex patterns of technology and product markets shift rapidly,[10] and because firms are reluctant to share information that might give competitors an advantage. Consequently, data limitations as well as the sheer number of technically distinct chemicals suggest that we focus on illustrative examples. Two have been chosen: ethylene and ammonia.

AMMONIA[11]   Until 1983, ammonia production (by weight) exceeded that of all other chemicals derived from premium fuel feedstocks, and until then,

[10] The petrochemicals industry involves complex branching, with many alternative pathways to produce the same end product and tens of intermediates as well as hundreds of final products traded as commodities. The best source of historical data has been the Annual Survey of Manufactures, but the Census Bureau has had a difficult time separating energy consumption for organic chemicals (SIC 286) from energy consumption for petroleum refining (SIC 29, see below), since large refining firms are often integrated downstream into chemicals.

[11] Ammonia combines three molecules of hydrogen to one of nitrogen (NH_3). Nitrogenous fertilizers account for 80% of total ammonia production. Resins, fibers, and plastics used

ammonia was among the top five chemicals overall (13, p. 35). Between 1981 and 1983, output declined by 30% because of a sharp fall in sales of agricultural fertilizer.[12] Domestic production is also strongly affected by imports from less developed nations, which convert natural gas into ammonia as the most profitable way to export plentiful natural gas resources.

Ammonia is made by direct synthesis of hydrogen from methane (natural gas) with nitrogen from the atmosphere (14, 17). The process occurs in three steps: 1. steam reforming of methane under pressure to make a synthesis gas of hydrogen and carbon monoxide; 2. CO shift and removal of CO_2 and other impurities by refrigeration and distillation; 3. compression (to 80–1000 atmospheres) and heating (750–1000°F) of pure H and N mixture (with catalysts) for conversion into liquid ammonia. In virtually all plants, natural gas is used for process heating as well as for hydrogen feedstock, with gas inputs accounting for about 60% of total production costs. In 1980, 0.67 quad of gas was used for ammonia (both energy and feedstocks), and a much smaller amount of electricity (15).

ETHYLENE   By weight (28.6 billion pounds in 1983) and by market value ($6 billion in 1983), ethylene exceeds by far any other basic petrochemical (13, 16).[13] Sixty-five percent goes into plastics; 10% into antifreeze; 5% into fibers; and 5% into solvents (16). As with most petrochemicals, production moves up and down with the economy, with percentage gains and losses that amplify GNP movements. For example, while the GNP declined by less than 2% in 1981–1982, and grew by less than 5% in 1982–1983, ethylene production fell and then grew over the same two-year period by about 17% (1, 13). Furthermore, although ethylene itself is in little danger from foreign competition (as a gaseous material, it is usually too expensive to ship long distances), a variety of derivative products have started to be produced for large-scale export from nations rich in oil and gas.

Ethylene is made together with other olefins (propylene and butadiene) and other coproducts (mainly benzene) via thermal (steam) cracking of feedstocks, which range from (light) ethane (C_2H_3) to (heavy) gas-oils (C_5 and higher) (16, 17). Each feedstock yields a different product mix; and to a lesser extent, product mix also varies with plant configurations and with

---

primarily in construction and automobiles account for an additional 15%, with the remaining 5% used in the production of explosives for mining and military applications.

[12] Ammonia is used as fertilizer either directly as anhydrous ammonia or with nitric acid as ammonium nitrate. Because of federal farm programs and generally low agricultural profits during the last two years, acres planted and fertilizer applications have both declined, reducing sales of nitrogen fertilizer.

[13] Although ammonia is derived from a hydrocarbon, natural gas, it is not considered a petrochemical since it contains no carbon.

equipment sold by different vendors. Most plants today have considerable feedstock flexibility, so the mix of petroleum-based ethylene feedstocks depends on the relative prices of alternative feedstocks and coproducts. Since thermal cracking is a basic operation in petroleum refining, olefin production is often integrated into large petroleum refineries, where feedstocks can be varied most easily with market conditions.

Ethylene is made by the reaction of steam and hydrocarbon feedstock, followed by thermal cracking. The resulting product mixture is cooled to $-150°F$ and compressed to 450–600 psi, after which ethylene is distilled from its coproducts. Because of the complex interaction of feedstock chemistry, process heat requirements, and coproduct accounting, the Btu implications of changing ethylene feedstocks are unclear. However, generic process cost calculations indicate that energy and feedstocks account for about 75% of total costs (18).

While it is impossible to summarize SIC 28 in a straightforward flow chart, these two examples illustrate the characteristic sequence of feedstock processing: (a) initial heating (under pressure or not) to break down feedstock into building-block molecules, (b) separation of building blocks and recombination in correct proportions, (c) forming product molecules under heat and pressure, and (d) cooling and distillation to separate product and by-product streams. Steam is commonly used for process heat and pressurization; for the entire industry, steam accounts for over 40% of total fuel use. Feedstocks account for the other very large share, 37%, with machine drive and direct process heat each at around 10% (8).

RECENT FUEL USE PATTERNS AND TRENDS   For the entire chemicals sector, the shares of oil and natural gas are each around 40% of total energy use (12). Oil consumption accounts for over 90% of feedstock inputs for petrochemicals.[14] Over 70% of gas consumption is for energy inputs, process heat, and steam. Electricity (9% of total Btu inputs) is used throughout the industry for motor drive, but it is also concentrated in a small number of inorganic chemical processes. Coal, which is used to raise steam, is also concentrated, in this case in large petrochemical and plastics plants.

To better illuminate fuel consumption prospects, SIC 28 can be disaggregated into four three-digit groups: 281, 287, 286 and 282, and the rest. Besides producing distinct products, each group has distinctive fuel services, fuel mix, and product growth patterns.

Industrial organic chemicals (SIC 281)   Industrial inorganic chemicals includes an extremely diverse group (4), but fuel use and product growth

---

[14] Because of the problems with census data noted above, the oil share may be several points higher than reported.

patterns can be generally described. Natural gas (41% of 0.6 quads of total fuel) is used for direct process heating and small boilers. On the other hand, electricity (38% of total energy) is consumed mainly in two distinctive processes, refining natural uranium into fissionable nuclear materials and electrolysis of salt brine to produce chlorine and caustic soda (NaOH). The production of fissionable nuclear materials has been growing much more rapidly (at about 6% annually from 1972 to 1981) than that of other inorganic chemicals.

Agricultural chemicals (SIC 287)   Natural gas dominates fuel consumption for agricultural chemicals (90% of 0.8 quad total) because nitrogenous fertilizers account for 87% of total fuel use, and the latter consumes natural gas almost exclusively for energy and material feedstock. (See AMMONIA.) Some gas is used also for miscellaneous material drying, but 53% of total fuel for SIC 287 is ammonia feedstock and 37% is heat and power for ammonia processing (15).

Since the hydrogen in methane and the nitrogen from the air are highly reactive, practically no efficiency gains can be expected in the future for feedstocks, but substantial energy-efficiency improvements would occur if new catalysts were developed to allow ammonia synthesis at atmospheric pressure (14). Perhaps more important than technical change, total fuel demand is likely to be strongly influenced in the future by fluctuations in US fertilizer demand and by the availability and price of ammonia imports (19).

Industrial organic chemicals, and plastics materials and synthetics (SIC 286 and 282)   Industrial organic chemicals (SIC 286) and plastics materials, and synthetics (SIC 282) are combined because the latter group is the primary consumer of organic chemicals. Together they account for 95% of the oil consumed by SIC 28. Oil accounts for 65% of total fuel use by these two chemical groups, most of it as raw material feedstocks (8, 12).

In the period 1972–1981, output of these two groups grew at an average annual rate of 2.4%, compared to 3.5% for the entire SIC 28. Although the particular end points of this period may give a downward bias, this relatively low growth rate nevertheless suggests that growth prospects for domestic petrochemicals and their material derivatives may be declining after several decades of leading the expansion of SIC 28 (20).

Other chemicals   The rest of SIC 28 chemicals consume less than 5% of total fuel but account for 46% of value added (Table 9). They include, however, biologically engineered products and other high value-added goods, all based on relatively new technology, which have the best growth prospects for chemicals.

Overall fuel efficiency trends   The diversity of products and the complexity of production operations make simple, transparent calculations of fuel-

Table 10 Chemicals and allied products: SIC code 28

	1972 percent of total	1982 percent of total
Electricity	21.24	29.27
Natural gas	49.73	41.08
Propane	0.06	0.06
LPG	0.07	0.05
Bituminous coal	8.64	10.17
Anthracite coal	0.18	0.09
Coke	0.21	0.17
Gasoline	0.03	0.03
Distillate fuel oil	1.08	0.51
Residual fuel oil	4.97	3.28
Petroleum coke	0.02	0.02
Purchased steam	3.52	3.14
Other	10.27	12.13
Total	100.00	100.00

Source: (21).

efficiency improvements impossible. However, other sources of information can be used. First, the Department of Energy's Office of Industrial Programs (DOE/OIP) publishes fuel use for energy data from a partial survey of chemical manufacturers (21).[15] Second, a variety of authoritative articles and opinions from industry experts are available to help at least sketch fuel efficiency trends.

The DOE/OIP has monitored energy efficiency improvements between 1972 and 1982 for a group of 120 large chemical manufacturers.[16] Over the 10-year period 1972–1982, the share of premium fuels of the reporting companies fell from 56% to 45% (Table 10), with a 24.8% improvement in overall fuel efficiency. This data is corroborated by one major company's internal assessment that the chemicals industry overall reduced the ratio of

[15] These data collection and reporting are authorized by Sections 371–376 of the Energy Policy and Conservation Act (EPCA) of 1975. Under EPCA, the DOE/OIP collects data from all corporations using over one trillion Btu per year in one or more of the two-digit SIC codes 20–39. That law also required that 1980 energy-efficiency improvement targets be established for the 10 highest energy using manufactures, including the three discussed here.

[16] Fuel consumption data for energy in chemical processing are collected for DOE mainly by the Chemical Manufacturers Association (CMA). Companies must report current, companywide fuel use, as well as using certain methodological guidelines to compute what fuel use would have been if technology and management had not changed since 1972 (or, for a small group of firms, 1978). The latter computation involves many complex judgments that make the comparison over time less accurate than if fuel use surveys had been taken at both the beginning and the end.

energy to feedstock inputs from 1.5 in 1972 to 1.1 in 1982 (22). Another major company reported using 97% of its 1972 fuel consumption in 1982, while constant dollar output increased by 37% (23).[17] A final fragment of data comes from a major builder of petrochemical capacity, which markets retrofits for olefin plants built in the early 1970s on the basis of projected energy savings of 30–40%.[18]

Improvements in feedstock-efficiencies have been much smaller, perhaps in the range of 5–7%, over the last decade.[19] This relatively small gain may be the result of process optimization prior to 1973 to minimize feedstocks, not energy inputs, because feedstocks generally account for a much larger cost share (see AMMONIA and ETHYLENE above). Although this is speculative, such a general explanation would explain the fact that we could find many substantial examples of energy-efficiency improvements in chemicals but only a handful of relatively small improvements in feedstock-efficiency.

While these data sources, other than the census, are not independent of the industry itself, we believe the data indicate at least the technical opportunities available over the past decade for energy-efficiency improvements. For example, raising furnace temperatures and shortening residence times have been reported by industrial contacts to yield comparably large ethylene savings. Furthermore, over the past decade, firms in this industry have had large financial incentives to exploit technical opportunities, because premium fuel cost shares are highest in this industry. Furthermore, despite the technical improvements made, incentives to increase premium fuel efficiency (for both energy and feedstock) and to switch fuels remain strong because fuel costs remain critical.[20]

Petroleum and Coal Products (SIC 29)

This discussion of petroleum and coal products (SIC 29) will focus entirely on petroleum refining because the latter accounts for 94% of fuels consumed in SIC 29 (4). The refining industry refines crude oil into eight major product categories (Table 11). The industry is dominated by large, vertically integrated firms that produce much of their own crude oil and sell

---

[17] In this companywide calculation, product mix adjustments as well as process improvements explain efficiency improvements.

[18] Private communication, S. B. Zdonik, Stone and Webster Engineering, Boston, Mass. Olefins include ethylene and propylene.

[19] Informal estimate by W. W. Reynolds, Shell Oil Company.

[20] W. W. Reynolds of the Shell Oil Company estimates that 70–80% of current cost-effective technical options for raising fuel-efficiency have been exploited by the chemicals industry. However, he believes that 50% of the efficiency improvement since 1972 was based on technologies developed since 1972, and thus he expects technical developments to continue to open up new opportunities for fuel savings.

Table 11 Petroleum refining product slate

Product manufactured	Percentage of total 1980 production	Definition of product and uses
Gasoline	39	A refined petroleum distillate, normally boiling within the ranges of 30° to 300°C, suitable as a fuel in spark-ignited internal combustion engines.
Distillate fuel oils	18	A general term meaning those intermediate hydrocarbon liquid mixtures of lower volatility than that of kerosene, but still able to be distilled from an atmospheric or vacuum distillation petroleum refining unit. Used as boiler fuel in industrial applications, and as home heating fuel.
Residual fuel oil	15	The material remaining as unevaporated liquid from distillation or cracking processes. Used mainly as boiler fuel in power plants, oceangoing ships, and so forth.
Aviation jet fuel	6	Specially blended grades of petroleum distillate suitable for use in jet engines. These fuels have high stability, low freezing points, and overall high volatility.
Petrochemical feedstocks	5	A broad term encompassing those refinery products, having typically low molecular weight and high purity (e.g. ethylene, propylene, and acetylene), which are used as feedstocks in chemical production of everything from food additives to textile fibers.
Liquefied petroleum gases	4	Light hydrocarbon material, gaseous at atmospheric pressure and room temperature, held in liquid state by pressure to facilitate storage, transport, and handling. Consists primarily of propane and butane. Used in home heating.
Kerosene	2	A refined petroleum distillate, intermediate in volatility between gasoline and heavier gas oils used as fuels in some diesel engines. Often used as home heating fuel.
Other products	11	Includes items such as petroleum coke, petroleum solvents, lubricating oils and greases, asphalt, and the like.

Source: (26, 27).

most of their refined products directly to final consumers. As noted above, large refining installations also typically produce petrochemicals.

Market trends in oil refining have been dominated by the availability and price of crude oil.[21] Until the early 1970s, domestic crude prices and refinery acquisition costs were held down by apparently elastic supplies of inexpensive imports. Around the time of the Arab oil embargo of 1973–1974, the price of imported crude skyrocketed and the domestic refiner-acquisition price of crude oil more than doubled (in constant dollars) (2, p. 131). The Islamic revolution in Iran precipitated another doubling of the acquisition price in 1978–1981. As a result of this large price increase, product demand has declined substantially. Production of refined products peaked at 17.3 million barrels per day in 1973; declined for two years to 16.3 million barrels per day; rose again to 18.9 million barrels per day in 1978; and then declined to 15.7 million barrels per day in 1984 (2, p. 101).

Furthermore, over the last decade the product mix has changed, in response to market demand and federal environmental regulation, which caused an increase in refining energy requirements. The market response involved the substitution of nonpetroleum fuels for residual fuel oil in large utility and industrial boilers. Since residual oil requires the smallest amount of molecular change in crude oil feedstocks, the reduction of demand for residual oil has forced refiners to do more cracking of large, heavy carbon molecules into smaller, lighter molecules. Furthermore, federal regulations call for the eventual elimination of lead-based gasoline additives as octane-boosters, and these have had to be replaced with highly refined oil-based additives. Regulations have also limited the amount of sulfur in fuel oils, which must be removed by an energy-intensive process of replacing sulfur with hydrogen.

A final market trend, involving the mix of crude oils available to domestic refineries, also has increased energy use in refineries. Crude oil resources, domestic and worldwide, are differentiated by weight and sulfur content. Light, low-sulfur reserves are easier to refine, so they tend to be produced first. As the resource base is depleted, the remaining reserves tend toward heavier crudes with relatively high sulfur contents. Although new discoveries are continually made, the depletion of known resources is clearly evident and the weight and sulfur content of crude supplies have been rising steadily. Both characteristics increase the amount of energy necessary to produce a product slate that, as discussed above, has an increasing share over time of light, low-sulfur products.

The process of refining crude oil is more complex than paper and steel

[21] All major refineries are located near domestic crude supplies or port facilities that can handle large crude oil carriers. No major new refineries have been built since 1965.

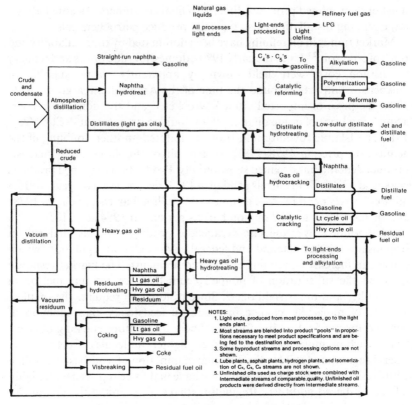

Figure 2   Complex petroleum refinery configuration. Source (3).

production but, unlike chemical manufacturing, the entire process can be illustrated in a sequential process diagram (Figure 2). The energy-using characteristics of a modern integrated refinery can be illustrated more easily (see Figure 3).

RECENT FUEL USE PATTERNS AND TRENDS   The fuel mix for the entire SIC 29 can be obtained by combining 1981 census data on purchased fuels with an independent estimate for captive fuels (6, 12). The main captive fuel is refinery or still gas, but a second, petroleum coke, is growing rapidly. Both are by-products or residuals from refining activities, and together they account for 62% of total fuel use. Still gas has limited value outside of the refinery complex because of its variable composition. Petroleum coke, on the other hand, is a premium solid fuel, sold mainly to specialty steel makers, but the bulk of it has greatest value at the refinery, where it has negligible transportation costs. Coke supplies are growing rapidly, as the

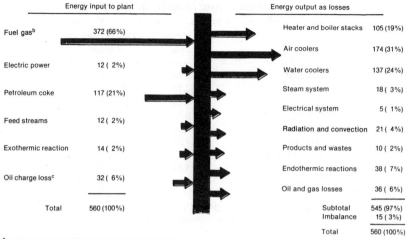

Figure 3   Modern refinery energy profile. Source (3).

product mix shifts toward lighter fractions produced by "coking" heavy bottoms. Natural gas accounts for the bulk of noncaptive fuels and 30% of total fuel use.

Since the product mix is changing over time, and since different products have different energy requirements, data available from independent sources do not allow definitive measurement of trends in fuel-efficiency. However, under the same legislative authority applied to all major manufacturers, the Department of Energy's Office of Industrial Programs (DOE/OIP) surveys the largest refiners in cooperation with the American Petroleum Institute (21). In 1972–1983, the DOE/OIP reports about a 10% drop in refining energy per barrel of crude processed. Taking into account the decreasing quality of crude feedstocks and the upgrading of the product slate over this eleven-year period, the large refiners in this sample estimate that the overall efficiency improvement exceeds 25%. Furthermore, the share of marketable premium fossil fuels declined over the same period from 44% of total fuel consumption to 32%. Electricity, on the other hand, increased its share from 6.5% to 9% (Table 12).[22]

[22] As always with energy data from different sources, we must be careful to distinguish the conversion factor used to convert kWh of electricity to Btus. The Census Bureau uses the actual Btu content of the electricity for the customer, 3412 Btu/kWh. The DOE/OIP also uses this factor for all industries except chemicals and petroleum refining. Since the Chemical Manufacturers Association and the American Petroleum Institute use 10,000 Btu/kWh, which includes generation and transmission losses as well as electric energy itself, the DOE/OIP made an exception and used this much larger conversion factor for just these two industries.

Table 12 Petroleum and coal products: SIC code 29

	1972 percent of total	1982 percent of total
Electricity[a]	6.46	8.94
Natural gas	33.16	22.09
LPG	1.25	1.48
Distillate fuel oil	0.82	0.46
Residual fuel oil	8.26	6.32
Petroleum coke	14.25	15.81
Purchased steam	1.12	0.74
Other	34.51	43.95
Total	100.00	100.00

[a] For SIC 28 and 29 only, DOE/OIP reports electricity at 10,000 Btu/kWh. All other sectors use conversion rates of 3,412 Btu/kWh. Source: (21).

SUMMARY—PAST TRENDS AND THE FUTURE

Most of our analysis on industrial energy use occurred in 1981 and 1982, when the industrial sector was severely depressed. Premium fossil fuel prices seemed to have stabilized, but at that time it seemed only a brief respite from long-term real inflation. In this circumstance, our analysis showed that industry was emphasizing energy-related equipment retrofits and managerial initiatives because they required small investments when capital budgets were extremely tight.[23] The largest investments identified, which can be directly related to energy costs, involved switching boiler fuels from #6 residual oil and natural gas to solid fuels (wood or coal). Other large investments were being made in petroleum refining and steel, with important energy implications, but energy costs were not their primary driving forces.

At that time and more recently, it is apparent in basic materials processing industries that the least expensive options for reducing energy costs have been exhausted, and that major improvements in energy efficiency or major fuel switches in the future will require large investments in new capacity. The outstanding questions now involve how fast new capacity will be built in these industries, and how fast the mix of products will shift away from energy-intensive commodities.

As oil prices continue to decline, a question also arises about the

[23] Petroleum refining was the exception. Vertically integrated with crude oil production and product sales, refiners had access to very large corporate profits derived from oil price increases.

attractiveness to investors of technologies that save premium fossil fuels. The answer would seem to depend upon the economic lifetimes of plant and equipment. While it may make sense for light manufacturers to make short-term investments that take advantage of falling oil (and gas) prices, it seems highly unlikely that manufacturers of basic industrial materials will replace or expand their energy-intensive capacity, which should operate for 15–20 years, without minimizing the long-term risks of increasing premium fossil fuel prices.

More complete answers to these questions, about the rate of investment and the response to falling premium fuel prices, remain problematic, but the preceding discussion has identified certain technological and market developments that will undoubtedly make new capacity in basic industry more energy-efficient than current capacity and also less dependent on premium fossil fuels. Examples include continuous casting and electric arc furnaces in steel, coke-fired boilers in oil refineries, coal-fired boilers in large chemical processing plants, and hogged fuel boilers and thermomechanical pulping for paper. Given long-term expectations of rising relative prices for premium fuels, these long-term investments will continue to be economical whenever new capacity is built.[24]

With regard to oil-fired boilers, we do not expect substantial conversion to coal as long as relative oil and gas prices remain stable or grow only very slowly in the near future. This type of coal switching on a large scale will require a substantially larger price incentive because it involves substantial capital expenditures (in materials handling and transportation facilities besides the boiler itself) and managerial initiatives that typically are not considered unless absolutely necessary to stay in business. Rapid acceptance of atmospheric fluidized bed combustion, particularly if driven by acid rain legislation and/or cogeneration prospects, may change this picture.

Certain product and industry growth trends also promise less dependence on premium fossil fuels because of independent technological and market forces as well as the delayed effects of energy price inflation since 1983. Such trends include 1. the life-cycle maturation of basic petrochemicals coupled with the outstanding growth expectations for biologically engineered products and other chemicals with relatively high value added and small fuel requirements, 2. continued improvements in automobile fuel-efficiency, which reduce growth in oil refining as well as steel (because less steel is used for lighter automobiles), 3. continued development and rapid

---

[24] Improvements in the energy-efficiency of these basic materials processing industries are also likely to continue because of a generation of engineers and corporate management made energy-conscious during the last decade, and because of built up experience with efficient process designs. This base of human resources has its own inertia that is likely to influence investment choices for many years.

market penetration of electronic information media, which can limit paper demand, and demand for other basic materials used in personal transportation, 4. continued market penetration of energy-intensive imports (e.g. petrochemicals and steel) from nations that have economic advantages in the related natural and human resources, compared to the US advantage in advanced technologies.

Listings of particular investment technologies and product market trends do not, however, do justice to the diversity of factors that will influence future fuel consumption. Future prospects can be substantiated more comprehensively in terms of two current models of industrial energy economics, EEA/ISTUM and EIA/PURHAPS. EEA/ISTUM (Energy and Environmental Analysis, Inc./Industrial Sector Technology Use Model) (8, 24) explicitly simulates industrial process technologies and investment decisions. EIA/PURHAPS (Energy Information Administration/ Purchased Heat and Power System) (25) projections are based on econometric functions estimated from historical relationships between fuel prices, fuel use, and industrial production (by two-digit SIC groups). Both models use macroeconomic projections for GNP, product demands, and energy prices as input assumptions in projecting industrial fuel consumption.

Compared to historical data, both models suggest that the decoupling of energy from output growth, which occurred after 1973, has substantially altered the relationship that existed before 1973 (Table 13). Both projections indicate that, if industrial growth continues its recovery from the stagnation of the past decade, then total industrial energy use will grow as well, but at a slower rate. Growth in consumption of premium fuels will or can be much slower. In other words, both models suggest that there remains considerable substitutability between fuels and capital in the industrial sector, both directly as process improvements and indirectly when markets move away from fuel-intensive products.

Table 13   Energy/output growth rates: the past and future projections

Time Interval/Model	Growth Rates (%)			
	Value added	Total energy	Premium fuels	Electricity
1950–1973	3.7	3	4.3	7
1973–1983	0.8	−1.7	−3.0	1.3
1985–1995 (EIA/PURHAPS)	3.1	2.0	1.3	4.3
1980–2000 (EEA/ISTUM)	2.7	1.0	0.1	2.4

Source: (1, 2, 5, 8).

Each model result requires considerable interpretation. ISTUM is relatively optimistic about fuel-efficiency improvements and fuel switching, because it assumes higher premium fuel prices and projects coal switching entirely on the basis of marginal cost calculations. Coal switching in this projection allows premium fuel consumption to remain relatively flat despite an assumed 4% growth in demand for organic chemicals. On the other hand, ISTUM simulation of industrial investment does not attempt to anticipate future technological developments that could substantially increase profits in fuel-efficiency and switching.

The econometric equations in EIA/PURHAPS may be more realistic because they are based on actual market behavior. Both models assume real price increases for premium fuels of 3–5%, but PURHAPS starts this relative growth from current prices, not the higher real prices that existed in 1980.[25] However, econometric estimation procedures have limited validity when based on short time series that include recent macroeconomic discontinuities, as well as sharp fuel price fluctuations. Consequently, the temptation is toward simple trend analysis, which can be misleading, especially when oil prices head downward after rising dramatically during the period when structural parameters were estimated.

Nevertheless, by pre-1973 standards, these two rather different models suggest a more relaxed relationship between future growth of industrial output and fuel consumption. There is definitely still a positive relationship, but in terms of public and private planning to meet energy needs, industrial fuel requirements definitely appear to be manageable.

This conclusion does not mean that industrial energy demand will never again grow in step with GNP. During periods of rapid economic expansion, industrial energy use could indeed grow rapidly as well since the leading activities of construction and capital formation involve energy-intensive, industrial goods and services. This possibility becomes the more likely as long as oil and gas prices decline or remain steady in real terms. A major uncertainty in all energy use projections are the prices of oil and gas.

On the other hand, these modeling projections reinforce the picture of technical and economic flexibility described in some detail in this article. Although there are important instances where industrial growth and technology change tend to increase energy intensity (such as rapid growth in demand for coated paper products, which use electricity, and increased refining requirements for converting heavier crude oils into lighter

---

[25] ISTUM process technologies were calibrated using 1976 Census of Manufactures data. Simulation begins in 1980 to minimize ad hoc calibrations to account for process changes that happened after the census.

petroleum products), industrial energy demand has in general considerable flexiblity to grow more slowly than industrial output.

This flexibility includes the premium fossil fuels, oil and natural gas. An important uncertainty involves fuels used for chemical feedstocks, but major technical opportunities exist to switch to coal in the production of steam, to switch to coal and electricity for direct process heat, and to increase process heating efficiencies when premium fuels must still be used. Beyond these substitutions, which typically involve heavy or otherwise expensive capital equipment, perhaps the most important opportunities (to reduce dependence on premium fossil fuels) arise from market driven product substitutions and process change.

Product mix trends were discussed in the first section of this paper because they indicate the broad, economic structure in which energy is just one among many important factors. Although this article cannot deal with the full range of technical and economic factors which determine the mix of products and the process technologies, the discussion is sufficient to suggest that the most rapidly growing industries and processes do not emphasize traditional, materials-intensive activities. Rather, they exploit new technologies (e.g. information processing, communications, and recombinant DNA) which use relatively little oil and natural gas. Continuation of this trend could increase the gap between the growth rates of industrial output and industrial fuel consumption.

Finally, to the extent that the US economy can indeed build upon new technologies, reliance on foreign sources for fuel-intensive goods (such as chemicals and refined petroleum products, which presently account for most industrial oil and gas consumption) may significantly reduce direct consumption of premium fuels. However, it also raises new issues of US dependence upon foreign trade and, since this is the direction in which the economy may be heading, future analysis of industrial energy requirements should include questions about the structure of international trade in fuel-intensive industrial commodities.

ACKNOWLEDGMENTS

The authors would like to thank the large number of contractors, advisers, and industry experts who contributed to the *Industrial Energy Use* report of the Congressional Office of Technology Assessment, the larger publication which provided most of the material for this article. We would also like to thank our colleagues at OTA who worked with us on that project. We would especially like to acknowledge the help of Samir Salama of Energy and Environmental Analysis, Mark Ross of the Alliance to Save Energy and the University of Michigan, Paul Werbos of the US Energy

Information Administration, and Joanne Sedor of the Office of Technology Assessment.

Literature Cited

1. US Dept. Commerce, Bur. Econ. Analysis, *Survey of Current Business*, April issues, and *The National Income and Product Accounts of the United States, 1929–76, Statistical Tables*
2. Energy Inf. Agency. 1983. *Ann. Energy Rev.* US Dept. Energy, April, 1983, and Marley, R. C. 1984. Trends in industrial energy use. *Science* 226:1277
3. US Office Tech. Assessment. 1983. *Industrial Energy Use*
4. US Bur. Census. 1972. *Standard Industrial Classification Manual.* Revised 1977
5. Energy Inf. Agency. 1985. *Annual Energy Outlook, 1984* (including unpublished tables), US Dept. Energy
6. Ross, M. 1984. *Industrial Energy Conservation*, Natural Resources Journal, pp. 369–404
7. US Dept. Commerce, Bur. of Econ. Analysis, Unpublished 2-digit data from the National Income and Product Accounts (NIPA)
8. Salama, S. Y., Forshay, P. H. 1982. Energy and Environmental Analysis, Inc., *Impacts of Four Legislative Initiatives on Energy Use in the Industrial Sector*, Report prepared for Pacific Northwest Laboratories, Battelle Memorial Institute, and the US Office Tech. Assessment
9. US Office Tech. Assessment. 1984. *Nuclear Power in an Age of Uncertainty*, Feb.
10. Energy Inf. Admin., *Annual Outlook for U.S. Coal, 1984*, US Dept. Energy
11. Am. Paper Inst. 1984. *1984 Statistics of Paper, Paperboard, and Woodpulp*, New York, Oct.
12. US Bur. Census, *1982 Census of Manufacturers: Fuels and Electric Energy Consumed*, Part 1, June 1983, *1981 Ann. Survey of Manufacturers*, April 1983 and *Preliminary Report: Industry Series*, July 1984
13. Facts and figures for the chemical industry. 1984. *Chem. and Eng. News*, p. 35
14. Young, J. K., Johnson, D. R. 1984. *Ammonia Synthesis, Energy-Use and Capital Stock Information.* Battelle Pacific Northwest Lab., July
15. Samsa, M. A., Hedman, B. A., Solomon, I. J. Status and Outlook for Natural Gas Use as Chemical Feedstock, *Gas Res. Insights*, p. 5
16. Greek, B. F. 1983. Basic petrochemical swing into long-awaited recovery. *Chem. and Eng. News*, p. 13
17. *Hydrocarbon Processing: Petrochemical Handbook '83.* 1983. Vol. 62, No. 11. Houston, Texas: Gulf Publ.
18. Zdonik, S. B., Meilun, E. C. 1983. *Cost and Implications for Feedstocks and Product Flexibility in Olefins Plant*, Stone and Webster Eng., paper presented at AIChE Spring meet.
19. Lyon, S. D. 1975. Development of the Modern Ammonia Industry, 10th Brotherton Memorial Lecture. *Chemicals and Industry*, Vol. 6, Sept.
20. Abshire, A. *Worldwide Olefins Production: Looming Threat*, and Soder, S. *World Feedstocks: Changing Patterns for Petrochemical Production.* Stanford Res. Inst., Inc., Energy Bureau Fall Petrochemicals Conf., September 1984
21. Office of Industrial Programs. 1984. *Annual Report to Congress and the President on 1982 Industrial Energy Efficiency Improvement*, US Dept. Energy, April
22. Reynolds, W. W. 1982. Shell Oil Company, *Feedstock Economics and World Petrochemical Markets*, Energy Bureau Fall Petrochemicals Conference, pp. 3–4
23. Workshop comments of James Borden, Energy and Materials Department, E. I. du Pont de Nemours, Inc., US Office Tech. Assessment
24. Lerner, M., Salama, S., Kothari, V. 1982. ISTUM-2. The Industrial Sector Technology Use Model. Final Report on the Industrial Energy Productivity Project. DOE/CS/40151-1–9
25. Energy Inf. Agency. 1984. *Documentation of the PURHAPS Industrial Demand Model*, Vol. 1, April
26. Am. Petroleum Inst. 1979. *API Data Book*
27. *Natl Pet. News*, Dec. 1980
28. US Office Tech. Assessment. 1984. *U.S. Vulnerability to an Oil Import Curtailment: The Oil Replacement Capability*

Ann. Rev. Energy. 1985. 10 : 201–16
Copyright © 1985 by Annual Reviews Inc. All rights reserved

MEASURED PERFORMANCE
OF PASSIVE SOLAR BUILDINGS[1]

Joel N. Swisher

Architectural Energy Corp., 8753 Yates Drive, Westminster,
Colorado 80030

INTRODUCTION

Passive solar buildings have become commonplace in the United States
during the last ten years, as designers have attempted to improve the energy
efficiency of new buildings. For the last five years, the US Department of
Energy has supported research to evaluate the performance of a variety of
passive solar designs. This article reviews the results of that work.

Passive Solar Principles

A passive solar building is one that directs natural energy flows to provide a
comfortable indoor environment with a minimum of purchased energy for
heating and cooling. Passive solar design requires application of fundamental solar and climatic relationships to orient the building and its
windows so that solar radiation is admitted during the winter and avoided
during the summer. Dense materials such as concrete are located where
they provide thermal storage mass by absorbing solar radiation and storing
it without severe temperature fluctuations.

Buildings that are well insulated and sealed against air infiltration
minimize heat loss to the outdoors and make efficient use of the captured
solar energy. Additional measures are frequently applied to reduce heat
losses, including movable insulation components that are installed on
windows at night.

---

[1] Abbreviations used: LANL, Los Alamos National Laboratory; SLR, solar-load ratio;
SERI, Solar Energy Research Institute; BPI, building performance index; PHR, passive
heating ratio; NSDN, National Solar Data Network; BNL, Brookhaven National
Laboratory.

201

0362–1626/85/1022–0201$02.00

Passive System Types

There are three basic types of passive solar system: direct gain, indirect gain, and isolated gain. These classifications refer to the geometric relationship between the building components that collect and store solar energy and the interior living spaces to which they deliver heat. In a direct gain design, sunlight enters directly into the living space. Energy is absorbed by a massive component, such as a slab floor, and later delivered back to the space through the same surface. Direct gain systems are simple and economical, but they often create large indoor temperature fluctuations, due to the lack of control over heat delivery.

Indirect gain systems include thermal storage walls, or Trombe walls, as well as water walls and roof ponds. Solar energy is absorbed at the outside surface of the massive wall or roof component and delivered to the living space through the inside surface of the same component. The wall thickness can be chosen so that heat delivery lags several hours behind the receipt of the radiation. This provides a measure of passive control and improves indoor thermal comfort.

An isolated gain system such as a sunspace collects solar energy in a space that is separate from the living space. Energy can be delivered by conduction through an intermediate wall or by natural or forced circulation of warm air from the sunspace through the living space. This flexibility allows greater control over heat delivery than in the other system types. A sunspace can contain the primary thermal mass, or it can rely on mass within the living space for additional thermal storage. A greenhouse is a sunspace in which the temperature is more carefully regulated to promote the growth of plants. This function requires more thermal storage and greater use of heat within the greenhouse, thus reducing the amount of energy delivered to the living space.

Of course, these distinctions can become rather ambiguous. A direct gain design can include a massive interior wall that resembles a Trombe wall or a heavily glazed space that is partially closed off from the living space, similar to the sunspace configuration. Also, many buildings contain both direct gain apertures and indirect gain or isolated gain components.

Design Parameters and Calculation Methods

The important design parameters in a passive solar building include the passive system type and its south-facing glazing area, usually expressed as a ratio to the floor area of the heated living space. Thermal capacitance is an important parameter, but it is often difficult to quantify. The building heat loss coefficient, which describes the thermal integrity of the building envelope, is a critical parameter in passive solar design and analysis.

At the research level, building thermal performance is analyzed using detailed hourly computer simulations. The designer, however, must often rely on simpler analysis procedures to evaluate prospective passive system designs. Numerous monthly and annual performance calculation procedures have been developed, based on correlations derived from extensive simulation work. The most widely used is the solar-load ratio method from Los Alamos National Laboratory (LANL) (1). This procedure uses the passive solar aperture area, the building heat loss coefficient, excluding the passive solar aperture, and monthly average weather data to establish the solar-load ratio (SLR). This ratio is used in a series of correlations for several different passive system configurations, to determine the monthly solar savings fraction (SSF), and the predicted annual auxiliary heating requirement. Recent performance monitoring results from passive solar buildings have given us the first opportunity to compare these design tool predictions to actual building performance, and to evaluate passive solar performance in general.

MEASURED PERFORMANCE RESULTS

SERI Monitoring Program

In 1980, the Solar Energy Research Institute (SERI), funded by the US Department of Energy, established three programs to evaluate the thermal performance of passive solar residential buildings at three levels of detail and expense, designated as Class A, B, and C. The Class A program was the most detailed, focusing on individual thermal processes in a small number of buildings. The Class B program, described below, represented the middle level of detail and expense. The Class C program was a noninstrumented evaluation of hundreds of buildings (2).

The SERI Class B monitoring program was a low-cost approach to evaluating the thermal performance of selected residential buildings throughout the United States (3). The goal of this program was to provide a consistent measure of the thermal performance of different types of passive buildings in different climates. Instrumentation was limited to that needed to calculate monthly building energy balances, separating passive solar heating from the other building energy flows. Thermal storage and other individual components were not monitored, and no attempt was made to determine a building's thermal processes in detail.

Instrumentation began in 1981; by 1983, more than 70 buildings had been instrumented. SERI published summaries of the results from these buildings in two volumes, along with comparisons between measured performance results and monthly and hourly design tool predictions (4–7). Additional results were produced by a similar program at the Bonneville

Power Administration (8). Other Class B program reports include documentation of instrumentation, processing, and analysis procedures (9).

The Class B approach features on-site data processing using a standardized microcomputer-based data acquisition system, which collects up to 22 continuous measurements every 15 seconds and processes these values in its microprocessor. The system is programmed to calculate daily and monthly performance factors in real time and print them on a daily basis. The real-time factors computed include the major building energy flows, weather variables, and indoor temperatures. The system also produces magnetic data storage tapes containing hourly averages of the raw continuous measurements, which can be used for graphic data presentation and building parameter estimation using regression techniques.

Building thermal performance calculations are based on a monthly building energy balance. The energy balance is calculated for a single control volume that includes the air of the living space, defined as conditioned space and space that always remains within 5°C of conditioned space temperature. This usually excludes attics and garages and includes basements and sunspaces only when the space is conditioned. The energy balance has four components: one heat loss component and three heat gain components: auxiliary heat, internal heat, and passive heat. Internal heat gains include heat produced by lights, appliances, and people.

The passive heat used by the building is found by subtracting the measured heat delivered by auxiliary and internal sources from the building heat loss, calculated from measured temperature difference. The passive heat includes the effects of all passive heat gains and losses, not just the direct solar gain through the south-facing glazing.

Measured building and heating system parameters are stored in the on-site computer memory and used during the real-time data reduction. The most important values are the long-term furnace efficiency and the building heat loss coefficient. The heat loss coefficient is measured using electric coheating, in which all heat gains are eliminated and the building is allowed to reach a steady thermal state, which is maintained by easily measured electric resistance heating (10). The heat loss coefficient is adjusted in real time, based on the calculated air infiltration rate and the positions of movable insulation components.

Summary of Results

Figure 1 shows the overall heating season performance for 70 buildings in the Class B program. Each bar represents the total building heating load divided into passive, auxiliary, and internal heating components. The energy quantities are normalized per unit floor area and degree-day (based

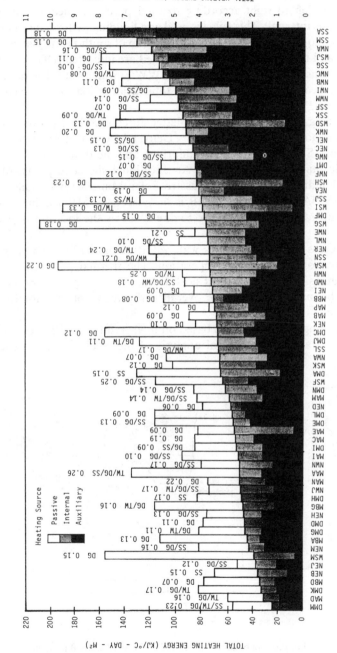

Figure 1   Normalized heating season energy summaries for 70 passive solar buildings.

on measured indoor-outdoor temperature difference). The buildings are ordered from left to right according to total purchased heating energy (auxiliary plus internal). This value is used as a building performance index (BPI). At the top of each bar, the building's ratio of south-facing glazing to floor area and the type of passive system used are indicated (DG = direct gain, SS = sunspace, TW = Trombe wall, WW = water wall). The ratio of the passive heating component to the total heating load is the passive heating ratio (PHR).

The buildings are identified by three-letter site codes. The first two letters denote the region of the country (DM = Denver, MA = Mid-America, NE = Northeast, SS = South, WS = California), and the third letter denotes the specific building in the region. An additional site code prefix, MB, identifies buildings in the SERI Passive Solar Manufactured Buildings Program (MBA is in Denver, MBB is in Wisconsin, MBD is in North Dakota, and MBG is in Virginia). Figure 2 shows the location of each building.

The 70 Class B sites represent a wide variation in building and passive system size, configuration, and operation. The thermal performance of these buildings also varies a great deal. However, some general observations can be made regarding the overall performance of the monitored buildings.

The following conclusions are based on interpretations of the results in Figure 1 and a simple statistical analysis of the results. The statistical

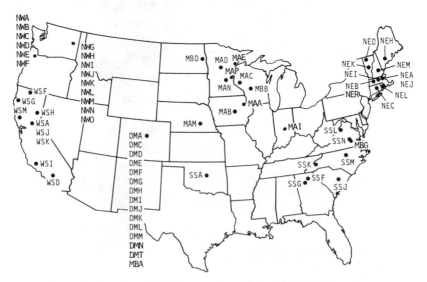

Figure 2 Locations of buildings monitored in the SERI Class B program.

approach involves averaging the results and grouping them according to various categories, such as passive system type and geographical region. The statistical results are shown in Table 1.

First, these buildings have low auxiliary heating needs. Figure 1 shows that the auxiliary heating is generally less than 60 kJ/°C-day-m^2 with an average of 49 kJ/°C-day-m^2. Conventional buildings typically use from 120 to 240 kJ/°C-day-m^2; the average for new construction is 100 kJ/°C-day-m^2; and the proposed Building Energy Performance Standard would allow 80 kJ/°C-day-m^2. The average BPI, which includes internal heat, was 75 kJ/°C-day-m^2. This indicates that internal heating was nearly equal to auxiliary heating on the average.

Solar performance is variable. The passive solar systems contribute a statistically significant portion (39%) of the total heating load, or 55% of the net heating load (total load minus internal heat). Figure 1 shows several buildings with large passive heating contributions, but this does not necessarily translate into energy savings. While many heavily glazed buildings perform well in terms of auxiliary heat (e.g. sites DMK, DMM, MAD, NEB, NEM, WSM), some are disappointing (e.g. sites NEA, SSA, SSM, WSJ).

Although thermal mass levels were not studied in detail, some trends were apparent. Designs with very little thermal mass exhibited large indoor temperature swings and disappointing energy performance. However,

Table 1 Thermal performance statistics

Category	Number of sites	Solar glazing/ floor area	BPI (KJ/°C-day-m^2)	PHR	Mean horizontal solar radiation (MJ/day-m^2)
System type					
Direct gain	37	0.13	78	0.37	9.4
Sunspace	22	0.15	69	0.33	8.2
Mass wall	11	0.19	75	0.39	8.6
Location					
Denver	15	0.13	59	0.46	11.5
Northeast	11	0.14	67	0.32	7.7
Mid-America	11	0.14	59	0.35	8.5
South	9	0.14	100	0.33	8.8
California	9	0.19	81	0.51	10.7
Northwest	15	0.14	89	0.25	6.6
Movable insulation					
Yes	26	0.14	73	0.35	8.9
No	44	0.14	76	0.38	8.9
Total	70	0.14	75	0.37	8.9

extremely massive buildings did not exhibit exceptional performance. The heavily earth-coupled buildings performed poorly, perhaps because an overemphasis on the utility of mass caused the designers to provide too little insulation between the building skin and the ground. Thermal mass effects were shown to be generally insignificant after more than one day (7, 11).

The passive solar houses are comfortable, with an average indoor temperature of 19.5°C. The time-averaged temperature of each site ranged from 12 to 23°C. Many of the buildings experienced large interior temperature swings, especially in direct gain areas. Approximately 20% of the monitored buildings experienced some winter overheating, with indoor temperatures above 27°C. These were mostly direct gain designs with ratios of solar glazing to floor area greater than 0.12. Severe summer overheating was infrequent, and some houses in the South stayed comfortable during the summer with no air-conditioning. Night ventilation was sufficient to prevent summer overheating in most northern sites.

Performance varied significantly by location. The PHR means for California and Denver were significantly larger (at the 95% confidence level) than the means for the other regions. The differences in measured incident solar radiation shown in Table 1 accounted for some of this difference in Denver, while larger solar glazing areas were more important in the California houses. The BPI means for California, the Northwest, and the South were significantly higher than those for the colder regions. There are two reasons for this: the buildings in the colder regions tend to be more heavily insulated; and there tends to be more venting of internal heat in the warmer regions.

None of the basic passive system types (direct gain, sunspace, and thermal storage wall) had significantly better or worse overall performance (at the 95% confidence level). It is interesting to note that several of the sunspace designs used very little purchased heat, but did not have an especially large passive heating contribution. This points out the advantage of using a sunspace for a thermal buffer to reduce heat losses, as well as for a solar energy collection device. As shown in Table 1, the PHR appears to depend more strongly on the ratio of solar glazing to floor area than on the system type. The sunspace average PHR was somewhat biased by the large number of sunspaces located in the least sunny region, the Northwest.

The energy saving effects of insulation and weatherization are critical. Modest solar designs on very tight buildings generally use less heat than more ambitious solar designs on very leaky buildings, and low heat losses increase the fraction of the heating load met by the solar components. Figure 1 shows that the largest purchased energy users (right side of the graph) have relatively high total heating loads; with the exception of a few very heavily solar-driven buildings in Denver and Northern California, the buildings with higher heat loss coefficients required more purchased heat. Some of the smallest users of purchased energy (left side of the graph) were buildings in the Northeast and Mid-America that had moderate passive heating and very low heat-loss coefficients.

National Solar Data Network

The National Solar Data Network (NSDN), which was established to monitor active solar heating systems, also monitored several passive buildings (12). The instrumentation was extensive, with up to 90 sensors at each site. However, building thermal characteristics, such as heat-loss coefficients, were not measured.

Most of the buildings were first generation passive designs with very large glazing areas. Although these buildings may not be representative of current design practice, the general conclusions of the NSDN work agreed with those of the Class B program. The importance of thorough insulation was stressed, and building performance was sensitive to proper occupant operation of components such as movable insulation.

Double-Envelope Results

The double-envelope configuration, a variation of the isolated gain concept, became popular in 1980. Many exaggerated claims were made about its thermal performance, comfort, and costs. A double-envelope house has two insulated roofs and north walls, a crawl space under the floor, and a large south-facing sunspace. A continuous air passage connects the sunspace, the interwall cavity, and the crawl space. This loop is supposed to distribute heat from the sunspace by thermocirculation, storing energy in the ground under the crawl space. Reverse flow is supposed to deliver stored heat at night.

Double-envelope buildings were tested in Rhode Island by Brookhaven National Laboratory (BNL), in Colorado by SERI, in California and Nebraska by the University of Nebraska, in Georgia by the Southern Solar Energy Center, and in Montana by Fowlkes Engineering (13, 14). These monitoring efforts showed low auxiliary heating use, around 40 kJ/ °C-day-m^2. This performance was attributed mostly to the excellent insulation and air-tightness of the buildings, rather than the predicted convection phenomena. Each study showed negligible useful energy quantities stored in the ground, and air flows that rarely reached 0.5 m/s. BNL blocked the air loop and observed no significant change in performance. Under normal operating conditions, thermal stratification resulted in a 10°C temperature difference between the upper floor and the lower floor. This indicates that the collected solar energy was not being effectively stored. The cost implications of double-envelope design are less certain, but it is reasonable to conclude that there is little justification for building two envelopes to obtain the same performance achieved by passive solar buildings with one envelope.

Los Alamos National Laboratory

Los Alamos National Laboratory (LANL) performed much of the early monitoring work, which was mostly concerned with small test cells. The data from this effort were used to validate a computer simulation program that was used to develop simpler design tools, such as the SLR technique. LANL also instrumented several full-scale buildings and focused on specific passive solar components (15). The performance evaluation methodology was not standardized, but instead was tailored to each specific situation.

Several of the LANL sites used a fan-charged rockbed to provide additional thermal storage mass. The rockbed receives excess heat from a sunspace and delivers heat by conduction through the floor of the living space. Rockbed performance was generally disappointing, with excessive conductive heat losses and nighttime convective air circulation between the rockbed and the sunspace.

Additional experiments were conducted at LANL to investigate convective heat distribution in passive solar buildings (16). It was found that a remote room could be adequately heated by two-way convective flow through a doorway to a solar heated area. Similitude experiments and full-scale tests led to a mathematical model of this process and showed that a doorway could deliver 500 W when driven by a 3°C temperature difference. LANL also investigated the effectiveness of convective loops within the primary building envelope, as distinguished from the external circulation suggested by the double-envelope concept. The results showed that natural convection driven by concentrated solar gains, as in a sunspace, could effectively distribute heat through a two-story building. The key design factor is the provision of a clear return flow path down the north side of the building and back into the solar-driven area. In the Class B program, several buildings with large quantities of direct gain glazing on the upper level demonstrated severe temperature stratification. Concentrating solar gains on the upper level establishes a stable temperature profile that prevents effective natural circulation.

Passive Cooling Results

Passive solar research and testing has concentrated on space heating because it is the largest building energy use, especially in the residential sector. Considerable progress has also been made in passive cooling systems, especially in the roof pond research conducted at Trinity University. Roof ponds are indirect gain systems in which the thermal mass is provided by water-filled bags in direct contact with the roof. The horizontal orientation is far from ideal for winter heating. However, it provides a clear view of the night sky for radiant summer cooling. For this purpose, the roof pond must be covered by insulating panels during the day while it absorbs heat from the building.

When supplemented by ceiling fans that deliver 0.6 m/s air motion, this system can provide year-round thermal comfort in residential buildings without mechanical cooling in nearly all American climates (17). The exceptions are Florida, Louisiana, Eastern Texas, and Southern Arizona. Air-conditioning loads are greatly reduced in these areas. Unlike mechanical cooling, passive radiant cooling removes only the sensible cooling load and not the latent load. Thus, any residual cooling load in a humid climate has a large latent component, making it difficult to treat efficiently with conventional air-conditioning. Also, the high relative humidity in a passively cooled building, although within the limits of comfort, can allow growth of mold and mildew. For passive cooling to be practical in humid climates, an efficient dehumidifier is needed (18). This conclusion has spurred the development of a new dehumidifier, based on a vapor compression air-conditioner with sensible cooling recovery by an air-to-air heat exchanger.

DESIGN IMPLICATIONS

Passive Solar vs Conservation

Considerable discussion has centered on the relative merits of passive solar design and energy conservation techniques such as insulation. The results of the thermal performance monitoring show that the best overall performance is achieved by buildings that employ both good solar design and good conservation practice. It can be argued that a so-called superinsulated building can perform as well as most passive buildings regardless of solar orientation. However, passive solar techniques allow much more flexibility in designing for energy efficiency. Passive design allows more liberal use of windows without thermal penalty. It creates warm daytime spaces and radiant wall surfaces that enhance thermal comfort and provide refuges from cold winter conditions. It allows the designer to select the most cost-effective combination of insulation, glazing, mass, and other materials. Recent studies have intended to derive economically optimal ratios of passive solar and conservation investments. These optima, however, are broad, and the designer really has many options that will provide significant energy and dollar savings. As a first step, sound energy conservation practice is essential, and careful solar design can further improve thermal performance while adding to the comfort, aesthetics, and flexibility of design.

Air Infiltration Testing

Pressurization tests were used to determine the effects of air infiltration on the heating loads in Class B buildings (19). Average infiltration rates varied from 0.1 air change per hour to more than 1.0. Some houses were so tight that air-to-air heat exchangers were needed to provide fresh air, and some were so leaky that it was difficult to pressurize them adequately. Air infiltration levels were generally no better than those measured in other American houses of similar ages. Air infiltration represented an average of 29% of the total building heating load, and it was the least certain component of the load because of its variability with weather and other factors. It is important to note that in some houses the infiltration load was as large as the passive heating contribution and was therefore a major determinant of overall energy performance. Greater attention to airtightness would have substantially improved these buildings' performance and increased the fraction of the heating load met by solar energy.

Pressurization testing is useful for comparing the air tightness of a variety of houses, and especially for diagnosing specific leaks in individual buildings. However, combining leakage characteristics with weather data to determine long-term infiltration flows is rather difficult and uncertain. Tracer gas decay tests can be used to measure infiltration rates directly, but these results are even more difficult to extrapolate to seasonal performance. Comparisons between the two methods were performed using the tracer gas test conditions as inputs to the leakage model derived from the pressurization tests. The results were typically higher for the pressurization results, especially at high windspeeds. While this does not significantly affect the seasonal energy balance calculations, it does suggest further work in modeling wind-induced infiltration.

Operable Components

It appears that the occupants' habits in using the building are critical to passive system performance. This applies especially to the use of operable components such as insulation, sunspace doors, and vents. Most of the Class B buildings with disappointing performance had problems with incorrect operation of such components. There are several reasons for this. In some cases occupants were simply inattentive. Some automatic components, such as thermostatically controlled fans, did not operate properly. Several occupants ignored operable components that were not sufficiently simple or convenient to use. This indicates that designers should continue to improve the simplicity and convenience of manual components and emphasize reliability in selecting automatic components.

While the monitored houses generally performed well, there was no significant performance difference between the houses with movable insulation and those without. As shown in Table 1, the two samples have identical ratios of solar glazing to floor area and average solar radiation available. Thus, movable insulation has no apparent performance advantage. This observation is surprising in that it contradicts the results of the extensive performance analysis of passive solar buildings, which predicts substantial benefits for all types of passive systems when movable insulation is added (1).

Individual sites with movable insulation varied in performance from nearly ideal to poorer in net performance than no night insulation at all. This disappointing performance was due to poor operation of the insulation components, not the quality of the materials themselves. Detailed data analysis showed movable insulation to be properly operated about 70% of the time, on average. However, it appears that at least 70% proper operation is necessary to obtain any net energy benefit.

Thus, if movable insulation is called for, the designer might consider motorized or automatic operation, but only if the system is very simple and reliable and the incremental cost is reasonable. It is also important to provide simple and clear instructions to occupants regarding operation of movable insulation and other building features.

Considering the disappointing performance of the movable insulation components that were monitored, consideration of alternative high-performance glazing systems is warranted. The simplest approach to reducing heat losses through windows without operable components is triple glazing. There are, however, newer glazing products available that promise still better performance. A layer of reflective film can cut window heat loss in half, although it also reduces solar transmittance significantly. This is an especially attractive alternative for nonsouth windows, where solar gains are less critical. Also, one or more layers of high-transmittance polyester can provide equal or better thermal resistance with only a slight reduction in transmittance. These products are promising for direct gain and sunspace applications, and they are far more practical than movable insulation in Trombe wall and water wall systems. Thermal storage walls are usually difficult to insulate on the inside, and exterior insulation presents practical problems, such as weathering, ice and snow buildup, vandalism, and space usage. An additional alternative for thermal mass walls is a selective surface such as black chrome, which can deliver equal performance compared to movable insulation with less inconvenience. The key advantage to all these products is that they are completely passive and require no operation by the occupant. They can therefore be expected to operate as designed in most applications.

Auxiliary Heating

As part of the Class B program, 23 brand new gas-fired forced-air furnaces were tested and found to have an average overall delivery efficiency of only 50%. Combustion efficiency was not the major problem; every furnace tested had a combustion efficiency near 80%. However, distribution heat losses from uninsulated ducts in uninsulated spaces had a significant effect on overall delivery efficiency in several systems. Also, furnace sizing has a major effect on delivery efficiency. Overall efficiency was found to be well correlated with heating system sizing. The severely oversized systems were significantly less efficient than the more accurately sized systems. Since accurate furnace sizing depends on accurate heat-loss calculations, infiltration and thermal testing on newly constructed houses may be warranted to facilitate energy-efficient sizing.

Design Tool Comparisons

Designers need simple design tools to quantify the potential performance implications of their design decisions. To use these tools with confidence, while realizing their limitations, the designers should understand the relationship between actual measured building performance and that predicted by the design tools. The Class B monitoring results were compared with predicted results from the SLR method, using measured weather and thermal input data (1, 6). On average, the difference between measured and predicted purchased energy values was only 1%. However, the mean difference for individual houses was 30%. Several corrections were applied to the predictions, using the LANL sensitivity curves, but the results showed no better agreement with measured values. Also, use of historical weather data in place of the on-site measured weather data introduced little additional difference between predicted and measured values. These results suggest that the SLR design tool is most useful in its simplest form. Sites with large indoor temperature fluctuations produced greater differences, and it is likely that other occupant interactions also contributed to the differences. Thus, it is probably unreasonable to expect greater consistency from a simple design tool. Since no bias is apparent in the performance predictions, their use provides an objective analytic aid to the design process. It is important, however, for designers to remember that significant discrepancies result when predictions for an individual house are compared to actual performance.

CONCLUSIONS

The following general conclusions can be drawn from the measured results available to date from passive solar buildings:

Passive solar buildings have very low auxiliary heating needs, less than half of those of conventional new construction, and less than one-fourth of those of the existing housing stock.

Thermal comfort in these buildings is acceptable, with occasional overheating in the heavily glazed direct gain designs.

Direct gain, sunspace, and thermal wall systems perform equally well.

Sound energy conservation practice, with thorough insulation and minimum air leakage, is critical to thermal performance.

Passive solar systems contribute a significant fraction of the building heating load in all US climates, although the contributions are higher in the sunnier regions.

Thermal storage mass is necessary to moderate indoor temperature fluctuations, but it does not reduce the need for insulation.

Operable components such as movable insulation are frequently used improperly, leading to disappointing thermal performance.

Natural convection can distribute heat effectively through a passive solar building if adequate air flow paths are provided.

On the average, measured energy performance agrees with predictions from the Los Alamos SLR calculation procedure, but for individual buildings the results differ considerably.

Passive cooling, using radiant roof ponds assisted by ceiling fans, can replace mechanical cooling in most US climates, but a high-efficiency dehumidifier is necessary for passive cooling to be practical in humid climates.

ACKNOWLEDGMENTS

Research work cited in this article was supported by the US Department of Energy. The author thanks the many people for their cooperative effort in the Class B program, especially Don Frey, Michael Holtz, Blair Hamilton, Mark McKinstry, Ralph Beckman, John Duffy, Sukhbir Mahajan, Charles Newcomb, Michael Shea, Bill Tolbert, Nancy Carlisle, Tom Cowing, Jim Boswell, and George Yeagle.

Literature Cited

1. Balcomb, J. D., Jones, R. W., McFarland, R. D., Wray, W. O. 1982. Expanding the SLR Method. *Passive Solar J.* 1(2):67–90
2. Solar Energy Res. Inst. 1980. *Program Area Plan: Performance Evaluation of Passive/Hybrid Solar Heating and Cooling Systems.* SERI/PR-721-788. Golden, Colo: Solar Energy Res. Inst.

3. Frey, D. J., Swisher, J. N., Holtz, M. J. 1982. Class B Performance Monitoring of Passive/Hybrid Solar Buildings. *Proc. ASME Solar Energy Div. 4th Ann. Conf.*, pp. 571–77. Albuquerque, N.M.: Am. Society of Mechanical Engineers

4. Swisher, J. N., Cowing, T. W. 1983. *Passive Solar Performance: Summary of 1981–1982 Class B Results.* SERI/SP-281-1847. Washington, DC: US GPO

5. Swisher, J. N., Carlisle, N., Oatman, P., Russell, K. 1984. *Passive Solar Performance: Summary of 1982–1983 Class B Results.* SERI/SP-271-2362. Washington, DC: US GPO

6. Duffy, J. J., Odegard, D. S. 1984. *Solar Load Ratio Design Tool Predictions Compared to Level B Monitoring Data.* SERI/STR-254-2251. Golden, Colo: Solar Energy Res. Inst.

7. Mahajan, S., Newcomb, C., Shea, M., Mort, D. 1983. *Performance of Passive Solar and Energy Conserving Houses in California.* SERI/STR-254-2071. Golden, Colo: Solar Energy Res. Inst.

8. McKinstry, M., Lambert, L., Busse, P., Gillen, P. 1983. Heating Season Results from the Bonneville Power Administration Class B Passive Solar Monitoring Program. *Proc. 8th Natl Passive Solar Conf.*, pp. 211–16. Santa Fe, N.M.: Am. Solar Energy Society

9. Frey, D. J., McKinstry, M., Swisher, J. N. 1982. *Installation Manual for the SERI Class B Passive Solar Data Acquisition System.* SERI/TR-254-1671. Golden, Colo: Solar Energy Res. Inst.

10. Sonderegger, R. C., Condon, P. E., Modera, M. P. 1980. *In-Situ Measurements of Residential Energy Performance Using Electric Co-Heating.* LBL-10117. Berkeley, Calif: Lawrence Berkeley Lab.

11. Balcomb, J. D. 1981. Heat Storage Duration. *Proc. 6th Natl Passive Solar Conf.*, pp. 44–48. Portland, Ore: Am. Solar Energy Society

12. Vitro Laboratories. 1982. *Comparative Report, Performance of Passive Solar Space Heating Systems.* NTIS Publ. DE82018506. Springfield, Va: Natl Tech. Inf. Serv.

13. Jones, R. F., Dennehy, G. 1982. The Mastin Double Envelope House: A Thermal Performance Evaluation. *Passive Solar J.* 1(3):151–73

14. Chen, B., Maloney, J., Wang, T. C., Demmel, D. 1981. Performance Results of the Demmel Double Shell Home. See Ref. 11, pp. 59–63

15. Jones, R. W. 1982. *Monitored Passive Solar Buildings.* LA-9098-MS. Los Alamos, NM: Los Alamos Natl Lab.

16. Balcomb, J. D., Yamaguchi, K. 1983. Heat Distribution by Natural Convection. See Ref. 8, pp. 289–94

17. Fleischacker, P., Clark, G., Giolma, P. 1983. Geographic Limits for Comfort in Unassisted Roof Pond Cooled Residences. See Ref. 8, pp. 835–38

18. Vieira, R. K., Clark, G., Faultersack, J. 1983. Energy Savings Potential of Dehumidified Roof Pond Residences. See Ref. 8, pp. 829–34

19. Hamilton, B., Sachs, B., Duffy, J. J., Persily, A. 1984. Measurement Based Calculation of Infiltration in Passive Solar Performance Evaluation. See Ref. 8, pp. 295–300

Ann. Rev. Energy. 1985. 10:217–49
Copyright © 1985 by Annual Reviews Inc. All rights reserved

INCENTIVES AND CONSTRAINTS ON EXPLORATORY DRILLING FOR PETROLEUM IN DEVELOPING COUNTRIES[1]

Harry G. Broadman

Resources for the Future, Washington, D.C. 20036

For some time it has been widely accepted that increasing indigenous supplies of petroleum in developing countries[2] is a worthwhile, if not imperative, goal. Fossil fuels in general, and petroleum in particular, are the predominant sources of commercial energy in these countries. Such fuels provide almost all commercial energy used in transportation and much of the energy used in the industrial, household, and electricity generation sectors. Agriculture too, depends increasingly on commercial energy for important inputs, such as chemical fertilizers, underscoring the growing importance of commercial energy relative to traditional fuels in economic development. As these economies expand, energy demand, particularly that for fossil fuels, will continue to grow. Indeed, most analysts believe that throughout this century petroleum demand will grow more rapidly in developing countries than in the industrialized world.

Most developing countries import all their petroleum. The tremendous rise in the real price of oil in the last decade has thus generated a massive transfer of wealth from domestic consumers in these countries to foreign oil producers. Many of these countries have experienced sizeable increases in

---

[1] A version of this article was presented at the annual meeting of the American Economic Association, Dallas, Texas, December 28–30, 1984.

[2] Throughout this article, unless otherwise noted, my focus is generally on developing countries that are not members of the Organization of Petroleum Exporting Countries (OPEC).

217

0362–1626/85/1022–0217$02.00

oil import bills and corresponding drains from foreign exchange earnings as they have financed their oil payments. Oil producing developing countries that are on balance importers have incurred similar economic costs, though on a smaller scale than their non-producing counterparts. Nonetheless, should oil consumption in these countries outpace oil production, their level of oil imports will grow, with the attendant economic costs.

It may be readily apparent that exploiting indigenous oil resources would be a valuable form of import substitution for developing countries, but it is less obvious that such moves would also benefit the developed nations. World oil production capacity is still relatively concentrated in the politically unstable region of the Middle East. Diversifying world production outside this area would enhance the energy security of all oil consuming nations.

There is considerable disagreement over how much oil might be discovered in developing countries, just as there is disagreement over the resource potential of any area. Resource assessment, rather than being an exact science, is based on imperfect information and partially subjective judgments. However, there is agreement that, despite the economic inducement brought about by the past decade's increase in oil prices, there has been less exploration activity than expected in many of these countries. Indeed, some of the areas seemingly ripest for exploration have seen disproportionately little (or no) activity; and certain countries have experienced notable declines in exploration.

Because increasing oil supplies in developing countries would bring about the benefits described above, there is concern about why exploration has languished in some of these areas over the past decade. Some seem to believe in a conspiracy of the major oil companies. Others think the international pattern of exploration is primarily the result of political or historical accidents.

A more credible explanation derives from the notion that, since most of these countries have insufficient risk capital and indigenous expertise to exploit their own petroleum resources and thus must rely on foreign investors to develop them, some factors constrain otherwise willing foreign capital from investing in such activity. The worldwide search for petroleum supplies, it can be argued, is a systematic process, and foreign risk capital will be less attracted to a geologically promising area where (a) some "market failure" exists, such as that giving rise to underdevelopment of the infrastructure required for hydrocarbon exploitation, (b) the contractual and fiscal arrangements governing exploration projects are unduly restrictive, or (c) excessive political risk makes foreign investment in the country unappealing.

In part, this is the view underlying a proposal made by the World Bank several years ago to create an "energy affiliate," through which the Bank would act as an "honest broker," bringing together developing country host governments and international oil companies, helping developing countries finance exploration and development projects as well as the infrastructure projects the Bank traditionally has promoted. Yet support for public sector intervention along the lines of an "energy affiliate" has been very mixed. Opposition to the idea by some governments, most notably the United States', and by some, though far from all, petroleum companies, derives partly from uncertainty as to the sources and the dimensions of the types of market failures and deterrents to investment upon which the proposal appears to be predicated.

It is not surprising that such uncertainty exists. There are plenty of anecdotes about why some company is active in country x, why another company withdrew from country y, and so on. However, little systematic empirical research has examined a broad array of developing countries, seeking to explain the relative importance of the various factors that may give rise to market failures and disincentives to investment.[3]

This article, and the larger body of research from which it is drawn (12), attempts to fill this gap. It reports on the development of a multivariate analytical framework that discriminates between the geologic, economic, institutional, and political determinants of exploration activity. While the research presented here pertains to developing countries, the framework is general enough to be applicable to a broader range of countries. For example, the framework has been used to assess the determinants of exploration activity in a large number of countries outside North America, including many developed nations and most members of OPEC.

The balance of this article is organized in four sections. The next section presents data on the scope of exploration activity for the period 1970–1982 in the sample of developing countries considered in the study and conventional measures of the petroleum resource base of these areas at the end of 1982. The following section discusses the pattern of exploration activity, which from country to country and over time is very uneven. It is not obvious from casual observation that this pattern can be explained by differences in resource potential alone. In the third section, I outline the elements of the framework I have devised for analyzing the determinants of international differences in exploration activity. Specifically, I identify the kinds of economic, institutional, and political factors likely to be important in the exploration investment process and develop a set of hypotheses that

---

[3] A number of recent studies (1–11) attempt to identify the range of factors that might act as constraints on oil investments in developing countries, but none assesses the relative importance of each factor for a wide array of countries in the manner pursued in this article.

suggest how these and other factors, including resource potential, relate to the extent of exploration in a given area. To test these hypotheses, in the course of my research I have transformed this conceptual framework into a number of multivariate pooled time-series–cross-section regression models and have fitted these models to data covering 1970–1982. In the fourth section, I report the results of some of these tests. Generally, the results suggest that factors other than resource potential can be important determinants of exploration activity in developing countries. The article concludes by drawing some observations for policymaking.

THE PATTERN OF EXPLORATION ACTIVITY AND PETROLEUM RESOURCE BASES IN DEVELOPING COUNTRIES

The Trend in Exploration Activity

Two important measures of petroleum exploration activity in developing countries are presented in Table 1.[4] Seismic surveying often precedes a more extensive commitment of capital for exploration. The first two columns of Table 1 show data on amounts of such activity, measured by the number of seismic party-months. Almost half of the countries experienced declines in the number of seismic party-months in 1979–1982 compared to 1970–1973. While Argentina, Brazil, Colombia, and Egypt dominate seismic activity in both periods, Peru, which was second only to Argentina in 1970–1973, has experienced a dramatic recent decline; so too have Angola, Malaysia, Papua New Guinea, and Senegal. On the other hand, Pakistan, Tunisia, Cameroon, and Bangladesh, among others, have all seen sizeable increases.

The third and fourth columns of Table 1 present data on the number of exploration wells drilled. In more than half of these countries, the number of exploration wells drilled in 1979–1982 was less than in 1970–1973. Casual inspection suggests that this pattern is similar to that in the data on seismic activity; that is, countries that experienced increases in drilling activity tend to be those that experienced greater seismic work, and conversely. In addition, the international distribution of the number of exploration wells drilled has changed. Of the five leading countries in 1970–1973, only Argentina and Brazil are among the leading five in 1979–1982 (but note how the dramatic increase in Brazil's drilling has changed its ranking with

---

[4] Unfortunately, the only data available measure physical exploration activity rather than expenditures. Of course, because cost conditions vary across countries, a given level of physical activity may well imply different levels of expenditure.

Table 1   Exploration activity in developing countries: 1970 to 1973 versus 1979 to 1982

	Seismic party-months (Annual averages)		Exploration wells drilled (Annual averages)	
	1970–1973	1979–1982	1970–1973	1979–1982
Angola	23.250	6.250	19.250	11.000
Argentina	292.500	204.750	128.250	103.750
Bahrain	2.500	5.750	0.500	0.250
Bangladesh	0.500	33.575	0.250	3.500
Barbados	0.250	1.375	0.750	0.750
Belize	1.750	1.375	2.000	1.750
Benin	0.500	0.250	1.750	0.000
Bolivia	30.000	14.250	6.500	13.750
Brazil	106.000	184.250	86.250	218.250
Cameroon	5.500	34.750	4.000	17.750
Chile	26.500	33.500	24.250	35.000
Colombia	55.250	78.250	18.250	52.250
Congo	1.000	9.250	3.750	9.000
Dominican Republic	0.250	4.250	0.500	1.500
Egypt	90.000	74.750	22.750	55.500
Ethiopia	11.250	0.000	2.250	0.000
Ghana	2.250	2.750	3.750	1.000
Guatemala	2.500	16.750	0.500	5.000
Guyana	1.000	4.000	0.250	0.500
Honduras	2.750	0.000	1.750	0.500
Ivory Coast	1.250	2.500	0.750	7.250
Jamaica	0.000	1.750	0.750	1.500
Kenya	4.250	0.750	0.750	0.500
Liberia	1.000	0.000	1.000	0.000
Madagascar	13.750	2.750	3.000	0.000
Malaysia	14.500	7.125	18.500	44.750
Mauritania	10.250	1.000	1.000	0.250
Morocco	12.500	17.750	5.000	10.500
Mozambique	7.750	4.000	4.500	0.000
Nicaragua	2.000	0.000	2.750	0.000
Pakistan	29.500	68.750	4.750	10.250
Panama	1.000	0.250	0.250	0.500
Papua New Guinea	18.575	5.250	12.750	1.000
Paraguay	1.500	6.000	4.000	0.750
Peru	107.500	11.500	26.250	11.250
Philippines	5.000	32.000	9.500	18.500
Senegal	14.750	3.000	3.000	0.250
Somalia	5.000	16.750	0.250	1.000
South Korea	5.500	0.900	1.500	0.750
Suriname	0.500	1.250	2.000	0.750
Taiwan	29.750	22.000	9.750	19.250
Thailand	0.250	31.800	2.750	18.500
Togo	0.000	0.000	0.500	0.000

Table 1 *continued overleaf*

Table 1 (*Continued*)

	Seismic party-months (Annual averages)		Exploration wells drilled (Annual averages)	
	1970–1973	1979–1982	1970–1973	1979–1982
Trinidad and Tobago	2.250	2.750	23.250	16.250
Tunisia	26.250	64.000	10.000	20.750
Turkey	29.250	41.000	31.500	22.500
Zaire	8.750	1.250	3.250	2.000

Source: Author's calculations using data from the *AAPG Bulletin* and other sources.

Argentina). Significant increases in the ranking have occurred in, among other countries, Colombia, Egypt, Malaysia, Tunisia, Taiwan, Thailand, and the Philippines; the reverse has been true for Turkey, Peru, Trinidad and Tobago, Angola, and others.

The Petroleum Resource Base

The petroleum resource base of the developing countries as a whole is relatively promising but, as Table 2 reveals, there are great disparities among countries. The first column in the table presents data on proved reserves at the end of 1982. Twenty-two out of the forty-seven countries under study have no reserves. Among those with reserves, the size distribution is considerably skewed; the top four reserve-holders, Malaysia, Egypt, Argentina, and Tunisia, account for about 54% of the total.

Data on annual crude oil production for 1982 (presented in column two) reveal an even more skewed distribution; the leading four producers, Egypt, Argentina, Malaysia, and Brazil, account for 59% of total crude production.

The third column shows the share of cumulative crude oil production at year-end 1982 accounted for by offshore fields in these countries. Of the 25 producing countries, 11 (or 44%) produce all of their oil from onshore areas. Importantly, all of the countries that obtain 100% of their crude oil from offshore fields are relatively new producers; thus, the first year of production for Cameroon was 1978; for Ghana, 1979; for Ivory Coast, 1981; for the Philippines, 1979; and for Zaire, 1975. Since exploitation of offshore fields tends to be considerably more costly than that of onshore areas, such a relationship between vintage of production and share of supplies coming from offshore areas would be expected, given the significant increase in the real price of crude oil over the period under observation.

Table 2 The petroleum resource base of developing countries in 1982

	Proved oil reserves (1000 bbl)	Oil production (1000 bbl/day)	Offshore share of cumulative oil production (%)	Total discovered recoverable resources[a] (1000 bbl)
Angola	1,635,000	122.000	74	2,344,426
Argentina	2,590,000	483.000	0	6,427,239
Bahrain	197,000	45.000	0	958,787
Bangladesh	0	0.000	—	0
Barbados	730	0.600	0	2.591
Belize	0	0.000	—	0
Benin	0	0.000	—	0
Bolivia	18,000	24.000	0	340,345
Brazil	1,750,000	252.000	17	3,141,420
Cameroon	530,000	109.000	100	637,675
Chile	760,000	41.000	12	1,059,368
Colombia	536,000	140.000	0	2,884,494
Congo	1,550,000	87.000	96	1,744,193
Dominican Republic	0	0.000	—	0
Egypt	3,325,000	667.000	70	5,985,116
Ethiopia	0	0.000	—	0
Ghana	5,400	1.500	100	8,685
Guatemala	50,000	6.400	0	56,095
Guyana	0	0.000	—	0
Honduras	0	0.000	—	0
Ivory Coast	110,500	9.000	100	116,340
Jamaica	0	0.000	—	0
Kenya	0	0.000	—	0
Liberia	0	0.000	—	0
Madagascar	0	0.000	—	0
Malaysia	3,325,000	306.000	90	4,447,267
Mauritania	0	0.000	—	0
Morocco	290	0.200	0	17,137
Mozambique	0	0.000	—	0
Nicaragua	0	0.000	—	0
Pakistan	196,300	12.000	0	311,548
Panama	0	0.000	—	0
Papua New Guinea	0	0.000	—	0
Paraguay	0	0.000	—	0
Peru	835,336	198.000	12	2,187,736
Philippines	35,600	7.000	100	49,470
Senegal	0	0.000	—	0
Somalia	0	0.000	—	0
South Korea	0	0.000	—	0
Suriname	0	0.000	—	0
Taiwan	6,700	3.000	0	21,077

Table 2 *continued overleaf*

Table 2 (*Continued*)

	Proved oil reserves (1000 bbl)	Oil production (1000 bbl/day)	Offshore share of cumulative oil production (%)	Total discovered recoverable resources[a] (1000 bbl)
Thailand	103,000	6.000	0	106,399
Togo	0	0.000	—	0
Trinidad and Tobago	580,000	182.000	41	2,731,279
Tunisia	1,860,000	106.000	23	2,410,502
Turkey	280,000	45.000	0	706,784
Zaire	139,000	22.000	100	193,385

[a] Defined in text.
Source: Author's calculations using data from the *Oil and Gas Journal.*

Finally, the fourth column of Table 2 presents data on total discovered recoverable resources, that is, reserves plus cumulative production, at the end of 1982. On this basis, the distribution of petroleum resources is about the same as it is for reserves or production alone; the leading four countries, Argentina, Egypt, Malaysia, and Brazil, account for about 54% of total discovered recoverable resources for all of the countries combined.

Of course, firms contemplating exploration in an area take into account not only how much oil has already been discovered, but also their hunches as to how much oil has yet to be discovered. Here resource assessment becomes almost as much an art as a science. In any event, estimates of the amount of total ultimate recoverable resources in each of these countries are not readily available in a format that would allow systematic comparisons. However, total ultimate recoverable resources of developing countries as a whole are thought to be relatively sizeable—although there is considerable disagreement over the precise amount of oil estimated to be ultimately discovered. Still, it is noteworthy that some observers believe these countries contain a significant portion of the undiscovered petroleum in the world. Grossling (13), for example, estimates that of the 30 million square miles of the world's sedimentary basins, the developing countries are likely to account for 13 million square miles. In a similar vein, a study commissioned by the World Bank (2) estimates that the remaining ultimate recoverable reserves in the developing countries stood at about 80 billion barrels in 1975, of which roughly 20 billion barrels were proved. This would be on the order of 4% of the prevailing consensus estimate of 2,000 billion barrels for ultimately recoverable world resources.

Observations

The preceding data reveal a very skewed pattern of exploration in developing countries during 1970–1982, both over time and across areas. Two observations seem warranted in this regard. First, while the increase in oil prices during the period would have been expected to bring about a relatively uniform rise in seismic surveying and exploratory drilling over time, not all the countries experienced robust trends in such activity; indeed, in a number of areas exploration activity declined appreciably. Second, there is considerable unevenness in the pattern of seismic and drilling activity among the countries at a given point in time, the reasons for which are not entirely obvious from conventional measures of resource potential alone.

DETERMINANTS OF EXPLORATION ACTIVITY: A CONCEPTUAL FRAMEWORK

There are likely to be several factors that determine the level of oil exploration in a given area. Here I formulate some hypotheses about what I believe to be the principal determinants, with particular reference to developing countries, to establish a conceptual framework that can be used to explain the patterns in exploration described above.

I discuss first the role of geologic promise, positing that its relationship with exploration can be characterized in a number of dimensions. But a high degree of geologic promise is likely to be only a necessary, and not sufficient, condition for undertaking exploration in a particular area. No hard and fast theory exists to explain what specific factors, beyond geologic promise, explain international differences in exploration. Various strands of literature and discussions with oil industry personnel, have yielded insights, however. I argue that three factors are important, and I consider each in turn: (a) the extent of infrastructure development in general, and that most relevant to hydrocarbon exploration in particular, (b) the nature of the contractual and fiscal system governing petroleum exploration; and (c) the degree of political risk associated with such activity. Because the roles of the latter two are sufficiently complex and not well understood, the balance of the foregoing discussion is weighted heavily toward them.

Geologic Promise

It is widely held that the worldwide petroleum resource base is unevenly distributed geographically. This simple fact of life implies that some countries are bound to have greater petroleum resource bases than others. But the amount of undiscovered petroleum deposits in a given area is not known with certainty *ex ante*, and it is on this basis that exploration investment decisions must be made in this uncertainty. It seems reasonable to suppose that, all other things being equal, if petroleum exploration is systematically related to the prospects for finding oil, then countries with better prospects should experience higher levels of such activity.

Apart from the sheer prospects that oil might be contained in an area, the size distribution of discoveries or fields is also likely to influence the level of exploration activity. Because exploration costs per barrel decline considerably as field size increases, investment is likely to be greatest in areas thought to have oil deposits in large fields. To be sure, there is a size continuum of oil fields in developing countries—just as there is throughout the world. It is worth noting, however, that most experts believe that the probability of discovering low-cost, large fields in these areas is quite low and that small discoveries are likely to be the rule (14).

Infrastructure

Even where petroleum deposits are known to exist, exploration could be forestalled by a lack of infrastructure in the forms of harbor facilities, pipelines, railways, inland roadway systems, storage facilities, communications systems, and power generation and distribution facilities. Developing countries in general are plagued by this problem, although it is a relatively modest problem in some areas. In extreme cases, harsh environmental conditions may make it very difficult to explore prospective areas.

If inadequate infrastructure is indeed inhibiting otherwise attractive exploration opportunities, this problem would be important to policymakers; for investment in infrastructure presents the classic problem of providing a "public good." Generally, the level of investment in such goods by any individual party tends to be socially insufficient because the benefits they engender are not privately appropriable. Accordingly, public sector investment is usually required. The recent experience in Thailand, in which a pipeline system financed with multilateral public support was viewed as a prerequisite to foreign investment in developing a gas field, provides evidence in support of this hypothesis.

Institutional Factors

Even if an area is rich in petroleum resources and has an adequate infrastructure, institutional factors may constrain exploration investment. Specifically, contractual arrangements and onerous tax provisions that provide for an unsatisfactory sharing of risks between the host government and foreign investors could offset incentives to invest in an otherwise attractive situation.

ECONOMIC RISK IN PETROLEUM EXPLORATION   Investment in petroleum exploration is an inherently risky economic activity.[5] Large costs are expended with generally small odds of a payoff. Uncertainty pervades the investment decisions and outcomes throughout the exploration process, which begins with locating petroleum resources through seismic surveys (and other geophysical tests) and drilling exploratory wells, and ends with drilling development wells and the extraction process (15). The risks of exploration derive largely from uncertainty about the resource potential of an area's geology—not only the existence of deposits, but their size, quality, and accessibility. The risks associated with development derive from uncertainty about these factors, which largely determine cost conditions, and also from uncertainty about commercial factors, such as the price of oil and that of competing fuels.

Petroleum firms allocate investable funds among exploration projects by ranking each project's risk-adjusted net present value, where risk is usually measured as the variance of the project's profit.[6] All other things being equal, the more risky the project the lower its risk-adjusted net present value. Economic efficiency requires that, under the conventional assumption of risk aversion, the reward structure of investment should match that of risk ; thus, a higher expected rate of return is required to induce a firm to invest in high-risk ventures than is required for less risky ones.

There are two dimensions of risk in this regard. One component, associated with all exploration projects, derives from economic forces (for example, the price of oil) that affect the outcome of all such investments. The other component is peculiar to each project, deriving from a variety of factors, for example, the geologic structure of the area in which exploration is contemplated, the degree of field accessibility, and so on. From the standpoint of an individual investor, all risk that is project-specific can be diversified away by investing in other exploration projects. However, project diversification cannot eliminate the portion of risk that is inherent in all exploration investments. It is this nondiversifiable risk with which we are concerned.

The allocation of the risks (and rewards) of an exploration investment between the parties to such a project—in our case the host government and a foreign oil company—is governed by institutional arrangements including contracts and taxes. The economic theory of principal-agent relations

---

[5] At the outset, it is important to note that the discussion here pertains to the economic risks inherent in petroleum exploration investments. Such investments may also be exposed to political risks, which I discuss shortly.

[6] This is a gross measure of risk. Below, I discuss how the variance of post-tax profits is affected by different taxation systems.

and optimal incentive contracting provides insights into how such arrangements can establish an efficient distribution of risk and create incentives for each party to perform according to the contract in an efficient manner (16).

Briefly, starting from the definition of an efficient allocation of risk as one in which one party cannot be made better off without making the other worse off, the theory suggests that such an allocation will arise only if a contract adequately reflects the differences in the parties' attitudes toward risk-bearing; in the specific case where all the parties to a contract are risk-averse, an optimal contract is one that provides for risk-sharing (in accordance with each party's relative degree of risk-aversion). In addition, the theory states that efficiency in contract performance will depend on the need for (and cost of) contract monitoring, and the degree to which there is "opportunistic" behavior (for example, reneging on an entire contract, or on specific provisions). Finally, the theory shows that the extent to which an efficient risk-reward structure can be negotiated into a contract will depend, in part, on the relative bargaining power of the parties.

Keeping in mind these observations from theory, let me now turn to examine the allocation of risk and the degree of efficiency embodied in the contractual and taxation arrangements actually governing petroleum exploration in developing countries.[7]

CONTRACTS   Here I will consider contractual arrangements as the institutional rules governing the allocation of risks and rewards of an exploration project as well as the distribution of management responsibilities for the actual operation of such a project. Five broad classes of contractual arrangements can be distinguished: (a) concessions, (b) production-sharing contracts, (c) nonrisk service contracts, (d) risk service contracts, and (e) joint ventures.

Concessions are arrangements under which a company obtains a lease from the government to explore in a well-defined geographic area and provides all the capital for exploration. If petroleum is discovered the company is free to develop and market it; in exchange, it usually makes a payment of royalties to the government for each barrel produced (thus royalty payments may well begin before the project becomes profitable). Traditional concession agreements, like the 1901 Anglo-Persian concession in Iran, give oil companies more discretion over rates of exploration and production than do recent versions, which often give them limited property rights and a strictly defined timetable of operations and set of obligations.

---

[7] The following two sections on contracts and taxation schemes draw on collaborative work with Joy Dunkerley (16a).

In general, however, concessions do not provide for direct participation by the government. All exploration risks are borne by the company, and the government shares only indirectly the development risks—in particular, the risk associated with the company's ability to market and hence extract discoveries.

Under production-sharing contracts, the oil company again provides all exploration and development capital. But if it makes a commercial discovery, rather than recovering its investment and operating costs by income it generates from sales, it does so by obtaining a share of the crude oil produced—so-called "cost oil." The balance of production—"profit oil"—is divided between the government (whose share is analogous to the royalty received under concession arrangements after the project becomes profitable) and the company according to the provisions of the contract. It should be obvious that, as in concessions, the company bears all exploration risks under production-sharing contracts. At the same time, the government shares more directly in the development risks than it does under a concession; if an exploration project results in a commercial discovery the government in effect compensates the company with a return commensurate with the exploration risk the latter bore.

Nonrisk service contracts are arrangements whereby the company provides an exploration and/or development service and is then compensated by the government for costs of the service rendered. For exploration services, compensation is generally a flat fee; for development services, compensation can be in cash or oil but is usually a flat fee per barrel of oil produced (and sometimes the right to purchase a share of the output). In either case, the contractor receives compensation regardless of the outcome of its activity. Under such arrangements all risks are borne by the government, and the company is guaranteed its fee regardless of the efficiency or success of its performance.

Risk service contracts combine features of production-sharing and nonrisk service contracts: the company provides the exploration outlays, and if a commercial discovery is made, its services to develop the field are contracted for by the government with a fee that is structured to compensate the company for both exploration and development costs. Under such arrangements, all exploration risks are borne by the company, but the government assumes the risks associated with development. In variant of this type of contract, the host country's state-owned oil company is responsible for the production phase of the project, but the foreign company still conducts exploration and development.

Joint ventures, wherein the government participates as an equity partner with an oil company, usually occur in development rather than in exploration. With the company assuming the exploration risks, these types of joint ventures usually provide for the company's exploration costs to be recovered first and then for the balance of the oil to be split in proportion to each party's equity interest in the project. The company, in proportion to its equity interest, is then usually compensated for any development costs it incurs, and subsequently shares any profits from the sale of the oil. Pure joint ventures, wherein both the host government and an oil company share in proportion to their equity interest the risks and rewards of a combined exploration and development project, are rarely formed in developing countries.

In the last two decades the developing countries have shifted away from almost exclusive reliance on concessions toward a mixture of contract forms. Concessions are still predominant, but production-sharing contracts are becoming increasingly common. As illustrated in Table 3, by 1982 the

Table 3   Contractual arrangements governing petroleum exploration in developing countries

Country	1970	1982	Country	1970	1982
Angola	C, JV	PS, JV	Liberia	C	C
Argentina	C	RS	Madagascar	C	PS, JV, C
Bahrain	C	JV	Malaysia	C	PS
Bangladesh	C	PS	Mauritania	C	C
Barbados	C	C	Morocco	C, JV	C, JV
Belize	C	C	Mozambique	C	C
Benin	C, JV	C, JV	Nicaragua	C	C
Bolivia	C	RS, PS	Pakistan	C	C, PS
Brazil	N/A	RS, PS	Panama	C	PS
Cameroon	C	PS	Papua New Guinea	C	C, JV
Chile	N/A	PS, RS	Paraguay	C	C
Colombia	C	JV	Peru	C	PS, JV
Congo	PS	PS	Philippines	C	PS
Dominican Republic	C	PS	Senegal	C	C
Egypt	C, PS, JV	PS	Somalia	C	C
Ethiopia	C, JV	C, JV	Surinam	C	PS
Ghana	C	C	Taiwan	C, PS	C, PS, JV
Guatemala	C	PS	Thailand	C	C
Guyana	C	C	Togo	C	C, PS
Honduras	C	C	Trinidad/Tobago	C	PS
Ivory Coast	C, JV	PS	Turkey	C	C
Jamaica	C	PS	Tunisia	C, JV	C, JV
Kenya	C	C	Zaire	C	C, PS
South Korea	C	C			

Note: These are the contractual arrangements most often offered by the governments of these countries to foreign petroleum companies.

Key: C = concession; PS = production-sharing; JV = joint venture; RS = risk service.

Source: Trade press reports; discussions with embassy personnel.

latter had been adopted by almost half the developing countries under study, including countries in Latin America, Africa, and the Far East. Service contracts—both the risk and nonrisk forms—are in practice often limited to countries with existing production, proved reserves, and a certain level of indigenous expertise in petroleum development.

If there is one overriding implication of the shift away from concession arrangements, it is that there is now greater government participation in petroleum investment activity in developing countries in general. Governments assume more risk (though primarily at the development rather than the exploration phase), have greater voices in management and operations, and hold greater ownership rights in crude oil.

However, it would be misleading to conclude that this increased government role has not or could not occur through other contractual arrangements as well. Not only has the number of contract forms increased, but also the likelihood of combining particular provisions of the different kinds of arrangements. Thus, it is possible that two otherwise different types of contracts can be structured to yield an identical risk-reward structure. For example, just as it is now customary for concessions to include provisions for host government participation, service contracts are increasingly often of the risk rather than nonrisk variety, providing for a greater share of risk to be assumed by the contractor.

TAXATION   There are three predominant types of taxation schemes for petroleum exploration and development in developing countries: (a) signature bonuses, (b) royalties, and (c) profit taxes.

Under signature bonuses, the company generally pays a lump sum to the government when an exploration contract is initiated. Such arrangements are most often used when exploration rights are obtained through competitive bidding. In this regard, they reflect a company's willingness to pay for the opportunity to explore for oil, which is derived from the company's evaluation of the value of the lease for which it is bidding and is conditional on the firm's risk profile. Signature bonuses are also used in conjunction with negotiated contracts, which, in contrast to contracts obtained through a competitive bidding system (such as that practiced in the United States), are more prevalent in developing countries. In any event, signature bonuses are not the primary tax instruments in these countries; they serve only a nominal revenue-raising function.

Royalties based on a fixed percentage of production used to be the main form of taxes, generally used in conjunction with concession arrangements. Under a traditional royalty system, the government collects cash from the contractor as soon as commercial production begins; thus such payments are a function of revenue rather than profits. This has important implications for the allocation of risk. Typically an exploration and development project begins earning revenues considerably before it begins earning a profit. As a result, production-based taxes such as royalties tend to allocate most of the development risk to the company.

Profit taxes are analogous to royalties except that they are determined not by output or gross revenues, but by profits. This means that the tax burden falls on the company later in the life of a project than it would under a royalty system. Moreover, relative to royalties, taxes on profits tend to result in greater risk-sharing.

Increasingly prevalent in developing countries are various types of profit taxes geared to earn the government revenue when there are windfall profits. One such tax scheme, based on the concept of "resource rent," usually involves subjecting the contractor's net income to a higher tax rate when the rate of return on a project exceeds a specified level (17, 18). Taxation of this type was first introduced in Papua New Guinea. Another type of profit tax is also aimed at capturing windfalls. Malaysia and Peru, for example, impose higher income taxes when international oil prices rise above a predetermined level.

There is another important difference between royalties and profit taxes. Income taxes paid to foreign governments can, in some countries, including the United States, be entirely offset against taxes owed to the company's home government. They are therefore less onerous per dollar collected to oil companies engaged in foreign investment than are royalties, which are deductible rather than creditable. Importantly, since 1976, taxes paid by a US oil company to a host government in the form of oil, as is often the case in production-sharing agreements, have been disallowed as a tax credit. As a result, some developing countries have renegotiated production-sharing contracts so that US companies' tax payments (in the form of oil) now appear in the form of taxes creditable against the firms' other US tax liabilities. Of course, in the countries where this has not been done, all other things being equal, the effective tax burden on US companies may well be considerably higher under production-sharing contracts than under other contract forms.

The overall trend in taxation on petroleum exploration in developing countries over the last decade and a half has been toward higher tax yields accruing to host governments (19, 20). The change has been almost uniform owing to a "demonstration effect." The increase in yields is the result of both higher levels of taxation and a change in the mix of instruments. Higher levels of taxation have been accomplished less by applying higher tax rates than by moving from specific to ad valorem taxes, reducing tax holidays, decreasing percentage depletion allowances, and scrutinizing the transfer prices governing transactions between foreign oil company exploration affiliates and their parent companies.

The most notable change in the mix of fiscal instruments has been a shift from flat royalties to graduated income taxes and, in a number of cases, taxes on resource rent. The greater reliance on profit taxes, particularly resource rent taxes, implies that, for the same level of tax yield, the structure of taxation is different. Tax structure is critically important in determining how efficiently petroleum resources of different qualities will be exploited. In this regard, a key question is how far does the typical tax structure in developing countries discriminate among fields of different size?

The tax structure in most developing countries tends to be uniform across oil fields of all sizes. But because, as noted earlier, exploration and development costs per barrel are negatively related to field size (and thus profitability is positively related to field size), tax systems that afford the contractor satisfactory returns for large, low-cost fields may not provide adequate returns for small, high-cost fields. An efficient tax structure allows the host government to capture its desired portion of economic rent but does not distort investors' incentives to explore for and develop oil resources of varying quality. A tax system that is progressive, in that tax rates increase with oil field profitability, meets this efficiency test. In theory, a progressive tax should reduce investor risk (that is, reduce the variance of returns) and increase the level of investment and host government take.

What this implies is that the high-cost, low-volume fields that may well be characteristic of developing countries could be made more attractive by introducing progressivity into the structure of taxes. Signature bonuses often amount to committing cash outlays even before an investment proceeds. They are regressive, and raise, not lower, the variance of returns. Traditional royalties, depending on how they are structured, may be regressive but could be proportional across fields of varying sizes. More generally, as shown by Kemp & Rose (21), production-based taxes tend to result in the contractor assuming a greater share of the superprofits generated from large fields than the share of profits generated from small fields that are only marginally profitable on a pretax basis. Resource rent taxes, on the other hand, are progressive and have the advantage of providing the investor with a specified rate of return, reduced exposure to risk, and quicker cost recovery on small fields. Although the main cause for the introduction of this type of profit tax was to capture windfalls for the host country, it also introduces progressivity into the structure of taxes.

The recent trend toward resource rent taxes is thus likely to make petroleum exploration in developing countries more efficient. Unfortunately, however, their adoption is still relatively circumscribed, for at least two reasons. First, they tend to shift economic risk to the host government, which (possibly myopically) may be reluctant to assume greater risk. Second, such taxes may be risky politically for the government, since petroleum tax revenues—which often rouse nationalistic feelings— will not begin to accrue until a project is well under way and profitability rises to the predetermined "windfall" level.

In any event, other changes in tax arrangements, such as differential royalties (for example, a two-part or sliding-scale tariff type of regime), also have been made recently and could help make small, high-cost fields more attractive to investors. In addition, some contractual arrangements have been modified so as to introduce progressivity into the risk-reward structure. For example, Bangladesh, among others, has structured production-sharing contracts such that higher shares of profit oil accrue to the government as a field's rate of production rises. Still, taken together these various changes have been made in a relatively small number of countries. Moreover, there is evidence suggesting that even the modifi- cations that have been made do not go far enough. For example, in simulations that assess progressivity in current taxation arrangements in Papua New Guinea, Malaysia, and Egypt (as well as in other areas of the world), Kemp & Rose (21) found that such systems still discriminate against exploiting high-cost, small deposits.

OBSERVATIONS ON THE ROLE OF INSTITUTIONAL FACTORS   The foregoing discussion suggests some hypotheses about the role of institutional arrangements in explaining the observed pattern of oil exploration activity across developing countries. Economic theory indicates that invest- ments involving parties that are risk-averse—perhaps a characteristic of both developing country governments and international petroleum com- panies—should be governed by arrangements that bring about risk- sharing, with the allocation of risk determined by each party's attitude toward risk-bearing. Thus, it is likely that where host governments are willing to bear more risk than provided for by their contractual and taxation arrangements, the level of exploration will be lower than it otherwise would be. Moreover, because risk-bearing preferences are likely to differ in the governments of different developing countries, an insti- tutional arrangement that appropriately governs the allocation of the risks and rewards of exploration activity in one country may not do so in another. Thus, two countries equally promising geologically but with different capacities for risk-bearing should offer different contractual and taxation arrangements to attract the same level of investment. The substantial degree of uniformity in institutional arrangements that has been characteristic of these countries for the past decade and a half, brought about perhaps by some sort of demonstration effect, may help explain the relatively low level of exploration investment in some of these areas.

Political Risk

Even if a country is blessed with petroleum resources, exploration could be constrained if investors perceived that they would face a high degree of political risk if they engaged in such activity. Political risk is a complex phenomenon, and its influence on investment in petroleum exploration is not well understood. Moreover, not only is political risk a slippery concept to define, as the numerous definitions of it in the literature testify, but its measurement is also problematic. The various methods petroleum firms use to incorporate political risk into investment decisionmaking, as well as the strategies they employ to reduce the costs it imposes, reveal these difficulties.

DEFINING POLITICAL RISK   Uncertainty is an inherent aspect of any business decision. As noted earlier, in oil exploration uncertainty arises because the exact outcome of drilling cannot be ascertained *ex ante*; for each possible outcome there is a probability of occurrence. In this context, it is useful to distinguish between the notions of uncertainty and risk. When the probabilities for all possible outcomes of drilling a well are known, there is risk associated with a particular outcome (for example, finding a commercial quantity of oil). Thus, the notion of risk implies knowledge of the entire probability distribution of outcomes of a repeatable event. While assessment of the economic risk associated with an exploration project can be achieved using this concept of risk (for example, by estimating the variance of expected profits), evaluating the impact of political factors on the various outcomes of undertaking the venture is far more difficult. This is because political risk includes elements of both uncertainty and risk. Estimating objective probabilities of the occurrence of political events is exceedingly complex. Therefore, it is not surprising that while research on political risk has shed light on the many elements that may determine the probability that a certain political event (for example, unilateral abrogation of an exploration contract, expropriation, etc) will occur, a unified theory of political risk does not exist. Instead, there are at least three definitions in the literature.

The first views political risk as a function of the stability of a country's underlying political and social system, that is, the amount of conflict that exists among groups or individuals in a society (22). This social-psychological concept of political risk might be appealing, but it is operationally cumbersome, since it suggests that gauging political risk requires intensive social and political study of each prospective host country.

Moreover, political instability may not be a reliable measure of political risk. Tensions in a country's underlying political and social spheres need not destabilize the conditions that directly face foreign investors, as the sustained presence of foreign oil companies in Angola, for example, testifies. This is the basis for the second definition of political risk which views specific political events in the business arena, such as expropriation or property damage, as giving rise to political risk (23). However, this definition suggests that the political risk is the product of some sort of irrational behavior, and in many cases this just doesn't square with the facts. Political risk is often the product less of irrational whim than of deliberate policy on the host government's part.

The third definition of political risk breaks away from looking at specific political acts, and conceives of it more broadly in terms of any un-anticipated discontinuity in the business environment associated with political events (24). This definition associates political risk with actions that are in substance rational but which may be applied in an ad hoc or selective fashion and thus perceived as arbitrary. In short, this definition views political risk as stemming more from changes in the rules of the game than from the rules themselves.

Conceiving of political risk in terms of such discontinuities offers the most meaningful, operational definition. One need only inspect casually the business press to observe that political risk goes well beyond individual, seemingly irrational, politically inspired acts against foreign corporations. In the oil industry and elsewhere, this broad view of political risk appears to be the most germane.

MEASUREMENT OF POLITICAL RISK   Just as there are many ways to define the concept of political risk, there exist a number of ways to measure it. Perhaps the most well known approach is to construct a quantitative index by polling experts for their assessments of a country's investment climate and aggregating their responses using what is known as the Delphi technique. One of the most widely available of these measures is the Business Environment Risk Information (BERI) index, in which a permanent panel of about 100 volunteers from industry, the financial community, government, and academia regularly rate countries on a broad range of criteria including political stability, prospects for nationalization, attitudes toward free enterprise principles, balance of payments positions, bureaucratic delays, and credit availability and terms.

While political risk is rarely measured on an industry-specific basis, it is worth mentioning one such approach developed by the Shell Oil Company (25). This approach gauges political risk according to the probability that an oil exploration and development contract, considered equitable to both parties at the time it is signed, will be maintained over a 10-year period without being changed unilaterally by the host government; the probability is estimated by interviewing a panel of experts with knowledge of the country in question. While this measure is perhaps the most suitable for this study's purpose, it is the product of a case-by-case approach, and in any event is not publicly available.

INCORPORATING POLITICAL RISK INTO THE FOREIGN INVESTMENT DECISION-MAKING PROCESS   Survey studies of company executives in petroleum as well as in a variety of other industries suggest that political risk, defined broadly, is a prominent factor in the foreign investment decisionmaking process. But they also show that despite the sophisticated, quantitative approaches to measuring political risk in the literature, corporate use of such techniques is not widespread. Indeed, one study (26) finds that the most common method by which US-based multinational corporations incorporate political risk into investment planning is "nothing more than a completely intuitive judgment made by line management (for example, the investment climate looks unfavorable or the country is against free enterprise or foreign investment)."

In assessing whether or not to invest abroad in a particular exploration project, international oil companies typically consider political risk at the very end of the decisionmaking process, that is, only after completion of their geologic and economic evaluations. Perhaps the most widespread practice in the industry in incorporating political risk into investment decisions is that followed by Exxon, which, "if it detects the probability of a radical change in government or tax policy, may add one to five percent to its required return on investment" (27). Adding a risk premium in this way is tantamount to restructuring a project for a more rapid payout, which is desirable in that uncertainty in general, and political risk in particular, looks worse over a long investment horizon.

AN OBSERVATION ON THE ROLE OF POLITICAL RISK   The role that a high degree of political risk is likely to play in deterring exploration activity in some developing countries is one of leading foreign investors to add some sort of premium to their cost calculations, implicitly or explicitly. Adding such a premium raises the so-called hurdle rate of return, which must be met before a given level of exploration funds is invested. Since this will tend to reduce the level of investment, we would expect that countries perceived as especially politically risky are likely to experience lower levels of exploration than countries with more favorable investment climates.

ECONOMETRIC ANALYSIS[8]

The Basic Model

The conceptual framework outlined in the preceding section yields a set of hypotheses that can be summarized as follows. For a given cross-section of countries, the level of exploration in any one area is likely to be greater:

(a) the more promising the geological prospects,
(b) the greater the development of the requisite infrastructure,
(c) the less onerous to risk capital the contractual and taxation system, and
(d) the lower the level of political risk.

These hypotheses of course do not exhaust all the possible hypotheses that could be adduced; however, taken together they probably form an analytical core for explaining any differences observed in exploration activity across a set of countries.

To lay the groundwork for empirical analysis of the determinants of exploration activity in the developing countries under study, this set of hypotheses can be transformed into a simple multivariate mathematical model.[9] Thus, the determinants of exploration activity across these countries can be expressed as a model whose basic functional form for the ith country in the tth time period is

$$EX_{it} = f(GEO_{it}, INF_{it}, CON_{it}, TX_{it}, POL_{it}), \qquad 1.$$

where:

EX = level of exploration activity,
GEO = geologic promise,
INF = extent of infrastructure development,
CON = type of contractual arrangements offered,
TX = nature of taxation system,
POL = degree of political risk.

To obtain a specification suitable for econometric analysis, this model, expressed as a pooled cross-section time-series regression model becomes

$$EX_{it} = a_1 + a_2 GEO_{it} + a_3 INF_{it} + a_4 CON_{it} + a_5 TX_{it} + a_6 POL_{it} + e_{it} \qquad 2.$$

[8] The data used in this analysis, the construction of the variables described below, and the econometric procedure employed are described in greater detail in an appendix available from the author.

[9] Readers familiar with the classic Resources for the Future study of the determinants of exploration wells drilled within the United States by Fisher (28) may note that this model is in many respects the international analog of that developed by Fisher. Siddayao (29) estimates a Fisher-type model using data on a small sample of South-East Asian countries.

where all the variables are defined as above, a_1 is a constant, and e_{it} is the residual term. There is no a priori specification of the functional form of this model. It is natural to consider a linear specification as a first order approximation to an unknown functional form. This specification implies that the independent variables have additive impacts on the level of exploration. That is, each parameter a_j $(j = 2, 3, \ldots, 6)$ equals $\partial EX/\partial X$ $(X = GEO, INF, \ldots, POL)$ respectively, and thus depicts the extent to which a change in the corresponding variable influences the level of exploration in the ith country, all other factors held constant.

Selection and Measurement of Variables

DEPENDENT VARIABLES   A variety of models has been estimated, each of which employs a different dependent variable to represent different aspects of exploration activity. Below I report some results in which the dependent variable is $EXNO_{it}$, the number of exploration wells drilled in the ith country in year t.

INDEPENDENT VARIABLES   Geologic promise is depicted by two variables. As the discussion in the previous section suggests, exploration activity is likely to be greater in areas where not only are the prospects for finding oil in commercial quantities greater, but where the size of discoveries is larger. To capture the first characteristic of geologic promise I have employed $SUCR3_{it-1}$, the success ratio of exploration wells drilled in the ith country calculated as a three-year moving average and lagged one year.[10] No single measure of discovery size is unequivocally superior conceptually. However, because the data required to construct a number of such measures are simply unavailable on an international scale, I did not have the luxury of choosing among competing measures. The variable employed, $DISZ3_{it-1}$, is calculated as a three-year moving average of

$$[(RESERVES_{it} + PRODUCTION_{it}) - RESERVES_{it-1}] \div EXNO_{it}$$

and lagged one year (where $RESERVES_{it}$ is the stock of proved crude oil reserves, and $PRODUCTION_{it}$ the flow of crude oil production, in the ith country in year t).[11] Thus, $DISZ3_{it-1}$ represents the net additions to reserves per exploration well drilled.

Choosing at a conceptual level the most suitable measure of infrastructure development raises the same problem as measuring discovery size. It is difficult to establish which of many possibilities is superior. Partly because

[10] A three-year moving average was used because a number of countries in the sample experienced no drilling at all in some years (sometimes for two years in a row) and therefore produced undefined success ratios when calculated on a single (or even a two-year) basis.

[11] A three-year moving average was used here for the same reasons cited in the previous footnote.

of the greater availability of data, and partly to save degrees of freedom that would be lost by employing several variables to capture the admittedly many dimensions of infrastructure development that are likely to be important, I have chosen a composite measure, $INFRS_i$, the percentage of the ith country's gross domestic product accounted for by manufacturing, averaged over the sample period.

Contractual arrangements have been measured by constructing the following dichotomous variables, which take on a value of 1 when the corresponding arrangement is the dominant contract form in the ith country in year t, and 0 when it is not: $CNCT_{it}$ (concessions); $PDSHCT_{it}$, production-sharing as well as concessions; and $NCPSCT_{it}$, neither concessions nor production-sharing (hence, risk service contracts, nonrisk service contracts, or joint ventures). To be sure, since (as noted earlier) otherwise different contractual forms may, in practice, yield similar risk-reward structures because of specific provisions in different contracts, these variables may depict only crudely the influence of different contractual arrangements on exploration activity. Nonetheless, this proved to be the only tractable approach for measuring these factors.

In developing a proxy for fiscal arrangements, I have had to settle on using $INCTX_{it}$, the (maximum) nominal tax rate on net income earned from exploration and development activity (which can, but need not, equal the general corporate tax rate on net income) in the ith country in year t. Needless to say, nominal tax rates may be a far cry from effective tax rates. Again, the lack of better data foreclosed the option of using alternative measures.

Less severe, though still not inconsequential, problems arise in measuring political risk. Conceptually, as argued above, a broad measure of political risk (i.e. one that goes beyond depicting the incidence of expropriations and the like) would be desirable, yet such a measure should also capture elements of political risk specific to foreign investment in oil exploration going beyond elements of political risk that pertain to foreign investment in general. Finding a measure with both characteristics proved impossible. There are considerably more data on the first, and for that reason I chose to measure political risk using the Operations Risk Index developed by BERI Inc., described earlier, which decreases in value as the degree of political risk rises. Because of problems involving missing data (described in an appendix), variants of an initial variable, PRK_i (an average for the ith country over the sample period), are used when different segments of the sample are analyzed.

The foregoing eight independent variables cover the primary determinants of international differences in exploration activity. To gauge accurately their explanatory power, it is essential to control for several other factors likely to influence such activity.

Two are variables that represent pure economic elements. The first is $DPTH_{it}$, the average depth of exploration wells drilled in the ith country in year t; it serves as a measure of cost of exploratory drilling (which tends to increase disproportionately with greater depth).[12] On the revenue side, an ideal measure would be the wellhead market price of crude oil produced in the ith country in year t. Needless to say, such markets are apt to be thin in developing countries, if they exist at all. In fact, most wellhead crude prices are determined through private negotiation between host governments and contractors, in which world oil prices are used as points of reference. Among the few alternatives available to approximate these prices, I chose to use variants of $PRICE_{t-1}$, the weighted average official OPEC crude oil price (in constant dollars) lagged one year. Among the variants used were $PRICE3_{t-1} = [PRICE_{t-1} + PRICE_{t-2} + PRICE_{t-3}] \div 3$ and $PRICE2_{t-1} = [PRICE_t + PRICE_{t-1}] \div 2$.[13]

Finally, two control variables are incorporated. $AREA_i$ is used to account for the ith country's total square area. (The presumption is that, all other things being equal, more wells are drilled in larger countries.) $NTLOC_{it}$, is a dichotomous variable that takes on the value of 1 if the ith country in year t has a national oil company (with jurisdiction over, though not necessarily active in, exploration) and 0 if it does not. This variable serves a dual purpose. It is used to try to discriminate between exploration wells drilled by foreign investors and those drilled by national oil companies, since the only data available for constructing $EXNO$ are not disaggregated in this fashion. In addition, it is employed to try to gauge, albeit crudely, the extent to which the presence of a national oil company has a stimulative or chilling effect on direct foreign investment.

THE ESTIMATED MODEL   Equation 3 summarizes the general specification of the model estimated:

$$EXNO_{it} = a_1 + a_2 SUCR3_{it-1} + a_3 DISZ3_{it-1} + a_4 INFRS_i$$

$$+ a_5 NCPSCT_{it} + a_6 PDSHCT_{it} + a_7 INCTX_{it} + a_8 PRK_{it}$$

$$+ a_9 DPTH_{it} + a_{10} PRICE_{t-1} + a_{11} AREA_i$$

$$+ a_{12} NTLOC_{it} + e_{it}. \qquad\qquad 3.$$

[12] In addition to well depth, exploration costs depend significantly on the location of drilling activity; in particular, whether a well is onshore or offshore. Unfortunately, there are insufficient data available on a well-by-well basis to construct an appropriate measure of these locational attributes.

[13] Contemporaneous as well as single-year lagged prices were also tried. The results reported here employ moving averages, which tended to yield the best fit. The use of moving averages can be justified on the grounds that the effective economic inducement to exploration activity provided by rising oil prices over the sample period is more accurately captured by a smoother upward trend in prices than what would be portrayed by using the dramatic year-to-year price charges that actually occurred.

A priori expectations about the signs of the estimated parameters are: $a_2, a_3, a_4, a_8, a_{10}, a_{11} > 0$ and $a_7, a_9 < 0$. There are no a priori expectations about the sign of a_{12}, and because of the crudeness of construction of *NCPSCT* and *PDSHCT* it is difficult to specify a priori expectations about the signs of a_5 and a_6.

SAMPLE CONSIDERATIONS   The sample was defined to include all developing countries that are not members of OPEC, and for which the requisite data were available on a standardized basis for the years 1970 to 1982. An initial sample contained over 100 countries; the lack of data for many of these countries led to a consolidated final sample of 47 countries (those countries listed in Tables 1, 2, and 3 above). These countries were then partitioned into two panels; one contains the 25 countries that have oil production, and the other, the 22 that do not. This was done to assess the extent to which the determinants of exploration differ in the two types of countries. The 13 years between 1970 and 1982 provide the most recent period that is both long enough to reveal systematic differences in exploration activity and has data available; of course they also cover the period for which there is the greatest interest in unraveling the determinants of such activity. Because of the lagged structure of several of the independent variables, data for the years 1970 to 1972 were truncated for the dependent variable and the other independent variables, and the regressions were run for the period 1973 to 1982.

Analysis of Results

A summary of the results is presented in Table 4. Perhaps the most striking overall result is that the hypotheses underlying the model generally receive stronger support for the panel of oil producing countries under observation than they do for the panel of nonoil producing countries. Moreover, the model itself attains a very high degree of statistical significance and explains about 60% of the variation in drilling for the oil producing countries, but it is barely statistically significant and explains about only 3% of the variation in drilling for the nonoil producing countries. These findings suggest that there are likely to be important differences in the determinants of exploration activity in these two areas.

Let me now comment on the performance of the principal explanatory variables individually. The estimated parameter for the success ratio is consistently statistically significant and has the expected sign, suggesting that the dimension of geologic promise represented by the success ratio is an important determinant of exploration. Although the estimated parameter for discovery size has the expected sign, it is not statistically significant; this may (*a*) stem from collinearity between *SUCR3* and *DISZ3* (although when the former is dropped from the model the increase in

Table 4 Cross-country time series regression models for exploration wells drilled in developing countries: 1973–1982

Sample: dependent variable:	Oil producers EXNO		Nonoil producers EXNO	
constant	−23.0618		0.2971	
	(−2.3329)***		(0.1650)	
SUCR3	4.8084	[0.0587]	1.1535	[0.0544]
	(2.7739)****		(2.5582)***	
DISZ3	0.23E-05	[0.0002]	—	
	(1.0378)			
INFRS	1.1982	[0.9407]	−0.0037	[−0.0589]
	(3.9722)****		(−0.1017)	
NCPSCT	−7.1375	[−0.0386]	—	
	(−3.4666)****			
PDSHCT	−3.680	[−0.0706]	0.6346	[0.1045]
	(−1.7181)*		(1.2602)	
INCTX	−0.1834	[−0.4161]	−0.0018	[−0.1126]
	(−1.7709)*		(−0.1129)	
PRKNDCP	0.2648	[0.5055]	—	
	(1.0706)			
PRKNDCNP	—		0.00389	[1.9326]
			(0.6194)	
DPTH	0.73E-04	[0.0309]	−0.00001	[−1.2632]
	(0.6865)		(−3.3680)****	
PRICE3	0.3562	[0.2276]	—	
	(3.7777)****			
PRICE2	—		−0.0046	[−0.1121]
			(−1.0394)	
AREA	0.0179	[0.7814]	0.00001	[0.0415]
	(3.620)****		(0.4236)	
NTLOC	6.0690	[0.1705]	0.3206	[0.0547]
	(5.1194)****		(0.6312)	
$\bar{R}^2$	0.6175		0.0266	
F	37.536****		1.664*	
n	250		220	

Note: One, two, three, and four asterisks denote statistical significance at the 0.1, 0.05, 0.02, and 0.01 levels, respectively. t-statistics are given in parentheses. Bracketed numbers are point elasticities calculated at the mean of each independent variable.

Key to Independent Variables: SUCR3 represents the "success ratio" of exploratory wells drilled; *DISZ3* represents discovery size; *INFRS* represents infrastructure development; *NCPSCT* represents the use of service contracts and joint venture arrangements; *PDSHCT* represents the use of production-sharing contracts; *INCTX* represents the income tax rate; *PRKNDCP* and *PRKNDCNP* represent political risk; *DPTH* is average well depth; *PRICE3* and *PRICE2* represent the world oil price; *AREA* is square area; *NTLOC* represents the presence of a national oil company. See text for a complete description of each variable.

Source: Author's calculations.

statistical significance of the latter's parameter estimate is insufficient) or (b) indicate that DISCZ3 is a poor measure of discovery size.

INFRS performs strongly and yields the expected sign for the oil producer panel, but for the nonoil producer panel its parameter estimate has the "wrong" sign and is insignificant. Taken together, these results suggest that either (a) infrastructure plays only a modest role in the exploratory drilling process or (b) there is some kind of "threshold effect," in that the degree of infrastructure development becomes an important influence on further drilling in countries where there has already been some payoff from such activity, but is a moot determinant of drilling in countries that have failed to yield discoveries.

The influence of contractual arrangements is illustrated by the strong and moderate performance of NCPSCT and PDSHCT, respectively, in the oil producer panel, and the very weak performance (and change in sign of the parameter estimate) of PDSHCT in the nonoil producer panel (none of the countries in the latter group offered contract forms other than concessions or production-sharing agreements). Bearing in mind the crudeness of these measures, the following interpretations of these results can be suggested. The relatively low level of significance of the parameter estimate for PDSHCT in the oil producer panel, and (despite the sign change) its insignificance in the nonoil producer panel, suggests that the effective difference between production-sharing and concession arrangements in influencing exploratory drilling is probably small.[14] This interpretation is consonant with the analysis above, in which it was indicated that, at least at the exploration stage, the contractor bears all the risk under both types of arrangements. On the other hand, the highly statistically significant parameter estimate for NCPSCT suggests that there might be a significant difference in the way foreign investors value nonrisk and risk service contracts, as well as joint venture arrangements, relative to concessions. That the coefficient is negative suggests that these contract forms are less attractive to a foreign investor, despite the greater sharing of risk they often entail. Perhaps the result is due to increased government control of exploration that usually accompanies these arrangements.

The sign of the parameter estimate for INCTX is as expected for both panels, though it is statistically significant (albeit at a lower level) only for the oil producer panel. Its low level significance is, in part, likely due to the fact that INCTX measures nominal rather than effective tax rates.

The parameter estimates for political risk, which is measured by

[14] This interpretation is based on the fact that CNCT is the omitted contractual arrangement dummy variable. Thus the coefficients on included contractual arrangement variables reflect the influence on exploration of the corresponding contract forms relative to that of concessions.

PRKNDCP and *PRKNDCNP*, have the expected signs, though they are consistently statistically insignificant. These results suggest that all other things being equal, an above-average degree of political risk, at least as measured here, does not appear to have been a strong deterrent to investment in exploratory drilling in the sample of countries in the time period considered here.

Let me now turn to the other included variables. The influence of *DPTH* is mixed. In the nonoil producer panel, its parameter estimate is negative (as expected) and statistically significant. In the oil producer panel, its parameter estimate is positive but not statistically different from zero. One possible interpretation of these results is that because some of the established oil producing countries in the sample already have been explored relatively intensively, with shallow fields discovered years ago, a greater proportion of exploration wells are being drilled to greater depths. Because the resource base of these areas is relatively well known, deep wells do not have a discouraging effect on further exploration efforts. On the other hand, in nonproducing areas, where the extent of the resource base may not be well known, deep wells are discouraging and perhaps yet another sign of wasted effort.

The price variables perform as expected in the oil producer panel, with *PRICE3* yielding the best fit. However, it is clear, as the theoretical literature (15) suggests, that the relationship between exploratory drilling and price is far more complex than that depicted by the relatively simple lag structures embodied in *PRICE3* or *PRICE2* (which, despite its negative sign, yielded the "least worst fit" between the two for the nonoil producing panel). Surely higher prices do not induce less drilling (as the results of the nonoil panel might suggest at face value); rather the past decade or so represented a radical departure from the historical trend of oil prices and modeling the structure of price expectations for that period is exceedingly difficult.[15]

Finally, a word about *AREA* and *NTLOC*. *AREA*'s parameter estimate has the expected sign and, at least in the oil producer panel, is statistically significant. The positive sign of the parameter estimate for *NTLOC* and its statistical significance in the oil producer panel suggests that in those countries either national oil companies account for a significant portion of drilling activity, the presence of a national oil company has a stimulative (rather than a hindering) influence on foreign investment, or there is some

[15] A detailed investigation into the cause of the negative sign on the price parameter estimate suggests that the relationship is indeed spurious (rather than causal), arising from the fact that there was a pronounced downward trend in exploratory drilling in the nonproducing countries over most of the sample period at the same time oil prices were moving in the opposite direction.

type of symbiotic relationship—whether brought about by host country laws or by choice—between foreign and national oil companies (perhaps formalized through joint ventures). For the nonoil producer panel, however, national oil companies apparently are not significant contributors to exploratory drilling nor do they influence foreign investment in such activity.

Overall, the results of the regression analysis suggest the following conclusion. The market for exploratory drilling in developing countries appears to be segmented in the sense that while factors unrelated to resource potential tend to exert a significant influence on the scope of drilling in oil producing areas, they appear to be considerably less important in nonproducing areas.[16]

CONCLUSIONS AND OBSERVATIONS FOR POLICYMAKING

The objective of this article has been to first establish and then test a set of hypotheses about the extent to which observed constraints on oil exploration in some developing countries are related to poor geologic prospects on the one hand and, on the other, economic, institutional, and political elements to reduce the net returns to capital invested in such activity. For if it is a question of these nonresource factors, there may be a role for policy; if the constraints relate to resource potential alone, then all the policy initiatives in the world will not help.

The conceptual framework developed here posits that the search process inherent in oil exploration is essentially systematic. It suggests that the prospects for investment in such activity in any given area depend on diverse factors, including geologic promise, economics, institutions, and politics. The empirical evidence adduced, both qualitative and quantitative, lends support to the hypotheses that are proposed, and suggests that there are a number of opportunities where productive policy actions could be taken to achieve a desired level of exploration in developing countries. Let me make three observations along these lines.

First, some host governments may be offering contractual arrangements

---

[16] Cross-section regressions generically tend to yield lower R^2s than time series regressions; hence the low levels of variation explained in the present exercise. This "argument," however, probably does not account fully for the very low R^2 in the nonproducer panel. Nor do I believe that the poor performance of the model in this case is due largely to a "missing variable" problem. Rather, I believe it is due largely to a significant amount of true "noise" in the equation and thus a large proportion of the variance of drilling in nonproducing developing countries defies explanation by a relatively simple econometric model such as the one employed here.

that, given payoff structures, provide insufficient risk-sharing with foreign investors, and this is deterring exploration. In cases where this is true and the host government is willing to bear more risk, the level of foreign investment could be enhanced if contracts were reformed to reduce the risk borne by the contractor. Moreover, because risk-bearing preferences are likely to differ across host governments, a contractual arrangement that appropriately governs the allocation of risk of investment in one country may not do so in another. Thus, two countries which are equally geologically promising but have different capacities for risk-bearing probably should offer different contractual arrangements to attract the same level of investment.

Second, and this is related to the first point, the level of taxation in some countries may be too high (given the amount of risk-sharing). The econometric analysis lends some support to the notion that high tax rates per se reduce exploration activity. An equally critical dimension of the relationship between taxation and exploration—the structure of tax rates across fields of varying size—was not incorporated in the econometric analysis (due to the lack of appropriate data). Nonetheless, as noted above, there is evidence that in many developing countries tax structure imparts a bias against exploiting the small-volume, high-cost fields likely to be the rule rather than the exception in many of these areas. A progressive tax structure will minimize this bias, but this is not to suggest that all countries should adopt the same tax system. Indeed, for the same reasons stated above, a taxation scheme that is optimal for one country need not be optimal for another.

In the case of onerous contractual and fiscal arrangements, host governments themselves clearly can enhance incentives for exploration. My third observation concerns two areas where this is less likely to hold. The first arises from the fact that there is some evidence that inadequate infrastructure may be a deterrent to exploration. Where this is true, and the capital investment required to develop the necessary infrastructure is large, host country governments may find it difficult to raise the requisite funds from indigenous sources and may need to obtain financial support from multilateral lending organizations. This suggests that the issue of multilateral financial support of petroleum investment, for example by the World Bank, may be less one of how much such support is appropriate than one of what form such support should take.

The second area where external assistance may be required concerns those countries with poor track records of discoveries. To be sure, for some of these countries there exist relatively complete accounts about the exploration activity that has taken place, and as a result, a relatively high degree of confidence about knowledge of the (extremely limited) oil resources with which these areas are endowed. But for other countries, information about the historical record of exploration activity is often either missing or does not exist in a systematic fashion, and this is likely to act as a deterrent to potential investors. Since many of these countries are likely to lack indigenous expertise to develop up-to-date prospectuses about this information, a third party, either from the public or private sector, could play a valuable role in fulfilling this function.

ACKNOWLEDGMENTS

Financial support for this article was provided by the US Departments of Energy and State. Opinions expressed here are the author's alone.

Literature Cited

1. World Bank. 1983. *The Energy Transition in Developing Countries*. Washington, DC
2. World Bank. 1980. *Energy in Developing Countries*. Washington, DC
3. US Treasury. 1981. *An Examination of the World Bank Energy Lending Program*. Washington, DC: Office of the Assistant Secretary for Int. Affairs
4. Natl Petroleum Council. 1982. *Third World Petroleum Development: A Statement of Principles*. Washington
5. Odell, P. R. 1981. Oil and Gas in Developing Countries: Prospects for and Problems of Their Development. *Nat. Resour. Forum* 5:1–46
6. Favre, J., LeLeuch, H. 1981. Petroleum Exploration Trends in the Developing Countries, *Nat. Resour. Forum* 5:327–46
7. Foster, J., Friedman, E., Howe, J. W., Parra, F., Pollock, D. H. 1981. *Energy for Development: An International Challenge*. New York: Praeger
8. Ball, B. Jr., Kaufman, G. M., Lingamneni, J. P. 1981. Alternatives for Accelerating Oil and Gas Discovery and Production in Oil Importing LDC's, MIT Work. Pap. MIT-EL 82-031WP. Boston: MIT
9. Parra, F. 1981. Exploration in the Developing Countries: Trends in the Seventies, Outlook for the Eighties. Presented at the Int. Petroleum Seminar sponsored by the Inst. Francais de Petrole, Nice, March
10. Ghadar, F. 1982. *Petroleum Investment in Developing Countries*, the Economist Intelligence Unit Special Rep. 132. London: Economist Intelligence Unit Limited
11. Blitzer, C. R., Lessard, D. R., Paddock, J. L. 1982. Risk Bearing and the Choice of Contract Forms for Oil Exploration and Development. *The Energy J.* 5(1):1–28
12. Broadman, H. G. 1985. An Econometric Analysis of the Determinants of Exploration for Petroleum Outside North America. Washington, DC: Resources for the Future. Manuscript in preparation
13. Grossling, B. 1976. *Window on Oil: A Survey of World Petroleum Sources*. London: The Financial Times Ltd
14. Nehring, R. 1978. *Giant Oil Fields and World Oil Resources*. Prepared for the Central Intelligence Agency. Santa Monica: Rand Corp.
15. Bohi, D. R., Toman, M. 1984. *Analyzing Non-Renewable Resource Supply*. Washington, DC: Resources for the Future
16. Shavell, S. 1979. Risk Sharing and incentives in the Principal and Agent Relationship. *Bell J. Econ.* 10 (Spring): 55–73
16a. Broadman, H. G., Dunkerley, J. 1985. The Drilling Gap in Non-OPEC Developing Countries: The Role of Contractual and Fiscal Arrangements. *Natural Resources J.* 24 (Apr)
17. Garnaut, R., Ross, A. C. 1975. Uncertainty, Risk Aversion and the Taxing of Natural Resource Projects. *Econ. J.* 85 (June): 193–201
18. Palmer, K. F. 1980. Mineral Taxation Policies in Developing Countries: An Application of Resource Rent Tax. *Int. Monetary Fund Staff Pap.* Sept., pp. 517–30
19. Gillis, M. 1982. Evolution of Natural

Resource Taxation in Developing Countries. *Nat. Resour. J.* 22 (July): 619–48

20. Mikesell, R. F. 1984. *Petroleum Company Operations and Agreements in the Developing Countries.* Washington, DC: Resources for the Future

21. Kemp, A. G., Rose, D. 1982. The Effects of Taxation of Petroleum Exploitation: A Comparative Study. Paper presented to the Ann. Int. Conf. Int. Assoc. Energy Economists and the British Inst. Energy Econ., Cambridge, England, June

22. Gurr, T. 1971. *Why Men Rebel.* Princeton: Princeton Univ. Press

23. Nehrt, C. L. 1970. *The Political Environment for Foreign Investment.* New York: Praeger

24. Robock, S. H. 1971. Political Risk: Identification and Assessment. *Columbia*

J. of World Bus. (July–August): 6–20

25. Gebelein, C. A., Pearson, C. E., Sillbergh, M. 1977. Assessing Political Risks of Oil Investment Ventures. *J. Pet. Technol.* May

26. Kobrin, S. 1982. *Managing Political Risk Assessment: Strategic Response to Environmental Change.* Berkeley: Univ. Calif. Press

27. Kraar, L. 1980. The Multinationals Get Smart About Political Risk. *Fortune* March 24, p. 88

28. Fisher, F. M. 1964. *Supply and Costs in the U.S. Petroleum Industry: Two Econometric Studies.* Baltimore: Johns Hopkins Univ. Press for Resources for the Future

29. Siddayao, C. M. 1980. *The Supply of Petroleum Reserves in South-East Asia.* Kuala Lumpur: Oxford Univ. Press

Ann. Rev. Energy. 1985. 10:251–84
Copyright © 1985 by Annual Reviews Inc. All rights reserved

DECOMMISSIONING OF COMMERCIAL NUCLEAR POWER PLANTS

J. T. A. Roberts, R. Shaw and K. Stahlkopf

Nuclear Power Division, Electric Power Research Institute, Palo Alto, California 94304

1. INTRODUCTION

Decommissioning of a nuclear facility is defined as the measures taken following the end of the facility's operating life to ensure protection of the public from residual radioactivity or other hazards. Active decommissioning normally starts when a nuclear power plant has reached the end of its service life (which may or may not coincide with the expiration of its initial operating license, generally granted for 35–40 years from the date of issuance of the permit that allowed the start of construction).

No plant is close to the end of its design service life yet, but there are several other possible reasons for closing a plant: (*a*) operation may become inefficient owing to decreased availability as a result of extraordinary increases in repair and maintenance requirements; (*b*) the need may arise for major and complex repairs, the expense of which would be unreasonably high; (*c*) requirements may be retroactively imposed on operating plants by the authorities, and bringing the plant to fulfillment of the revised safety status may involve unjustifiable expenditure; or (*d*) a major accident may have occurred, which has led to consideration of closure for economic, technical, and/or safety reasons.

A number of commercial power plants are facing closure: Shippingport, Dresden-1, Humboldt Bay-3, Indian Point-1, and Three Mile Island-2 (TMI-2) in the United States and Gundremmingen A and Garigliano in Europe. Accordingly, interest in decommissioning has been growing over the past six years, and this interest has been manifested in numerous articles and at least three major international meetings, in Vienna in 1978 (1), Seattle in 1982 (2), and Luxembourg in 1984 (3).

251

0362–1626/85/1022–0251$02.00

All aspects of decommissioning are being addressed in studies by government and industry. Issues include regulations governing decommissioning costs, approaches to long-term financing, approaches to decommissioning, associated technology, and environmental impacts. The emphasis is on commercial nuclear power plants, because of the sizes of these projects. However, test reactors, nuclear fuel processing plants, and other nuclear facilities are included in some of these studies, implicitly or explicitly.

The question facing the industry today, and the focus of this paper, is just what these studies have achieved. Do we now better understand the financial requirements for decommissioning, and is the technology available for moving ahead safely to decommission the aforementioned reactors? The question is a timely one in that in February 1985 the US Nuclear Regulatory Commission (NRC) staff recommended to its commissioners amendments to their Decommissioning Criteria for Nuclear Facilities, and several decommissioning projects are either in the advanced planning stages or have already begun.

We are concerned in this paper with the decommissioning of large light water reactors (LWRs) of the boiling water reactor (BWR) and pressurized water reactor (PWR) types. Other facilities will no doubt have a few unique needs, but in general decommissioning criteria and technologies established for power plants will be directly applicable to these other situations.

This paper proceeds from the background of early policy and small plant decommissioning experience to today's activities on regulatory, economic, and technological fronts. It identifies issues settled and those still open, and recommends topics for future study.

2. BACKGROUND

The process whereby decommissioning can be accomplished depends on the existing regulatory framework, the possible decommissioning alternatives, and the information base from previous decommissioning experiences.

2.1 *Regulatory Framework*

Decommissioning in the United States is covered in various parts of Title 10 of the Code of Federal Regulations (Parts 20, 30, 40, 50, 51, 70, and 72).

The principal thrust of the regulations is to establish a plan for ensuring that funds will be available when property is removed from nuclear service, for disposing of all radioactive material, and for reducing the level of any residual radioactivity remaining on the property after decommissioning, to allow unrestricted use of the property.

In March 1978, the US NRC announced its intention to reevaluate its regulations. Additionally, in June 1979 the NRC responded to a petition for rulemaking concerning decommissioning financial assurance (filed by the Public Interest Research Group in June 1977) by granting that issues and funding alternatives raised by the petitioners would be considered in the context of this decommissioning rulemaking.

The policy reevaluation included the development of an information base; a series of studies on the technology, safety, costs, and financing of decommissioning various types of nuclear facilities (4–6); and the preparation of a draft generic environmental impact statement (7). The nuclear power industry, under the auspices of the Atomic Industrial Forum (AIF) and Edison Electric Institute (EEI), also initiated independent studies of the costs and financial aspects of decommissioning (8).

The various industry, NRC, and industry "watchdog" group activities converged in mid-1984 with a series of meetings and presentations by these parties before the NRC Commissioners (9–11) and the proposed amendments to decommissioning criteria recommended to the Commissioners in September, 1984 (12). The specifics of the amendments and the contrasting opinions are discussed in Section 3.

2.2   Decommissioning Alternatives : Pros and Cons

The decommissioning process leads from the termination of power operation (at which time the facility license is changed to one of "possession only") to an objective of achieving a condition in which there is no significant hazard to the public and workers (at which time the facility license is terminated). To achieve this objective, three alternatives have been used.

The three approaches to decommissioning considered worldwide are:

1. Immediate dismantlement (DECON)[1]—The unit is decontaminated, and the radioactive materials are removed. Upon completion of the procedure, the nuclear license is terminated and the property is released for unrestricted use.
2. Safe storage, with later dismantlement (SAFSTOR)[1]—The radioactive materials and contaminated areas are decontaminated or secured, and the structures and equipment to be dismantled later are securely maintained to ensure protection of the public from residual radioactivity. During the period of safe storage, few, if any, machinery systems are active, that is, the facility is in a passive state. Use of the property is

---

[1] The acronyms in parentheses, used by the NRC, are becoming standard nomenclature in the United States.

controlled by the conditions of the amended nuclear license. Eventual dismantlement is to be conducted for unrestricted release and license termination.

3. Entombment (ENTOMB)[2]—All areas that are to be accessible in the future are decontaminated. The remaining radioactive materials are confined in a monolithic structure with the integrity to ensure protection of the public from the entombed radioactivity for long enough to permit the decay of the radioactivity to unrestricted release levels. During the period of entombment, the property is guarded as necessary and remains restricted in use.

Of the three alternatives, it is generally accepted that the first two raise fewer open issues (12). For example, studies indicate that occupational doses from DECON of LWRs would be about 400 man-rem per year (1200–1900 man-rem over 4–5 years for large reactors). This is generally less than current annual doses at operating reactors. The long-term SAFSTOR alternative will reduce both the occupational dose and the amount of radioactively contaminated waste. Based on the half-life of Co-60 and the volume reduction in contaminated waste that would result from its decay, 30- to 50-year SAFSTOR periods appear reasonable.

The overall impacts of the first two alternatives are generally similar, with the lower occupational dose and waste with SAFSTOR compensating for the costs and uncertainties of controlling the site for a long period. The choice of alternative in individual cases will depend on a number of factors specific to the particular reactor, site, and time of decommissioning.

With regard to the ENTOMB alternative, long-lived activation products in reactor internals, such as Nb-94 and Ni-59, would probably preclude entombment for power reactors unless the reactor internals were first removed. If reactor internals were removed, it would be necessary to find a way to demonstrate that the entombed radioactivity would decay to levels permitting release of the property for unrestricted use within about 100 years.

2.3 Decommissioning Experience

Considerable experience exists in decommissioning nuclear reactor facilities, although this experience has typically been with small reactors, of less than 200 MW(t). To date decommissioning has been performed on 23 reactors, 14 with ratings greater than 10 MW(t). Of these 14 plants, 10 were power production units and four were principally test facilities. Table 1

---

[2] The acronyms in parentheses, used by the NRC, are becoming standard nomenclature in the United States.

summarizes the decommissioning experience. About 50 research reactors have also been decommissioned, typically by dismantlement. From Table 1, it can be seen that the amount of work done at each unit varied considerably, and most of the units are in some form of safe storage (SAFSTOR). At only one LWR, Elk River, has immediate dismantlement (DECON) been performed (Figure 1).

3. DECOMMISSIONING RULEMAKING

As noted in Section 2.1, the US NRC has actively pursued a change in its decommissioning criteria since 1978. Industry groups organized by AIF and EEI have conducted simultaneous independent studies to evaluate their positions.

The industry position is straightforward. It is that current technology and existing regulations provide adequately for safe decommissioning of nuclear power plants with acceptable costs after the nuclear fuel has been removed from the facility. The industry has taken the position that regulations should focus on the end objectives of no hazard to the public and license termination, leaving planning and financing to existing managerial and rate regulation practices. In their briefings of the NRC commissioners both the Utility Decommissioning Group (UDG), consisting of EEI and 16 nuclear utilities, and the AIF Subcommittee on Decommissioning, representing a cross section of industry, have emphasized that a rule specifically requiring a demonstration of the capability to finance decommissioning is not necessary for electric utility licensees (10). Nearly all electric utilities are specifically authorized by rate-making authorities to recover estimated decommissioning costs through current rates. In cases in which the power plant is the sole asset of the utility, then external mechanisms for ensuring funding are appropriate, in the industry view, but their establishment is the responsibility of the rate authorities.

On the other hand, the NRC has been under pressure from "watchdog" groups to ensure that adequate financing is available when decommissioning activities begin. The Public Citizen group, for example, suggests that at least 11 reactors have gone through one-third of their operating lives having collected no funds from ratepayers for decommissioning, and that only a total of $600 million has been collected to date of the possible hundreds of billions needed for decommissioning (11). Furthermore, Public Citizen points out, almost 70% of licensed reactors today are tentatively scheduled for immediate dismantlement; the group does not believe that the technology exists to do this job. It is particularly concerned about problems such as radiation exposure to workers and the quantity of waste generated (11).

Table 1   Decommissioned nuclear power units

Experience	Cost ($)	Cost per kW(th) ($)	Comments
Safe storage (SAFSTOR)			
Hanford production reactors (8)	175,000/yr for each reactor		Limited decontamination, about 8000 Ci sealed in each reactor; intruder alarm system installed
CVTR	250,000		Limited decontamination; sealed reactor area
Pathfinder	3.7×10^6	63	Removed some piping and components; reused turbine with fossil boilers
Saxton	2.5×10^6	106	Welded security enclosure
SEFOR	—		Welded security enclosure
Fermi	6.95×10^6	35	Locked doors and security fence
GE EVESR	—		Continuous security
Peach Bottom 1	—		Continuous security
VBWR	—		Safe storage with steam plant conversion; continuous security
Westinghouse test reactor	—		Continuous security with locked doors
SRE	—		Planning and preparation for dismantling started in 1974
Entombment (ENTOMB)			
Hallam	3.148×10^6	12	300,000 Ci (200 irradiated items) sealed in the reactor and underground vaults; reactor building superstructure removed, area was weatherproofed and topped with soil
Piqua	$12 \times 10^6 +$ 5000/yr	267	260,000 Ci sealed in the facility; entire floor surface weatherproofed and reinforced concrete poured over; reactor building superstructure left intact, remaining facilities used for offices and warehouses
Bonus	1.7×10^6 (1970$) $+ 10,000$/yr	34	50,000 Ci sealed in the facilities. (Objective was to leave facility in as-built condition, for subsequent use as public display.)

Table 1 (*continued*)

Experience	Cost ($)	Cost per kW(th) ($)	Comments
Dismantlement (DECON)			
Elk River	More than 5.73×10^6 (1974$)	79	Reactor vessel and other components cut for shipment; reactor building completely removed, other buildings removed to grade level; no radiological monitoring required
Sodium Reactor	16.6×10^6 up to 1983	830	Reactor vessel and other components cut for shipment; contaminated soil and bedrock removed; building superstructure retained and released for unrestricted use; project cumulative dose was 89 man-rem

The NRC position is embodied in the proposed amendments to 10 CFR, Parts 30, 40, 50, 51, 70, and 72, Technical and Financing Criteria for Decommissioning Nuclear Facilities, which were published in the Federal Register in February, 1985, for public comment by May 13, 1985 (12).

The proposed rules address the major issues of financing decommissioning; approaches to decommissioning (including alternative methods); planning (at licensing and just prior to decommissioning); and timing. An additional issue, the acceptable levels of residual radioactivity for release of property for unrestricted use, is being addressed in a separate set of proposed rules as an amendment to 10 CFR Part 20.

The key elements of the proposed rules are:

1. Decommissioning financing must be ensured. For electric utilities' applicants and licensees, the amount of funds assured can be based either on an amount prescribed in the regulations, proposed to be $100 million (1984 dollars) adjusted for inflation at a rate two times the change in the Consumer Price Index, or on a facility's specific cost estimate. The methods of providing financial assurance are also indicated. They fall into three general types: (*a*) deposit of funds (either at startup or over plant life) into accounts separate from licensees' assets and outside their

Figure 1 The Elk River Reactor was completely dismantled and removed from its site in 1974. The operation is complete in the bottom photograph (courtesy UNC Nuclear Industries).

control; (b) holding of funds by licensees as part of their assets and under their control (referred to as internal reserves); or (c) some form of guarantee of funds (such as insurance or sureties). The licensees must indicate which form of financial approach is to be used and must demonstrate good intentions.

2. Decommissioning alternatives. More than one alternative method of decommissioning may be acceptable depending on the type of radioactivity and contamination present at shutdown and other factors. The proposed rule indicates that deferring immediate dismantlement following cessation of operations would be acceptable in cases where sufficient benefits can be demonstrated. Possible benefits include such things as reduction in occupational exposure or waste volume. The alternatives considered are essentially the same as those that have been used in the past, but they have been redefined to include all activities leading to termination of the license, in keeping with the definition of decommissioning in this proposed rule. The approaches of complete dismantlement and supervised storage are preferred, with the latter being limited to 50 years. Entombment is not a preferred approach, and accordingly the licensee must demonstrate that the entombed radioactive materials will decay to acceptable levels by the end of the surveillance period, considered to be about 100 years.

3. Planning. Detailed decommissioning plans must be submitted two years after operations cease or one year prior to operating license expiration, whichever occurs first. Decommissioning plans must contain sufficient detail to demonstrate that decommissioning can be accomplished safely. Additional guidance on the standard format and content of decommissioning plans will be made available in planned regulatory guides. The plans must, however, address the decommissioning approach proposed and must justify the proposal. They must contain plans for disposing of radioactive waste (including the availability of waste burial ground); for controls and limits on procedures and equipment to ensure occupational and public safety (e.g. a quality assurance program) and to protect the environment during decommissioning; for final radiation surveys. They must also include updated cost estimates and plans for ensuring that adequate funds are available.

4. Timing. The proposed amendments also address the time that would be appropriate to allow for completion of decommissioning, including the entire period between final shutdown and license termination. The primary consideration here is the decay of radioactivity, which can result in reductions in both occupational exposure and quantities of waste for disposal. Also, the proposed amendments will require facility owners to maintain detailed records on topics that affect decommissioning, such as

as-built drawings, specifications, and significant events during operation.

5. Residual radioactivity. The amendments will reference the residual radioactivity analysis that will be established in 10 CFR Part 20, indicating that actual levels attained should be as low as reasonably achievable (ALARA). An upper bound of 50 man-rem per year is being considered.

At the time of this writing, official industry comments had not been released; however, based on past presentations before the NRC commissioners (10), it is likely that the industry position will be that the rule is unnecessary in the case of electric utility licensees, but that, if the regulation on decommissioning financing is adopted, it (a) should acknowledge that the choice among funding methods is for rate regulators, not the NRC, to make, (b) should not specify a generic cost of decommissioning, and (c) should not impose regulatory requirements regarding financial assurance as a condition of operating licenses. However, the recommended rules, as currently written, are not too far from the industry position.

4.   DECOMMISSIONING COST ESTIMATES

The cost of decommissioning a nuclear power plant has been the subject of intensive study for five years. This effort was initiated by the NRC's assigning to Battelle's Pacific Northwest Laboratory the task of estimating the decommissioning costs of a variety of facilities that would provide a basis for rulemaking (4–6). In addition to rulemaking by the NRC, estimates are needed by utilities to support requests to their rate regulating commissions to permit collecting funds for decommissioning. Collecting these funds in advance is based on the idea that the consumers of electricity from these facilities are responsible for paying for their closure (when the time arrives).

4.1   Estimating Methods

In general, there have been three approaches to estimating decommissioning costs. The first is to conduct, as with construction estimates, a detailed step-by-step assessment of the tasks and activities necessary to decontaminate, secure, place in storage, and/or dismantle a specific plant. This is the approach used independently by Battelle and AIF, using a specific plant model with generic costs factors (4, 5, 8). The second method is to take an existing estimate, such as the Battelle work, and adjust it for regional effects, differences in the plant, and inflation in costs with time. The third, used mostly prior to the Battelle work, is to scale costs from actual decommissioning experience; in particular the Bonus and Elk River projects.

A survey in 1983 indicated that 26 of 34 respondees had conducted site-specific estimates (10). This number would be higher today. These estimates use a combination of the above three methods.

One of the most significant plant-related factors for site-specific estimating is the type of nuclear reactor plant; that is, whether the plant is a pressurized water reactor (PWR) or a boiling water reactor (BWR). The BWR has a more extensive system exposed to radioactive fluids and, in general, will require more extensive decontamination of materials that are to be released as nonradioactive. However, the radiation level induced in the reactor vessel of a BWR is somewhat less than that in a PWR. Other important factors are:

1. Regional variations in labor rates.
2. Amounts of reinforcing in concrete structures.
3. Sizes of plants.
4. Assumptions regarding operations during plant lifetimes.
5. Assumptions about future availability and location of waste disposal sites.

The industry group AIF has sponsored the development of a set of guidelines that can be used uniformly to generate estimates (13). The project provides a reference work plan that estimates the unit quantities of labor and materials for specific tasks, nonroutine cost factors, and collateral costs such as insurance, the value of scrap, and sensitivity of total costs to waste disposal costs. This work should provide a useful basis for making different estimates readily comparable.

4.2 Cost Estimate Update

In preparing this article, we sponsored the Battelle effort to update its previous estimates to reflect 1984 costs and other changes (14). We believe Battelle's methodology is both sound and comprehensive; however, because site-specific features can significantly influence costs in the various categories, we include two other recent site-specific estimates for comparison.[3]

The models for the NRC-sponsored Battelle studies (4, 5) are the Washington Public Power System's No. 2 plant (a BWR); and Portland General Electric's Trojan plant (a PWR). In those studies Battelle's approach was to develop detailed work plans based on the specific designs

---

[3] Site-specific information on BWR obtained from A. Weinstein, Stoller Corp.; on PWR from J. F. Risley, TLG Engineering, Nov.–Dec. 1984.

of the reference plants, and on expected experience. For each of the three alternatives, the Battelle studies estimated labor, design, and supervision manpower requirements, and requirements for equipment, energy, supplies, and specialty contract labor, for decontamination and removal of activated and contaminated equipment and piping. They also calculated the waste disposal costs, including transportation and disposal, based on the expected quantities of radioactive material. Finally, they estimated nuclear insurance and licensing fees to derive a total estimated cost for decommissioning by each alternative.

The basic approach in the present study (14) was:

1. The cost information contained in the two original decommissioning reports prepared for the NRC by Battelle (4, 5) and in their addenda was adjusted to reflect cost escalations from January 1978 to January 1984.
2. With the new 1984 cost base, a number of factors that could add to costs were evaluated: (a) The cost of additional staff employed to ensure that staff radiation exposure averages less than 3 rem per yr (5 rem per yr and 1 rem per yr scenarios were also examined). (b) The additional cost of using an external firm as the decommissioning contractor, compared with the utility's acting as its own decommissioning contractor. (c) The additional cost of the predecommissioning planning, engineering, and licensing activities that are now necessary but were not included in the 1978 evaluation activities. (d) The additional fees payable to the NRC as the result of the Commission's change to full cost recovery for decommissioning licensing activities.

A selection of the results of these evaluations is summarized in Tables 2–4. The first two tables compare generic and site-specific estimates for a PWR and BWR, respectively. The third table illustrates the differences in cost of the three decommissioning methods for a PWR.

4.3 Variations in Estimate Assumptions

It is characteristic of estimates for different applications that assumptions will be different. To illustrate one example, Figure 2 shows the variations in staff labor costs that would result from different assumptions of average worker exposure during decommissioning. In this case, lower average annual doses are achieved by increasing staff, and therefore costs. Other approaches to lowering doses are available, such as extra shielding or the use of robotics. Thus the reader may see estimates elsewhere that vary considerably from these. Nevertheless, decommissioning costs (either DECON or SAFSTOR) are on the order of $100 million per plant, which is the value used in the rules recommended by the NRC staff as a nominal basis for fund accumulation.

Table 2 PWR cost estimates for immediate dismantlement (DECON)

	Cost (millions of dollars)			
	Battelle reference 1100 MW(e)		Site-specific 1100 MW(e)	
Cost category	Est. 1984	% total	Est. 1984	% total
1. Disposal of radioactive material				
Neutron-activated	6.0		14.7	
Contaminated	15.4		27.5	
Process waste	1.6		3.8	
Total disposal cost	23.0	22.1	46.0	40.9
2a. Staff labor	14.4			
b. Additional staff needed to reduce average annual radiation dose to 3 rem/yr	16.1			
c. Total	30.5	29.4	22.2	19.7
3. Use of external decommissioning contractor	12.7	12.2	12.5	11.1
4. Energy	9.1	8.8	3.7	3.3
5. Tools and equipment	1.2	1.2	2.3	2.1
6. Predecommissioning engineering (external contractor)	7.3	7.0	6.6	5.8
7a. Speciality contractors	0.9	0.9	—	—
b. Demolition (nonradioactive)			14.3	12.7
8a. Miscellaneous supplies	2.3	2.1	3.5	3.1
b. Supplies for extra staff	2.7	2.6	—	—
9. Nuclear insurance	1.1	1.0	1.2	1.1
10. License fees	0.1	0.1	0.1	0.1
Subtotal	90.9		112.4	
25% Contingency	13.0[a]		28.1	
Grand total	103.9		140.5	

[a] Contingency only for items 1, 2a, 4, 5, 7, 8a, 9; other items have contingency included.

Table 3   BWR cost estimates for immediate dismantlement (DECON)

Cost category	Cost (millions of dollars)			
	Battelle reference 1100 MW(e) Study (16)		Site-specific 1100 MW(e)	
	Est. 1984	% total	Est. 1985	% total
1. Disposal of radioactive material				
Neutron-activated	4.5		12.3	
Contaminated	15.9		22.4	
Process waste	3.0		6.7	
Deep geologic disposal of highly activated materials	—		4.6	
Fuel channel disposal (shallow burial)	—		2.4	
Total disposal costs	23.4	17.6	48.4	36.2
2a. Staff labor	28.1		33.0	
b. Additional staff needed to	16.9		—	
reduce average annual radiation dose to 3 rem/yr				
c. Total	45.0	33.9	33.0	24.6
3. Use of external decommissioning contractor	20.7	15.6	—	—
4. Energy	10.7	8.0	11.0	8.3
5. Tools and equipment	3.0	2.3	3.1	2.3
6. Predecommissioning engineering (external contractor)	7.3	5.5	—	—
7. Specialty contractors	0.6	0.4	0.5	0.4
8a. Miscellaneous supplies	2.8	2.1	2.8	2.1
b. Supplies for extra staff	0.6	0.4	—	—
9. Nuclear insurance	1.1	0.8	10.0	7.5
10. License fees	0.1	0.1	0.1	0.1
Subtotal	115.3		108.9	
25% Contingency	17.4[a]		24.7[b]	
Grand total	132.7		133.6	

[a] Contingency only for items 1, 2a, 4, 5, 7, 8a, 9; other items already have contingency included.
[b] Contingency not added to item 9.

Table 4 Summary of reevaluated decommissioning costs for a PWR (14)

| | | Estimated costs in millions of 1984 dollars | | | | |
| | Immediate dismantlement | Deferred dismantlement | | | Entombment (internals removed) | |
		Prep. for safe storage	50-years safe storage	Dismantlement after 50 years	Prep. for entomb.	100 years of surveillance
Base case estimated						
Decommissioning costs:						
1978 dollars[a]	31.0	9.5	3.8	21.1	24.7	3.9
1984 dollars[a]	65.0	17.9	6.5	41.4	46.7	5.9
Possible additional costs						
Additional staff needed to reduce average annual radiation dose to 3 rem/yr	16.1	3.7		—	10.0	—
Use of external decommissioning contractor	12.7	4.5		11.9	11.2	
Predecommissioning engineering: external (contractor)	7.3	4.4		5.8	7.3	
Supplies for extra staff	2.7	0.4		—	1.7	
NRC fees	0.1	0.1		0.1	0.1	1.0
Subtotals	103.9	31.0	6.5	59.2	77.0	6.9
Total estimated cost (1984 $)	103.9		96.7		83.9	

[a] These costs include a 25% contingency.

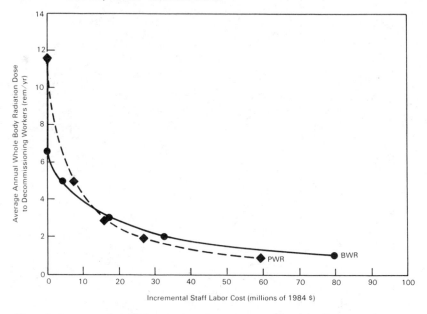

Figure 2   Incremental staff labor cost to reduce average worker dose. (Source: Ref. 14.)

5.   CURRENT DECOMMISSIONING PROJECTS AND ASSOCIATED ACTIVITIES

Several nuclear power plants are either already shut down for planned decommissioning, or may be shut down. In this section we describe the approaches being taken in two current major projects and the various technologies available to accomplish the work.

5.1   *Current Projects*

The power plant decommissioning projects at Gundremmingen A in West Germany and Shippingport in the United States are proceeding at a fair pace (Table 5). They illustrate the range of activities associated with decommissioning. These activities depend on the objective and on the pace and difficulty of working conditions.

GUNDREMMINGEN A   Gundremmingen A is a 250 MW(e) BWR at a power station with two other large (1000 MW(e)) BWRs that recently started operation. In January 1980 the decision was made to decommission the unit. Different decommissioning alternatives were examined, and the one finally chosen was to place the reactor building in safe storage and to

Table 5   Status of two current decommissioning projects

Project	Gundremmingen A	Shippingport
Objective	SAFSTOR (mothball)	DECON (dismantle)
Radiation	Low	Low
Contamination	Medium	Low
Degree of dismantling	Medium	High
Status at end of current project	Containment and fuel isolated, turbine building decontaminated	Nuclear island removed

decommission the turbine building for unrestricted use.[4] Figure 3 summarizes the approach.

Physical barriers were identified to prevent the release of radioactivity to the environment. The reactor pressure vessels, biological shield, steel containment, and concrete walls of the containment and auxiliary building were all left intact. The outer concrete structure of the containment defines the limits of the SAFSTOR area. All the radioactive materials are being stored in this area, including the spent fuel, stored in nodular cast iron casks; highly radioactive waste, to be evaporated and stored in drums; and low-level waste, to be compacted and stored in drums.

BWRs build up some radioactivity in their turbine generator systems, and the turbine building is the focus of a major effort to remove all the components and return the building to unrestricted use. To date, about 100 tons of material have been dismantled and decontaminated using conventional demolition technology, such as plasma arc torches and metal saws for cutting, and low-pressure warm water spray decontamination. About half the materials were electropolished (see Section 5.2) for final decontamination prior to release for unrestricted scrap use. One problem was what to do with 35 tons of asbestos insulation (now being stored in drums in the building). The disposal of material that is only slightly contaminated but otherwise toxic is an issue that has to be addressed for large-scale decommissioning.

The metal decontamination program is very straightforward. All materials are decontaminated to the point at which they can be released as scrap. Research on the optimal combination of decontamination techniques so far has shown that the best combination is ultrasonic cleaning followed by electropolishing to reduce the radiation to below the allowable maximum limits. Prebrushing with steel brushes sometimes cuts elec-

---

[4] Information based on private communication with W. Stang, October 1984.

tropolishing time in half. Two electropolishing baths are in operation on what is essentially a production line basis. As a further cost-saving measure, an in situ melting furnace is planned to reduce scrap volume for transport.

After work at the turbine building is done, piping and other low radioactive components in the reactor building will be decontaminated. About a thousand tons of steel will be removed from the reactor building as nonradioactive scrap.

In summary, for Gundremmingen A the decision to place the reactor building in safe storage and decontaminate the turbine building for unrestricted use was prompted by consideration of the long-term future use of the site, principally the fact that two large BWRs on the site were about to start up. The entire decommissioning effort is being implemented in a straightforward way, using existing technology with a minimum number of personnel.

SHIPPINGPORT ATOMIC POWER STATION   Shippingport, the first commercial nuclear central electric generating station in the United States, was dedicated on May 26, 1958. The reactor plant is a four-loop 72 MW(e) PWR. It will be decommissioned by its owner, the US Department of Energy. The approach is to dismantle the reactor building completely. The balance-of-plant belongs to the utility, Duquesne Light Company, and will remain under its ownership. No other nuclear power plant to be

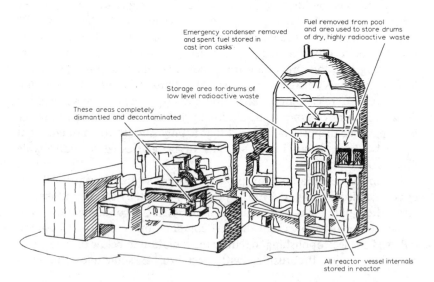

Figure 3   Sketch of Gundremmingen A, illustrating its decommissioning plan. (Source: W. Stang.)

decommissioned has operated such a long time (25 years) and produced so much electricity (over 7.4 billion kWh). Therefore Shippingport provides an interesting contrast to the activities at Gundremmingen A.

Decommissioning work will be managed by the General Electric Company. Decommissioning started in January 1985 and is to be completed in March 1990. The decommissioning logic diagram and schedule (Figure 4) shows the sequence of events from the start of training to site restoration and preparation of the final decommissioning report (15).

Two noteworthy features of the Shippingport decommissioning plan are that (a) no primary system decontamination will be undertaken, and (b) the reactor pressure vessel will be removed in one piece. Because of the low radiation levels in the reactor containment chambers, gross reactor systems decontamination is not required. The estimated occupational exposure for the entire project is approximately 1000 man-rem. It is estimated that at the peak workload approximately 250–350 people will be employed in the decommissioning effort. It has been estimated that the one-piece removal of the reactor pressure vessel will save about $7 million, reduce personnel radiation exposure for this task by about 100 man-rem, and reduce the total decommissioning schedule by about one year compared to segmentation of the vessel and internals. The total estimated cost of the Shippingport station decommissioning project is $98 million (1986 dollars).

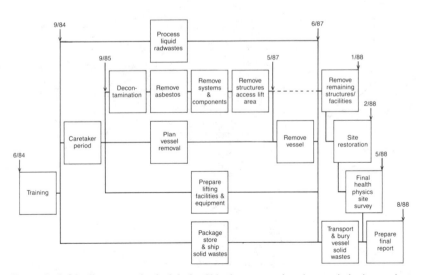

Figure 4 Logic diagram and schedule for Shippingport station decommissioning project. (Source: Ref. 15.)

5.2 *Associated Activities*

Decommissioning activities include the removal of radioisotopes (decontamination), the disassembly of components of the power plant, and the processing and packaging of radioactive materials for disposal (Figure 5). The current status of the technologies involved is described below.

DECONTAMINATION   Radioisotopes exist in nuclear power plants in two key forms, activated material and contaminated material. Activated material includes the reactor vessel and its internals, which have been

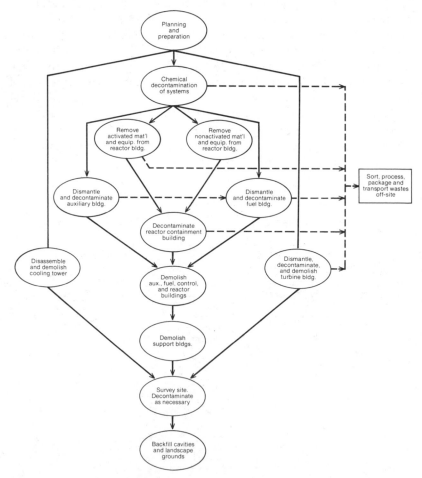

Figure 5   A typical sequence of major decommissioning activities for DECON mode. (After Ref. 4.)

exposed to neutrons and thus have radioisotopes incorporated in their constituent metals. These components cannot be decontaminated, and must be disposed of as radioactive materials. The preferred method, if allowed, would be to bury these materials in low-level waste burial grounds.

The second category, contaminated material, results from the distribution of radioisotopes during plant operation. These radioisotopes, mainly cobalt-60, are in the surface films of the metal in pipes, pumps, valves, and other components exposed to radioactive fluids. Contaminating radioisotopes in this case can be removed, leaving the component in a nonradioactive state, and amenable to disposal as normal scrap as in the Gundremmingen A approach. The amount of radioactivity contained in the contaminated material, on the order of 1000–10,000 curies, is much less than that in the activated material, which holds on the order of 1–10 million curies.

Decontamination processes can reduce the amount of material that must be disposed of as radioactive waste. The processes fall into two general categories: (a) chemical decontamination, in which deposits and oxide films are dissolved, and (b) nonchemical (mechanical) processes, using force to remove these films.

A number of chemical processes have been used recently to decontaminate components for continued operation (16). Some of these processes use low concentrations of acids for dissolution and chelates to hold the dissolved ions in suspension (e.g. low-oxidation-state metal ions (LOMI), CAN-DECON). Because of the low concentrations of the chemical solutions, the dissolved ions can be removed from solution by ion exchange.

Concentrated mineral acids, such as hydrochloric or hydrofluoric acid, are much more aggressive than low-concentration solvents; and are more likely to remove nearly all of the contaminating radioisotopes, leaving the material suitable for recycle as scrap. For economic reasons these high concentration solutions are usually processed by evaporation rather than ion exchange.

An alternative to mineral acids is electropolishing, a process used successfully by Battelle and adopted by many utilities and vendors for decontaminating ferritic and stainless steels and copper-based materials for scrap (17). Electropolishing applies electric current to the metal surface through an acid bath to remove the surface layer of the base metal, releasing any oxides along with the contained radioisotopes. Use of a dilute acid process like LOMI in combination with electropolishing could provide a decontamination system to yield unrestricted scrap with minimal liquid radioactive wastes to process for disposal.

Strippable coatings containing sequestrants, complexing agents, oxidiz-

Table 6 Nonchemical ("mechanical") decontamination techniques

Technique	Description	Comments
Pressurized water 10,000 psi	Low-pressure water sprayed through nozzle onto surfaces; used to remove sludges and deposits. Manual application.	Used at TMI-2 (20).
35,000 psi	High-pressure water sprayed through nozzle onto surfaces; used to remove oxide films, sludges, and deposits. Remote equipment application preferred for better control.	Used at TMI-2 (20, 21). Westinghouse added abrasive grit to speed removal of surface decontamination (19).
Mechanical devices	Rotating brushes, hones, cutters, scrapers, and pigs used to clean inside surfaces of pipes; surface grinders (scarfing), scabblers, pneumatic drills, and chippers used to remove the upper layer of a concrete surface; saws used to cut sections of floors or walls.	Selected for use at TMI-2. See (18, 22). See Figure 7.
Abrasive blasting	Dry (using sand, garnet, etc.) and wet (abrasives are introduced into the stream of a high pressure water jet) application used to remove paint and underlying surfaces. Requires use of a vacuum pickup attachment.	Dry ice crystals were used as the abrasive in recent applications in an effort to reduce the volume of waste generated; not as effective as conventional abrasives.
Steam-vacuum	Superheated water at <200 psig pressure loosens the contamination; high capacity vacuum then removes the water/steam/contaminant from surfaces.	Used at TMI-2 and other reactor sites.

ing or reducing agents, and wetting agents have been tested at the TMI-2 and Sequoyah plants with success. They decontaminate by migrating into the pores of a surface and attaching to the contaminants. When the coating dries, the contaminants are locked into a solid matrix and the film is stripped off to produce a solid contaminated waste.

Foams and gels represent relatively new methods for decontamination. They are potentially useful in decontaminating internal surfaces, large areas, and components with complex geometries. A major advantage is that they cover much larger volumes per unit weight than liquids.

A wide variety of nonchemical decontamination techniques are also available (18). They are listed and described in Table 6; all are reasonably well established and available commercially from a number of vendors. Both metal and concrete surfaces have been effectively decontaminated in these ways (18–22). The success of these processes is nowhere better illustrated than by the exposure management program for the cleanup of TMI-2. Decontamination using pressurized water, scabbling (removal of the top one-sixteenth to one-eighth inch of the concrete surface by cutting), and steam vacuuming has resulted in significant reductions in worker exposure. Figure 6 shows a strongly decreasing average exposure to workers inside containment throughout the last four years of the program (23). At present routine surveillance work entries can be made without the use of closed breathing apparatus.

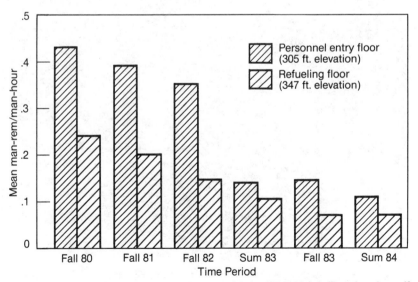

Figure 6   Average worker dose rate as a function of time at TMI Unit 2. (Data based on self reader results.) (Source: Ref. 23.)

The drawbacks to these techniques are lack of control and uniformity of decontamination and the large volumes of waste produced. Current development work is addressing these issues with very encouraging results, particularly in the area of automation (see Figure 7, for example).

DISMANTLING   Major components such as pipes, valves, pumps, and tanks must be dismantled in preparation for shipment as either scrap or radioactive waste. Particular care is required in dismantling the pressure vessel and its internals due to the high radiation fields emanating from these materials. They most likely require underwater and/or remote dismantling, and must be segmented or cut using arc saws, plasma arcs, oxygen burners, explosive cutting, lasers, or mechanical nibblers.

Demolition of the large concrete structures in a nuclear power plant can be accomplished by using controlled blasting, wrecking balls, rams, cutting, sawing, chipping, scarifying, grinding, and water cannons. Among the less traditional techniques are the application of heat or chemical expanding agents to cause loss of strength and disintegration of concrete structures.

Controlled blasting is the concrete demolition method recommended for all concrete more than two feet thick, provided noise and shock in adjacent

Figure 7   Remote scabbling system developed for EPRI by Pentek Corp. (system eliminates worker fatigue and radiation dose and provides a more uniformly decontaminated surface than do manual scabbler systems). Unpublished Information. EPRI November 1984.

occupied areas are not limitations. The process is well suited to demolition of heavily reinforced concrete, because proper selection of blast parameters can achieve a high degree of fragmentation. The exposed reinforcing bar may then be cut with an oxyacetylene torch or other cutter. Explosive cutting techniques are available for precision cutting rather than massive demolition.

The Elk River reactor dismantling program used controlled blasting to demolish the eight-foot-thick, steel-reinforced radioactive biological shield. A blasting mat (composed of automobile tire sidewalls tied together), was placed over the blast area. Continuous fog sprays of water were used before, during, and after the blast to hold down dust.

A recent development in concrete cutting is the use of ultrahigh-pressure water (> 35,000 psi). The water is directed through precision nozzles to produce a clean cut; standoff distance is precisely controlled. Recent EPRI-funded developments show that it is feasible to cut through eight feet of concrete pavement using this technology.

REMOTE SYSTEMS ("ROBOTS")   Remote systems are indicated in a number of applications to reduce (or eliminate) worker's radiation exposure and stress, but also to ensure reproducibility and uniformity of operations. Robotic devices have already been used in decontamination, dismantling, and repair activities in nuclear plants, and the trend is toward expanded use (24).

Single-purpose devices—such as remote cutters, welders, and inspectors—are used daily in nuclear plant repairs such as BWR pipe replacement and PWR steam generator tube plugging. One such device developed by Westinghouse, called ROSA, is a robotic arm that can perform all the functions involved in inspection and repair of steam generator tubes (Figure 8). The ROSA might be used in the defueling work at TMI-2 Combustion Engineering and Babcock and Wilcox have similar devices.

Several mobile "robotic" devices are in use in nuclear facilities, principally in surveillance tasks. A six-wheel remotely controlled device dubbed FRED was outfitted with a high-pressure water spray and used to decontaminate the walls and floor of a cubicle in the TMI-2 auxiliary building basement. FRED weighs 400 lb (181 kg), and its mechanical arm can lift 150 lb (68 kg) and extend to a height of 6 ft (1.8 m). More recently, a remote surveillance vehicle (ROVER) developed by General Public Utilities, DOE, EPRI, and the state of Pennsylvania at Carnegie Mellon University made two inspections of the TMI-2 reactor building basement, sending back television pictures of excellent resolution in radiation fields that ranged from 35 to 1100 rad/hr.

Several developments currently under way should result in even better decommissioning capability. For instance, a six-legged, free walking machine known as ODEX is being "hardened" for nuclear plant application (24). ODEX represents a breakthrough in its strength-to-weight ratio, as it can lift more than $5\frac{1}{2}$ times its 370 lb (168 kg) weight. Applications include both decontamination and dismantling tasks.

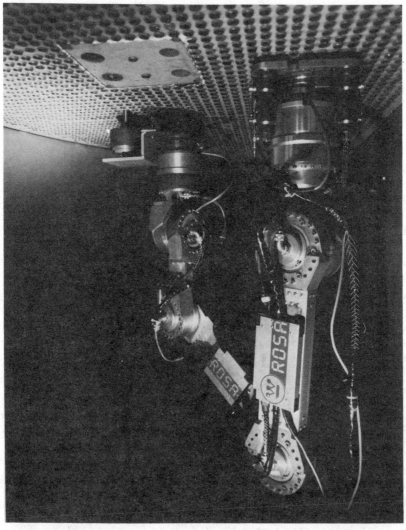

Figure 8 ROSA manipulator arm fixed to underside of steam generator tubesheet, working on tube sleeving task (courtesy Westinghouse).

VOLUME REDUCTION OF WASTE FOR DISPOSAL   Two types of waste are generated as a result of decommissioning activities, radioactive and nonradioactive. The amount of the latter will be determined by "de minimis" standards, which determine the amount of radioactivity below which materials constitute no public health hazard and can be considered releasable as trash or recycled. Cost-benefit analyses must be applied to decontamination processes to determine when it is desirable to use de minimis standards and when it is more cost-effective to treat the material as radioactive.

The cost of decontamination depends strongly on the volume of low-level waste (LLW) shipped, and a number of volume reduction techniques have recently become available (25). These include incinerators for dry active waste (and in some cases for resins); acid digesters for resins; liquid waste processing by evaporation, demineralization, crystallization, or extruder evaporation; and volume reduction of dry active wastes by high-pressure super compactors. A number of vendors now offer some of these processes in mobile facilities, which can be brought to the plant site for LLW processing.

In preparation for delivery to the burial sites, these LLWs, with the possible exception of the dry active wastes, will require either solidification or placement in high-integrity containers. Cement, bitumen, or polymerization processes are generally used for solidifying of liquid wastes. High-integrity containers are frequently used for high-activity resins or for activated materials.

6.   OPEN ISSUES

Several issues can be identified as requiring resolution, including regulatory constraints, the terms of operating licenses, assurance of funding for decommissioning, and uncertainties in cost estimates.

6.1   Regulatory Constraints

At this writing, the NRC's proposed rules for financing decommissioning, terminating operating licenses, and determining allowable levels of residual radiation for unrestricted access are out for comment. When these rules are promulgated, they will establish the legal basis for proceeding to fund and plan decommissioning projects for nuclear power plants. It is anticipated that these rules will be issued in late 1985.

6.2   License Terms

The operating license term for most nuclear plants begins on the date when the construction permit is authorized. Thus, the initial licenses for many of today's operating plants will expire around 2010. These expirations

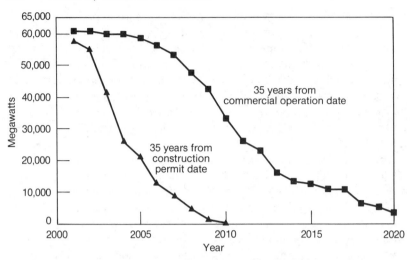

Figure 9 Decrease in nuclear capacity (based on license termination dates of plants operational in 1984) and effect of dating licenses from first commercial operation date. The lost power (i.e. area between two curves) is equivalent to eight 1000 MW(e) plants operating for 35 years. (Source: Unpublished information, J. Carey, EPRI, November 1984.)

potentially represent a loss to the nation of 60 GW of power in a very short interval (see Figure 9) in addition to whatever nonnuclear power units might terminate operation in the same period.[5] Furthermore, the amounts of capital required to replace this generation capacity in a short period will be tremendous. The cost of replacing these units would be in addition to the resources needed to decommission the retired plants.

There are two approaches to easing this potential financial burden. Because of the stretchout of the construction period for many of these plants, not all started operating in the same time interval. For the first approach, the license term for these plants should start at the date when power operation was authorized. As shown in Figure 9, this approach, if it were instituted industrywide, would yield a tremendous energy resource, equivalent to some eight large power plants for their full lifetimes.[5] There is precedent for this approach at two plants, and there are indications that the industry will approach the NRC to correct this anomaly generally. To do so is important to the planners of future power generation, and to the nation.

The second approach is to operate some plants beyond their initial operating license terms. This approach has been studied generically (26), and efforts are being initiated by several utilities and EPRI to determine its practicality in several plant-specific cases by identifying the associated

[5] Unpublished information, J. Carey, EPRI, November 1984.

safety and reliability concerns. The study will define the technical issues that must be addressed in providing operating history and material condition data for the equipment and structures. Future utility planners and regulators will need such data to make life-extension decisions.

6.3   Financial Issues

In several cases, state regulators and the Federal Energy Reliability Council (FERC) are allowing cost recovery for decommissioning at rates less than those corresponding to full accumulation of the estimated decommissioning cost. This is not necessarily imprudent; for example, if life extension is implemented in the future, the cost recovery period would be greater.

In some cases, though, "balloon" costs may be incurred when decommissioning times arrive. Nevertheless, even if this cost is half again of the original estimated cost of decommissioning, the financial exposure in a nuclear-plant-owning utility is not great in terms of the utility's total resources and annual budget. However, intervenors and utilities disagree about whether this cost should be borne by electricity users or utility stockholders.

6.4   Uncertainty in Cost Estimates

COST ESTIMATING   Cost estimation for decommissioning has been extensively studied, to the point at which the scope of the work is well defined. Unit costs for labor, materials, energy, and other elements of task definition have been established based on experience in demolition, plant modification, decontamination, and new construction. Industry efforts (13) will contribute to estimating consistency.

WASTE DISPOSAL COSTS   The cost of disposing of LLW has increased significantly in the last few years (Figure 10), and the future is uncertain. The cost is generally based on both the volume of the packages and the amount of activity in a particular shipment. This will be resolved only when the activities to establish regional low-level waste burial grounds are substantially complete. For now, utilities that have conducted site-specific analyses estimate that 30–40% of the cost of decommissioning will go to waste disposal (refer to Tables 2 and 3 for examples).

COST TO MINIMIZE RADIATION DOSES TO PERSONNEL   Another area is the additional labor required to minimize average personnel exposure. For example, removing the Surry plant's steam generators was estimated to require 1155 man-rem and actually required 946. However, 302,500 labor hours were expended for the job, compared to the 96,400 estimated (27).

The revision of the original Battelle cost estimates, reported (for the first time) in Section 4, provides an evaluation of the extra costs associated with

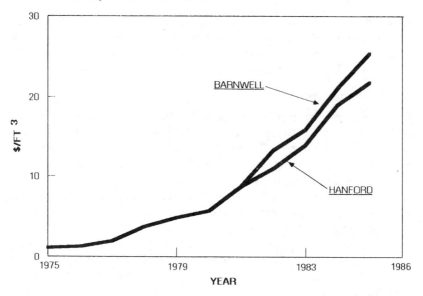

Figure 10 Radwaste burial cost history and forecast (55 gal. drum, <50 mr/hr); includes base, taxes, and perpetuity. (Source: Unpublished information, P. Deltete, November 1984.)

tighter management of personnel radiation exposure. Experience at TMI-2, where personnel exposure is very tightly controlled (as discussed in Section 5.2), will also permit better understanding of these costs.

6.5   Uncertainty in Personnel Exposure Estimates

The general range of estimates of personnel exposure for decommissioning is 1200–1900 man-rem. However, the Surry experience cited above, other repair and refurbishments, and the TMI-2 experience (noted with caution because of the special circumstances of high radiation levels) indicate that this estimate may be low. The experience to be gained at Shippingport will remove some of this uncertainty. However, the Shippingport approach of removing the reactor vessel as a unit may not be applicable to larger plants, whose larger vessels will have to be cut up for removal.

Compensating for these uncertainties are developments in remote technology for decontamination and dismantling operations. TMI-2 is a proving ground for some of these developments. When that project is substantially completed, because of the very careful control and documentation of worker exposure, there will be a substantial data base on what exposure savings result from remotely operated equipment. The TMI-2 project will also yield ways of setting decontamination priorities that are sharply focused on personnel dose avoidance.

6.6   *Method of Decommissioning*

Many existing nuclear power plant sites, especially multiple-unit sites, will always be candidate sites for electricity production, and these sites will be closely regulated by the government. This observation suggests the need to review the decommissioning option of permanent entombment (ENTOMB) in some cases. The advantages of lower cost, waste disposal volume, and radiation exposure to personnel, possibly with small environmental impact, if any, warrant such a review.

Many assume a priori that such a mode of decommissioning should not be allowed. It is not clear, however, that this assumption is based on a systematic evaluation of costs and risks to the public and environment. In any event, future consideration will undoubtedly be on a case-by-case basis.

6.7   *Waste Disposal*

Nuclear power plant materials other than fuel will generally be classified as low-level waste (LLW). Such wastes will include activated materials, contaminated materials, wet ion-exchange resins, liquid wastes, and dry radioactive materials. These must all be stabilized for burial and delivered by truck, rail, or barge to a burial ground.

The packaging appropriate to LLW will usually be defined by the burial grounds. This definition will depend on the type and amount of radioactive isotopes in the LLW package. Many states are developing compacts or agreements with neighboring states for the development of burial grounds. Consequently, the rules associated with the delivery of LLW to the three presently established burial sites in the United States will likely undergo significant changes in the near future, as new sites are developed.

This question of where the wastes from dismantling and decontamination activities will go is being solved, though somewhat slowly, as a result of initiatives in response to the Low Level Waste Policy Act of 1980. It is anticipated that regional burial grounds will be established within the next five years.

Some higher-activity waste from the internals of the reactor may fall into a waste disposal category not clearly defined by regulations. Some of this material is not candidate for the spent-fuel repository being established under the authority of the Department of Energy, but is higher in radioactivity than currently allowable for land burial as LLW as defined by 10 CFR 61. The quantity of such material is not great, and there may be ways to reduce its volume further. This issue can be addressed by defining an additional category of waste and specifying fixing methods that ensure indefinite isolation from the environment.

Two other technical areas related to waste disposal are receiving attention: the disposal of waste from the decontamination of metal surfaces with chelating agents, and the volume reduction of radioactive metals. The ultimate objective in both cases is to optimize the disposal of waste materials in terms of volume, stability, personnel exposures, costs, and conservation of burial ground space and natural resources.

7. CONCLUSIONS

The questions raised at the beginning of this chapter can now be addressed. We conclude, as many others have before, that the technology to decommission a nuclear power plant does exist. There is, however, room for improvement. Specifically, the continued development of remotely operated equipment for decontamination and dismantling will prove beneficial in reducing costs and minimizing personnel exposures.

The programs and projects being initiated to define the constraints for extending the lives of nuclear power plants also represent technological progress. The ultimate benefit of life-extension programs will go to the consumer of electricity by providing a basis for using valuable resources for longer periods of time and deferring the need for decommissioning.

The costs and financing of decommissioning are much better understood than they were five years ago. Estimating procedures have evolved to the point where one can be confident of their completeness, and there is a trend toward codifying the estimating procedure. Large-scale repair, decommissioning, and decontamination projects will provide a larger data base for use in estimating costs. Included are projects such as dismantling Shippingport, placing Gundremmingen A in safe storage, decontamination and decommissioning of Dresden 1 and Humboldt Bay 3, decontaminating TMI-2, replacing steam generators at Surry and other PWRs, and replacing major reactor piping systems at several BWRs. These projects provide a wide spectrum of experience in scope and endpoint of project, labor intensity, and work conditions. These projects also represent an evolution of process sophistication, techniques for dose management, and remote operations.

The remaining open issues related to decommissioning are institutional and political. It appears that the rules to be proposed by the Nuclear Regulatory Commission are reasonable and will resolve some of these issues. These rules, when adopted, will provide much of the definition needed to allow firm planning by utilities, state regulators, and other institutions responsible for deciding what to do when a plant reaches the end of its useful life.

ACKNOWLEDGMENTS

The authors wish to recognize the following people who contributed to the preparation of this article: R. I. Smith, G. J. Konzek, E. S. Murphy, and H. K. Elder, who developed the updated Battelle cost study for Section 4; J. F. Risley and A. A. Weinstein, who provided additional site-specific cost data for Section 4; S. Lefkowitz and K. Schwartztrauber, who contributed to Section 5.2; and C. Negin, who contributed to several sections.

Literature Cited

1. *Proc. Int. Symp. Decommissioning of Nuclear Facilities.* 1978. Vienna, Austria
2. *Proc. 1982 Int. Decommissioning Symp.* Oct. 10–14. Seattle, Wash. CONF-821005 (DE83008702)
3. *Proc. 1984 Int. Decommissioning Symp.* May 21–25, Luxembourg
4. *Technology, Safety, and Costs of Decommissioning a Reference Pressurized Water Reactor Station.* NUREG/CR-0130. June 1978, and 0130 Addendum, August 1979. Pacific Northwest Laboratory for US Nuclear Regul. Commission
5. *Technology, Safety, and Costs of Decommissioning a Reference Boiling Water Reactor Power Station.* NUREG/CR-0672. June 1980. Pacific Northwest Laboratory for US Nuclear Regul. Commission
6. *Facilitation of Decommissioning Light Water Reactors.* NUREG/CR-0569. December 1979. Pacific Northwest Laboratory for US Nuclear Regul. Commission
7. *Draft Generic Environmental Impact Statement on Decommissioning of Nuclear Facilities.* NUREG-0586. Jan. 1981. US Nuclear Regul. Commission
8. Subcommittee on Decommissioning of AIF Committee on Environment, 1983. *An Overview of Decommissioning Nuclear Power Plants.* Atomic Industrial Forum, Inc.
9. Status Report on Decommissioning Issues, Open Meeting of NRC Commissioners, June 11, 1984
10. Statement of Position of Utility Decommissioning Group and the Edison Electric Institute on Proposed Decommissioning Rule, Open Meeting of NRC Commissioners, Sept. 20, 1984
11. Totten, M. 1984. Comments on the Nuclear Regulatory Commission's Proposed Rule on Decommissioning. Public Citizen's Critical Mass Energy Project, Open Meeting of NRC Commissioners, Sept. 20
12. Decommissioning Criteria for Nuclear Facilities: Proposed Rules. 1985. *Federal Register* 50(28): 5600–25
13. Atomic Industrial Forum/National Environmental Studies Project, "Guidelines for Producing Commercial Nuclear Power Plant Decommissioning Cost Estimates." Available from M. Renner, Project Manager, Atomic Industrial Forum
14. Smith, R. I., Konzek, G. J., Murphy, E. S., Elder, H. K. 1984. Costs for Decommissioning Nuclear Power Facilities, Estimated in 1984 Dollars, Research Project 2558-5, Dec. EPRI Special Report prepared by Battelle Northwest Labs
15. Crimi, F. P. 1984. Decommissioning Program for the Shippingport Atomic Power Station, AIF Fuel Cycle Conference
16. *Proc. ANS Executive Conf. Decontamination of Power Reactors, The Costs, Benefits and Consequences.* Sept. 17–19, 1984. Springfield, MA
17. Hamilton, R. G. 1984. Decontamination for Volume Reduction and Materials Disposal. See Ref. 16
18. Gardner, H. R., Allen, R. P., Polentz, L. M., Skiens, W. E., Wolf, G. A. 1982. *Evaluation of Nonchemical Decontamination Techniques for Use in Reactor Coolant Systems.* EPRI NP-2690, Project 2012-2, Topical Report. Oct.
19. Sejvir, J., Davison, P. H. 1982. Evalua-

tion of Abrasive-Grit-High Pressure Water Decontamination. EPRI NP-2691 Project 2012-4, Final Report. Oct.

20. *Gross Decontamination Experiment Report.* 1983. GEND 34. Prepared for US Dept. Energy, TMI Operations Office

21. Gardner, R. H., Polentz, L. M., Bateman, D. B., Scott, J. L., Canlibe, H. L. 1984. High-Pressure Water Systems for Flushing Loose Fuel Debris from the Reactor Underhead Area at TMI-2. EPRI NP-3510 Project 2012-6 Interim Report. June

22. Gardner, R. H., Polentz, L. M., Bateman, D. B., Allen, R. P. 1984. Laboratory evaluations of mechanical decontamination and scaling techniques, EPRI NP-3508, Project 2012-6, Interim Report, July

23. Flanigan, J. A. 1984. Three Mile Island Unit II recovery update, Presented at *Westinghouse Radiation Exposure Management Seminar.* Pittsburgh, Pa. Oct.

24. Robots join the nuclear workforce. 1984. *EPRI J.* 9(9): 6–17

25. Naughton, M. D. 1984. Radwaste volume reduction economics: An overview. *Proc. Am. Power Conf.* 46: 591–97

26. Negin, C. A., Goudarzi, L. A., Kenworthy, L. D. 1979. Planning study and economic feasibility for extended life operation of LWR plants. EPRI TPS 78-788 Technical Planning Study. Sept.

26b. Negin, C. A., Walker, R. S., Shantzis, S. B., Worku, G. 1982. Extended life operation of light water reactors: economic and technological review. Vol. 1, EPRI NP-2418, Project 1253-6, Final Report, June

27. Virginia Electric Power Co. Progress Reports to USNRC Docket Nos. 50–280 and 50–287, Aug. 1978–Nov. 1978

Ann. Rev. Energy. 1985. 10 : 285–315

A REVIEW OF THE RESIDENTIAL CONSERVATION SERVICE PROGRAM[1]

Dr. James A. Walker

Core Ventures, Incorporated, 1225 Eighth Street, Suite 285, Sacramento, California 95814

Theador N. Rauh

California Energy Commission, 1516 Ninth Street, Sacramento, California 95814

Karen Griffin

California Energy Commission, 1516 Ninth Street, Sacramento, California 95814

INTRODUCTION

In 1977, a presidential initiative launched a national program to improve the energy efficiency of existing American homes. Embodied in law and regulation, the Residential Conservation Service (RCS) was intended to make home energy audits widely available to homeowners through local utilities. The program's rationale was that, given adequate information on the availability and advantages of home energy conservation options, US consumers would significantly increase their conservation efforts. Early in 1985, after five years of program operation, key program requirements were being allowed to lapse, minimizing the federal program mandate.

[1] The California state government has the right to retain a nonexclusive royalty-free license in and to any copyright covering this paper.

During its five years of full-scale operation, RCS generated controversy and criticism regarding its cost-effectiveness, its impact on the conservation marketplace, and the proper federal, state, and utility roles in energy conservation. It resulted in a dramatic expansion of the number of energy audits performed in US homes, but may not have contributed significantly to the achievement of national energy goals. In this article we shall examine a number of questions regarding this ambitious and complex program: What is RCS? How successful has it been? What problems did it encounter? What sources of information are available for evaluating it? What general lessons and conclusions can be drawn from the national experiment that RCS represented?

PROGRAM HISTORY

The Residential Conservation Service had its origins in the Carter administration's National Energy Plan (1). This plan was released as a major policy initiative in April 1977, and was embodied legislatively in the National Energy Act. Portions of this legislation were considered and acted upon by Congress by the end of 1978, among them the National Energy Conservation Policy Act (NECPA) (2). NECPA became law on November 9, 1978. Along with amendments in the Energy Security Act of 1980 (ESA) (3), NECPA formed the basis of the federally mandated RCS program.

In retrospect, the energy programs of Presidents Nixon, Ford, and Carter share many features. Federal energy policy initiatives of this period were driven by the political, economic, and national security vulnerability exposed by the first oil crisis. They also shared an active, even interventionist, role for government in energy markets. This was evident in the imposition of oil price and allocation regulations, as well as in the aggressive funding for energy research and development and commercialization programs at the federal level. The general public policy rationale was that government needed to ensure that actions were taken to reduce energy vulnerability. It was felt that market forces alone could not be relied on to induce these actions quickly enough.

At the time of its release, the Carter plan's emphasis on conservation and renewable energy sources provided a significant contrast with prior presidential energy messages. Carter's own introduction noted, "We can rediscover the ingenuity and the efficiency which have made our nation prosper, rather than deepening our dependence on insecure imports and increasingly expensive conventional energy supplies" (1).

While the Carter administration was searching for initiatives to include in the National Energy Plan, the postembargo flood of forecasts and analyses identified (a) the residential sector in general, and home retrofit in particular, as a significant target of opportunity for mid-term (five-year) conservation potential, and (b) electric and gas utility home energy conservation programs as a possible delivery vehicle. An estimated 74 million residential dwelling units in the United States in 1977 used the energy equivalent of 8.3 million barrels of oil per day, or about 23% of the nation's energy consumption. Savings potential in the mid-1980s was projected at up to 500,000 barrels per day. This potential became embodied in a goal-oriented residential sector program "designed to bring 90 percent of all residences and many public buildings up to minimum Federal standards by 1985" [(1), p. 41].

The residential program included:

1. Federal tax credits for residents who purchased and installed energy conservation measures.
2. Expanded funding for low-income weatherization programs.
3. Authorization for the federal secondary mortgage markets to purchase energy conservation loans.
4. A rural home conservation program.
5. Direction to state public utility commissions to require their regulated utilities to "offer their residential customers a 'turnkey' conservation service, financed by loans repaid through monthly bills. Utilities would also inform customers of other available conservation programs, and advise them how to obtain financing, materials, and labor to carry out conservation measures themselves" [(1), p. 41].

It was this last element that became the RCS program. RCS was built on the "rational actor" concept. Utilities were chosen as the vehicle for disseminating the information that would allow customers to act rationally (i.e. to conserve energy) because (a) utilities have direct contact with virtually all residences; (b) existing programs in various utilities provided prototypes for program design and scale-up; and (c) state commission jurisdiction over the utilities provided a mechanism for large-scale program implementation without the need for direct federal outlays other than those for central program direction. The state commissions even provided an intermediate level of program oversight that would limit the need for oversight at the federal level.

As proposed to Congress, the RCS program required large electric and gas utilities to offer a variety of conservation services. The centerpiece was a home energy audit tailored to the individual residence, to identify appropriate conservation features and provide estimates of likely costs and savings. The technology for such audits was in itself a rapidly evolving field. Virtually nonexistent before the embargo, by the late 1970s, it could draw on extensive research in computer-aided building energy modeling and engineering analyses of various conservation measures. Utilities were also to offer to install recommended measures directly or through independent contractors, and to finance installations directly or through outside lenders.

The potential magnitude of the original RCS program and the role it implied for utilities vis a vis other participants in the energy conservation marketplace became significant topics of discussion in the congressional hearings on the Carter proposals (4). The administration estimated that the cost of inspecting 40 million one- and two-family homes (80% of the US stock of these types) could be between $800 million and $1.6 billion. This penetration was planned to occur within five years, an ambitious program goal. The Edison Electric Institute estimated that 320,000 auditors and 2.5 million insulation installers would be needed to accomplish the administration's goal. The same factors of size, administrative infrastructure, and ubiquity that made utilities attractive as a sort of rapid deployment force for home retrofit raised concerns about the possible anticompetitive effects of utility involvement. The Federal Trade Commission testified that

> Unless there is a maximum degree of competition at each level of the program, consumers may find that the utility monopoly, however well-regulated, has expanded to include the retrofitting business. The requirement that public utilities arrange for home appraisals, installation of energy-saving materials or devices, and financing of weatherizing costs may drive smaller businesses that lack the economic resources of utilities from the market. . . . The result may be increased concentration, a greater potential for development of oligopolies or monopolies in individual geographic regions, and related problems of overcharging for supplies, services, and capital. On the other hand, direct entry by utilities into the retrofit business in some markets may increase competition with existing businesses, large and small [(5), pp. 34–35].

This ambivalence about the role of utilities remained a major policy issue at all program levels throughout the life of the program. At the federal level, the original enabling legislation for RCS dictated that utilities not be allowed to finance or install major conservation measures, such as insulation, except under preexisting programs covered by a grandfather clause. Only two years later, Congress decided to modify these prohibitions when it passed the Energy Security Act of 1980. The ESA authorized utilities to install any energy measure, so long as it was done through an independent contractor, and lifted restrictions on utility financing.

The net result of NECPA, embodied in federal regulations first issued in November 1979 and later amended by ESA, was to create an RCS program with the following major requirements (6):

1. State Plan. Through the governor or a designated agency, each state was required to submit a state RCS plan implementing the federal regulations and providing further guidance to the state's utilities. Covered utilities included gas utilities with sales over 10 billion cubic feet per year and electric utilities with annual sales over 750 million kilowatt-hours,

although smaller utilities or other fuel distributors could be added. Municipal utilities could join the state plans or submit their own.
2. Energy Audit. As the centerpiece of the program, each utility was required to offer a home energy audit to anyone who either owned a dwelling or lived in a multiunit dwelling of less than four units. The basic, or "Class A," audit included an on site inspection of the home by a qualified utility representative. The utility auditor evaluated conservation measures specified in NECPA, such as caulking, storm windows, furnace modifications, insulation, and low-cost/no-cost conservation practices. The auditor also provided estimates of costs, energy savings data, and recommendations on which measures to implement. In some cases consumer information on available tax credits and financial assistance programs was also to be provided.
3. Arranging Installation and Financing. The utility was required to assist eligible customers by presenting state-compiled lists of qualified installers, equipment suppliers, and financial institutions for the measures involved, and by helping to obtain bids and other arrangements. A minimum one-year warranty was required for inclusion on the installer or supplier lists. Utilities were also required to establish procedures for resolving customer complaints arising out of RCS program activities.

Several other program provisions were significant. The ESA set an upper limit of a $15 charge to any customer receiving an audit. This proved to be well below the average cost of $118 per audit, determined by a 1983 national program evaluation (7). Moreover, many states chose to keep the audit fee at less than $15 to encourage participation. This implied that most RCS program costs would be recovered from a broad utility customer base independent of participation rates. Second, the requirement for utilities to make offers of program audits expired January 1, 1985, providing an effective sunset to the RCS program. Finally, although no formal evaluation program was established in the law, annual program reports to the Department of Energy (DOE) were required, providing a data base for monitoring and evaluation.

The ability to achieve a high degree of program "leverage" per federal dollar or bureaucrat also raised questions as to how much flexibility to allow states and utilities in program implementation. As discussed later, the RCS program was characterized by extremely detailed specifications in the legislation and in the original federal regulations, which were reflected and often compounded in state RCS plan requirements. As DOE's 1983 evaluation report stated, "Congress and DOE were attempting to respond to a number of issues raised by the Program, including anticipated utility compliance problems, climatic diversity, introduction of new conservation technology, and protection of the consumer and small business" [(8), p. 2-1]. DOE recognized this and issued revisions to its regulations in 1980, 1981, and 1982, designed in part to streamline the program and its reporting requirements.

The first state RCS plan was approved by DOE in October 1980. Most states began full-scale implementation of RCS in 1982 or 1983, resulting in less than five years of program operation prior to the January 1, 1985, expiration of key requirements. Early in the Reagan administration, DOE expressed its intention to fund RCS program activities only minimally. In anticipation of the January 1985 expiration date, DOE sponsored a series of internal and contractor evaluations of the RCS program during 1983 and 1984. The conclusions of these studies reinforced the Reagan administration's intent to allow RCS requirements to expire.

While this effectively ended RCS as a national program, many utilities and states can be expected to continue RCS in one form or another. Several states have enacted legislation at the state level authorizing continuation of RCS-type activities. Closely related activities continue even at the federal level under the Commercial and Apartment Conservation Service Program (CACS). CACS was enacted by the 1980 Energy Security Act and extends energy audit and related utility-delivered services to apartments and small commercial buildings. Since CACS is scheduled to run until January 1, 1990, it is hoped that some of the experience in RCS can aid in implementing CACS as well as in the continuing state and utility efforts to promote residential conservation.

OVERVIEW OF NATIONAL PROGRAM RESULTS

In 1983, DOE initiated a series of evaluation studies of the RCS program, which were later summarized in a document referred to as the *RCS Evaluation Report* in January 1984 (8). This series of reports and their governmental distillations represent the most accessible sources of information on the results of RCS, including program statistics, costs, energy savings estimates, cost-effectiveness, and qualitative implementation issues. The principal drawback of these analyses (and it is a significant one) is that their main data bases were the 1982 and 1983 state annual reports on RCS. This meant a cutoff date of April 1983 for basic program statistics. Many state and utility programs were just beginning to gather momentum in late 1982 after 1981 start-ups and mid-course corrections to take advantage of DOE's 1982 revisions of the regulations. Thus, these reported results cannot be considered the final word on the program. Moreover, in the area of cost-effectiveness in particular, limitations on data and methodology raise further questions. Overall, however, the DOE-sponsored reports serve a useful function in giving an overview of RCS in midstream and frankly outlining implementation issues and policy viewpoints in a way that rings true. The results given below are drawn principally from these DOE-sponsored studies.

Participation by States

By early 1984, 40 states had plans approved by DOE and operational RCS programs; 37 had conducted at least some audits prior to the April 1983 data cutoff for the DOE evaluation. At least two of the states not reporting any RCS audits at that time had aggressive home audit programs, however. Utilities in New York State, under its 1977 Home Insulation and Energy Conservation Act, had audited over 400,000 homes and had made over 22,000 conservation loans through 1982 (9). The Tennessee Valley Authority, headquartered in a state not participating in RCS, had made nearly 500,000 residential surveys and loaned $185 million to 200,000 consumers by September 1981 (10). Tennessee was one of several non-participant states abiding by the spirit of NECPA more than some half-hearted participants. Nonetheless, it is an interesting commentary on the limits of federal authority that one-fifth of the affected states could still be nonparticipants more than five years after passage of NECPA. As of the evaluation date, six states accounted for 56% of RCS audits conducted, but a number of smaller states had respectable audit tallies exceeding 10% of eligible households.

Audit Statistics

Table 1 reproduces one of the central statistical summaries developed in the DOE evaluation process. Compiled by DOE and its consultants from 1982 and 1983 annual reports from 37 states, it reveals much about program results up to April 1983. The state reports contained some recognized data problems, such as double-counting of customers served by both electric and gas utilities in states such as Pennsylvania, which overstated eligible customers. DOE estimates that the total number of eligible customers for the 37 reporting states is closer to 50 million than the 61.7 million shown in Table 1. With 2 million audits performed in this two-year period, this implies a 4% cumulative penetration rate (8). DOE also compiled data correlating customer requests for audits with the number of "unconditional offers" made by a utility (an unconditional offer is the RCS term for a direct contact with a customer clearly stating the availability of an audit and describing how to request one). They found that 5.6% of those receiving such offers in 1982 requested audits (8). This is slightly higher than the ratio of audits to eligible customers, since utilities were allowed to space out their audit offers. The customer response and audit penetration rates fell below

Table 1   RCS audits and program costs as of April 1983

State[a]	First audit performed[b]	Eligible customers[c]	Number of households[d]	Number of audits performed[e]	Utility expenses[f] ($)	State expenses[g] ($)
NORTHERN STATES						
Alaska			131,000			
Colorado	4/81	600,000	1,061,000	43,553	1,100,000	500
Connecticut	10/80	844,000	1,094,000	79,859	10,500,000	203,454
Delaware	3/80	90,418	207,000	9,060	900,000	18,874
D.C.			253,000			
Idaho		384,871	324,000	2,301	600,000	41,077
Illinois	10/81	5,159,982	4,043,000	63,255	7,000,000	178,000
Indiana	7/81	2,091,795	1,927,000	32,299	3,900,000	60,298
Iowa	6/81	770,000	1,053,000	29,808		
Kansas	2/83		872,000			
Kentucky	11/82	821,870	1,264,000	125	400,000	50,283
Maine	11/81	408,099	395,000	9,122	800,000	18,388
Maryland	9/81	1,392,147	1,461,000	28,598	3,000,000	163,000
Massachusetts	12/81	1,876,730	2,033,000	165,000	15,400,000	88,000
Michigan	7/81	3,000,000	3,195,000	334,000	46,000,000	
Minnesota	9/81	1,187,000	1,447,000	46,517	5,200,000	184,000
Missouri			1,794,000			
Montana			284,000			
Nebraska	10/81	517,424	571,000	9,535	1,700,000	
New Hampshire	11/81	259,000	323,000	1,786	200,000	21,182
New Jersey	3/81	2,158,657	2,548,000	18,517	3,600,000	81,000
New York			6,340,000			
N. Dakota			228,000			
Ohio	7/81	3,374,215	3,834,000	16,056	3,700,000	42,723
Oregon	4/81	429,650	991,000	32,597	1,000,000	64,500
Pennsylvania	8/81	5,640,049	4,221,000	22,146	4,000,000	883,402
Rhode Island	1/81	300,000	339,000	20,300	2,200,000	132,989
S. Dakota	10/81	255,832	242,000	5,133	900,000	49,565
Utah			449,000			
Vermont	7/82	147,000	178,000	17,000		932,727
Virginia			1,863,000		100,000	

Washington	2/83	273,134	1,540,000	55,235	5,600,000	201,443
W. Virginia	2/82	436,754	686,000	1,454	1,400,000	51,900
Wisconsin	9/81	1,500,000	1,652,000	280,416	3,500,000	
Wyoming			166,000			
NORTHERN TOTAL		33,918,627	49,009,000	1,323,682	122,700,000	3,467,297
SOUTHERN STATES						
Alabama	11/81	850,000	1,341,000	21,990	2,200,000	30,300
Arkansas	3/82	999,470	816,000	5,060	1,200,000	90,250
Arizona			957,000			
California	3/82	6,500,000	8,630,000	311,262	54,700,000	529,600
Florida	3/81	4,345,700	3,741,000	244,876	15,200,000	234,094
Georgia	4/81	1,717,247	1,872,000	23,241	2,900,000	41,919
Hawaii			294,000		100,000	5,000
Louisiana	7/81	1,685,000	1,411,000	3,914	400,000	5,500
Mississippi	3/81	490,650	827,000	587	300,000	58,430
Nevada			304,000			
New Mexico	7/80	570,457	441,000	7,212	500,000	52,240
N. Carolina	9/81	1,500,000	2,041,000	6,685	1,300,000	139,700
Oklahoma	6/81	1,547,465	1,119,000	22,869	4,200,000	203,060
S. Carolina	5/81	766,012	1,030,000	8,083	1,100,000	125,000
Tennessee			1,618,000			
Texas	3/81	6,880,060	4,929,000	52,213	9,100,000	102,700
SOUTHERN TOTAL		27,852,061	31,371,000	707,992	93,200,000	1,617,693
NATIONAL TOTAL		61,700,688	80,380,000	2,031,674	215,900,000	5,084,990

[a] Data for Table 1 are presented for each state, for southern states, for northern states, and for the nation.

[b] This is the date of the first audit performed by any covered utility in the state under the RCS program as reported in the state annual reports. Some states started providing audits earlier than the official RCS Program approval date.

[c] The data presented are as reported in the state annual reports for 1982 and 1983. There is some double counting, resulting from failure to account for overlapping electric utility service districts.

[d] The data are taken from the 1981 U.S. Statistical Abstract and are based on the 1980 US Census.

[e] This is a cumulative total of the number of audits performed as reported in the 1982 and 1983 state annual reports. Some states cannot distinguish between RCS and other audits and thus probably overreport the number of RCS audits performed. Data are not available for some states, causing the totals to be understated.

[f] This is a cumulative total of utility expenses as reported in the 1982 and 1983 state annual reports. Uncertainty exists about the accounting procedures for these expenses. For example, reported utility expenses may or may not include program administration costs in addition to the costs of performing the audit. Often the reported costs per audit multiplied by the number of audits performed is far less than the reported utility expenditures.

[g] These are a cumulative total of the state expenses as reported in the 1982 and 1983 state annual reports. Uncertainty exists as to each state's procedures for accounting for these expenses. For example, the expenses may or may not include personnel and overhead costs, or costs reimbursed to the state by utilities. The national total is likely to be understated.

Source: DOE Evaluation Report (8).

the expectations of program planners, lying at the low end of the range of forecasts in the 1982 DOE Regulatory Impact Analysis (7) and well below the 7% annual penetration envisioned in DOE's 1979 program projections for free audits. DOE found a high variability in response rates among different utilities and regions, with the number of audits per eligible household varying from 0.1% to 31% (9). It also found audit response rates averaging 20% for RCS programs coupled to financial incentives, compared with 5% for programs without such incentives (9). For the two-year reporting period, only Delaware, Michigan, Vermont, Washington, and Wisconsin had audited more than 10% of their eligible customers. All of these were Northern states, consistent with an overall tendency observed for states with higher heating loads to have higher participation rates (7).

Various possible reasons for the shortfall in RCS participation rates compared with early expectations are identified in the DOE reviews:

1. Data samples from the early stages of program implementation, which may not reflect the ultimate potential for participation.
2. Inability or unwillingness on the utilities' part to market the program effectively.
3. Overly restrictive program requirements that inhibited flexible and innovative marketing.
4. Narrower than expected demographic appeal of the RCS audit with lower socioeconomic groups especially underrepresented.
5. The possibility that much of the easy conservation "cream" had been skimmed in the RCS eligible group by individual efforts or in response to tax credits and other programs.
6. Lack of economic incentive for homeowners in areas of milder climate and lower utility costs.
7. Declining concern about energy as an issue.

Unfortunately, the information bearing on the reasons for the overall participation rates is largely qualitative and anecdotal. It is therefore difficult to determine whether there was a problem with the product being offered (basically home audits by local utilities), with the way the product was being marketed and distributed, or perhaps simply with the pre-introduction market penetration projections and their interpretation. Nonetheless, the unexpectedly modest early penetration rates in all but a few states, such as Michigan, contributed to DOE's and others' sense that RCS was far from a runaway success. At the same time, with 2 million RCS audits performed in the first two years of operation, it was evident that home energy conservation services were being delivered on an unprecedented scale. Expectations were that time, experience, and relaxed federal and state requirements would improve the program, and preliminary cost-effectiveness analyses for RCS-type audits were by no means uniformly negative (11, 12). Penetration rates thus became but one of several factors figuring in assessments of the program's overall success and desirability.

Characteristics of Participants

Virtually all studies of the demographics of participants in RCS and comparable residential audit programs (12, 13) reach conclusions similar to DOE's: "Ideally, RCS would serve all demographic groups living in RCS eligible dwellings with incomes above the poverty level, i.e. those who are not already served by other federal programs such as the Weatherization Assistance Program. Results indicate, however, that RCS reaches an affluent, highly educated, home-owning subset of the general population" (9). Typical of the results referred to (which were from a few states that conducted demographic surveys) was that for RCS participants in Massachusetts; 93% owned their homes, 58% were college graduates, and 55% had incomes over $30,000, versus 81%, 31%, and 31% for the state's average utility customer. Other studies showed that participants tended to have greater energy awareness and to have installed more conservation measures in their homes prior to the audit (14). This suggested that RCS audits appealed to the very people who are likely to have already implemented the more cost-effective measures. Whatever effect the skewed demographics of RCS may have on program impact and cost-effectiveness, they contributed to the perceptions among some consumer and other groups that RCS had regressive income-transfer effects.

Program Costs

As shown in Table 1, DOE estimated utility and state program costs at nearly $221 million for the first two years of operation. Including federal program costs for this period brings the total to an estimated $240 million. Since there was little effort to ensure that different utilities and states used consistent accounting techniques, this must be considered only an approximate figure (8). DOE divided this by the 2,000,000 audits performed to arrive at an estimated cost per participant of $120, including $10 per participant in federal costs. Only about $22 million, or an average of $11 per participant, was recovered directly from audit fees paid by participants. The bulk of program costs were spread over the nonparticipant ratepayer and general taxpayer populations. The main cost component of the program was the direct cost of providing a trained energy auditor in the participants' home for a one- to two-hour Class A audit, with associated computer link for energy analysis calculations. For one California utility typical of those providing a detailed Class A audit, these direct costs were over 78% of total utility program costs (15). These cost characteristics led many utilities to seek ways to shorten and simplfy the RCS audit without losing the face-to-face, in-home opportunity to promote conservation that the RCS Class A audit provides.

Energy Impacts and Cost-Effectiveness

Energy savings and their cost-effectiveness are central to any evaluation of RCS. Such data are, naturally, much harder to obtain than raw audit and program cost figures. At the national level, DOE had to rely mainly on an analysis of evaluations designed and executed at the state level, since federal RCS rules provided little direction to the states in this regard. Only 23 states had data on differences in rates of implementing conservation measures between participants and nonparticipants, and these data were primarily the unverified results of telephone surveys. DOE found only five states with usable data on incremental energy savings resulting from RCS participation compared to nonparticipants (8). While recognizing explicitly many weaknesses in data and methodology, DOE felt that some conclusions could be drawn, including the following:

> . . . the differences in the rates of adoption of conservation measures between RCS participants and non-participants are small. Where participants installed more measures than non-participants, the percentage differences are: caulking 13%, weatherstripping 7%, ceiling insulation 10%, duct insulation 6%, and clock thermostats 1%. In the five states where data are available, the RCS program has increased conservation investments by about 80% relative to all non-participating households. (From about $250 to $475 per household.) The average incremental energy savings in these states analyzed for participants versus non-participants is about 4 million Btus per year. (Nine million Btus in savings for participants versus 5 million Btus for nonparticipants.) For the two million audits conducted in the first two years of the program, this would represent savings to consumers of approximately 12 trillion Btus, over the two-years of program operation . . . this represents about 17 trillion Btus in primary energy use [(8), pp. 4-1 to 4-3].

With respect to overall program cost-effectiveness, the DOE *RCS Evaluation Report* provides an estimate of national incremental cost-effectiveness (the dollar value of additional energy savings attributable to the RCS program divided by RCS program costs) ranging from a high of 0.9 at a 5% discount rate to 0.5 at a 15% discount rate, commenting only that, "A ratio greater than one indicates that the additional savings attributable to the RCS Program exceed the additional costs generated by the program" (8). The principal contractor study on which DOE drew for its cost-effectiveness figures was far less coy:

> An evaluation of the Residential Conservation Service (RCS) Program, as it is currently implemented, suggests that the program may not be cost-effective. This conclusion is derived from the base case analysis, that is, the best estimate of the RCS Program's impacts. As the sensitivity analyses below indicate, conditions would have to change significantly in order for the primary conclusions to be found invalid. This evaluation is

primarily from a national or societal perspective considering the investment in energy conservation made by consumers, the administrative costs of the program, and the energy savings for consumers. Conversely, this evaluation does not extensively consider the Program's benefits to utilities or government, or the indirect benefits to society. Based on limited available data over about two years of program implementation experience, participation rates have not matched expectations and, for the average audited customer, energy savings do not match the total conservation investment costs required to produce those savings. . . .

There are at least three possible explanations for the poor cost-benefit results presented above. First, many homeowners have already implemented energy conservation measures prior to receiving the audit and the additional measures recommended by the audit may be at the low end of the cost-benefit spectrum. This explanation is the most likely possibility. In a recent GAO [Government Accounting Office] report, selected gas utilities reported that residential customers currently use 25 to 30 percent less energy than in the mid-1970s. Surveys in California and Massachusetts as well as many utility officials suggest that audited customers have made major improvements prior to the audit. Analysis of the use of the federal residential energy tax credit indicates between 5 and 6 million households were annually investing about $700 in energy conservation for the years 1978, 1979, and 1980, that is, before the RCS Program began in all but a few states. Second, audit advice based on engineering estimates may not reflect the real world. For example poorly installed insulation may have an effective R-value that is only a fraction of its nominal R-value. Finally, homeowners may be raising their thermostats and improving their quality of life after implementing the audit recommendation. That is, people may decide that once their homes are energy efficient, warmth is more important than dollar savings [(7), pp. iv–v].

While strongly put, these conclusions should not be viewed as final or definitive assessments of RCS cost-effectiveness, if for no other reason than that the final years of program operation are not reflected. These "national" assessments certainly do not imply that all state or utility RCS programs would fail an incremental cost-effectiveness test. The Connecticut, Michigan, Minnesota, and California programs, for example, appear to meet this criterion under some conditions [(8), p. 4-5]. Most RCS programs, it should be noted, are clearly cost-effective for participants (8). The DOE figures are in agreement with other findings that it is much more difficult to make a case for RCS being cost-effective for the nonparticipant. Crudely speaking, audit costs average $120, 90% of which costs are borne by nonparticipants, and annual incremental energy savings, while respectable, averaging $35 per year per participant (8), are not dramatic enough to have large spillover benefits to nonparticipants.

IMPLEMENTATION ISSUES—A CALIFORNIA FOCUS

The national "numbers game" summarized above, even if expanded upon to reveal the variation in results among different RCS programs, tells only a small part of the RCS story. To extract the full benefit of the massive experiment that RCS represented, one must look at implementation issues that arose at a variety of levels and from a variety of perspectives. DOE recognized this to an extent in its evaluation efforts, by sponsoring a series of case studies (8) and by using focus groups and other means to identify recurring issues and their potential solutions (16). In this section we will look selectively at several of the more interesting of these implementation issues, from the perspective of experience operating one of the largest state RCS programs—that of California. We hope that this will also reveal some of the insights obtainable from a more detailed look at individual programs.

Penetration Rates and Utility Regulation

In the prior section, the disparity between the DOE target of 7% per annum participation rates and actual participation rates was discussed. While the 7% figure was never an official goal, it had become an unofficial benchmark after being included in the DOE regulation impact analysis. Actually this analysis estimated a range of penetrations for different audit fees: 7% for a free audit and 1.5% if the audit charge was $15 (17). These estimates were supported by a study of 12 pre-existing audit programs with average response rates of 7.7% for eight free audit programs and 0.9% for four programs with low but nonzero audit fees (18). Penetration rates can be determined in a number of ways. Among the first year's reporting states and utilities, 1.18 million audit requests were made out of 66.8 million eligible customers, for a response rate of 1.77% [(19), p. 5]. However, measured another way, 1.18 million audit requests were made by the 10.8 million customers who received an actual audit offer, for a 10.9% response rate. Counted the first way, program accomplishments seem meager; counted the second, they were quite respectable.

California argued that a more informative measure of response was the rate of audit requests by eligible customers who expressed interest during a preprogram survey. Fifty one percent of eligible customers responded affirmatively in a survey of interest in the program taken in March 1981 (20). As Table 2 shows, over time the response rate rose. In California, the response rose from 4.5% of those interested and eligible in year one of the program to 13.7% in year three. In the early summer of 1985, California was projected to perform its millionth audit. (RCS audits are free in California.) By this measure, the program has achieved substantial penetration of the 6.5 million eligible customers.

In the discussion of national penetration rates it was noted that there is little consensus about the reasons for the observed penetration rates. In California at least, an institutional situation existed which had a direct bearing on penetration rates. Federal regulations required utilities to provide an audit to any customer who asked for one. However, utilities

Table 2 California program statistics

	Eligible customers	Interested customers	Number of audits requested	Requested eligible (%)	Requested interested (%)	Costs		
						Utilities ($)	State ($)	Per audit performed ($)
Year 1	6,459,500	3,294,345	147,871	2.3	4.5	27,166,640	331,900	279
Year 2	6,459,500	2,648,395	302,719	4.7	11.4	26,369,748	165,500	126
Year 3	6,459,500	2,648,395	363,060	5.6	13.7	28,845,937	79,985	95
Year 4/1st half	6,459,500		197,078			14,469,094	30,000	74

Source: RCS annual reports and quarterly operating statistics, California Energy Commission.

were allowed to pace their workload by nondiscriminatory spacing of the unconditional audit offers. DOE's original requirements for information to be included in utility program announcements, which were supposed to generate customer interest, made the announcements more information booklets than effective marketing tools. As a result, utilities could use other marketing and advertising, coupled with the allowed spacing of unconditional offers, to influence the rates of audit requests. Initially, auditor training and other program scale-up considerations set the optimal pace for audit requests. After the first year, however, RCS program budget constraints became the crucial factor. As noted earlier, a high percentage of utility RCS program costs are directly related to the number of audits conducted. (Table 2 shows per audit costs declining over time, but within each year costs were roughly constant.) In California, each participating investor-owned utility was given a line item RCS program budget by the California Public Utility Commission (CPUC). While these allocations were not absolute ceilings, it was the clear intention of the CPUC that utilities observe these allocations whenever possible. Not the least of the factors involved was that while the CPUC controlled the RCS purse strings, substantive program control over RCS had been assigned by the Governor to a separate state agency, the California Energy Commission (CEC). The CPUC and CEC were both anxious to promote cost-effective utility conservation programs, but there was a degree of interagency rivalry. On several occasions the CEC argued before the CPUC for RCS budgets substantially higher than those finally approved, while the CPUC staff argued for lower allocations. Regardless of the bureaucratic details, however, annual penetration rates in California were strongly influenced by funding rates set by the state ratemaking body. Although the reviewers have not specifically researched other states in this regard, it would not be surprising to find penetration rates in other RCS programs to be a function as much of budget limits on the supply side as of market demand.

RCS and Consumer Protection

When Congress enacted RCS, it included a number of consumer protection elements in the program. These included having the utility prepare lists of qualified installers, suppliers, and financial institutions, arrange for installation, provide inspections, and mediate complaints. DOE eventually found these provisions not to have been used much by participants, except in a few programs in which audits were linked to the availability of financial incentives, such as loans or state tax credits (8). Congress also sought to protect both consumers and independent conservation contractors from the threat of utility marketing power by putting limits on how far the utility could proceed in arranging for or performing the installation of conservation measures.

In California, the establishment of lists of qualified suppliers, lenders, and installers became one of the most controversial operating issues. The state sidestepped the logistical problems of developing a retail suppliers list by creating a generic list of vendors (e.g. air-conditioner dealers, appliance dealers, and hardware stores). The state attempted a full-blown implementation of the installer listing provisions, spending over $100,000 to invite the state's 110,000 licensed contractors to become "RCS listed." Early DOE regulations would not allow the state's contractor license list to be used, since it did not include provision for a one-year written warranty, a delisting procedure for RCS offenses, or conciliation procedure agreements. This, in turn, alienated contractor organizations and the existing Contractors State License Board, who saw the DOE requirements as cumbersome and unnecessary. In the end, only about 1,600 contractors chose to be listed on the computer-generated matrices distributed to RCS customers.

The state, contractors, and utilities all agreed that the cumbersome RCS lists were of little use to customers. (Some smaller states culled all contractors out of their local phone books for use as their lists.) The contractor list envisioned by DOE would have been specific by product and by market area. It was impossible to separate out relevant market areas in densely populated areas (Los Angeles) and the resulting lists often contained contractors who were two or more hours away. Also, utilities did not want to be intermediaries between customers and contractors; they did not want to pay for the service, resolve complaints against contractors, or allow contractors access to the names of audited customers (20). In 1983, DOE relaxed the rules and allowed temporary exemptions for listing. California immediately applied for and was eventually granted authority to use licensure as the accepted listing procedure.

By not accepting existing, if imperfect, state regulatory mechanisms and instead insisting on conceptually more perfect consumer protection procedures, the RCS program overstepped practicality. It tried to use one program to radically restructure existing markets and relationships among parties who fundamentally did not trust one another. Without providing any money to ease the pain, DOE ordered that states, utilities, contractors, community organizations, and consumer groups would cooperate and establish a new consumer conservation service. Part of this order worked— utilities did create a new customer service that frequently dwarfed all their other customer programs. The audits, in turn, generated a whole new industry of auditing techniques and measurements. However, one utility program could not carry the weight of reforming an entire market. The hostility aroused by this attempt weakened the entire program and consumed a great deal of political and administrative effort (20). In California, RCS became an organizing issue for trade associations. During the 1982 recession, when many small businesses were having trouble staying afloat, it became convenient to blame the state, utilities, and RCS for disrupting the marketplace for conservation services. To the contractors, the costs and constraints imposed by RCS were all too apparent. One often-cited problem was the need to wait for a utility to conduct an RCS audit before closing a sale to a customer for certain measures in order to qualify for California tax credits. The promised benefits of referrals to RCS-listed contractors from audited customers never materialized on a large scale.

Audit Calculations and Auditor Training

At the working level of RCS, approving audit methodologies and auditor training were the key state controls over program content. These tasks were feasible if a state had an independent modeling capability. States that did not, as the GAO report found, wound up with widely disparate payback and energy savings information being reported to customers of different utilities (17). Operational issues included estimation of energy savings for specific products, the useful lives of products, product prices, and the trade-off between an audit's accuracy and its costs.

In the process, a national mini-industry of conservation engineers and consultants emerged to answer these questions. In the latter years of RCS, indications were that even highly technical audits could make only approximations of actual energy to be saved (21). This called into question one of the program's basic theses—that accurate, house-specific predictions of energy and dollar savings could be provided to consumers through such audits.

The most common reasons for requesting an audit in California were found to be the following: to learn ways to conserve from an energy professional (49%), to reduce utility bills (43%), and to save money generally (8%) (13). RCS participants rated the auditors highly on knowledge and courtesy and gave mid-range responses for the believability and utility of recommendations. Sixty seven percent expected to save some energy from their audits, but 30% did not expect to save, and 3% expected to save a lot.

Combining customer responses with marketing research, California revamped its audit to increase the role of the auditor as a persuader and decrease the technical component by prepackaging many of the computer calculations. Average audit length was cut in half, from 3 hours to $1\frac{1}{2}$ hours. Costs dropped by a third. Such an overhaul would not have been possible without the DOE change to a minimalist regulatory philosophy. Even so, it took much creative wordsmithing and liberal interpretations to be able to use operational feedback in redesigning the program. One of the state's key concerns, that the program be cost-effective, was not an explicit requirement of the law; nor did it appear, from the state's perspective, to be a high-priority design criteria among DOE's program overseers. This was reflected in the overspecification of audit content and method, which inhibited efforts by states and utilities to develop more cost-effective approaches. For example, in California, amending the state RCS plan to take advantage of the increased flexibility in DOE's 1982 rules became an extensive process in itself, with preparation and hearings at the state level taking over six months, followed by additional time for DOE approval (22).

Broadening the Market for RCS

One difficult implementation issue, for California and many other states, was resolving the question of what was the appropriate target audience for RCS. It was known in advance who the prime market would be for an in-house, ultility-sponsored informational audit. Various surveys had identified the potential clientele as middle income and above, well-educated, middle-aged homeowners (11). As reported earlier, the experience of the first two years of RCS verified these tendencies. This came at a time of deepening recession, high unemployment, and sudden rate increases for customers of some major California utilities in early 1982. The growth in utility expenditures on federal- and state-mandated conservation programs was clearly visible in the utility ratemaking process. Conservation programs in general, and RCS in particular, began to come under critical scrutiny by public interest groups concerned about the lack of participation by the elderly, the poor, and renters (22). This contributed to the emergence of cost-effectiveness to nonparticipating ratepayers as an explicit and often the dominant criteria for utility program cost-effectiveness analyses (23). This in turn was a factor in the CPUC's reluctance to fund RCS at the levels proposed by the utilities and the California Energy Commission. (There was a certain Catch-22 aspect to this, since higher funding rates would presumably have meant more participants and fewer nonparticipants.)

Concern for access to conservation services by nonparticipants also led to modifications and additions to California RCS programs to serve these populations better. Strategies included (a) marketing the program more intensively to underrepresented groups; (b) redesigning the program to be more attractive and useful to them, and (c) sponsoring companion programs more appropriate to these clienteles. California experimented with the first two options and invested most heavily in the third, principally by expanding government and utility-funded direct weatherization programs. Local governments and community groups lobbied for a role in marketing RCS programs to underrepresented groups, and several were employed by utilities on a contract basis to increase penetration rates. Perhaps the most interesting combined approach developed is the experimental, door-to-door program by the city of Santa Monica. The city felt that its demographic characteristics (80% renters) and climatic conditions did not fit the standard RCS mold, and it also wanted to be directly involved in community conservation activities. The program Santa Monica developed is different from conventional RCS activities in at least four ways: (a) every resident is eligible; (b) audits are delivered by a door-to-door neighborhood canvass; (c) results are precalculated for standard housing types, and; (d) low-cost installations are made at the time of the audit at no charge to the resident. First and second quarter results are impressive:

1. 34–37% of eligible customers were reached.
2. At least one installation was made in 94% of houses.
3. Low-income residents, senior citizens, and renters were reached in proportion to their presence in the population.
4. Surveys show 95% satisfaction by participants, with 73% giving the program the highest possible rating (24).

Intended as a model for what a community can do within RCS to fashion a program to meet its citizens' needs, the Santa Monica program is a promising and innovative effort. Ironically, the fact that it took over two years of intense negotiations to get the program approved within RCS, despite active support from the state RCS lead agency, may end up discouraging other less resolute communities.

The implementation issues addressed in this section are typical but by no means all-inclusive of those encountered in putting RCS into operation. Many of the problems that consumed the most effort at the operational level could have been avoided or lessened had the program been designed with more flexibility from the beginning. Flexibility seems to be particularly important as a design principle in programs such as RCS where there are several links in the program delivery chain (DOE-state-utility-contractor), regional and geographical differences, and close ties to other programs or objectives (tax credits, low-income weatherization, consumer protection).

RCS AND EVALUATION

A common and often justified lament of analysts and public policy makers is the lack of adequate, timely program evaluation. The RCS program has generally been perceived as suffering from this malaise. One of the RCS's observers, Eric Hirst, wrote in 1984, after reviewing available RCS evaluation studies, "Unfortunately, little is known about the performance of the RCS program, either at the local or national levels. Questions remain concerning energy savings that can be directly attributed to the program; cost-effectiveness of the program to participants, nonparticipating rate-payers (who may pay for the program through gas and electric rates), utility stockholders, and society as a whole; and alternatives to the RCS that might stimulate greater residential energy-efficiency improvements at lower costs" [(25), G-29]. Virtually all RCS evaluation studies feel compelled to include strong caveats with regard to data quality, unresolved meth-odological issues, and the tentativeness of their policy recommendations (8).

That the RCS evaluation studies have limitations both individually and collectively cannot be denied, but should not obscure the overall contri-butions of this body of work. First, the evaluation literature provides the most accessible and useful source of information on RCS. The nature of the program has not required or inspired a large technical literature. The methodological issues, while significant to the program, have not been so weighty or longstanding as to result in new analytical cottage industries (like weapon system cost-benefit analyses or nuclear risk assessment). Many of the descriptive studies done at the state or utility level also include considerable evaluative components.

Second, the RCS evaluation literature is at least partly successful in illuminating the central issues and results relevant to policy making in RCS.

Third, the RCS evaluation activities are themselves an important part of the RCS "story," with implications for future policy making.

The RCS evaluation literature can be organized into four main groups by source:

1. The DOE evaluation studies.
2. The Oak Ridge National Laboratory (ORNL) study series.
3. Studies by several states with significant evaluation efforts.
4. Miscellaneous analyses of individual state or utility programs.

The DOE Studies

The principal findings of the series of reports commissioned by DOE as part of its 1983–1984 RCS evaluation activities were described earlier. The data sources consisted primarily of the annual state RCS reports submitted to DOE, augmented by contractor reviews of the few more in-depth evaluation efforts and a series of utility case studies (7, 16, 26–28). For the most part, the DOE studies did not seek to break new methodological ground, but rather synthesized and summarized the state report data into a national progress report on RCS, while pushing as far as possible toward some useful cost-benefit quantification. The Centaur Associates cost-benefit evaluation provided an additional useful set of sensitivity analyses of the preliminary cost-benefit results (7). The sensitivity analyses indicated that substantial changes in input assumptions (e.g. 100% increase in final costs or 50% reductions in RCS program costs) would be needed to make the program cost-effective on an incremental basis. The output of the DOE-sponsored reports was ably summarized by DHR, Inc. (16) before being further distilled into DOE's own reports (8, 9). Collectively, the reports are useful compilations of state-level statistics and also provide a good summary of the principal program implementation issues that had emerged by the RCS program's middle years. To students of bureaucracy and federal programs, they also could provide material for an interesting case study of public policy evaluation. The major limitations of the DOE studies stem from their dependence on data from only two start-up years of program operation, data that were already highly aggregated and had passed through uncertain filters at the utility and state levels. The problem was not so much the uncertainties introduced into the overall cost-benefit analyses. In fact, the overall assessment of average program cost-effectiveness appears quite robust. What was lacking was the ability to identify, among the many program variants that were being developed, especially after the 1982 program revisions, those features that had the potential to improve program results significantly.

The Oak Ridge National Laboratory Studies

A group of staff members from the Oak Ridge National Laboratory (ORNL) under the direction of Eric Hirst, has produced an extensive series of evaluation-oriented studies of RCS and similar home energy audit programs (11, 12, 19, 25, 29–33). The ORNL studies have made significant methodological contributions to the evaluation of RCS-type programs (29–32). They have also provided evaluations of major home audit programs in Wisconsin, Minnesota, Connecticut, and the Bonneville Power Authority (BPA) with a consistent outlook and increasing sophistication. ORNL is a DOE national laboratory, with a history of home audit evaluation efforts going back prior to 1980; its work has been the reference point for most serious evaluation efforts at both the national level and below. Any citation analysis of the conservation evaluation literature would give ORNL high ranking.

The ORNL studies collectively provide a clear presentation of the methodological issues faced in RCS evaluation, along with concrete programs for dealing with them. The basic RCS evaluation issue is whether RCS is a cost-effective means of achieving significant energy efficiency improvements in the residential sector. Residential users may opt to take all or part of any energy efficiency improvements in the form of higher comfort levels, instead of energy savings. Neither the ORNL studies nor any of the other studies reviewed do more than note that this could lead to an underestimate of RCS program benefits if energy savings are used as a proxy for efficiency improvements. Energy savings attributable to RCS (or the associated stream of dollar savings when overlain by energy prices) are reasonable benefit proxies, especially for interprogram comparisons within the residential sector. They also have the advantage of being quantifiable through a variety of means.

One approach is to use engineering estimates of energy savings from particular conservation measures in conjunction with survey data on the frequency with which each measure is implemented. Estimates of this kind appear frequently in the RCS literature because such estimates can be readily derived from the engineering estimates used in the audits and records of audit recommendations, as well as the pre- and postaudit customer surveys (29). Unfortunately, discrepancies between engineering estimates of such measure surveys and actual savings in the field have been shown to be substantial, with actual savings usually less than engineering estimates (34). Unverified phone or mail-in surveys of conservation measures implemented suffer from the tendency of respondents to provide "socially desirable" answers, again overstating savings.

Oak Ridge and other serious evaluation efforts rely on statistical analyses of utility billing data to derive savings estimates. In addition to some fairly standard adjustments for weather effects, analysis of utility billing data to detect RCS program impacts presents some special challenges. The main problem is that of approximating a controlled experiment where none actually exists (e.g. comparing savings with and without an RCS program when RCS is a fact.) Since RCS is a voluntary program consistently shown to appeal to people with higher levels of income, education, and interest in energy issues, the potential for a "self-selection" bias exists (29), particularly when overall audit penetration rates are low. Interestingly, it is not clear whether the self-selected RCS participants are more likely to respond to an RCS audit and make installations, perhaps because of their interest and financial capability, or less likely to do so because they may already have taken the basic conservation steps. To minimize the effects of any self-selection bias, Oak Ridge, in its evaluation plan for BPA and elsewhere, employed matched samples of participants and nonparticipants (29).

Implementing its evaluation plan for BPA, the Oak Ridge team found some interesting results, including "This surprising finding–that the [audited] group did not save more electricity than the [nonparticipant] group" [(12), p. 63]. Customers who had an audit and took advantage of a financial incentive (loan) program, however, did show statistically significant savings (12). A later ORNL study found that including a second postaudit year in the analysis showed a measurable increase on the audit-only and audit-plus-loan groups relative to nonparticipants (32). The fact that adding second-year effects made the audit look much better illustrates the early developmental stage of this sort of program evaluation as well as the advantages of an ongoing multiyear analytical program like that at Oak Ridge.

Going from estimates of incremental energy savings to incremental cost-effectiveness analysis for RCS adds at least two sorts of complexities. First, as Hirst points out in a recent survey article (25), cost-effectiveness results are sensitive to site-specific factors such as climate and marginal fuel costs. He found RCS looked much more cost-effective in Michigan, with marginal gas costs above average year costs, than in Minnesota, where the reverse was true. In addition, cost-effectiveness depends strongly on the perspective considered. In Minnesota, for example, Hirst estimated RCS was cost-effective for participants, but "economically unattractive to nonpartici-pants and economically neutral to the state" [(25), p. G-36].

Current evaluation efforts and techniques have not provided much help in another task of program evaluation—designing improvements to the program. As Hirst succinctly notes: "Several alternatives to RCS have been proposed. These alternatives seek to reduce the prescriptive nature of the present program, reduce program operation costs, and increase participa-tion and subsequent retrofits. Unfortunately, even less empirical evidence concerning the advantages and disadvantages of these alternatives (except for subsidized loan programs, which show much larger energy savings than do audit-only programs) is available than exists on the RCS program itself" (25). The Oak Ridge studies suggest that conscientiously applied, available program evaluation techniques can provide "ballpark" estimates of overall savings and cost-effectiveness for a program like RCS, but they are far from being tools either to fine-tune such a program or to make fine discrimi-nations or final judgments on individual operational programs.

State Studies

There are several states the RCS evaluation efforts for which are frequently cited in the literature (Connecticut, Michigan, Wisconsin, Minnesota, California, and Washington), largely because of their efforts to quantify energy savings (25). We will illustrate these state efforts by briefly reviewing evaluation activities in Michigan and California.

In Michigan, an independent RCS evaluation effort was established in an agency separate from the program operating agency. Dr. Martin Kushler and John Saul of the Michigan Energy Administration have produced a series of evaluation reports dating back to December 1981 (35–39). Impetus was added to these efforts because the initial response rates to RCS in Michigan greatly exceeded expectations. The Michigan RCS program got off to an early running start, with good government and utility cooperation. Penetration rates were about 5% per year, with over 500,000 audits completed by mid-1984 (36). The state's evaluation efforts were comprehensive, including basic costs and participation statistics, customer attitude surveys, energy savings estimates extending to two postaudit years, matched samples of participants and nonparticipants, and simple cost-effectiveness calculations from participant, utility, and societal perspectives (38). Taken together the Michigan studies are a valuable documentary of one of the most successful RCS programs. A minor criticism is that they afford little specific insight into how another state might recreate the same favorable conditions, such as strong utility marketing of RCS.

The California evaluation plan for RCS set four objectives:

1. To test the hypothesis that the RCS program reduced energy consumption.
2. To determine RCS cost-effectiveness and means for improving it.
3. To provide feedback on the interaction between RCS and conservation incentive programs.
4. To assess the effectiveness in terms of whom the audits reach, installation rates, and the expenses of different audit types.

The approach was to use quarterly utility reports, biannual consumer surveys of participants and nonparticipants, and matched energy savings from billing data.

The requirements for RCS evaluation were generally defined in the state plan and were developed in a cooperative venture between the state and the utilities. A task force, chaired by a utility, was established to design the questionnaire and the utilities jointly funded a professional marketing research firm (40) to conduct the three customer surveys (Benchmark, Follow-up I, and Follow-up II). The utilities paid for all the surveys and for the collection of metered data, while the state paid for its own analysis of the statewide and utility-specific data. Each utility was allowed to add specific questions to its sample, in order to make the data collection more useful to it in other projects. This cooperative venture was very successful in providing a common baseline for evaluation throughout the state.

The Benchmark survey (14), designed to set a preprogram baseline, was a stratified random sample of utility customers prior to the introduction of RCS. Conducted in March 1981, the survey collected information from 2457 customers on their awareness of conservation measures and practices, their attitudes towards energy conservation, and their participation in conservation activities. Survey findings included (a), that RCS would need to raise awareness of utility conservation services from 36–50% up to at least 70%, (b) that there was a need for special marketing to older persons and lower-income individuals, since these groups were least interested in RCS type services, and (c) respondents tended to overreport conservation measures, when compared to other CEC appliance saturation data. Therefore, there was an expectation that the Follow-up surveys would have to be checked against other penetration estimates (14).

The Follow-up I survey (13) collected statistics on the first year of California's RCS program. The survey sampled 3177 households of both program participants and nonparticipants in order to assess the program impacts. It validated many of the projections of the Benchmark survey regarding the demographics of program participants and their reasons for wanting an audit. The RCS audit did invite action. Seventy-two percent of those who received audits took conservation actions either by making an installation or adopting a practice. RCS participants made more installations and adopted a greater number of conservation practices than nonparticipants. Even low-income participant households took conservation actions as a result of the audit.

Based on the survey results and engineering estimates of energy savings, the program appeared to be slightly cost-effective to society and very cost-effective to the participants. Matched sets of pre- and postaudit billing data were collected so that the engineering results could be improved by measured savings. Two utilities did publish conditional demand evaluations based on their billing data, but the state as yet has not completed an evaluation of the statewide data. This is a deficiency in the overall evaluation plan.

A similar Follow-up II survey was conducted in 1983 to document the third year's program results. Interest had stabilized somewhat over that period, and awareness had finally reached the 70% minimum marketing target. Again, participants reported more conservation activities and practices than nonparticipants, but not by wide margins.

One of the problems with the evaluation was that the program was changing so rapidly that customer surveys and billing data results were "old news" by the time data could be collected and analyzed. The state used crude aggregations, anecdotal data, and the professional judgments of utilities, CEC staff, and advisory groups in much of its program modification activities. There was also a serious methodology problem in that RCS was a component of several large consumer programs including state tax credits, utility loans, and utility rebates. It was almost impossible to identify what results should be attributed to which program.

Having utility-specific evaluations with common methodology, common interview techniques, and common reports vastly increased the usefulness of the data within California. The program was the only statewide utility program that benefited from this cooperative approach.

Other Evaluation Studies

As the RCS program was gearing up in 1981 and 1982, surveys of evaluation efforts were conducted at the RAND Corporation (41), Oak Ridge (11), and in the state of Michigan (38). A typical finding was that "in 1981 only one-half of the 38 utilities with active RCS programs had performed any evaluations" [(41), p. 11]. Though dated now, these surveys provide benchmarks for the evaluation effects at that time. Other more recent compendiums of RCS program and evaluation studies include the proceedings of the 1984 summer study of the American Council for an Energy-Efficient Economy (42), particularly the section by the panel on Federal and State Programs, and the proceedings of a conference on Energy Conservation Program Evaluation held in Chicago in 1984 (43). Reports on individual state and utility RCS programs too numerous to be reviewed here [e.g. (44)] are well represented in these two conference proceedings in addition to being available in individual reports [e.g. (45–49)].

Evaluating RCS Evaluation

As noted at the beginning of this section, the tone of many RCS evaluation studies is quite self-critical, apologizing for inadequacies in data, analyses, and results. Evaluation was not designed into the original legislation, and most states did little formal or quantitative evaluation beyond reporting of basic program operating statistics (38). Taken together, however, the DOE, ORNL, state, and utility evaluations constituted a serious self-examination effort. When compared to the paucity of good empirical data on the incremental cost-effectiveness of other conservation programs with much larger economic impact, such as federal and state conservation tax credits and new building efficiency standards, the RCS efforts appear even more creditable. The RCS evaluation efforts were a useful real-world application of program assessment techniques with more general usefulness, including the combination of utility billing data, attitudinal surveys, construction of experimental designs incorporating approximations of control groups, and incremental cost-benefit analysis from participant, nonparticipant, and other perspectives. Moreover, even where formal quantitative evaluation was lacking or incomplete, a feedback loop into program operations often existed. This was reflected in the major plan revisions in 1982 in California and other states to take advantage of the more flexible DOE regulations.

SUMMARY AND CONCLUSIONS

What can be said of the national experiment that RCS represented, and what lessons can be learned from this experience? There is little question that the program objective—increasing residential energy efficiency—was laudable and achievable through cost-effective means. Per household energy use had been increasing an average of 2.7% a year from 1960 through 1972. Subsequently, per household use dropped 20%, with most of the decline taking place from 1979 to 1981 (25). The potential for cost-effective residential conservation has been borne out by experience, in addition to innumerable analyses. Did RCS make a significant contribution to this national trend? Probably not. Hirst notes that "at current participation rates and energy savings, the total impact after five years of RCS operation would be a reduction in residential energy use of 0.2% (25). As a contribution to national energy security, "RCS energy savings are small" [(25), p. G-38]. Was RCS cost-effective? At the start of the program, there was little empirical evidence on this. With evaluation and hindsight, RCS must be placed in the marginal category, neither a dramatic winner nor loser with respect to cost-effectiveness, with much local variation.

What lessons can be learned from how RCS was created and administered? Pre-RCS utility home audit programs—pilot efforts for RCS, in effect—were sufficient to show that such programs were technically and administratively feasible on an individual basis. RCS sought to scale these up to a nationwide level within a short one or two years. The implementation problems proved to be immense. Perhaps the most serious flaw in RCS was that these problems were not adequately anticipated, nor was the administrative capability in place to respond to the challenges of introducing a new consumer service to the national market. Viewed as a national energy program that could be implemented through states and utilities with little federal overhead or expense, RCS looked like an attractive initiative. Viewed as an effort to market a new product through a diverse, multilevel, untrained, and often resistant distribution network, with no national marketing assistance or overhead funding and little profit potential, it is remarkable that RCS succeeded as well as it did (even more so in view of the restrictions on product design and marketing that NECPA and the DOE regulations imposed). With so many actors and organizational interfaces (e.g. DOE, states, utilities, and contractors), additional pilot testing and evaluation, perhaps in several states, might have exposed many implementation problems and calibrated the program's overall promise. In a program of this institutional complexity, a significant effort is justified to anticipate, analyze, and monitor organizational and institutional problems above and beyond the traditional legislative, rulemaking, and ratemaking processes. Combining the quantitative and qualitative observations on RCS, programs with the high administrative complexity and risk of RCS should preferably be pursued only when their potential programmatic rewards, in terms of absolute savings and cost-effectiveness, are commensurably high.

As for the future of RCS, the lapse of certain DOE requirements in January 1985 will mean a significant reduction in RCS-type activities in many areas. However, many utilities and state and local government are committed to maintaining home audit programs of some form. The increased flexibility and diversity are likely to provide some promising prototypes for future programs should a national interest develop, while providing increasingly cost-effective residential conservation services at the local, regional, or state level.

Literature Cited

1. US Govt. 1977. *The National Energy Plan.* Washington, D.C.
2. US Congress. 1978. *National Energy Conservation Policy Act*, PL 95-619. Washington, D.C.
3. US Congress. 1980. *Energy Security Act*, PL 96-294. Washington, D.C.
4. US Congress. 1977. Hearings before the House Commerce Committee. Washington, D.C.
5. Calif. Public Utilities Commission. 1980. *Energy Efficiency and the Utilities: New Directions.* A Symposium held at Stanford University, San Francisco, Calif.
6. US Dept. Energy. 1982. *Residential Conservation Service Program, Final Rule*, Federal Register 47: June 25. Washington, D.C.
7. Centaur Associates, Inc. 1983. *1983 RCS Highlights: Cost-Benefit Evaluation of the Residential Conservation Service Program.* Washington, D.C.
8. US Dept. Energy. 1984. *Residential Conservation Service Evaluation Report.* Washington, D.C.
9. US Dept. Energy. 1984. *Analysis of Options for the Residential Conservation Service and Commercial and Apartment Conservation Service.* Washington, D.C.
10. Tenn. Valley Authority. 1981. *Program Summary.* Chattanooga, Tenn.
11. Berry, L., Soderstrom, J., Hirst, E., Newman, B., Weaver, R. 1981. *Review of Evaluations of Utility Home Energy Audit Programs.* Rep. ORNL/CON-58. Oak Ridge, Tenn. Oak Ridge Natl. Lab.
12. Hirst, E., Bronfman, B., Goeltz, R., Trimble, J., Lerman, D. 1983. *Evaluation of the BPA Residential Weatherization Pilot Program.* Rep. ORNL/CON-124. Oak Ridge, Tenn. Oak Ridge Natl. Lab.
13. Ainsworth J. 1983. *RCS Follow-up Survey Analysis.* Staff Report. Calif. Energy Commission. Sacramento, Calif.
14. Ainsworth, J. 1982. *Residential Conservation Service Benchmark Survey Analysis.* Calif. Energy Commission. Sacramento, Calif.
15. Calif. Public Utilities Commission. 1984. *Energy Conservation Branch Analysis of Southern California Edison Company's Conservation Adjustment Request for 1983 Actual and 1984 Proposed Residential Conservation Financing Program and Residential Conservation Services Program.* San Francisco, Calif.
16. DHR Inc. 1983. *National Costs and Benefits of the RCS Program.* McLean, Va.
17. US Gen. Account. Office. 1982. *The Residential Conservation Service: Issues Affecting the Program's Future.* EMD-82-78. Washington, D.C.
18. Rosenberg, M., Lanore, R. 1980. *The Residential Conservation Project: The Costs and Benefits of Utility-Operated Residential Conservation Programs.* Technical Development Corp. Boston, Mass.
19. Morris, L. E., Frogge, L. M., Ehrenshaft, A. R., Riordan, J. M. 1983. *Residential*

314   WALKER, RAUH & GRIFFIN

Program: Summary of the Second Annual Reports Submitted to the Department of Energy by States and Non-Regulated Utilities. Rep. ORNL/CON-121, Oak Ridge, Tenn. Oak Ridge Natl. Lab.

20. Calif. Energy Commission. 1982. *Residential Conservation Service, Amending the State Plan: Phase I.* Revised Staff Report. Sacramento, Calif.

21. Merrill, L., Potter, T. 1982. *Quality Control in Energy Audits Based on Observation of Five Residential Energy Audit Programs.* Presented at Am. Council for an Energy-Efficient Economy (ACEEE) Summer Study. Santa Cruz, Calif.

22. Calif. Energy Commission. 1982. *Residential Conservation Service, Amending the State Plan: Phase II.* Sacramento, Calif.

23. Calif. Public Utilities Commission and Calif. Energy Commission. 1983. *Standard Practices for Cost-Benefit Analysis of Conservation and Load Management Programs,* Calif.

24. Egel, K. 1984. *Santa Monica RCS Program Progress Report,* (First and Second Quarters, 1984), Santa Monica, Calif.

25. Hirst, E. 1984. *Evaluation of Utility Home Energy Audit (RCS) Programs.* Presented at Am. Council for an Energy-Efficient Economy (ACEEE) Summer Study. Santa Cruz, Calif.

26. Energy and Environmental Analysis, Inc. 1983. *Statistical Analysis of Energy Conservation Trends Using the Annual Housing Survey.* Washington, D.C.

27. DHR, Inc. 1983. *The RCS Program. Status of States and States' Utilities.* McLean, Va.

28. Synergic Resources, Corp. 1983. *The Impact of the Residential Conservation Service Program on Natural Gas and Electric Utilities,* Rep. to US Dept. Energy. Washington, D.C.

29. Hirst, E., Berry, L., Bronfman, B., Johnson, K.-E., Tepel, R., Trimble, J. 1982. *Evaluation Plan for the Bonneville Power Administration Residential Energy Conservation Programs Volumes I and II.* Rep. ORNL/CON-94. Oak Ridge, Tenn. Oak Ridge Natl. Lab.

30. Hirst, E., Goeltz, R., Thornsjo, M., Sundin, D. 1983. *Evaluation of Home Energy Audit and Retrofit Loan Programs in Minnesota: the Northern States Power Experience.* Rep. ORNL/CON-136. Oak Ridge, Tenn. Oak Ridge Natl Lab.

31. Tonn, B. 1983. *Selection of States for Evaluation of the Residential Conserva-* tion Service. Rep. ORNL/CON-139. Oak Ridge, Tenn. Oak Ridge Natl. Lab.

32. Hirst, E., White, D., Goeltz, R. 1983. *Comparison of Actual Electricity Savings with Audit Predictions in the BPA Residential Weatherization Pilot Program.* Rep. ORNL/CON-142, Oak Ridge, Tenn. Oak Ridge Natl. Lab.

33. Hirst, E., White, D., Goeltz, R. 1984. *Energy Savings Due to the BPA Residential Weatherization Pilot Program Two Years After Participation.* Rep. ORNL/CON-146. Oak Ridge, Tenn. Oak Ridge Natl. Lab.

34. Regional Econ. Res. 1983. *Estimates of the Realized Savings from Selected Conservation Measures for San Diego Gas and Electric Company,* San Diego, Calif.

35. Kushler, M., Saul, J. *The Results of a Survey Evaluation of the Michigan RCS Program.* Lansing, Mich.

36. Kushler, M., Saul, J. 1981. *The Results of a Preliminary Survey Evaluation of the Michigan R.C.S. Energy Audit Program.* Lansing, Mich.

37. Kushler, M., Saul, J. 1983. *Fuel Consumption Impacts of the Michigan RCS Program: First Year Results.* Lansing, Mich.

38. Kushler, M. 1982. *RCS Program Status and Evaluation Activities in the 50 States.* Lansing, Mich.

39. Kushler, M., Witte, P., Crandall, G. 1984. *RCS in Michigan: Positive Results to Date—Where Do We Go From Here?* Presented at Am. Council for an Energy-Efficient Economy Summer Study. Santa Cruz, Calif.

40. Marylander Marketing Research, Inc. 1984. *1984 RCS Follow-up Survey,* Volumes I and II. Encino, Calif.

41. Pease, S. 1982. *An Appraisal of Evaluations of Utility-Sponsored Programs for Residential Energy Conservation.* Rep. N-1925. Santa Monica, Calif. The RAND Corp.

42. Am. Council for an Energy-Efficient Economy. 1984. *Proc. from the Panels on Federal and State Programs.* Proc. from 1984 Summer Study. Santa Cruz, Calif.

43. *Energy Conservation Program Evaluation: Practical Methods, Useful Results.* 1984. Proc. of a Conference at DePaul University. Chicago, Ill.

44. Polich, M. 1984. *Minnesota RCS: Myths and Reality.* Presented at Am. Council for an Energy-Efficient Economy (ACEEE) Summer Study. Santa Cruz, Calif.

45. Ganza, J., Hamilton, B., Lepage, A., Sachs, J. 1982. *Evaluation of the Vermont Energy Extension Service Home Energy Audit Team Program.* Prepared for the Vermont State Energy Office. Newport, Vt.

46. Hannigan, S., King, P. 1982. *Residential Conservation Programs at Pacific Power and Light Company: Models, Forecasts, and Assessments.* Pacific Power and Light Company. Portland, Ore.

47. Wisconsin Power and Light Company. 1981. *Home Energy Check Evaluation.* Madison, Wis.

48. MASS-SAVE, Inc. 1981. *Survey of Audited Customers.* Boston, Mass.

49. Thomas A. Heberlein and Associates. 1980. *Assessing the Effects of the Wisconsin Power and Light Company Energy Audit Program.* Lodi, Wis.

Ann. Rev. Energy 1985. 10:317–39
Copyright © 1985 by Annual Reviews Inc. All rights reserved

OIL PRICES AND INFLATION[1]

H. G. Huntington

Energy Modeling Forum, Stanford University, Stanford, California 94305

INTRODUCTION

The oil shocks of the 1970s caused severe adjustment problems for oil-importing nations. Rapidly increasing oil prices created widespread joblessness and sharply higher consumer prices in these economies. These conditions confronted policymakers with a particularly pernicious trade-off. Pressure on consumer prices could be eased by adopting restrictive policies that slowed down aggregate economic activity and worsened unemployment. Alternatively, policymakers could try to protect jobs at the expense of possible increased inflation.

During the past several years oil prices have been falling, while inflation has weakened considerably. This more recent experience has reinforced the general public's perception that oil prices and inflation are very much interrelated. Monthly reports on inflation are usually accompanied by the latest estimates of the change in energy prices.

This paper reviews several key aspects of the relationship between oil prices and inflation. There has developed over the years an extensive literature on inflation that has been adequately reviewed elsewhere.[2] Moreover, several other sources lucidly describe the interactions between oil prices and the general inflation process.[3] Rather than duplicate these efforts, this paper will emphasize the available estimates of the effect of oil price increases on inflation in the United States.

---

[1] Abbreviation used: EMF, Energy Modeling Forum.

[2] Two such reviews include Trevithick & Mulvey (2) and Santomero & Seater (3). Moreover, see *Economic Report of the President* (4: pp. 32–47), or for a useful textbook treatment, Dornbusch & Fisher (5: pp. 419–65).

[3] See Congressional Budget Office (6) on the subject of energy shocks and inflation. Solow (7) and Poole (8) appraise macroeconomic policy for responding to inflation and unemployment following the 1973 OPEC shock.

317

0362–1626/85/1022–0317$02.00

These estimates are derived from detailed macroeconomic models of the United States economy. These models were initially developed for explaining fluctuations in economic activity over the business cycle. Based on responses to economic conditions observed during the period following World War II, they focus on the determination of prices, expenditures, and income in the national economy. Since the 1970s, they have been extensively revised and expanded to incorporate important energy variables in order to study the consequences of oil shocks.

We develop in the next section several key conceptual points about the general inflation process. This discussion provides the appropriate background for the following section's consideration of available estimates of the effect of oil prices on consumer prices during the 1970s. Next, we focus on some more recent estimates of this relationship available from an Energy Modeling Forum (EMF) study on the macroeconomic impacts of energy price changes. This study focused on the effects on output, inflation, and unemployment during the four years after a hypothetical oil price change in the 1980s. It compared the responses of 14 prominent models of the aggregate economy (13 US models and 1 Canadian model) to energy price changes and to policies for cushioning the loss in real gross national product (i.e. adjusted for inflation) that results from sharp price increases.

The EMF results are discussed in two parts. We emphasize initially the broad conclusions about oil shocks and consumer prices that emanate from the comparison of models. In a second section we relate differences in these estimates among models to differences in observed characteristics of the participating models. A final section summarizes the key findings from this review of available estimates of the relationship between oil prices and inflation.

INFLATION

An appreciation of the estimated effect of oil prices on consumer prices requires some rudimentary knowledge of the general inflation process. We address only a few salient aspects of a complicated issue in order to highlight these important points: 1. economists make a sharp distinction between an increase in the price level for consumer goods and services and inflation (or the rate of increase in such prices); 2. the inflation experienced by an economy will depend on the amount of excess capacity and the aggregate economic policies (monetary, tax, and government spending) adopted by the government; and 3. due to the role of expectations, consumer prices should respond more strongly to expansions and contractions in economic activity over time. The third point becomes important for understanding the response of consumer prices to oil shocks over time.

Relative Prices and Inflation

Economists define inflation as a persistent tendency for a wide range of prices in the economy to rise over time. This definition excludes several phenomena.

First, inflation is not simply a rise in the price of one commodity while other prices remain unchanged. For example, the price of sugar or coffee can suddenly increase without inducing price increases in a wide range of commodities purchased by consumers. This increase in sugar or coffee prices relative to other prices would have very small effects on the general price level.

Second, even if other prices are significantly influenced, as happens during a widespread crop shortage or an oil disruption, this increase in the general price level will not necessarily continue for many periods after the initial jump in prices. Although measured inflation as represented by a variety of price indices will rise temporarily, economists do not consider these price increases as inflation unless the process continues.

It is sometimes argued that a sharp increase in oil prices will not raise the general price level, since other prices will fall over time. This statement that only relative prices are changing is seriously misleading on several accounts. In particular, many prices are sticky in a downward direction in the short run, and many others may actually move upward in sympathy with the oil price shock. Moreover, even with full employment and perfectly flexible prices, an oil price shock will result in a higher general price level. The energy shock will reduce total output by reducing energy use and capital formation. As output and income fall, households and businesses will demand less money than before. If the government does not alter the money supply (in current dollars) in response to the oil shock, the supply of money will exceed its demand. Prices must rise to maintain the demand for money balances (in current dollars) to satisfy the fixed supply of money.[4]

Thus, oil shocks will raise the general price level but will not necessarily lead to continued higher inflation rates.

---

[4] A simple monetarist explanation of this result would employ the basic "equation of exchange" relating nominal expenditures to the money supply, $MV = PY$, where M is the stock of money, V is its velocity or rate of turnover, P is the price level, and Y is real GNP. If both the money stock (M) and its income velocity (V) are fixed, reductions in real output (Y) would result in increases in the price level (P). Ott & Tatom (9) develop this monetarist perspective more fully and explain why they think that the income velocity (V) will be constant during an energy shock.

Inflation and Excess Capacity

Although inflation will be emphasized in this paper, this issue is closely associated with economic growth and unemployment. This point deserves some explicit consideration before we discuss estimates of the impact of oil prices on inflation.

Prices tend to rise more rapidly when there is less excess capacity and unemployment. Thus, a government can fight a burst in inflation by adopting more restrictive policies that limit economic activity and induce greater unemployment. Alternatively, it can accommodate the burst with more expansionary policies that will cushion the real GNP loss during the first few years. However, the latter policy will put greater inflationary pressure on the economy. As a result, the government faces a particularly pernicious trade-off between inflation and economic activity under these conditions.

This trade-off can be represented as a "reaction" function that relates the percentage change in money wages to the excess supply of labor. Since wages are a dominant factor in determining product prices, this relationship forms an important link to inflation. The degree of excess supply of labor is often measured as the unemployment rate, even though the latter has well known imperfections for this purpose.[5]

This "Phillips curve," shown in Figure 1, is not significantly different in concept from the OPEC price reaction functions that some world oil analysts have employed.[6] Both reaction functions show the response of suppliers, who have some discretion over setting prices, to the degree of slack in the markets in which they sell. With no further embellishments, this framework applied to the aggregate economy implies that a government could adopt an expansionary policy that lowered unemployment but increased inflation (moving from an original point A to point B) or a restrictive policy that lowered inflation but increased unemployment (moving from point A to point C).

An oil price shock will worsen both measured inflation and unemployment in the short run, pushing the trade-off outward from the origin. The economy's new position might be at D along curve II, as shown in Figure 1. The government might try to use aggregate demand policies to return

---

[5] Measured unemployment reflects decisions on entering and leaving the labor force as well as shifts in labor demand. Some economists have tried to overcome this problem by measuring job vacancy rates.

[6] Many of the world oil models discussed by the Energy Modeling Forum (10) used such reaction functions, relating the change in oil prices to the amount of excess productive capacity for oil.

temporarily to the lower unemployment rate at A, but inflation will be worsened as the economy moves to E.

Expectations

Over time, expectations become important and reduce the scope for expansionary policy for mitigating unemployment problems. Workers are concerned with their real wages (adjusted for inflation) rather than their money wages. Like OPEC, they seek a higher real price when high aggregate demand tightens the market. When the government embarks on an expansionary program, both money wages and expected future prices begin to rise. Workers base their new wage demands not only on the lower unemployment rate (as in Figure 1), but also on the higher anticipated inflation.

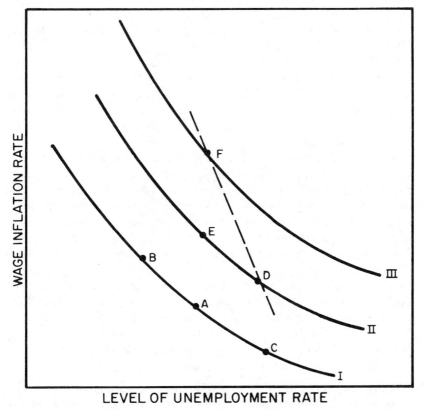

Figure 1   Short-run trade-off between wage inflation and level of unemployment rate (excess capacity).

Each time the anticipated inflation rate rises, workers demand a higher money wage rate at the old level of unemployment. These conditions are represented as an upward shift in the Phillips curve from II to III in Figure 1, implying that maintaining the old unemployment rate requires greater inflationary pressures. With the expansionary policy, the economy moves from point D to point F (rather than point E). Thus, over an extended period, the longer-run trade-off between inflation and economic expansion (represented by the dashed line) can become substantially steeper because anticipated inflation is increasing both actual inflation and unemployment in each period (through shifts of the curve upward and to the right).[7]

Expectations can also complicate the economy's response to restrictive policies for lowering inflation. Reductions in current output and labor demand should moderate wage demands, providing some inflationary relief. However, workers may believe that the government will not commit itself to a long-run strategy of high unemployment and that it will ultimately reverse itself by adopting an expansionary program in the future. The effect of these expectations may be buttressed by multiyear wage contracts and informal agreements and practices that avoid sharp wage reductions during slack times. Over time, wages and prices will adjust, although some economists believe that government policy must become very restrictive and unemployment very widespread (as happened during 1981–1982) before inflationary expectations are weakened enough to moderate wage demands.

The trade-off summarized by the Phillips curve is explicitly or implicitly imbedded in the macroeconomic models used to generate the estimates of inflation's response to oil shocks. Different assessments about the nature of this trade-off can be an important source of variations in the estimates to be discussed in the next several sections. One model may represent this trade-off as vanishing very quickly over time, i.e. the Phillips curve becomes steeper within a short time period. Another model may consider this trade-off as lasting longer.

In addition, this discussion emphasizes that these estimates are sensitive to the underlying aggregate economic conditions and policies during the oil shock. While very illuminating, the historical estimates from the 1970s and the representative ones for the 1980s may be inappropriate under a significantly different set of economic and policy conditions.

---

[7] These shifts result in a long-run Phillips curve that is considerably steeper than its short-run counterpart. In fact, when expectations correctly anticipate the experienced inflation rates, the Phillips curve becomes completely vertical, implying no long-run trade-off between inflation and unemployment.

PAST OIL SHOCKS

Declining world economic growth and mounting inflationary pressures characterized the 1970s after the oil price shocks of 1973 and 1979–1980. The United States' stagflationary experience is reviewed in Table 1, which shows economic growth, inflation, and unemployment during four distinct phases of this decade. Economic growth was considerably depressed and inflation more severe during 1973–1975, and again in 1979–1980, than in the other periods.

A number of factors contributed to the fluctuations in economic growth and inflation over this period. Economic activity was particularly brisk in 1971–73 because the economy was recovering from the 1969–70 recession. Price controls were artificially restricting the increase in measured price inflation, even though inflationary pressures were mounting due to expansive monetary and fiscal policy. By 1973, these pressures were becoming visible, while the economy was being shocked by increases in prices for food and imported nonoil materials as well as petroleum. Economic growth slowed dramatically during the severe recession of 1973–1975; the growth rate in real gross national product (GNP) declined from 3.6% for the four previous years to −0.9% for this two-year period. As a result, the real GNP level was 7–8% lower by 1975 than it would have been if it had increased by 2.5–3% per year during these years. Inflation as measured by the GNP deflator increased from about 5% to 9% per year during this same period.

Past studies of this 1973–1975 period assign a significant part of this poor

Table 1   Economic growth, inflation, and unemployment during 1969–1980

	1969–1973	1973–1975	1975–1978	1978–1980
Real GNP growth rate (%)	3.6	−0.9	5.3	1.2
Inflation rates (%)				
GNP deflator	5.1	9.1	6.1	9.0
Consumer price index				
All items	4.9	10.1	6.6	12.4
Excluding food and energy	4.4	8.8	6.7	11.1
Excluding food, energy, and				
housing	4.0	8.2	6.3	8.2
Unemployment rate (%),	4.9	8.5	6.1	7.1
end of period				

Source: *Economic Report of the President*, 1983.

economic performance to the first oil price shock during the last quarter of 1973. Previous simulation studies with macroeconomic models suggest that the 1975 GNP level was reduced by about 3.0–5.5% by the oil shock alone.[8] These same models attributed about 2 percentage points of the incremental inflation to this shock as well. According to these estimates, the economy's performance would have been significantly improved if the first oil shock could have been avoided.

Rising inflation rates and slower economic growth rates again characterized the period during and after the second oil price shock. Once again, the oil shock was accompanied by other factors that complicate a simple interpretation of the figures for the period. The growth rate of real output in 1975–1978 was far higher than its long-term trend, reflecting the economy's rebound from the 1973–1975 experience. Moreover, inflation was increasing throughout this period, reaching 7.4% (as measured by the GNP deflator) by 1978. Finally, major shifts in US macroeconomic policy and the substantial appreciation of the US dollar complicate interpretation of the period after 1980. Accounting for some of these other factors, two studies of the second OPEC oil price shock again suggest that the oil shock did considerable damage to the US economy. It was estimated that the level of real GNP would have been about 2–4% higher and prices 1–3% lower by the second year if the second oil price shock had not happened.[9]

This review of the 1970s experience suggests that each oil shock raised prices throughout the US economy by 2% or more, but that measured inflation subsided in later years. Expansionary macroeconomic policies in the mid-1970s (after 1975) kept inflation from returning to its pre-1973 level. While these increases were significant enough to cause hardships and widespread public concern, they appear to be moderate compared to the underlying inflation rate that was increasing over time. Inflation in the United States rose from about 1% per year in the early 1960s to 4–5% by the end of that decade. After the first OPEC oil price shock it returned to 6–7% through the 1976–1978 period. By 1980–1981, it peaked at 9–10% following the second oil price shock, beginning in 1979. Later, annual inflation rates fell to 3–4% during the last three quarters of 1983. This break in the inflation rate demonstrates that inflation can be curtailed if the country is willing to allow substantial and widespread unemployment. An important policy dilemma is how much economic activity should be foregone to achieve this lower inflation rate.

[8] Dohner (11) derives this conclusion after reviewing simulation studies performed with the MIT-PENN-SSRC (MPS), Michigan, Data Resources (DRI), Mork, and Fair models.

[9] The lower estimates for the second OPEC shock are from simulations with the DRI model reported by Eckstein (12); the higher ones are reported for the Mork model by Mork & Hall (13).

THE ENERGY MODELING FORUM STUDY

In 1982–1983 the Energy Modeling Forum undertook an effort to compare the responses of macroeconomic models of the United States to hypothetical future oil price shocks and to policies for mitigating the adverse real GNP effects of such events. The participating models and specific assumptions of the 10 scenarios simulated in that study are described at length by the Energy Modeling Forum (1). A complete list of the participating models appears in Table 2.

Importantly for this discussion, the working group specified oil shock simulations in which there were no conscious government efforts to mitigate either the higher prices for all goods and services or the reduced real GNP. The estimated inflation effects would have been lower than reported here if restrictive policies for holding down inflation, such as a tighter monetary policy, had been assumed. They would have been higher if expansionary policies for stabilizing real GNP had been included. The working group also simulated the effects of expansionary or accommodating policies in separate policy scenarios, which are reported in the Energy Modeling Forum (1) but not here.

The estimated impacts in the study are also sensitive to the underlying economic trends that were assumed to prevail in the absence of a change in oil prices. Specifically, considerable excess capacity and unemployment remained throughout the four-year horizon of the study, although the economic slack decreased over time, as has happened during the 1980s. Had the models simulated an economy with less excess capacity, the estimates would have shown greater pressures on prices in the economy.

Oil Price Increases

The EMF results indicate that an oil price shock would produce an immediate short-run burst of inflation, which is considerably weakened by the third and fourth years. For a sustained 50% increase in the oil price, the prices of goods and services purchased in the United States would rise by about 2% in the first year and by another 1% in the following year. These results are similar to the previously reported effects caused by the historical oil shocks of the 1970s, which were roughly comparable in size to the 50% shock in terms of increased expenditures for oil imports. While these impacts are significant, they are not overwhelming in the context of inflation rates experienced during the past decade.

The price level for all goods and services, as measured by the implicit price deflator for GNP in Figure 2, is raised in the early years in all models, but these effects begin to level off by the third year in many of them. The upward pressure on prices of goods and services begins to subside over time

Table 2 Features of participating models

Model	Stochastic equations	Parameter estimation period	Data frequency		Endogenous oil import		US macro forecasting and policy	Linkage
			Q	A	Price	Quantity		
Large US								
BEA Quarterly Econometric	150	1959–1979	×				×	
Chase Quarterly Macroeconomic	250	1958–1980	×			×	×	
Data Resources Quarterly	424	1960–1980	×			×	×	
LINK	3000[a]	1955–1980	×					International Macro
Michigan Annual Econometric	64	1954–1979		×	×		×	
MPS (MIT-PENN-SSRC)	130	1946–1980	×			×	×	
Wharton Quarterly	432	1955–1980	×			×	×	
Small US/International								
Claremont Economics Institute Domestic	10	1970–1980	×				×	
FRB Multi-Country (MCM)	500[b]	1960–1975[c]	×			×	×	International Macro
Hickman-Coen Annual Growth	60	1949–1979		×				
Hubbard-Fry	80	1960–1980	×		×	×	×[d]	Energy-Macro
MACE (Macro Energy)	23[e]	1955–1980		×		×		Canadian
Mork Energy-Macroeconomic	15[f]	f		×		×		Energy-Macro
St. Louis (FRB) Reduced-Form	3	1955–1981	×		×			Energy-Macro

[a] Total for 32 countries.
[b] 100 for each of five countries (Canada, Germany, Japan, United Kingdom, and United States).
[c] Oil and exchange rate equations are estimated through 1980.
[d] Long run as well as short run.
[e] Macroeconomic sector only.
[f] Parameters are judgmental rather than estimated.
Source: EMF (1).

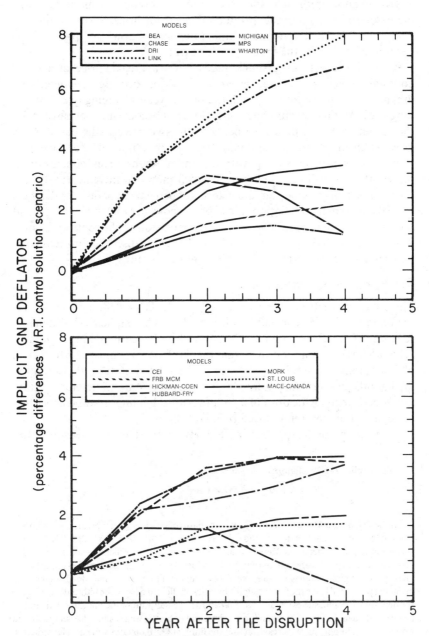

Figure 2   Effect of 50% oil shock on aggregate price level by year.

as growing unemployment and excess productive capacity moderate wage increases and other variable production costs for firms. Several models—Chase, MPS, Hickman-Coen, and Michigan—show the price effects becoming smaller by the fourth year or earlier.

A one-time energy price shock increases the aggregate price level for all goods and services in all models. However, it does not alter the underlying inflation rate, or the rate at which the price level is changing over time, as long as discretionary economic policy does not become more expansive. By the second year, the difference between the two price paths has stopped growing in most models, implying that the inflation rate has begun to return to its rate prevailing without an oil shock. This conclusion is represented more directly in Figure 3, which shows the incremental effect of the oil shock on the inflation rate in terms of percentage points. By the third year, the incremental inflation rate subsides considerably in all models and appears to be not much above zero. This conclusion applies only for a once-for-all increase in the price of oil without changes in discretionary economic policies, such as the growth rate of the money supply.

Measurement of Inflation

The inflation effects have been measured by the GNP deflator because it is the one price index reported by all modelers. The implicit GNP deflator represents the ratio of nominal GNP to real GNP measured in 1972 dollars. It has an important limitation in that the GNP deflator measures the prices of final goods and services produced in the United States rather than the final goods and services purchased in the United States. Therefore, it does not directly include the price of oil imports (or other imports) and hence will understate the inflation induced by an oil shock.[10]

Three other indicators in Table 3 represent different measures that

[10] The implicit GNP deflator is calculated as:

$$\frac{PC \times C + PI \times I + PG \times G + PX \times X - PM \times M}{C + I + G + X - M}$$

where C, I, G, X, and M are respectively real purchases of goods and services for consumption, investment, government, exports, and imports, and PC, PI, PG, PX, and PM are the corresponding price deflators. Import prices enter negatively in order to exclude the value added by foreign producers from the national product. Assuming no escalation of import price increases through the domestic price structure to augment domestic value added, an increase in import prices will leave the GNP deflator initially unaffected despite the induced increases in the deflators for domestic expenditures and exports. Actually, domestic markups will tend to be maintained in the face of higher import prices, and domestic wages and prices may spiral in response to an import price increase, so that the implicit GNP deflator will rise. Nevertheless, by the arithmetic of its construction, the implicit GNP deflator will tend to understate the initial rise in domestic prices of final goods from import price shocks.

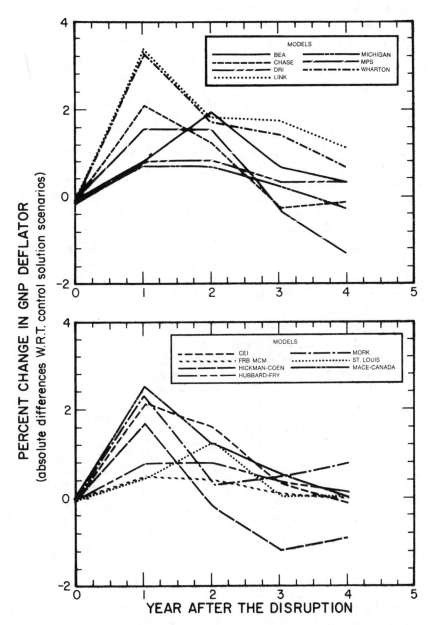

Figure 3 Effect of 50% oil shock on inflation rate by year.

overcome this problem by measuring prices of goods consumed in the United States including imports. These include the price deflators for gross domestic purchases and personal consumption expenditures and the consumer price index. The last measure represents the cost of goods and services purchased by consumers in some particular base period; the importance of each item to the overall index does not change. The first two indices allow changes in the mix of items. The deflator for personal consumption expenditures measures the price of goods and services purchased by households. The deflator for gross domestic purchases incorporates the prices of goods and services purchased for investment and for the government as well for consumption.

The main difference among the various inflation measures occurs in the first year of the shock, when the inflation rate for the GNP deflator is substantially smaller. After the first year, however, the different indices display similar inflation effects, because the price effects are predominantly indirect results of the previous rise in the price of imported oil. Nevertheless, the absolute level of the GNP deflator remains permanently lower than those of the other indices. This occurs because the oil price shock has no direct effect on the GNP deflator but does directly increase the three other measures.

As a result, the inflation effects shown in Figures 2 and 3 will be similar to those based on other price measures, except that the median inflation rate during the first year could be 0.5 percentage points higher if the deflator for gross domestic purchases was used or 0.8 percentage points higher if the

Table 3 Comparison of different inflation measures for 50% oil shock case (Median Model results) for US models, excluding LINK

Inflation measure	Absolute difference in inflation rates with respect to control solution scenario for year:			
	1	2	3	4
Deflator for				
Gross national product	1.18	1.00	0.23	−0.07
Gross domestic purchases[a]	1.72	0.65	0.17	−0.15
Personal consumption expenditures[b]	2.00	1.07	0.17	0.10
Consumer Price Index[c]	2.68	1.08	0.12	0.12

[a] Could not be calculated for Claremont, Mork, and St. Louis.
[b] Not reported by FRB Multi-Country, Mork, and St. Louis.
[c] Not reported by Claremont, Hickman-Coen, Hubbard-Fry, Michigan, MPS, and St. Louis.

deflator for personal consumption expenditures was used. From Table 3, we conclude that inflation from the consumer's perspective would increase by about 2 percentage points in the first year and by an additional percentage point in the second year.

Direct and Indirect Price Effects

Oil price shocks have both direct and indirect effects on the prices paid by consumers. The direct effects occur as crude oil prices raise the price of gasoline and heating oil used by households. In addition, crude oil prices will indirectly raise the prices of final products that use energy.

A very significant part of the first-year effect on the general price level originates with the price increases experienced directly by households. A 50% increase in crude oil prices will translate into a substantially lower percentage price increase for petroleum products purchased directly by households, because of the difference in price levels between crude oil and products. Table 4 shows that the 1983 crude oil price was approximately 60% of the product price level and that motor gasoline and heating oil combined accounted for slightly more than 5% of total personal consumption expenditures by households. A somewhat simplistic but revealing benchmark estimate of the direct price effects can be derived from these expenditure shares by assuming that a 50% crude oil price shock raises product prices by 30%. Under such conditions, the table highlights that the combined effect of higher oil prices on the price deflator for total personal consumption expenditures would be about 1.6%, or 80% of the first-year effect reported in Table 3.

The indirect price effects play a more dominant role by the second year. The estimated 1.6% of direct effects represents about 60% of the total combined inflation impacts over the first two years. A significant part of the indirect effect can be accounted for by wages, which rise by about 0.9% for the median model result by the second year. (Total employee compensation represents about 75% of total national income, which excludes capital

Table 4 Estimate of direct price effects of 50% oil shock on households

	Ratio of crude to product price	% of total personal consumption expenditures	Estimated direct effect of 50% oil price increase
Gasoline	0.60	4.2	1.3
Fuel oil[a]	0.64	1.0	0.3
Combined	—	5.2	1.6

[a] Expenditures are for fuel oil and coal.

depreciation allowances.) Changes in labor productivity can also be important, because firms are viewed as setting prices based on unit labor costs in these models. Unit labor costs rise with wage increases but fall with gains in labor productivity.

Oil Price Reductions

There has been recent interest in the effects of oil price reductions on inflation and the US economic recovery. Some say that the economy can only benefit from an oil price reduction, while others think that financial constraints and adjustment costs could be harmful to the economy.

When the economy is experiencing significant unemployment, the economic gains from an oil price reduction are equal but opposite to the losses induced by an oil price increase of comparable size. The model results suggest that, by the second year after a 20% reduction in the world oil price, real GNP would be about 1.2% greater than otherwise, while the inflation rate as measured by the GNP price deflator would be about 0.4% lower in each of the first two years. Table 5 summarizes the principal impacts on economic activity and inflation in terms of each $2 per barrel reduction in the world oil price. The models show unambiguously that an oil price reduction will have a net benefit to the economy. Oil price increases during the last decade contributed to the stagflationary conditions of greater short-term inflation and lower economic output; conversely, an oil price reduction in today's economy should ease short-term inflationary pressures and aid the economic recovery.

Table 5 Economic impacts of future oil price reductions: based on results of the EMF study

	Impacts resulting from a $2 per barrel reduction	
	Impact in year[a]:	
	1	2
Incremental effect on:		
Real GNP level (%)	0.15	0.33
Price level/GNP deflator (%)	−0.12	−0.22
Unemployment rate	−0.06	−0.14
Change in inflation rate:		
GNP deflator	−0.12	−0.12
Consumer prices	−0.24	−0.12

[a] Table should be read as follows: During the second year after an oil price reduction of $2 per barrel, the GNP level will be 0.33% higher, the price level as measured by the GNP deflator will be 0.22% lower, and the unemployment rate will be 0.14% less than otherwise. Inflation rates would be 0.12% lower.

All these developments represent an unqualified gain for the US economy. Of course, sudden price changes, rising or falling, will create adjustment problems for certain firms and individuals. These problems are not represented in the models, which focus instead on the aggregate relationships of economic activity. Moreover, for large reductions in the world oil price, it is possible that a crisis in the international financial system could develop and affect all economies by depressing world economic trade. While these caveats are important to note, they do not invalidate the EMF conclusions that moderate oil price reductions are beneficial for the US economy.

DIFFERENCES AMONG MODELS

The previous results emphasized a strong consensus among participating models in describing the dynamic response of inflation to oil price changes. A short-term burst in prices throughout the economy was eventually weakened by the third and fourth years. However, the models did show noticeable differences in the effect of oil prices on the aggregate price level for goods and services during the first two years. In this section we probe the differences between models by concentrating on changes in the aggregate price level rather than inflation rates.

Important differences in estimated inflation effects of an oil shock can be attributed to varying model structures and assumptions about key parameters. Hickman (14) used the results from the oil shock and policy scenarios simulated in the EMF study to derive certain key parameters associated with the participating models. The full analytical framework, outlined in detail by Hickman (14) and more briefly by the Energy Modeling Forum (1), is based on an analysis of aggregate supply and demand conditions in the economy.

Size of the Energy Shock

One important parameter that determines the estimated responses to oil shocks is rather straightforward. The models demonstrated a striking difference in the effect of oil prices on the aggregate price of energy. The latter is measured at the secondary level in most models, as the wholesale price index for fuel and power. It incorporates the wholesale prices of refined petroleum products, natural gas, coal, and electricity, as well as the average crude oil price to refiners.

The differences between models on the magnitude of the energy price shock is a little surprising, because the EMF study design attempted to standardize for the price responses associated with other energy sources. Some models used an input–output approach to determine the initial impact of the oil price shock on the producer price index for fuel and power. If oil prices accounted for 50% of the total value of energy costs, a 50% oil price increase would translate into a 25% increase in the producer price index for energy. In general, the energy price index changes by a similar amount across these models. Other models used a historical regression of the producer price index on the cost of oil and other variables. These models indicated energy price changes that differed from each other as well as from those in the first group of models. While there was some criticism of the input–output approach on the grounds that the weights were from 1967 or 1972 input–output tables and hence might be out of date, at least one modeling group decided to revise its results by rescaling its energy price shocks to be consistent with the models based on the input–output approach. This revision, however, was not uniformly accepted, and differences in the treatment of linkages between oil and energy prices remain.

This issue remains an important source of differences between models. Ironically, it reflects different approaches for estimating the link between energy prices rather than variations in the key macroeconomic relationships represented in the models.

Price Stickiness

A second significant parameter is the response of the aggregate price level for all goods and services to the excess capacity or unemployment rate. Prices in the economy are usually represented in these macroeconomic models as set by firms on the basis of markups above the average unit costs of variable inputs like labor, energy, and other materials. Labor costs form a large share of these costs. As discussed previously, the money wage rate negotiated by labor can be quite sensitive to the unemployment rate. As firms curtail their planned levels of output, they decrease their demand for labor, putting downward pressure on money wages and final prices of goods and services. In addition, the markups used by firms can also be sensitive to the degree of excess capacity, decreasing somewhat at higher levels of unemployment. Both factors contribute to reduced pressure on the general price level for goods and services when falling economic activity increases unemployment during an oil shock. These effects partially counter the upward movement in prices generated by the increase in oil prices.

Cyclical changes in labor productivity will tend to weaken but not reverse the response of the aggregate price level to the unemployment rate. During a recession, output falls more rapidly than labor demand because firms try to maintain their labor forces in anticipation of better times in the future. Immediate layoffs would require costly retraining when the economy begins its recovery. As a result, a reduction in economic activity accompanied by a less than proportional decline in employment lowers output per worker immediately after an oil shock. [This temporary decline is separate from any longer-term deterioration in labor productivity, as discussed by Dohner (11).] This cyclical deterioration in output per worker raises unit labor costs, which are passed through by firms to the final prices of goods and services.

The effect of output reductions on the aggregate price level in the economy is often summarized in terms of an aggregate supply curve for the economy. The shape of this function represents the price elasticity of output, which measures the proportional decline in output associated with a 1% decrease in prices. Higher price elasticities denote greater insensitivity of prices to changes in output.[11]

Aggregate Demand Responses to Price

The estimated effects of an oil shock on the price level in the various models are dominated by the two preceding model characteristics: the effect of oil prices on the wholesale price index of fuel and power and the price elasticity of aggregate supply. In fact, a simple analysis based only on these characteristics is extremely effective in explaining much of the difference between models. However, the macroeconomic effects depend conceptually on a third parameter, the price elasticity of aggregate demand (the response of total spending to a higher price level for goods and services).

When the aggregate price level rises, real output can be directly discouraged through several mechanisms: (a) The fixed nominal supply of money will be reduced in real terms, resulting in higher interest rates and reduced investment and purchases of consumer durables; (b) real consumption may be reduced by a decline in the real value of assets held by households due to higher prices; (c) government purchases may decline to the extent that they are not indexed or adapted to the higher price level; and (d) real net exports will be reduced as higher domestic prices make domestic goods less attractive and foreign goods more attractive, under the assumption of fixed foreign prices. In all these cases, the initial declines in aggregate demand due to rising prices will induce further reductions in consumption and investment expenditures through multiplier and accelerator effects.

The price elasticity of aggregate demand was not as important for explaining model differences in the EMF study, because it tended not to vary greatly from unity in most models. With an unchanged nominal

[11] Conversely, output is more responsive to price for higher price elasticities. Energy economists often use a similar concept to measure the flexibility in energy demand when energy prices are changing.

supply of money, a rising price level reduced real spending on goods and services by an approximately equal proportion. By the second year, nominal GNP (the GNP deflator times real output) was not substantially below its reference levels in many models.

A simple supply–demand model of the aggregate economy was developed during the study to try to explain the key differences between the models. This framework allowed one to develop some basic relationships among prices, economic output, and energy price changes. The key parameters identified previously were derived from the policy and oil shock scenarios by Hickman (14), who discusses extensively the important assumptions and caveats necessary for interpreting these estimates correctly.

Figure 4 emphasizes how closely the reported model results for the change in the aggregate price level during the 50% oil shock conform to this simple supply–demand framework. The figure compares the percentage increases in the GNP deflator reported by the models with those calculated from the simple analytical framework, using Hickman's estimates of the model parameters. The analysis was restricted to the US models that reported a variable for the change in the wholesale price of fuel and power.[12] Further discussion of the approach is included in the appendix of the full EMF report (1984).

All models lie either on or near the 45° line that indicates when the calculated impacts equal those reported by the models. Moreover, the simple model preserves exactly the ordering of the reported price responses, from the model showing the highest impact to the one revealing the smallest effect. The effects on the aggregate price level depend positively on: (a) the size of the shock in terms of its effect on the wholesale price for aggregate energy and (b) the degree of inertia in wages and prices with increasing unemployment; and negatively on (c) the responsiveness of total spending to changes in the aggregate price level.

The Wharton results show the largest effect on the aggregate price level because oil prices in the Wharton model influence energy prices more, while wages and prices are less responsive to unemployment than in other models.[13] For the opposite reasons, the Data Resources, St. Louis, and

[12] Three other models report the change in primary energy price, which excludes the price of electricity. Since the latter is relatively stable during an oil shock compared to other energy prices, the change in the primary energy price will understate that in the secondary or wholesale energy price that includes electricity. We did not attempt to include these models in the present calculation because it would have required some arbitrary assumptions about the relationship between the two energy prices.

[13] The aggregate price level in the Wharton model also depends on the cost of capital, which increases when interest rates rise in response to the oil shock. This specification, which is not found in the other models, makes prices appear less responsive to increased unemployment.

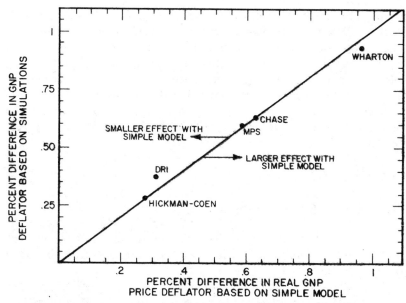

Figure 4 Comparison of second-year effects on aggregate price level with estimates from basic supply–demand analysis.

Hickman-Coen results reveal smaller impacts on the general price level. While the estimated response of prices in the economy to oil price shocks will vary across models, these differences can be attributed to structural characteristics that can be clearly identified.

As explained earlier, the general price level begins to moderate over the four-year period considered in the EMF study. The oil shock triggers a wage–price spiral that puts upward pressure on the prices of all goods and services. However, over time these prices become more sensitive to the unemployment that accompanies the initial spurt in prices. This charac-

Table 6 Average price elasticity of aggregate supply as revealed in EMF income tax reduction scenario

	Year:			
	1	2	3	4
Estimated price elasticity of aggregate supply[a]	20.0	2.61	1.04	0.55

[a] Measures the proportional decline in economic output for every 1% reduction in the general price level, when aggregate demand falls.

teristic of the models provides an important explanation of why the initial spurt in prices is not maintained and translated into a permanent inflationary spiral. In the absence of accommodating policy (e.g. an expansionary monetary policy), reductions in aggregate output will place increasing downward pressure on wages and prices. This response is summarized in Table 6, which shows that the average price elasticity of aggregate output, as estimated from the EMF policy scenario of a reduction in income tax rates, decreases markedly over the four years of the simulation period.

CONCLUSIONS

There have been frequent discussions of the role of energy prices in inflation. Some believe that the inflation of the 1970s was largely caused by the OPEC price increases and other exogenous factors. The same view would hold that significant inflationary relief has resulted from the recent oil price reductions.

The EMF results indicate that a once-for-all oil price shock induces only a short-lived burst of inflation. The prices of goods and services purchased in the United States would rise an additional 2% during the first year and 1% during the second year as a result of the 50% oil price increase. Moreover, the inflation effects in the 50% shock case do not last; they moderate over time and become negligible by the fourth year. A one-time increase in the price of imported oil cannot permanently alter the underlying inflation rate so long as economic policy, particularly the growth rate of the money supply, is not changed in response to the oil shock.

These results suggest that a prudent policy for future oil market disruptions should avoid combatting the short-lived burst in measured inflation by purposely reducing economic activity (through restrictive monetary, tax, and expenditure policies). While painful during the first year or two, these price increases would shortly be weakened by the slowdown in economic activity that results from the oil shock itself. Adopting more restrictive macroeconomic policies, as was done after the 1973 OPEC embargo, appears to be both unnecessary and extremely costly under these conditions. A more difficult question, which is not addressed here but considered in the EMF study (1), is to what extent discretionary economic policy should be used to offset declines in economic activity, with possible adverse effects on inflation.

By analogous reasoning, we should not expect long-term relief from inflation to materialize from one-time decreases in world oil and other energy prices. The oil price reduction in 1983 and softening domestic energy prices have lowered the general price level for all goods and services below what it would have been otherwise, thereby temporarily reducing measured inflation. However, these energy price reductions will not permanently lower the underlying inflation rate of the economy during the 1980s, unless fuel and electricity prices continue to fall throughout the decade.

ACKNOWLEDGMENTS

This paper is based upon a particular study of the Energy Modeling Forum, Stanford University, conducted by the EMF 7 working group, chaired by Bert G. Hickman. It draws on several sections of that group's report (1) for its principal conclusions on inflation and on model differences. I would like to acknowledge the significant contributions of the EMF 7 working group and the many comments and suggestions offered by EMF sponsors and affiliates, EMF staff, and members of our senior advisory panel. However, I alone bear the responsibility for the views and opinions expressed in this paper.

Literature Cited

1. Energy Modeling Forum. 1984. *Macroeconomic Impacts of Energy Shocks*, EMF 7.2, Draft Sept. Stanford Univ., Stanford, Calif.
2. Trevithick, J. A., Mulvey, C. 1975. *The Economics of Inflation*. Glasgow Soc. Econ. Res. Stud. 3. London: Martin Robertson
3. Santomero, A. M., Seater, J. J. 1978. The inflation–unemployment tradeoff: a critique of the literature. *J. Econ. Lit.* 16(2): 499–544
4. *Economic Report of the President*. 1981. Washington, DC: USG PO. 357 pp.
5. Dornbusch, R., Fischer, S. 1981. *Macroeconomics*. New York: McGraw-Hill. 738 pp.
6. US Congressional Budget Office. 1981. *The Effect of OPEC Pricing on Output, Prices, and Exchange Rates in the United States and Other Industrialized Countries*. Washington, DC: Congr. United States
7. Solow, R. M. 1980. What to do (macroeconomically) when OPEC comes. In *Rational Expectations and Economic Policy*, ed. S. Fischer, pp. 249–67. Chicago: Univ. Chicago (NBER Rep.).

293 pp.
8. Poole, W. 1980. Macroeconomic policy, 1971–75: an appraisal. See Ref. 7, pp. 269–84
9. Ott, M., Tatom, J. A. 1982. Are there adverse inflation effects associated with natural gas decontrol? *Contemporary Policy Issues* 1: 27–46
10. Energy Modeling Forum. 1982. *World Oil*, EMF Rep. 6. Stanford Univ., Stanford, Calif.
11. Dohner, R. S. 1981. Energy prices, economic activity, and inflation: a survey of issues and results. In *Energy Prices, Inflation, and Economic Activity*, ed. K. A. Mork, pp. 7–41. Cambridge, Mass: Ballinger. 171 pp.
12. Eckstein, O. 1981. Shock inflation, core inflation, and energy disturbances in the DRI Model. See Ref. 11, pp. 63–98
13. Mork, K. A., Hall, R. E. 1981. Energy prices and the U.S. economy in 1979–1981. *The Energy J.* 1(2): 41–53
14. Hickman, B. G. 1984. *Macroeconomic Effects of Energy Shocks and Policy Responses: A Structural Comparison of Fourteen Models*, EMF 7.4, Sept. Stanford Univ., Stanford, Calif.

Ann. Rev. Energy 1985. 10: 341–405

EXPLAINING RESIDENTIAL ENERGY USE BY INTERNATIONAL BOTTOM-UP COMPARISONS[1]

Lee Schipper and Andrea Ketoff

Lawrence Berkeley Laboratory, University of California, Berkeley, California 94720[2]

Adam Kahane

Energy and Resources Group, University of California, Berkeley, California 94720[2]

INTRODUCTION

In 1982, the residential sector accounted for approximately 20% of energy use, 20% of oil use, and over 30% of electricity use in OECD countries.[3] This review examines the structure of energy use in the residential sector and identifies the factors that have driven changes over the past 25 years, particularly since the 1973 oil shock.

There are several possible approaches to explaining residential energy use. The one taken here is to compare use in a number of countries and look

---

[1] The US Government has the right to retain a nonexclusive royalty-free license in and to any copyright covering this paper.

[2] Opinions are those of the authors and not those of the sponsoring institutions or those with which they are affiliated.

[3] OECD: The Organisation for Economic Cooperation and Development. Energy consumption statistics for many OECD countries fail to correctly disaggregate the "other consumption" sector into residential, agricultural, commercial, and public components. Data used in this report were developed by the authors and their associates at the Lawrence Berkeley Laboratory and at VVS Tekniska Foreningen (the Association of Swedish Heating and Ventilation Engineers), Stockholm. Residential energy use is defined in this study to include all energy consumed in full-time places of residence, including farmhouses.

for patterns. International comparisons show the variation possible in household consumption over a wide range of economic, demographic, and policy conditions. More than single-country studies, these comparisons allow us to examine the responses of consumption under a wide range of changes over time in variables like energy (and other factor) prices, fuel availabilities, household incomes, appliance saturations, climates, dwelling types, building traditions, household and building sizes, and policies.

A clear picture of the factors determining the variation in household energy consumption gives us an understanding of the possibilities for future consumption change, including policy-induced change, within a single country. International comparisons reveal the effects of extremes of efficiency or life-style on energy use, allowing us to test whether proposed policy-induced end-use patterns are reasonable and to evaluate the possible influence of different technologies or policies on energy use.

Part 1 outlines the bottom-up methodology used in the international comparisons and discusses the primary data sources and associated problems. Part 2 gives a snapshot picture of residential energy use in nine countries in 1982, accounting in detail for differences in consumption among these countries. Part 3 explains the changes in residential consumption in these countries from 1973 to 1982. Our conclusions are given in Part 4.

1.   THE BOTTOM-UP APPROACH

With demand analysis more than any other energy field, the method adopted for analysis determines the quality of the results. We use the bottom-up approach, which explains changes in energy use by analyzing the changes in the underlying energy intensities of individual end uses. Our comparison among countries concentrates on these components of consumption, rather than on total consumption.

1.1   *The Method*

The bottom-up approach uses information on the residential capital stock (including housing characteristics, heating systems, and fuel and electric appliances) and measurements or estimates of unit energy consumption (by end use and fuel) to determine total energy use for each end use and fuel. This approach can be used in two ways: to disaggregate gas and electricity use by end use where total residential consumption is relatively well known, or to build up total use from its components for solid and liquid fuels where total consumption is poorly known. Meyers (1) described this approach in his analysis of US residential consumption patterns, and showed that the totals obtained by building up the components were close to sales totals independently reported for electricity and gas.

Other authors have compared energy use among countries by analyzing differences in total energy deliveries reported to the residential sector (or even the OECD "other consumption" sector), using prices and incomes as explanatory variables on econometric regressions. Regression analysis can also be applied to data derived from the bottom up (2). The analysis which follows demonstrates the importance of producing reliable and comparable data series and analysis for both total consumption and the underlying components.

1.2   Information Required for the Bottom-Up Analysis

The bottom-up method requires three kinds of data. The first are data on the structure of energy-using household capital stock, including building and appliances characteristics. The second are measurements or estimates of unit consumptions for different purposes—or combinations of purposes—by fuel. The final are data on deliveries of energy to the residential sector, which are used to check the totals derived from multiplying stocks by unit consumption.

1.2.1   THE STOCKS OF ENERGY-USING EQUIPMENT   Stocks of equipment are estimated using detailed household surveys carried out for different purposes: censuses, surveys of electric appliance ownership, or estimates of the potential for converting users of one fuel to another. In our analyses (3–15) we found these surveys to be of varying quality, with sample sizes ranging from a few hundred households to entire census populations. The most important requirements for making comparisons are:

1. that stocks be defined according to the main fuel used for space heating, water heating, and cooking
2. that single-family and multifamily dwellings (henceforth SFD and MFD) be characterized separately
3. that space heating systems be characterized according to their type, separating central heating from stoves (henceforth CH and non-CH)
4. that hot water equipment be identified separately and according to how it is combined with space heating equipment
5. that characteristics of dwellings themselves, such as size and vintage, and data on the characteristics of families be identified
6. that the consumption of each fuel over a well-defined period be recorded.

The main problem we encounter here is the multitude of definitions for each important datum. Dwelling type, for example, is not defined the same way in each country, because of differences in classifying buildings with 2–4 units, in whether to include farmhouses, and in whether to include principal dwellings in primarily nonresidential buildings. This problem also arises when different surveys of the same population or complementary populations are fused in order to have a more complete description of the entire stock. We have found, however, that by using original survey results whenever possible, we can adjust input data to reduce uncertainties to a small fraction of the intercountry differences.

1.2.2 UNIT CONSUMPTIONS Consumption data are collected from households in surveys or panels. Some surveys collect data on most or all fuels (16–19). In France, for example, specific fuel panels complement the periodic structure surveys, in an effort to match quantities of energy delivered with those calculated by multiplying the numbers of each kind of consumer times the unit consumption of each fuel for each purpose (20). In other cases we combine unrelated sources that monitor individual fuels. The Oil Panel in Germany (21) measures oil use monthly in several hundred dwellings. Another survey covers electricity use in Germany (22). Other countries regularly carry out similar efforts for electricity (23, 24) and gas (25).

Since most survey data give only the overall consumption of each fuel for each household, we must partition consumption of each fuel into its end-use parts. This introduces a number of uncertainties. Space and water heating, for example, are produced by the same fuel in 65% of all households. Since space heating dominates, the uncertainty introduced by any method designed to isolate fuel used for heating water will be small for space heating but large for hot water. If hot water is not separated from space heating, however, comparisons of fuel used for heating among homes using different fuels for hot water will be incorrect.

While measurements of space heating energy intensity are made relatively frequently, few studies meter hot water use or cooking use separately (26) because of the difficulty and the high cost of the procedure. Fortunately, the composition of the stocks in most countries allows us to compare fuel use in SFD that have space and water heating based on the same fuel with that in SFD or MFD that have the same space heating fuel but a different water heating fuel. This comparison shows how much more of the first fuel is used to provide hot water in the combined system, and how much of the second fuel is used to provide hot water in the separate system. Since average weather conditions, family incomes, and family sizes vary, these comparisons only yield approximate results, but the uncertainties are small compared to those caused by leaving hot water consumption combined with space heating.

There are other methods of partitioning consumption among end uses. Using data on stocks of equipment in the households surveyed, Latta (27) used multiple regression of consumption data on equipment stocks to decompose US residential consumption for 1981, as did the German Vereinigung Deutscher Elektrizitatswerke (VDEW) (28) for German appliance use. Fels & Goldberg (29) used regression of time series data from oil-heated homes to identify the hot-water components and to analyze changes in oil use. Meyers (1) compared the cohorts of homes in the first Residential Energy Consumption Survey (RECS) (cf 16) with different combinations of equipment using a given fuel to estimate the consumption of that fuel for each purpose. A series of German studies (30, 31) used engineering estimates of current consumption; although they do not apportion these estimates between SFD and MFD, they provide a plausible time series partition of each fuel to each end use over more than 25 years. Schipper used a combination of techniques to establish energy uses for Sweden (32) and Denmark (10).

1.2.3 TOTAL CONSUMPTION   Total residential consumption is used to check the bottom-up analysis. Electricity and gas consumption are reported reliably by supply measurements because tariff categories are usually well-defined to exclude nonresidential consumers and because most consumption is metered for each dwelling. Where gas and electric companies differentiate between heating and nonheating customers or offer special rates for heating consumption (which is metered separately), the end-use partition is simpler to determine. Finally, gas and electricity are not stored, so sales reflect current consumption. In all, the uncertainties in gas and electricity consumption are small compared with the major differences among countries or the changes over time.[4]

The data for liquid and solid fuels are much less reliable. Liquid fuels— heating oil, kerosene, and liquefied petroleum gas (LPG) can be stored by the consumer, obscuring the distinction between consumption and sales. Carlsson (33) and Schiffer (34) showed that this difference can amount to over 5% of consumption in any one year. In addition, few oil companies or national energy data bodies have attempted, until recently, to separate deliveries accurately between residential and commercial sectors. Oil use is therefore impossible to ascertain without recourse to bottom-up surveys of oil heating users and their average yearly consumptions.

The identity of the final consumer is also uncertain for solid fuels. Sales

---

[4] The situation for district heating with hot water, is similar to that of gas and electricity. We found, however, that in Scandinavia bottom-up surveys were necessary to arrive at total district heating consumption, because the deliveries of local companies were not broken down by customer categories that allowed a summation of only residential use at the national level. The same thing happens for some gas and electricity apartment customers taking collective deliveries and paying nonresidential tariffs; the companies do not know how many dwellings are involved.

**RESIDENTIAL WOOD USE
IN SWEDEN**

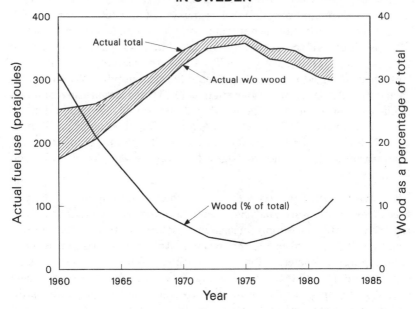

Figure 1 Residential wood use in Sweden, 1960–1982. The scale at left shows absolute use with and without wood included, the scale at right the share of wood in total delivered energy. The shaded area shows the use of wood.

records are even less complete for wood—which has reemerged as an important fuel (by itself or as a complement to oil or electricity) in Scandinavia, Canada, and the northern United States, because it is often free rather than purchased. Failure to include wood would cause consumption undercounting in several countries, as Figure 1 shows for Sweden. If wood use were omitted, the rate of change in total delivered residential energy use in Sweden would have been nearly twice as great between 1960 and 1973 as it actually was. Similarly, the rate of decline after 1973, and particularly after 1978, would be seriously overstated. The situation is similar in Norway and the United States; in all three countries (35–37) wood use is now surveyed. Only in Canada is wood use large but information almost totally unavailable.

1.3 *From Survey Data to Energy Analysis*

Once we have allocated energy use to each dwelling type and end use by fuel, we sum uses over each fuel and over each end use. This is illustrated in detail in Table 1 for France in 1981. This type of matrix, the basic unit in our data base, shows how consumption is broken into stocks and unit consumption. For some countries, we can split heating accurately into its four components of SFD and MFD and central heating (CH) and noncentral heating (non-CH). For other countries, the noncentral-heating figures may be rough, or all four categories aggregated. Most energy uses after 1972, however, were concentrated in central heating, hot water, and electric appliances, so that the uncertainties from cooking and from heating in noncentral-heating (or supplementary) systems result in a small total uncertainty.

After completing this matrix for a country, we complement it with data on incomes, energy prices, the age of the dwelling stock, family size, climate, and other variables. We then aggregate energy uses into three totals that express energy consumption differently. *Delivered energy* includes energy crossing the building boundary, and is also known as purchased energy (plus wood), or gross energy. The individual data in Table 1 are presented this way. Delivered energy, which represents the quantities consumers buy, counts electricity (or district heat) and fuels with the same weight, distorting the picture of energy use by omitting the losses associated with production and distribution of electricity and district heat.

Primary energy is a measure of delivered energy as well as the losses associated with transforming and distributing electricity (1.89 times actual consumption is the average value we use for all countries, although it varies depending on the proportion of hydroelectric and nuclear production) and district heat (0.33 times actual consumption). Because it includes these losses, primary energy represents more clearly the total energy used by society.

Useful energy, also known as net energy, is the energy reaching the consumer after estimated combustion losses in the home are subtracted. Our convention is to count solid fuels at 55% efficiency, gases and liquids at 66%, and electricity and district heat at 100%. More complicated schemes are suggested by Leach (38), and Gronheit [in (19)] estimates the changes in conversion efficiencies over time for Denmark. Without such detail for each country over time, we have assumed a single value for each conversion efficiency in all countries and at all times.

Because of the increasing share of electricity and district heating in the energy picture of each country, we use useful energy as a measure when aggregating energy use on a household basis. Useful energy allows a more equal estimate of the energy services—space heating, water heating, and cooking—offered by fuels as compared with electricity. Uncertainties in the "true" conversion efficiencies are much less important than the distortions that arise in comparing countries with different mixes of fuels and electricity. However, we find it important to give delivered or primary

Table 1 Residential energy use—France—1981

	Space heating					Hot water			Cooking	Appliances	Total
	Central heating			Stoves (SFD+MFD)	Total heating	From CH boiler	Independent system	Total hot water			
	SFD	MFD	Total CH								
Oil, PJ	286	174	459	72	532	85	0	85	0.5	0	618
Stock, 10³	3094	3222	6316	1576	7892	3713	0	3713	117		
Unit cons., GJ	92	54	73	46	67	23	0	23	4		
LPG, PJ	16	2	18	4	22	3	11	14	40	0	76
Stock, 10³	214	46	260	479	739	160	2159	2319	8848		
Unit cons., GJ	76	40	70	9	30	16	5	6	5		
Gas, PJ	111	113	223	9	232	44	18	62	37	0	331
Stock, 10³	1405	2461	3866	461	4327	2606	3532	6138	7419		
Unit cons., GJ	79	44	58	19	54	17	5	10	5		
Solids, PJ	28	15	43	91	134	3	0	3	2	0	139
Stock, 10³	351	259	610	2295	2905	175	0	175	319		
Unit cons., GJ	80	56	70	40	46	17	0	17	7		

Dist ht, PJ	6	28	33	0	33	12	0	12	0	0	46
Stock, 10³	73	508	581	0	581	482	0	482	0		
Unit cons., GJ	77	55	58	0	58	26	0	26	0		
Elec, PJ	35	19	54	10	64	0	37	37	17	107	225
Stock, 10³	926	888	1814	695	2509	0	4969	4969	5091	19354	
Unit cons., GJ	38	21	30	14	26	0	7	7	3	6	
TOTAL, PJ	482	350	832	186	1018	147	66	213	97	107	1435
(%)	34	24	58	13	71	10	5	15	7	7	100
GJ/dw[a]	79	47	62	34	54	21	6	12	4	6	74
Primary, PJ	550	395	945	205	1150	151	136	287	129	309	1875
(%)	29	21	50	11	61	8	7	15	7	16	100
GJ/dw[a]	91	54	70	37	61	21	13	16	6	16	97
Useful, PJ	329	246	573	116	689	101	56	157	70	107	1023
(%)	32	24	56	11	67	10	5	15	7	10	100
GJ/dw[a]	54	33	43	21	36	14	5	9	7	6	47

[a] Indicators per dwelling calculated on stock described in column.

Sources: LBL French Data Base, Ketoff (1985); based on figures from l'Observatoire de l'Energie, Agence Francaise pour la Maîtrise de l'Energie (AFME), and from yearly surveys by the Centre d'Etudes et de Recherches Economiques sur l'Energie (CEREN).

energy use in some contexts, and when we discuss a single fuel we always refer to delivered energy. Once the energy uses have been established and aggregated, we normalize by dividing by the number of dwellings, people, or dwelling area. Other normalizations are also possible, and are discussed below.

All values for energy consumption given herein are corrected to average yearly weather conditions (as defined below). The years for which data are available are not the same for each country; in general, 1972 and 1973 are considered together, as are 1981 and 1982.

2.   RESIDENTIAL ENERGY USE IN 1982

Table 2 shows, for nine countries, the basic quantities from Table 1 aggregated into total use per household as delivered, useful, and primary energy; fuel shares and choices; end uses; and parameters that describe the housing stock, climate, and other aspects of residential energy use.

Total energy use per dwelling varies considerably among the study countries; the differences depend on which accounting method we use, since the shares of electricity and district heating are different. For each important end use—space heating, water heating, cooking, and electrical appliances (including air-conditioners)—we examine the determinants of consumption and fuel choice. This disaggregated approach allows us to analyze the differences among countries on as equal a basis as possible by comparing like fuels, dwellings, and systems for these end uses.

2.1   *Space Heating*

Space heating, which consumes the largest portion of household energy, is displayed in Row 17, Table 2. We count only energy from the heating system, excluding contributions from other sources (e.g. people, the sun, other appliances), the significance of which has been illustrated by Elmroth & Levin (39). The great differences in useful space heating energy per dwelling are the results of differences in climate; dwelling size, type, and thermal properties; heating services demanded; and heating equipment efficiency.

2.1.1   CLIMATE   Climate is the key to understanding household energy use for heating and cooling. Although windspeed and humidity (39) are very important in determining the amount of heat needed in a house to maintain given internal comfort conditions, data limitations forced us to settle for the comparative analysis of degree-days, the integral over the heating season of the difference between the actual outdoor temperature and a reference indoor temperature. Our figures for the long-term average number of degree-days for each country, as well as the original figures supplied by local authorities, are given in Table 3. We use 18°C as our base temperature for every country, although the base temperature actually depends on indoor temperature, heat from all sources besides the heating system, and the thermal integrity of the building. To account for year-to-year variations in degree-days we note the yearly deviation for each country and correct our calculated space heating consumption proportionally. Schipper (32) describes several alternative methods that lead to under- or overproportional correction.

The figures in Table 3 give an objective measure of climate because they use the same base temperature (18°C) and are not varied to reflect actual free heat, actual thermal integrity, or actual indoor temperature. Row 15 of Table 2 (reported or measured indoor temperature) suggests, however, that these three factors vary greatly from country to country. The true climate indicator therefore depends on consumer behavior as well as on technical aspects of the building stock. Nevertheless, we found the adjusted degree-day figures in Table 3 sufficient to represent most of the differences in climate among the study countries.

Although this climate measure varies among these countries by a factor of two, it is striking that countries with similar climates have very different levels of per dwelling space heating energy use (Row 17 in Table 2). Scott (40) presented evidence that actual consumption does not quite follow outdoor temperature, i.e. that an underproportional correction is indicated. In a multinomial logit regression analysis developed within our study, Chern et al (2) found that long-term heating energy use per household was proportional to the square root of the number of degree-days.[5] Parikh & Rothkopf (41) and Sjölund (42) predicted this behavior, for a given energy price, on the basis of theoretical considerations: the colder the climate, the greater the incentive to insulate. Comparing Canada and Sweden, with similar climates but great differences in heat demand (cf. Table 2, Row 17), shows that climate does not completely explain all the difference; we must also consider the structural and economic factors.

2.1.2 DWELLING TYPE AND SIZE   Another large difference among countries is related to dwelling size and type. Dwelling type includes single-family dwellings (detached, semidetached, or attached, and mobile homes in North America, all abbreviated SFD) and multifamily dwellings (MFD, including dwellings in nonresidential buildings, referring to the individual

[5] Chern et al (2) counted actual consumption and degree-days, so their results could have picked up the year-to-year variation, but the range of intercountry variation (Table 3) is greater, by almost a factor of ten, than the maximum year-to-year deviation from the long-term average.

Table 2 International comparison of residential energy use

	Canada 1981	Denmark 1982	France 1982	Germany 1982	Italy 1980	Japan 1979	Norway 1981	Sweden 1982	United States 1982
Delivered energy									
1. —total, PJ	1237	173	1384	1852	986	1132	—	335	10120
2. —climate-corrected,[a] PJ	1176	176	1424	1880	976	1183	130	334	10190
3. —per capita, GJ	48	35	26	30	17	11	32	40	44
Energy per dwelling, GJ									
4. Delivered	143	82	72	76	56	36	84	90	122
5. Primary[b]	222	111	83	100	72	54.4	184	145	179
6. Useful[c]	109	63	51	54	41	27	72	73	90
7. Occupied dwellings, 10^6	8.0	2.2	19.7	24.7	17.5	33.2	1.54	3.7	83.2
8. Single-family dwellings (dw)									
% of occupied dw	67	57	51	49	40	65	68	46	69
9. Dwelling size,[f] m^2	100	86	79	77	78	80	84	90	150
10. People per dw	2.95	2.34	2.75	2.50	3.27	3.20	2.65	2.25	2.77
11. Area per capita, m^2	33	36	28	31	24	25.1	32	40	47
Stock vintage[d]									
12. % Built before 1949–1951	23.4	51	56	25	43	—	34	37	40
13. % Built after 1974–1975	18	13	11	12	9	—	17	12	15
14. Central heating,									
% of occupied dw	95	94	69	78	58	4	60[e]	99	86
15. Ave. indoor temperature, °C	—	17–18	17–19	18–20	18–20	13–15	16–18	21	19

16. Space heating fuel choices, %									
Oil	34	54	41ᶠ	50	53ᵍ	—ʰ	25	44	17
LPG	1	↑ʲ	4	1	→	—	0	0	5
Piped gas	41	2	22	24	28	—	0	0.5	55
Electricity	21	6	13	7	7	—	50	21	19
District heat	→	36	3	8	0.1	—	0.5	30	0
Coal/Coke		2	17	9	12	—	→	→	→
Wood	3	1	↑	1	↑	—	25	5	2
Space heating, useful									
17. —climate corr., GJ/dw	68	40	31	39	27ⁱ	8	48	44	53
18. —KJ/dd-m²	148	149	161	160	159ⁱ	50	147	122	136
Hot water,									
19. —useful GJ/dw	22	13	8.9	8.3	4.8ⁱ	8.3	13	16.6	17.0
20. —useful GJ/capita	7.3	5.5	3.2	2.9	1.5ⁱ	2.5	4.9	7.4	6.1
21. Cooking, GJ/dw	4.1	2.5	5.3	2.7	4.6	4.9	2.1	2.3	4.9
22. Appliance electricity, kWh/dw	4420	2165	1540	1360	1560	2055	2350	2800	5130

ᵃ These and all subsequent values include climate correction of the space heating consumption. Average degree-day values given in Table 3.

ᵇ With district heating counted at 75% total efficiency and electricity counted at 34.6%. (Actual consumed DH multiplied by 1.33; consumed electricity multiplied by 2.89.)

ᶜ Oil and gas consumption for heating and hot water multiplied by 0.66, solids by 0.55.

ᵈ Based on samples taken in 1980 (Sweden), 1981 (Canada), 1982 (Denmark, France, Germany, United States), with the last year in the class varying within the range given. For construction after 1974–1975, all data run through 1982.

ᵉ Includes central boilers 11% and direct electric 48%.

ᶠ French fuel shares refer to 1981.

ᵍ Italian space heating fuels penetration based on 98% of dwellings, since 2% are nonheated. LPG is accounted with gas.

ʰ Japanese space heating fuel shares impossible to determine because of the widespread use of multiple fuels.

ⁱ Calculated based on the number of dwellings with the facility, i.e. 98% for space heating, and 87% for hot water.

ʲ An arrow pointing up (down) means that the fuel source in this row is grouped with the fuel source in the row immediately above (below) the row for this country.

Sources: Compiled in our individual country papers (1, 3, 5–7, 10, 13) and in references therein, as well as (18–20, 30, 31, 51, 57, 90).

354   SCHIPPER, KETOFF & KAHANE

Table 3   Degree-days in study countries[a]

	Official (Base temp.)	This study (18°C)
Canada	—	4581
Denmark	2987 (17°C)	3122
France	2450 (18°C)	2450
W. Germany	3613 (20°C)[b]	3163
Italy	2140 (18°C)	2140
Japan	1000 (14°C)	1800
Norway	3680 (17°C)	3910
Sweden	3760 (17°C)	4010
United Kingdom	2380 (15.5°C)	2917
United States	—	2600
(Oil)	—	2950[c]

[a] For each country the long-term average is used; derivation is given in Ref. (12).
[b] 20°C is indoor reference temperature, but heating days are counted only when the daily average is less than or equal to 15°C.
[c] Based on the distribution of dwellings with oil as principal heating source in 1980 and the average values of degree-days (dd) in those areas. In 1960, the distribution gave 0.5% more dd, in 1970 0.5% less.

Sources: Canada, Shell Oil Co., Canada; Denmark, Energistyrelsen, based on Teknologisk Institut without adjustment for sun (225 heating days, based on monthly averages); France, Agence Francaise pour la Maîtrise de l'Energie (AFME); Germany, Deutscher Wetterdienst and Esso (250-day season); Italy, Consiglio Nazionale delle Ricerche (CNR); Japan, Japan Institute for Energy Economics; Norway, Statistik Sentralbyraa, Meteorologisk Institut (225-day season assumed); Sweden, Oeverstrelsen foer Ekonomiska Foersvar and own calculations (7), (250-day season calculated from VVS monthly heating day averages); United Kingdom, Building Services Research. United States, LBL calculations (1) based on population weighted distribution of oil-heated dwellings.

units, not the entire building). Table 2 (Row 8) shows that the share of SFD falls between 40 and 70%.

Building and system characteristics determine the intensity of energy use in terms of both engineering and the potential behavior of the occupants. Apartments have fewer outside walls per unit of floor area, suggesting they should have lower heat losses per unit of floor area than SFD, for equivalent heating systems. The evidence we have reviewed from every country suggests, however, that this is true only when each individual apartment has its own central system, as is common in France, Germany, and recently in Italy (43). Because of economies of scale, building-size heating systems have lower energy intensities for the same comfort level than systems in each apartment (44, 45), but since these MFD are generally not metered individually for heat, they reach intensities that are close to those for SFD on a square meter or per capita basis, actually exceeding those of SFD in Sweden (46). Also, apartments offer less living space per occupant than

SFD, and they have fewer occupants; this results in further variation between apartments and detached dwellings. MFD use less energy per dwelling than SFD, but more MFD are required to house a given number of people.

The ownership structure can also result in less efficient energy use in apartments than in SFD (47). The problem of metering heat from collective heating/hot water systems in apartments makes it unclear who should invest in more efficient structure and equipment. In addition, there are important differences between SFD and MFD in heating fuel structure; in no country we studied are MFD distributed the same way as SFD among fuels, because MFD are more likely to lie in denser areas (where gas and district heating are more likely available) than are SFD. For these reasons we keep data on structure and consumption in MFD and SFD separate whenever possible.

Although the energy consumption properties of SFD and MFD are very different, we aggregate all building types by dividing energy use by average floor area per dwelling. Recent values of net dwelling area are listed in Table 2, Row 9.[6] The variations reflect large differences in the area of SFD, from 150–160 m^2 (the United States) to 100–110 m^2 (Germany, Italy, and France), with Sweden, Canada, and Denmark intermediate. The SFD/MFD share also affects this average, although MFD in every country (except the United States) cluster around 60–70 m^2. Since, in the first approximation, heating needs are proportional to living area, the differences in home size explain much of the difference in heating use (see Row 17).

2.1.3 SPACE HEATING FUEL CHOICE   Use of useful energy measures to aggregate different fuels does not account for all of the differences in unit consumption caused by differing fuel mixes. Further investigation therefore requires explaining space heating energy use by fuel; Table 4 shows that space heating fuel choice varies markedly among countries. The share of oil-heated dwellings in the United States is significantly lower than in other countries, and the gas share significantly higher. District heating, insignificant in North America and very small in Germany and France, is important in Denmark and Sweden. The electricity share also varies, showing that Canada, Sweden, and the United States have the highest electric heating

---

[6] This variable is not well defined for all countries over time. For some years, we estimated area from changes in average number of rooms per dwelling over time and the relation between rooms and area for a year when both are known. Also, different countries measure inner area (net), outer area (gross), or heated area, which may be greater than net area because of certain areas that are not defined as part of living area (such as storage rooms, cellars, garages, washrooms).

Table 4 Principal space heating fuel choices[a]

Country Year	Dwellings 10⁶ units	Oil	LPG	Piped gas	Electric	District heat	Coal/ coke	Wood	Other
					(% of dwellings)				
Canada									
1961	4.6	59	↓[b]	19	0	0	22	↓	—
1971	6.0	57	1	31	6	0	1	4	↓
1977	7.2	45	1	38	14	0	0	2	—
1981	8.0	34	1	41	21	0	0	3	—
Denmark									
1965	1.60	46	↓	1	0	19	33	↓	—
1972	1.86	62	↓	2	1	30	6	↓	—
1977	2.02	59	↓	2	3	32	3	1	—
1982	2.2	54	↓	2	6	36	2	1	—
France									
1962	14.6	12	3	6	10	0.7	68	↓	↓
1973	17.3	50	3	12	4	1.6	26	↓	3
1978	18.6	45	3	18	10	3	18	↓	3
1983	19.9	36	4	25	17	3	7	8	—

Germany								
1960	15.4	14	→ᵇ	1	0	1	84	↓	—
1972	21.4	48	↑	11	4	5	32	↓	—
1978	23.4	53	1	15	7	5	19	1	—
1983/4	24.7	50	1	24	7	8	9	1	—
Norway								
1960	1.08	13	0	0	16	0	4	63	—
1973	1.37	47	0	0	31	0	1	21	—
1980	1.53	34	0	0	42	0	1	23	—
1983	1.58	25	0	0	49	0	0	26	—
Sweden								
1963	2.8	57	0	1	1	4	13	24	—
1972	3.3	69	0	2	6	17	1	6	—
1978	3.6	57	0	1	15	24	0.5	4	—
1982	3.7	39	0	0.5	24	32	0.3	4	—
United States								
1960	53.0	32	5	43	2	0	12	4	—
1973	69.3	25	6	55	10	0	1	1	—
1978	77.2	21	5	55	16	0	1	1	—
1981	83.2	17	5	55	19	0	0	2	—

[a] For Canada, France, and the United States, the figures reflect occupied dwellings; for other countries, all dwellings. Other includes unknown, or combinations of three or more fuels and systems; in the case of France the figures refer to the share of nonheated dwellings.

[b] An arrow pointing left (right) means that the fuel source in this column is grouped with the fuel source in the column immediately to the left (right) of the column for this country.

Sources: Government and private surveys in each country, LBL Reports. See the reference list to Table 2, and (6).

penetrations. In France, electric heating is diffusing very rapidly in both new and existing homes. Wood use has increased since 1975 in Sweden, Norway, Canada, France, and the United States, and is used in smaller quantities in other European countries, judging by the reported ownership and use of wood burning stoves.

Fuel choices in different countries have also changed appreciably over time, but with contrasting patterns. Note, in Table 4, the change from solids to liquids in the 1960s, continuing to gas and electricity after 1973. In North America, liquids were yielding to gas even before 1970. In all countries, electricity gained at the expense of all other fuels in the 1970s, particularly in new homes (electricity is used in over 45% of new homes in Canada, the United States, and France).

An additional phenomenon that became more prevalent after 1978 is the use of supplementary fuels for space heating. Until recently, this practice, allowing occupants to turn central systems off or down, was uncommon in homes with central heating. Use of secondary fuels could account for a significant drop in the intensity of one fuel without necessarily representing increases in efficiency or changes in comfort. The use of secondary fuels may change rapidly with changing prices, however, since such use requires continual activity on the part of occupants. Accounting for secondary fuel use is therefore important.

While data are difficult to compare, we can give some indications of the importance of secondary fuels. In France, electricity and wood served as secondary sources in more than 40% of all homes in 1982 and in 25% of all homes in Canada in 1981. One quarter of all oil-using homes in the United States, 33% in Denmark, and 45% of all oil-heated SFD in Sweden used secondary systems in 1982. In Germany there is widespread use of small electric heaters and a large capacity to use coal as a secondary fuel. Italy is seeing increasing use of secondary fuels—mostly electricity—as the CH systems become individually metered with the introduction of natural gas. Wood is important in rural dwellings, and combinations of electricity, kerosene, and liquefied petroleum gas (LPG) are common in homes without central heat. Norway and Japan represent special cases, where use of two heating fuels is the rule.[7] In Japan, with central heating in only about 4% of all homes—those in the far north—about 75% of all households used small kerosene heaters as principal or secondary heat sources throughout the 1970s, although the fraction began to dip as LPG and liquefied natural gas (LNG) became more popular (48). Almost all Japanese homes possess at least one *kotatsu*, or electric foot heater. Norway represents the European

---

[7] This was also the case in the United Kingdom until natural gas entered the heating market in the early 1970s.

equivalent; while more than half of all homes used oil or kerosene throughout the 1970s, the combination of wood, oil, and electricity characterized most SFD and even many MFD there (49). These combinations make it very difficult to trace the structure of oil use in Japan and Norway.

Few estimates are available of the real quantities of principal fuels displaced by secondary fuels, although there is evidence that secondary fuel use is concentrated in homes using oil and electricity as principal fuels. In Sweden, 10 PJ of wood and 3 PJ of electricity were used in 1982 in SFD, which also used about 79 PJ of oil as the principal fuel; by comparing homes using only oil with those using oil and other fuels, we estimate that the secondary fuels replaced about 20% of the oil used in those homes with secondary fuels. For Norway, the penetration of secondary wood and electricity is even higher. These countries represent likely upper limits on the use of secondary fuels in Europe and North America.

2.1.4 SPACE HEATING INTENSITY   Differences in space heating use per dwelling due to differences in climate and dwelling area were discussed in Sections 2.1.1 and 2.1.2. What remains to be explained is the differences among countries in space heating intensities, defined in kJ/degree-day/square meter and shown in Row 18 of Table 2. These can be attributed to differences in the actual heating services required (indoor temperatures and the portions of living areas actually heated), and in the engineering efficiency of space heating systems (which includes the effect of fuel choice, discussed in 2.1.3 and below, and of housing vintages).

Heating services required   Some of the variation in space heating intensities can be explained by differences in indoor temperature. Table 2, Row 15 shows the considerable variation of indoor temperatures. These might be considered life-style differences, although incomes and energy prices certainly shape the cost and value of comfort. The low space heating intensities in Japan (50) and the United Kingdom (not shown), for example, are partly attributable to exceptionally low indoor temperatures, and to the fact that only parts of homes are heated in both countries. However, an important feedback of efficiency to fuel choice through changed running costs is noted by Scott (40). The marginal change in heating costs is proportional to the change in temperature times the effective transmission (k-) value of the home times the price of the fuel. This means that Swedes, whose homes have the lowest transmission values (typical values of k are 33–50% lower than in other countries), save less money (at a given fuel price) than anyone else by lowering indoor temperatures. They can therefore be expected to respond less than others if heating prices increase, and in fact have the highest indoor temperatures in the OECD. Higher indoor temperatures, however, justify increased insulation investments, which in turn makes the Swedes considerably more interested in efficiency than people in Norway, Japan, or the United Kingdom, countries with the coldest indoors, and less likely to reduce heating use in the short run with increasing prices (32).

Space heating efficiencies   These differences in heating services required do not, however, explain all the variation in space heating intensities. The Swedish space heating intensity, for example, is the lowest of those shown (except for that of Japan), even though the indoor temperatures and CH penetrations are highest in Sweden. More detailed studies (51) confirm that Swedish space heating was in fact far more efficient than that in any country.

Differences in consumption per square meter reflect differences in lifestyle, incomes, and housing standards, as well as differences in the efficiency of the space heating systems. The presence of central heating, for example, varies greatly around the OECD, as shown in Table 2, Row 14. Its penetration has increased with income in every country.[8] Heating stoves heat less space than central boilers using any fuel, and households using heating stoves tend to heat to lower temperatures than those in centrally heated homes. This is discussed in (31), and in virtually every other national study of energy use.

Efficiency has three components. The first is relative fuel efficiency, the ratio of energy contained in the heating fuel to that supplied to the living space. Solid fuels are converted to heat with somewhat lower efficiencies, typically 45–55%, than liquid or gaseous fuels (typically 60–70%). The second is equipment efficiency, which depends on central heating penetration as well as technical characteristics. Central heating equipment (or a room stove) converts fuel into heat with varying efficiency, depending on its age and design and whether air or water is the medium (52). Little information is available on the differences in the technical characteristics of heating equipment among countries or over time.

The third component of efficiency is the thermal properties of the home, insulation and air infiltration rates. The latter vary from well below 0.5 air changes per hour in new homes (and 1.0 in existing) in Sweden to above 3.0

---

[8] By central heating we mean a heating source (furnace, boiler, or connection to an outside source of heat) that distributes heat via circulating air or hot water, or electric radiators (including storage radiators) placed in virtually every room. In Germany and the United Kingdom storage systems are not counted as central heating, and direct, unrestricted systems used as the principal heating source are uncommon. In Denmark direct acting systems are common but are not officially counted as central heating. In this study we count all these systems as central heat.

in the United States and Holland (39). Insulation levels are partly related to building materials, which vary considerably among countries, with wood dominating in Sweden, Norway, Canada, and much of the United States, and masonry more important elsewhere. Insulating wooden structures costs considerably less than insulating masonry.

Both heat system efficiencies and building thermal properties have evolved over time, so dwelling vintage must be taken into account. In general, the older the dwelling, the more leaky it is (with lower insulation levels), and the less efficient the heating system is, although local surveys and audits bring evidence that the more wasteful buildings were built in the booming postwar years (1946–1960) (53). The average age of the dwelling stock varies considerably among countries (and within each country between SFD and MFD and among fuel types): Table 2 (Row 12) shows the significant differences, among countries, in the shares of homes built before 1949, and after 1974 (Row 13). These data yield significantly different average ages (8); with France having the oldest dwellings and Canada the youngest.

In every country, heating system fuel and appliance ownership are related to house vintage, partly because house vintage is related to location and partly because it is related to the income of the occupants. The relationships are not simple, however. In Germany, the oil-heated SFD are young, with two thirds built after 1960, while in the United States two thirds were built before 1960. In Sweden, electricity dominates the new homes and oil those built between 1940 and 1965, while all fuels (including wood) are found in those built before 1940. Comparisons of homes using different fuels must therefore take into account probable differences in thermal properties.

Table 5 shows typical k values for walls, compiled by the European Insulation Association (Eurima) for homes built before 1973. The great variation among countries reflects the greater costs of insulating masonry walls and the differences in climate more than any other factors, since homes built before 1973 in most countries tended to use oil, which cost approximately the same everywhere. The low average k values for Sweden are outstanding and unequalled (on average) elsewhere. A recent Swedish survey (54) of the transmission (k-) values of the stock in 1984 confirms this trend: k-values for walls in SFD built before 1940 were 0.54 ± 0.05 W/m^2/°C, for those built between 1941 and 1960 they were 0.52 ± 0.06, for those built between 1961 and 1975 they were 0.37 ± 0.02, and for those built between 1976 and 1981 they were 0.28 ± 0.03. Other components of SFD and MFD showed the same trend.

Consumption data by vintage for homes using a given fuel tend to agree with the improvements in thermal integrity noted above. Figure 2 shows the 1981 consumption of the oil-heated stock for Sweden by year of con-

Table 5 Typical or maximum permitted transmission coefficients (k-values in W/m²°C)

Structural parts	Phase[a]	Belgium	Denmark	Finland	France	Germany	Holland	Italy	Norway	Sweden	United Kingdom
Wall	1 (pre-1973)	–			1.00	1.57	1.67	1.39	0.58	0.58	1.70
	2	0.75/2.00	0.42	0.41/0.70[b]	0.70	0.81	0.68	1.39	0.43	0.35	1.00
	3	0.60/2.00	0.30/0.40[b]	0.29/0.35[b]	0.70	0.57/0.90[c]	0.68	0.43/0.68[c]	0.43	0.30	1.00
	4	0.42	0.22	0.20	0.31	0.38	0.50	0.30/0.49[c]	0.23	0.20	0.40
Roof	1 (pre-1973)	–			1.10	0.81	0.97	2.03	0.47	0.47	1.42
	2	0.51/3.00	0.37	0.35/0.47	0.55	0.69	0.68	2.03	0.33	0.25	0.60
	3	0.45/3.00	0.27	0.23/0.29	0.55	0.45	0.68	0.48/0.87	0.33	0.20	0.60
	4	0.30	0.14	0.15	0.19	0.32/0.27[d]	0.50	0.33/0.61[c]	0.17	0.15	0.25

[a] Explanation of the phases: 1. Pre-1973 figures refer to practice except for Sweden, Germany, Norway, and the United Kingdom, although values in Sweden were much lower; 2. Generally values required or used in the mid-1970s; 3. Values required by codes introduced in 1977–1979; 4. Eurima's recommendations in 1980, which were close to practices in Finland, Sweden, and Denmark, but lower than codes or practices elsewhere.
[b] Dependent on the type of construction.
[c] Dependent on the shape of building.
[d] Sloping roof above ventilated crawl space.
Source: Evolution of Regulations and Practice on Thermal Insulation in Residential Buildings. EURIMA (European Insulation Manufacturers' Association), Roskilde, Denmark. See also Uitenbroeck, J., and Carpentier, G., 1981. Thermal Insulation Requirements in E.E.C. Countries. Situation 1980. Brussels: Centre Scientifique et Technique de la Construction.

SPECIFIC OIL CONSUMPTION FOR HEAT AND HOT WATER IN OIL-HEATING HOUSEHOLDS IN SWEDEN

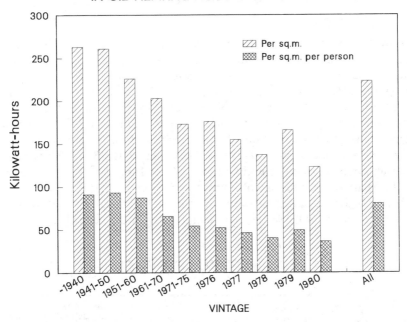

Figure 2 Specific oil consumption, for heat and hot water, in oil-heating households (single-family dwellings with central heating, by vintage) in Sweden, 1981. Given is energy use in kWh (1 kWh = 3.6 MJ) per square meter of heated area and per occupant, according to the year of construction.

struction. Other countries' stock shows similar behavior: Denmark (55), Germany (56), and the United States (27). The most recently built homes use the least energy per m² for heating (and hot water). Older homes have, however, likely been retrofitted and renovated, while those built in the 1950s and 1960s still enjoy their original equipment. This means that occupants of the most recently built homes will have had less interest in retrofitting their homes to save heat than those in the older ones. House vintage is therefore important in explaining differences among countries as well as differences in each country's response to higher energy prices over time.

In all, we have seen that large differences in heating consumption and intensity exist among the OECD countries. Climate, central heating penetration, home area—the latter two being structural factors influenced by incomes in the long run—explain much of the difference. Efficiency, including the effects of fuel choice, equipment, and insulation levels, as well as indoor temperature, all of which are affected by energy prices and also incomes, explain the rest. Housing type is less important.

2.2 Hot Water

Because there are few measurements of energy use for water heating that can be generalized to any country's entire housing stock, we have accepted the evaluations of each country's experts in extracting the hot water proportion of energy used in combined heating/hot water systems (57) or in free-standing systems that use gas or electricity. In spite of these uncertainties, there are large differences among countries that are robust against changes in assumptions about intensity by fuel. The choice of fuels used in six countries in 1981–1982 is shown in Figure 3. Useful energy consumption per dwelling is given in Row 19 of Table 2.

Two patterns emerge. In North America, Sweden, and Denmark, regions with the highest incomes and central heating penetration, hot water energy consumption in 1981–1982 was considerably higher than elsewhere, about

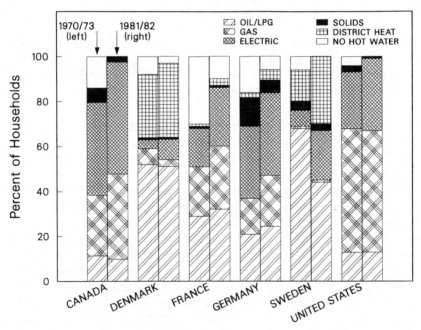

Figure 3 Hot water fuel (percentage). Data for six countries in two periods, with the use of two fuels indicated.

13–22 GJ/dwelling. In all other countries the figures are considerably less, a far greater difference than can be explained by differences in occupancy or fuel choice. If we normalize the figures in Table 2 to household size, however the differences among these groups diminish somewhat (Row 20).

A number of effects are important in explaining this and other differences. First, hot water consumption differs among countries. This is due to differing bathing habits, as well as the higher share of washing and dish-washing machines that heat their own hot water when needed in conti-nental Europe and Britain (which are thus not included in the hot water figure). Household size is also important (58). In addition, ambient (and output) water temperatures differ among countries, as well as seasonally.

Second, there is an important difference in systems between the two groups of countries. In North American and Scandinavia, central systems dominate in over 95% of all homes, either with free-standing tanks (gas, electricity, LPG, and some oil), or in combination with heating boilers (gas, oil, wood). In Germany, by contrast, only 44% of homes in 1982 had centrally distributed hot water, while 51% had point of use systems that provided hot water on demand or storage systems used only for a single bath (56). The figures for the United Kingdom (38) and continental Europe are similar, and in Japan almost all hot water is locally produced (51). Some of these systems ("geysers") have no storage, and therefore no standby losses, as well as virtually no distribution losses. Moreover, they produce hot water at a limited rate, which may limit total use. These noncentral systems were introduced in Europe when central heating, whose penetra-tion was lower than 15% in the United Kingdom and on the continent in 1960, was an exceptional luxury, and continued in popularity even in buildings whose heating systems were converted to central supplies, in part to make each household responsible for water heating costs.

Finally, the connection between water and space heating systems has an impact on water heating energy consumption. The high use in Canada, Denmark, and Sweden is due in part to the fact that almost all oil-heated homes in Sweden and Denmark and nearly 50% of those in Canada have hot water produced in the same facilities as the space heating. Outside of the seven-to-nine–month heating season, oil boilers providing only hot water are only around 25% efficient (59), increasing oil use for hot water considerably.

2.3   Cooking

Cooking in stoves and ovens accounts for a declining share of household energy use, and has decreased in absolute terms in almost every country. As we noted above, it is usually possible to estimate cooking use by isolating households using gas only for cooking, or by comparing electric consumption in homes with similar electric appliances except for the presence or absence of electric cooking. These judgements, made by each country's utilities, form the basis for our figures in Table 2, Row 21. Cooking fuel choices are shown in Figure 4. The high values for delivered energy for France, the United States, and Canada reflect the large penetration of gas cooking, while lower values for Denmark, Sweden, and Germany reflect the dominance of electric cooking.

One reason energy use for cooking has declined is the decreased time spent by adults and by children in homes, i.e. the increasing number of meals eaten away from home by all family members. In addition, small specialized electric appliances that each perform one task have proliferated in every country. In the Federal Republic of Germany in 1981–1983, 36% of all households had an electric grill, 74% a coffee maker, 34% an egg cooker, and 73% a toaster (22, 60); the saturations were similar for the United States (61), Sweden (32), and Denmark (62). Each of these devices replaces gas or

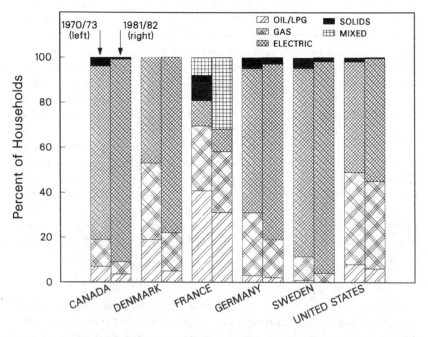

Figure 4   Cooking fuel (percentage). Data for six countries in two periods, with the use of two fuels indicated.

electricity consumption in large stoves.[9] Gas stoves in Europe almost never use pilot lights, while their North American counterparts made before the mid-1970s always had pilot lights, which significantly raises consumption.

2.4   Electric Appliances

So far we have reviewed energy uses that can be met by either fuels or electricity. We classify electric appliances separately because they generally fulfill needs met almost exclusively by electricity, although hot water in washers and hot air in clothes dryers can be heated by fuels, and gas or kerosene refrigerators are not unknown. Their consumption is estimated as a residual, after space and water heating and cooking have been estimated. Electricity use for electric appliances grew more rapidly than other energy uses in every country we studied, but consumption per household today still varies by 3 to 1 among the study countries, as Table 2, Row 22 shows.

Air-conditioning, included in the US figures [2000 kWh per household in 1981 among the 57% of households with cooling (1)] and the Japanese figures [310 kWh per year in 1979 (13, 51)] is important, but not enough to account for the difference between these two and the other countries. Moreover, ownership of the major appliances is now high in every country, with refrigerators (Figure 5a), washers, and televisions virtually saturated, freezers (Figure 5b) and dishwashers (Figure 5c) between 30% and 60%, and only clothes dryers (Figure 5d) below 25% in most countries. Estimates of electricity use for lighting vary from a few hundred kWh per household in Germany, France, the United Kingdom, and Japan (13), to around 1000 kWh per household in the United States (1) and Canada (13), with Denmark (10) and Sweden (7) at around 500–700 kWh per household.

Although differences in appliance saturations explain some of the differences in total consumption, differences in unit consumption and service offered are also important. Top loading washers need more hot water and energy than front loaders (63, 64). Appliance size and performance varies also; 150–250-liter refrigerators dominate Japan and Italy, 200–350-liter refrigerators in France and Germany, 350–450-liter refrigerators in Scandinavia, and refrigerators of about 400–600 liters in North America. Scandinavian and North American models are more likely to have freezer chests and automatic defrost cycles. At the same time, there is considerable evidence of actual efficiency variation across as well as within countries (52, 65, 66). Finally, life-style, i.e. actual utilization, undoubtedly

---

[9] One time series of cooking energy use for apartments in Stockholm using city gas for cooking only shows a drop from 4 GJ per apartment in 1960 to around 2 GJ per apartment in 1980, reflecting all of these changes as well as some increases in efficiency as older stoves were replaced with newer ones (32).

ELECTRICAL APPLIANCE SATURATION:
Refrigerators

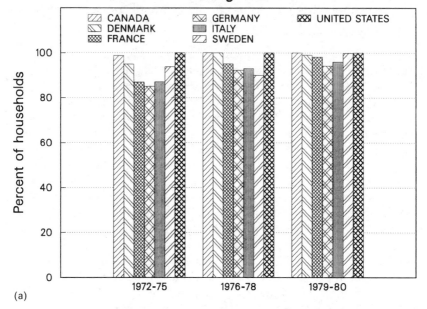

(a)

ELECTRICAL APPLIANCE SATURATION:
Freezers and Combinations

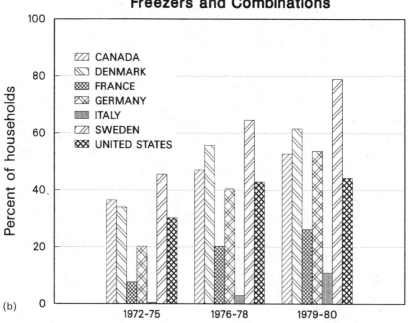

(b)

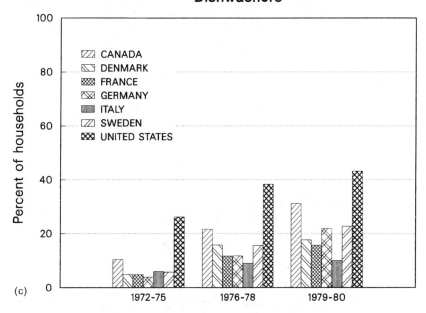

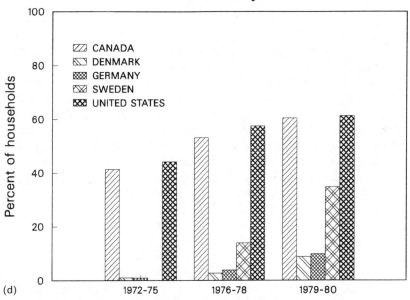

Figure 5 Percentages of households owning (*a*) refrigerators; (*b*) freezers, including combined freezers/refrigerators; (*c*) dishwashers; and (*d*) clothes dryers.

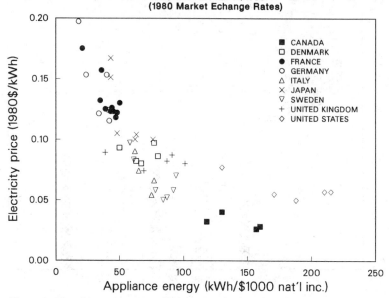

Figure 6   Electricity prices vs appliance energy/national income. The vertical axis shows electricity prices (US$ per kWh, 1980 exchange rates), and the horizontal axis gives appliance electricity use in kWh/US$1000 national, 1970 income in US$ at 1980 exchange rates.

varies among countries, although this principally affects smaller appliances and those that heat water, not refrigerators or freezers.

While a careful comparison of appliances among countries has not yet been carried out, the work to date explains in rough terms the gross differences in appliance energy use. Appliance ownership, utilization, and efficiency all share in explaining differences in appliance energy use among the OECD countries. As ownership levels in most of the countries approach saturation, differences in consumption are roughly equally attributable to utilization (including of features) and efficiency. This is best summarized in Figure 6, displaying price and consumption per US dollar of real income for nine countries over the period 1960–1982. Not surprisingly, the countries with the highest incomes and lowest electricity prices have the highest consumptions.

2.5   Economic and Demographic Determinants

We showed above how each energy end use was related to structural factors such as equipment ownership and characteristics, or fuel choice. Certain questions, however, cannot be addressed within the factor-analysis framework we have used in most of our work. In particular, the response of consumption to changes in income and factor prices (either between countries or over time) is obscured by disaggregation. The plausible effect of an increase in income, for example, is divided into: increases in average dwelling size (including perhaps a shift from MFD to SFD), increases in appliance ownerships, increased average appliance and building shell efficiencies (through both purchases of new, more efficient units and retrofits), and an increased demand for energy-consuming services (more discretionary appliance use). Each of these effects must be considered in order to understand the change in total consumption.

Energy prices are an important determinant of energy use in the short and long term [(2) and references cited therein]. Income level is a diffuse factor that affects the choice of homes and energy-using equipment, as well as the purchase of energy to use the equipment. Although exchange rate variability makes exact comparison of countries difficult, higher real incomes in North America and Scandinavia have permitted residents of those countries to own more (and larger) energy-related capital goods. The effect of income on these structural factors (house type and size, central heating ownership, and appliance ownership and sizes) is measured by the long-term income elasticity. Chern et al (2) reviewed recent findings and found values between $+0.44$ and $+1.77$. Once the penetration of an energy-using device approaches saturation, however, its ownership and energy consumption stops growing, even as incomes increase. Thus the income elasticity of energy use should be a decreasing function of income, once important energy-using devices are in place.

Demographic differences such as household size also have an important impact on energy use. Household size reflects not only the number of children in a family, but also the age at which children leave home, the divorce rate, and the number of retired parents living with their children. Another factor is occupancy, or the number of people per hour actually occupying the home. This is related to the age of children (and whether they are in school or child care facilities), employment of a second parent, and the age of grown occupants. Diamond (67) measured energy use in a retiree community of homes and found patterns different from those of other age groups, influenced strongly by the fact that most occupants stayed home most of the time.

Household size should be an important factor in explaining energy use, particularly hot water use (58). This factor was measured by Lundström (68), and Gaunt & Berggren (69), who monitored household energy use and all water use. These studies, as well as the Twin Rivers Project (70), examined many behavioral and attitudinal variables, confirming the importance of family behavior and life-style (from house occupancy and the number and age of children to the number of showers), in accounting for the variability of energy use in households, even in physically similar structures.

Demographic differences among countries depend partly on overall national as well as individual income levels as well as personal choices. We have not been able to account for all these factors in our work, but we point to their potential importance in explaining differences in energy use among countries. From our analysis we conclude that household size changes, as described herein, will have an impact on energy use in the long term. As to other aspects of life-style, several major international meetings have discussed influences of behavior and life-style on energy use (71), but a complete multinational comparison of these factors and their influences on energy use has not yet been made.

2.6   Fuel Shares

We have compared total energy consumption among countries by aggregating over all fuels, ignoring differences in fuel shares. These differences can be obtained by combining unit consumption and fuel choice data. We show fuel shares for 1972–1973 and 1981–1982 in Figures 7 and 8. The variation among countries is clear, with the oil share in the United States significantly lower than in the other countries and the gas share significantly higher. District heating, insignificant in North America and very small in Germany, France, and Italy, was important in Denmark and Sweden in 1972 and even more important in 1982. The electricity share is highest in countries with high electric heating penetration, i.e. Canada, Norway, Sweden, and the United States. Moreover, the ownership of electric appliances (as well as hot water heating and cooking equipment) has increased steadily in every country, so that the overall market share of electricity in residential energy use has increased even without consideration of changes in the home heating market. Wood use has increased since 1975 in Sweden, Norway, Canada, and the United States, and is even used in small quantities in other European countries, judging by the reported ownership and use of woodburning stoves. In Sweden and Canada, woodburning boilers account for most wood use, while in the United States and Norway, heating stoves are the most important equipment, usually in combination with another source or as a complement to oil or electricity use. Note that in all countries shown except the United States (and Holland and the United Kingdom, not shown), oil accounted for over 50% of energy use in 1972–1973, reaching as high as 70% in Italy, Sweden, and Denmark; by 1982 it was under 50% except in Denmark.

Fuel prices explain much of the differences in fuel choices and shares (2).

BREAKDOWN OF DELIVERED ENERGY BY FUEL
1972-73

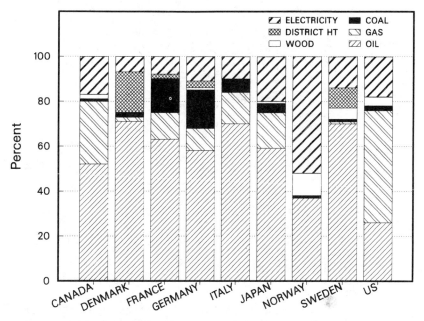

Figure 7   Breakdown of delivered energy by fuel, 1972–1973. Years shown are 1971, Canada;
1972, Denmark, Germany, Sweden; 1973, France, Norway, United States, Japan.

In Canada, the United States, the United Kingdom, and the Netherlands,
natural gas was generally available and less costly than oil before 1973. Seen
in cross section, gas dominates where it is less expensive, and increased its
market share after oil became expensive. Electricity has the highest shares
in Norway, Sweden, and Canada, where it is cheapest in absolute terms, and
in the United States, where air-conditioning needs make heat pumps
attractive over direct heating in the 50% of new homes choosing electricity.

After 1973, oil prices in every country shot up to nearly the same levels,
although relative change was greater in some countries (Sweden) than in
others (the United States, Canada). Gas and electricity prices, although
increasing almost everywhere, reached very different levels. These fuels
therefore vary in price compared to oil and each other, shaping different
choices in different countries (8). These variations have been reflected, in
large part, by differences in fuel mix, and are summarized in the changes
between Figures 7 and 8.

BREAKDOWN OF DELIVERED ENERGY BY FUEL (1981-82)

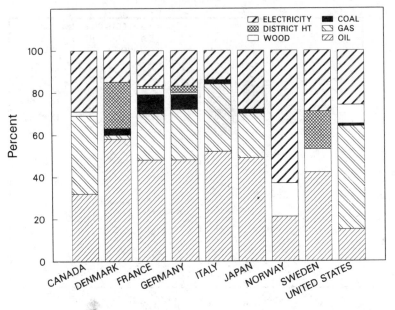

Figure 8   Breakdown of delivered energy by fuel, 1981–1982. Years shown are 1979, Japan; 1981, Canada, France, Norway; 1982, Denmark, Germany, Sweden, United States.

Figure 9 shows the relative cost of electricity and oil for heating, ignoring actual end-use efficiency. The large increases in oil prices caused movements away from oil (6), while the improved cost relation of electricity (or gas, not shown) to oil promoted increased choice of those fuels over oil in new homes, as well as conversions of existing homes in some countries.

We have not looked at equipment or housing costs per se, even though the investment costs for various systems that use each fuel must be weighed against fuel prices. We do know, however, that among central systems, direct electric resistance heating has the lowest installation cost in every country; boilers or hot air furnaces come next, followed by electric heat pumps, then gas or dual fired (electric/fuel) heat pumps. In every country, however, decisions about equipment in new homes are split between owners, occupants, and builders, rarely the same individual. In existing SFD, however, changes in fuel and equipment are decided largely by owners. We found (6) that only a great price differential (or subsidy) motivated fuel switching, suggesting that owners consider both capital and running costs when contemplating a fuel switch.

ELECTRICITY HEATING PRICE TO FUEL OIL PRICE RATIO FOR O.E.C.D. NATIONS

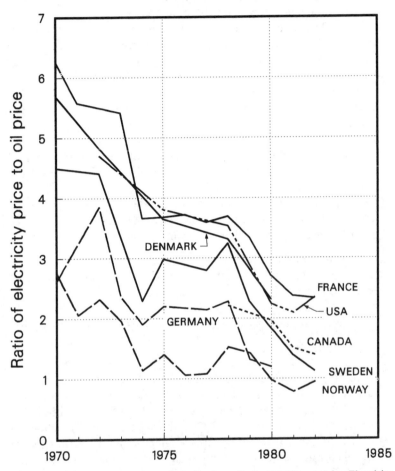

Figure 9   Electricity heating price to fuel oil price ratio for OECD countries. Electricity is counted at 3.6 MJ/kWh, and oil counted at its thermal content, approximately 35.59 MJ/liter for #1 oil, 36.6 MJ/l for #2 oil.

2.7   *Conclusions from the Cross-sectional Comparison*

1. Space heating dominates energy use in every country, followed by hot water, electric appliances, and cooking.
2. Climate, home size, efficiency, fuel mix, and life-style (indoor tempera-

tures, hot water use, and other aspects of family life) affect energy use, particularly that for space heating. The latter four are responsive to energy prices in the short and long term, and consumers can even change the areas of their homes they heat. Relative energy prices are useful in explaining the fuel choices and shares of different fuels.

3. Incomes affect mainly energy consumption in the long term through changes in the equipment stock (structure), home size, and appliance characteristics. The role of incomes acting through differences in structure must be taken into account in any causal study of differences or changes in household energy use.

4. Technologies, including the makeup of the building stock, have a big role in explaining the levels of household energy use and the prospects for changing it. Technologies change with prices and incomes; in general, all technologies have become more energy-efficient over time in response to higher energy prices, although improvement also occurred when energy prices were falling.

5. Demography is an important factor in household energy use in the very long run. However, we see no evidence that family size or other household characteristics change because of energy concerns.

Below we examine the changes in household energy use over time, and then return to see whether these five findings hold when we follow energy use over time.

3.   CHANGES IN RESIDENTIAL ENERGY USE SINCE 1973

The bottom-up method used to compare countries in 1982 can also be used to study changes in individual countries across time. In this section we examine changes in household energy use from 1973 to 1982. Although the level of disaggregation of the underlying data is the same as for the 1982 data presented above (using snapshots similar to that in Table 1 for each country and for five or more years between 1960 and 1982), here we discuss the effects of more general changes in structure (home size and type, heating system fuel and type, appliance penetrations) and note changes in key energy-use intensities.

We can illustrate this use of the bottom-up method by examining changes in heating oil use in the Federal Republic of Germany from 1960 to 1983. This is shown indexed to its 1960 value in Figure 10. Oil used reaches a maximum in 1979 and then falls. Below the curve for oil use we have plotted the evolution in the number of dwellings heated with oil, which increased almost fivefold over the period. The average size of these dwellings was larger in 1983 than in 1960, both because the share of SFD in the stock increased, and because in both SFD and MFD average floor areas grew by about 15%. Nearly 85% of these oil-heated dwellings were centrally heated in 1983, compared with only 45% in 1960. Furthermore, the fraction of these homes also using oil for hot water increased dramatically, as "hot water" indicates. In spite of these increases in substructure and hot water, total oil use per dwelling, represented by the "Intensity" curve in Figure 10, was about 5% lower in 1982 than in 1960.

The same type of time-series bottom-up analysis for all residential energy use in the Federal Republic of Germany is shown in Figure 11. We show both delivered and primary energy use, which diverge as the share of electricity (principally for appliances and water heat) increases. As with oil heated homes, increases in both the total number of dwellings and the area per dwelling drove up energy use through the late 1970s. The percentage of homes with central heating (CH) and with hot water (HW) systems rose significantly as well.

We see a similar pattern in all the countries we have studied, although the importance of individual structural factors varies. Prior to 1973, total primary and delivered energy use per dwelling increased steadily, although the rates of increase were slowest in North America, the United Kingdom, and Sweden, where levels were highest. Up to 1972, structure was the dominant force in pushing energy use upward. After 1972, however, the course changed markedly.

We show useful energy per dwelling during the 1970s and early 1980s in several countries in Figure 12. After 1973, energy use per dwelling dropped in every country (not shown in the figure but evident from our yearly data), driven principally by reductions in oil use, but rebounded by 1978–1979. This rebound occurred both because temporary conservation in 1974 and 1975 was reversed as real oil prices stabilized and fell, and because structural growth resumed in the continental European countries and Japan. However, the drop recurred in 1979–1981, with deeper cuts, affecting even gas, electricity and district heating. Most of the change appears to be in space heating; we really measure space heating plus hot water in most countries, but the drop in oil (or gas) use in homes without oil- or gas-based hot water is great and rapid enough so that it must have arisen principally from space heating.

3.1   Structural Changes

The effect of most of the structural variables after 1973 was to continue to push up total energy use per household, although at a slower rate than before. In general, the structural differences between countries narrowed.

First, living area per household increased significantly between 1973 and

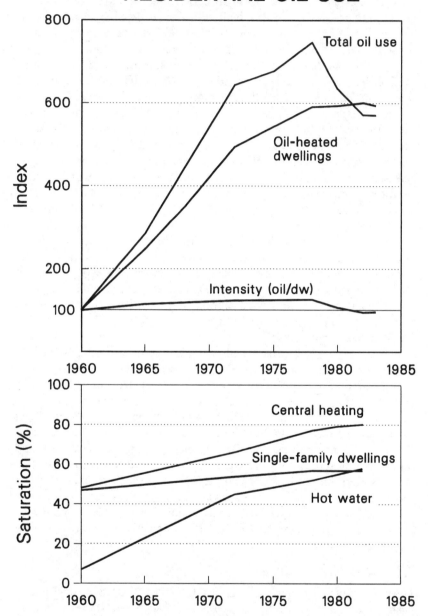

FACTORS DRIVING GERMAN RESIDENTIAL OIL USE

Figure 10 Factors driving German residential oil use [1960 values (= 100)]. Shown below are penetration of central heating, hot water, and the share of single-family dwellings in the oil-heated stock, in percentages.

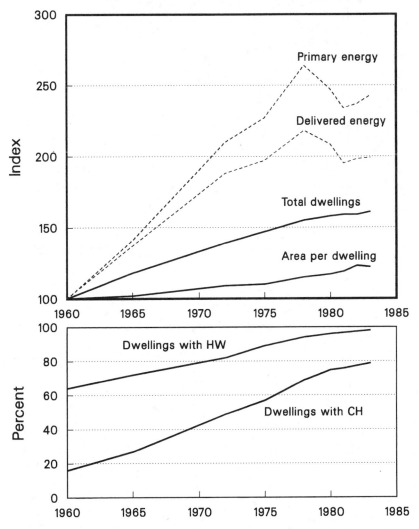

Figure 11 Factors driving German residential energy demand [1960 values (= 100)]. Primary energy, delivered energy, total number of dwellings, indexed to their 1960 value. The percentages of households with hot water (HW) and central heating (CH) are shown below.

USEFUL ENERGY PER DWELLING

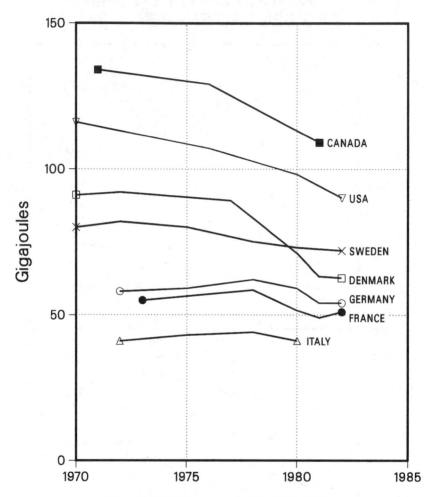

Figure 12   Useful energy per dwelling (1970–1982).

1982 (1–2% per year), both because of the increased share of SFDs in the housing stock (by 1980, new construction in almost every country was 60–70% SFD and only 30–40% MFD), and the continued increase in SFD size. Numbers of people per dwelling also decreased, so that area per capita increased (Figure 13) even more than area per dwelling.

Second, central heating penetration increased notably in countries with low levels (shown in Figure 14), except in Japan. Third, appliance stocks continued to grow, although more slowly than in the previous decade. This is illustrated with an index of the average penetration of five major appliances shown in Figure 15; for the European countries (except Sweden and Denmark) this index nearly doubled between 1960 and 1970 (not shown in the figure), with growth slower after 1970.

DWELLING AREA PER PERSON

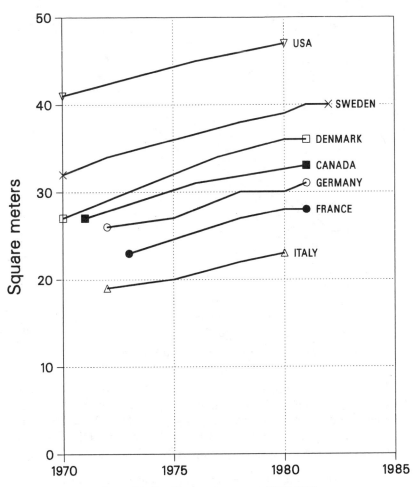

Figure 13 Dwelling area per person (1970–1982).

CENTRAL HEATING PENETRATION

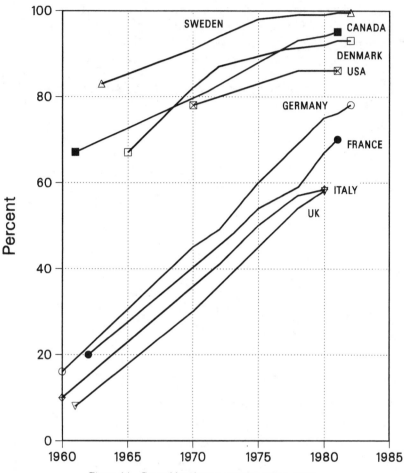

Figure 14   Central heating penetration (1960–1982).

These structural changes would, in isolation, have increased energy use per dwelling between 1972 and 1982 by about 10–15% in North America, Denmark, and Sweden; 15–20% in Norway, the United Kingdom, and Germany; and more than 20% in France, Japan, and Italy.[10] In most countries, however, energy use per dwelling did not grow this quickly

[10] This is done by estimating increases in energy use between 1972–1973 and 1981–1982, caused by changes in structure, holding the 1972–1973 intensities constant.

APPLIANCE OWNERSHIP INDEX
(AVERAGE OF SATURATION PERCENTAGES
FOR FIVE MAJOR APPLIANCES)

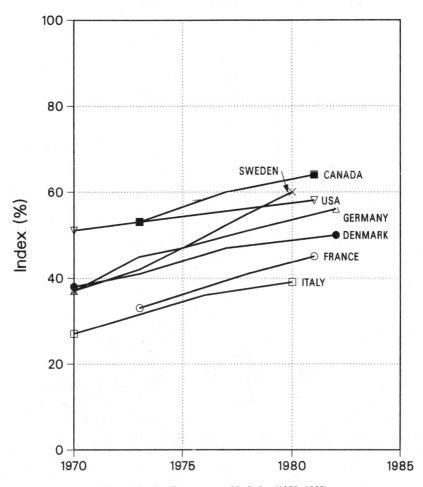

Figure 15   Appliance ownership index (1970–1982).

because of large simultaneous decreases in energy intensities which offset the changes in structure.

3.2   Changes in Energy Intensities

The slowdown in the growth of the principal structural factors does not explain all of the slowdown in the growth of total delivered energy per dwelling. The energy intensities of individual fuels and end uses also declined with the energy price shocks of the 1970s.

Heating intensities fell in all countries between 1972 and 1982 because of higher stock-average insulation levels, improved equipment maintenance (or replacements), reduced heated areas, shorter heating seasons, and lower indoor temperatures. Figure 16 shows these changes on a per dwelling

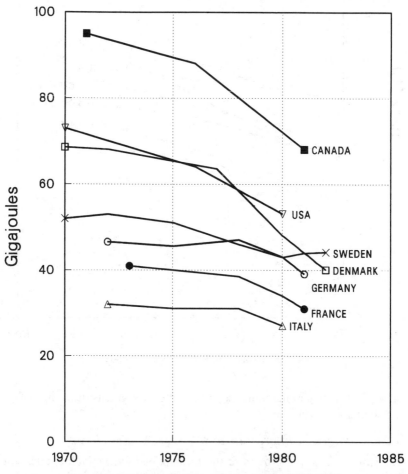

Figure 16   Space heat. Useful energy per dwelling (1970–1982).

basis, averaged over all dwellings, fuels, and systems. Figure 17 shows changes for oil-heated SFD only, in kilojoules per degree-day per square meter. Similar patterns emerge for oil-heated MFD, suggesting that our aggregation in Figure 16 is basically correct. Figure 18 shows intensity, in kJ/dd/m^2 useful energy, averaged over all fuels and systems. The drop is

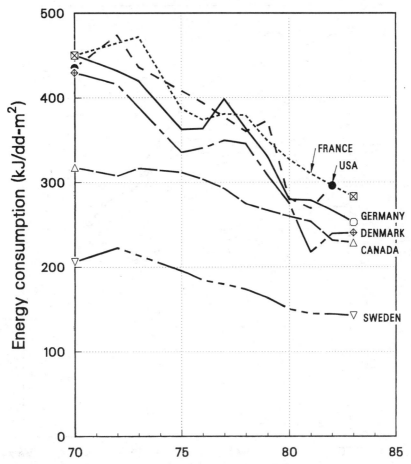

OIL HEATING INTENSITY
SINGLE FAMILY DWELLINGS
Space Heating: kJ/dd/m^2

Figure 17   Oil heating intensity, single family dwellings, space heating: kJ/sq. meter/deg-day (1970–1982).

greater in the latter two figures because dwelling size has increased in every country over the study period. The drop in oil intensity is greater than in that for all fuels, reflecting greater efforts to save oil, as oil prices generally increased more than prices for electricity and gas.

The gradual shift toward electricity and district heating reduced

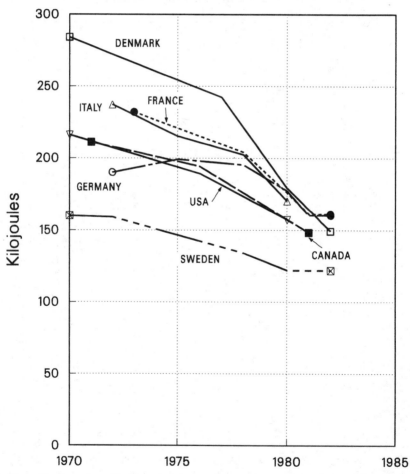

Figure 18 Space heat. Useful energy per dwelling per degree-day per square meter (1970–1982).

delivered energy more than useful energy, while primary energy use fell less than either useful or delivered energy, because of increased use of electricity for appliances. Trends in total energy used in water heating and cooking per dwelling were similar, but changes in these figures are very uncertain. Useful energy consumption for heating and hot water combined—a measure that eliminates the uncertainties over separating hot water from heating in the six countries shown in these figures—dropped at average annual rates of between 1.2% in France and 1.5% in Sweden and 4.5% in Denmark (8), during the 1972–1982 period.

We can break these changes into components for some countries. Schipper, Meyers, and Kelly (51) found that for SFD in Sweden, the intensity of useful energy for heating and hot water was 760 GJ/m^2 in 1970 and 475 GJ/m^2 in 1981 in the SFD built in 1970 or earlier, a 37% drop from the 1970 value. (The comparable intensity in SFD built during this period, about 25% of the 1981 stock, was 42% below the 1970 value.) Of the drop in the pre-1970 homes, about 5% was caused by the drop in the number of people per dwelling and 10% by the additional heat from electric appliances (not counted in heating and hot water totals), and 15–20% was caused by lower temperatures (still 21°C in Swedish homes in 1982) and lower hot water use. The remaining drop—65–70%—was attributable to retrofits of the building and boiler improvements.

The drop in total useful energy use per dwelling was greater in the United States and Denmark, and less in other countries, where more structural growth occurred. The drop in Sweden was slow and steady (cf Figures 16–18), while it occurred largely in two brief periods (1974–1976 and 1979–1981) in the other countries. This pattern suggests that more of the drop in intensity is likely to be of a permanent nature in Sweden (6, 8) than in the other countries.

New building shell heat loss rates have been reduced in all countries, partly because of stricter building codes. The Eurima data in Table 5 indicate a significant improvement in k-values by 1980 over their pre-1973 values in every European country.[11] New codes are even stricter in Denmark (after January 1, 1979) (72), the Federal Republic of Germany (after January 1, 1984) (73), France (for 1985) (74), and most strict for homes using electric resistance heating in Sweden (after January 1, 1984: shown under the Fourth Phase in Table 5) (75). Still, there is considerable spread between the Nordic countries at the lowest range of k-values, and the rest of Europe, with the highest values.

[11] Sweden had the lowest k-values, both in 1973 and in 1982, and values there were always lower than code requirements (51). Typically, Swedish walls fall under $k = 0.15$ W/m^2/°C, Danish walls fall close to $k = 0.25$ W/m^2/°C, and k-values of German walls lie between 0.40 and 0.50 W/m^2/°C, figures earlier typical for masonry construction in Denmark.

Measurement of changes in actual consumption, where they have been done, show reductions related to the improved building shells. The Swedish household survey shows that homes built after 1976 and using electric resistance heating—the dominant form—used about 15% less electricity in 1981 or 1982 than those built between 1970 and 1976 (7), a savings that roughly corresponds to the improvements in k values.[12] US data indicate that newer homes also use less energy for space heating; those built between 1975 and 1979 using electricity used 20% less electricity per square meter per degree-day in 1980 than those built between 1970 and 1974; for gas the reductions were 10%. Oil-heated homes built after 1974 used 44% less than the average for all oil-heated homes, although the sample size of post-1974 homes is very small (27). These figures confirm that there is measurable improvement in space heating use in at least two countries.

Changes in per household appliance electricity use were also significant. Appliance efficiencies increased everywhere, but this effect on consumption was somewhat offset by larger unit sizes in some countries and greatly increasing saturations. The Japanese report a 60% drop in energy use for refrigerators (of 170 liters) sold in 1982 compared with models on the market in 1973, and a 30% drop in air conditioner and television energy intensity (65). Similar, though somewhat smaller, improvements—10–30% —have been noted in Denmark (76, 77), Germany (78), Sweden (79), and the United States (80). The improvements in appliances actually sold are, however, not as great as those in the best available models. Nevertheless, the increases in Figure 19 are slowing down in the 1980s. In the United States, Germany, Sweden, and Denmark efficiency is responsible for much of this slowdown, but slower growth in ownership has also become important in all countries.

Table 6 shows strong evidence of this slowdown in Denmark. We show electricity use in new single-family dwellings (without electric heat or hot water), arranged to show consumption in 1980, 1982, and 1983 for homes built in successive years. According to the utility collecting the data, family size and appliance ownership are roughly comparable. The reduction in use per home with decreasing house age is apparent, as is the conservation that occurred after 1980 after a significant real price increase for electricity occurred. These data are real evidence of the improvement in efficiency that is occurring in electric appliances. The 1982 RECS survey (81) shows a decline in total electricity use per household in the United States, led by decreases in space heating and appliance energy use; similar evidence is apparent from the Federal Republic of Germany (VDEW, unpublished data for 1978–1982). It appears that electric appliances are becoming more

---

[12] Less than half the electricity in these homes goes to the heating radiators.

efficient everywhere, offsetting or reversing growth in energy consumption caused by increased penetration.

We can summarize changes in the most important energy uses with the following indicators (9–13). Space heating intensity, in kJ/m²/°C (Figure 18), demonstrates that space heating has become more efficient in every country since 1973. Growth in heating fuel use before then was caused

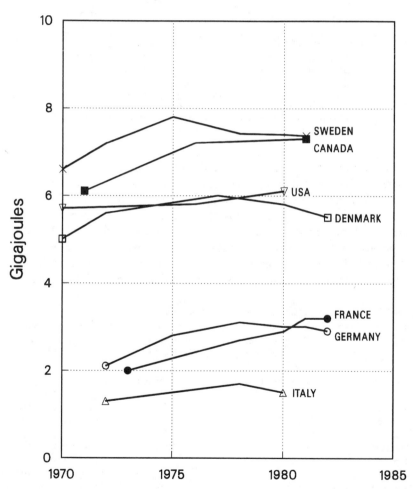

Figure 19 Water heat. Useful energy per person (1970–1982).

Table 6 Electricity use for appliances and home vintage in Denmark

Year home completed	Consumption (kWh)[b]		
	1980–1981	1982–1983	1983–1984
1974	5080	4761	4647
1975	4502	4530	4128
1976	4859	4758	4639
1977	4257	4229	4199
1978	4220	4204	4246
1979	3959	3882	3864
1980	—	3371	3614
1981	—	—	3423
Electricity price[a]			
Current Danish krone	0.50	0.72	0.88
1970 Danish krone per kWh	0.20	0.27	0.26

[a] Prices reflect year-end values for 1980, 1982, and 1983. Source: Energistyrelsen.
[b] Data on consumption are from the region supplied by SEAS, a large electric utility on Sjaelland, Denmark, as communicated by Danske Elvaerkersforening.

mostly by increases in floor area and central heating penetration, while there was little evidence of growth in use in homes already equipped with central heating. The hot water energy use indicator (Figure 19) shows the differences among the two groups of countries narrowing somewhat, but the growth in each group has slowed. The final indicator (Table 7) relates appliance electricity use to disposable income. This indicator grew rapidly through 1972 (11), but its growth after that time has been slower, reflecting both saturation and conservation. To eliminate uncertainties over ex-

Table 7 Appliance electricity per unit of real disposable income

Country	Changes before and after the oil embargo Annual percentage change		
	Pre-oil embargo	Post-embargo	Years covered
Canada	2.1%	1.1%	1961–1971; 1971–1981
Denmark	5.5%	0	1965–1972; 1972–1982
France	—	3.2%	—; 1973–1981
Germany	5.7%	2.6%	1960–1972; 1972–1981
Japan	—	7.3%	—; 1973–1979
Sweden	5.4%	1.4%	1963–1972; 1972–1982
United Kingdom	5.1%	5.7%	1961–1971; 1971–1978
United States	—	0.5%	—; 1970–1980

change rates we have shown the indicator indexed to real local disposable income. This slowdown is also seen in the individual country points, plotted for a constant exchange rate in Figure 6.

Figure 20 shows 1972–1973 useful energy use per dwelling for all purposes, for five countries, compared to 1981–1982. We estimate the amount of energy that would have been consumed in 1981–1982 if intensity had remained at preembargo levels and the structure had evolved to its 1981–1982 levels (i.e. increases in house size, central heating penetration, hot water penetration, and appliance penetration). This theoretical 1981–1982 use is shown by the middle bar in Figure 20. The bar farthest to the right shows the actual 1981–1982 levels. The actual 1981–1982 use,

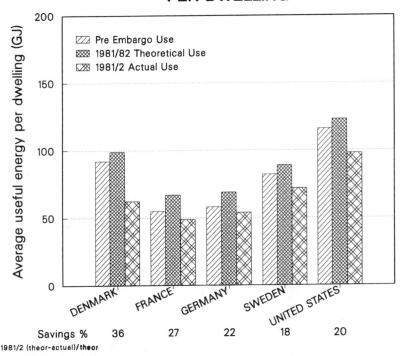

Figure 20 Residential energy use. Savings per dwelling. Lefthand bar is preembargo useful energy per dwelling. Middle bar is theoretical useful energy consumption, calculated using 1971–1972 intensities and 1981–1982 structure. The righthand bar is 1981–1982 useful energy per dwelling. The preembargo year is 1972 for all countries except the United States, for which the year is 1970, and France, for which the year is 1973. The postembargo year is 1981 for the United States and France, and 1982 for Denmark, Germany, and Sweden.

compared with actual 1972 use, was lower in every country, but the actual drop is smallest for France and Germany. The effect of this theoretical increase, however, is greatest for France and Germany; their levels of theoretical savings, measured as the ratio of the difference between 1981–1982 theoretical use and 1981–1982 actual use to 1981–1982 theoretical use, resemble those in other countries more closely. Increases in the standards of living have therefore pushed up energy use, while energy savings have pushed it down.

What about the other countries? In Norway growth in residential energy use accelerated after 1978, which Ekelund (82) has attributed to income growth–induced higher temperatures and higher hot water use. In the United Kingdom (11), there was little drop in energy use per dwelling after 1973 because natural gas was falling in price and rapidly entering the heating and hot water market, replacing inefficiently used solid fuels, but offering greater heating possibilities. Gas prices began to increase in 1979, causing a reduction in intensity for heating. A similar situation occurred in Holland (25), and was reversed by price increases toward the end of the 1970s and an aggressive insulation program. In Holland and the United Kingdom; energy use per household in 1982 was approximately 10–15% below its value in the mid-1970s. In Canada, oil and gas prices were approximately controlled and increases were moderated (6, 8), but standard increases were only moderate, and energy per dwelling was lower in 1981 than in 1971. In Japan (11, 12), use per household was higher in 1979 than in 1973 because of structural growth. In both Japan and Italy structural growth appears to have significantly offset conservation.

Figure 20, and our other data, show that the largest energy savings per dwelling occurred in Denmark, a 32% reduction in useful energy (and 34% in delivered energy). Decreases were greatest in Denmark because 1. it had the highest oil dependence; 2. pre-1972 dwellings were leaky (in contrast with Sweden); 3. there were few alternative fuels for existing oil users; 4. personal incomes were stagnant during the late 1970s. No other country faced all of these problems at the same time.

Could the reductions in energy use as a result of the reductions in intensities be reversed by a sustained fall in the price of energy? We found (6) that the combined residential oil use of seven countries (Canada, Denmark, the Federal Republic of Germany, France, Norway, Sweden, and the United States) fell by 33%, or 2000 PJ, between 1972–1973 and 1981–1982. Of this reduction, approximately 40–45% resulted from conversions away from oil or other investments which will not be reversed even if oil prices fall. The remainder of the reduction resulted from rapid drops in unit consumption whenever prices increased, changes that could be reversed if prices fell. While the fall in unit consumption from investment seems to have slackened, conversions away from oil continued in 1983 and 1984. Therefore, the share of permanent in total oil savings should increase for some time. Indeed, oil heating intensities remained nearly constant in 1982 and 1983 (6), in spite of stable or falling prices.

The decreases in heating intensities were less dramatic for other fuels than for oil (4), and can be interpreted as smaller-than-anticipated increases because newer homes and equipment were more efficient than older. Decreases have also been gradual for electric appliance efficiencies, as illustrated by Table 6 for Denmark; the United States registered some decrease in use per dwelling over the 1978–1982 period (81).

3.3 Conclusions from the Time Series Analysis

At the end of our cross-country comparison we listed the five important factors (or groups of factors) driving or explaining residential energy use. Our time series analysis supports the conclusion that these are the most important factors, here grouped in the same way.

1. Space heating still dominates energy use in every country, but its share has diminished, because of both conservation and structural growth in hot water and appliance energy use. Heat from hot water and appliances contributes to heating homes. Increased energy use for these purposes thus reduced energy needs for the space heating system somewhat.

2. Home size increased during the period, partially offsetting energy saving from increased efficiency. The fuel mix changed radically, with oil's share (as well as that of coal/coke) dropping, while electricity, gas, and district heating increased in importance. This change meant that delivered energy decreased more than, and primary energy decreased less than, useful energy.

   Indoor temperatures and hot water use have declined, and family size has decreased slightly. The reduction in family size (#5 in the conclusions of the cross-sectional comparison) led to reductions in energy use per dwelling but an increase in the number of dwellings. Thus per capita residential energy use did not decrease as rapidly as per dwelling use.

3. Prices affected energy use rapidly in 1973–1974 and again in 1979. The effect of these shocks is still being felt in decisions about insulation and equipment in new and existing dwellings. The rapid rise in oil prices explains the switch to other fuels for heating and hot water (the oil share in new homes had all but evaporated by 1982), as well as the drop in oil heating intensities. Heating intensities for other fuels dropped less than oil intensities in existing equipment, but dropped as much as oil intensities in new equipment.

Incomes increased, leading to higher living standards, but the stagnation in the early 1980s may have retarded replacement of inefficient equipment by more efficient equipment. The effects were less important in North America, Sweden, and Denmark than in countries with lower incomes, indicating declining elasticities of energy use with respect to income.

4. Technologies continued to be important in explaining the levels of and changes in household energy use. Swedish houses have even lower heat losses in 1982 than in 1972, although all countries' homes have improved; appliances have also improved in all countries. Proliferation of appliances and even housing technologies (50) appears likely to spread energy savings more rapidly among countries.

4. PROSPECTS FOR THE FUTURE

4.1 *Overview*

Moving from analysis to prediction, the disaggregated approach we have taken in this study multiplies the number of forecast variables. Nevertheless, our understanding of the trends in the underlying components is sufficiently better than our understanding of trends in the total to overcome this added uncertainty. Furthermore, the bottom-up method has the advantage that it deals with components that are related to energy-policy concerns, and aids in directing decisions about residential energy policy. In general, an understanding of the factors driving consumption allows a better targeting of policy designed to affect consumption, and a clearer evaluation of the impacts of existing policies.

For most of the countries we have studied closely, particularly those in North America and Scandinavia as well as the Federal Republic of Germany, structural growth is slowing because of heating equipment saturation. In addition, new homes are not large enough to significantly increase the average size of the stock. This means that the major uncertainties on the structural side in these countries are the population growth rate, the rate of household formation, and the rate of turnover (83) or improvement in the housing stock. In Sweden, the latter factor has become the centerpiece in the conservation program (84) because the impact of new construction will be very small. The situation is similar in Germany, Denmark, Norway, and, to a lesser degree, the United States. For the European countries, Canada, and Japan, structural growth will be stronger because of new construction, and in some cases there is still room for increases in central heating penetration.

An additional uncertainty is the impact of a greatly aged population. Will we live in institutions, in single family dwellings built when families were larger, or in a new generation of smaller apartments, where appliance ownership levels are lower than they are in single family dwellings? Will this effect cause a continued diminishing of household size, which, other things being equal, will push energy per capita up while reducing energy per household? Or will the elderly move in increasing numbers to collective care centers, as in Sweden?

Whatever the impact of demographic changes, electric appliance ownership growth will continue in all countries, but will be more important in Japan, the United Kingdom, and continental Europe than in Scandinavia and North America. Judging from the way household equipment penetrations approach similar values near saturation, the greatest uncertainty is the time when a given level is approached. Given the difference in energy efficiency among new and existing refrigerators, freezers, stoves, etc, it appears that energy use per household could decrease or hold steady through the end of the century in North America and Scandinavia, and increase only slightly in Germany and France. Thus, electric appliance efficiency improvements could offset structural growth.

As Klingberg & Wickmann (83) pointed out in an international comparison, construction rates may be an important determinant of the rate of decrease in energy intensity; we would add that housing and equipment vintage are important as well, since these influence the rate of rehabilitation and replacement. The building rate was slower in the 1970s than before because of sluggish economic growth and slower household formation. This slowdown affects replacement of older homes. The share of the 1982 building stock built since 1975 (i.e. after the first oil shock), for example, is smaller than 13% except in France, the United States, and Canada. Half the homes using oil today in France and Germany were built between 1960 and 1975, and these are too recent to be in need of major rehabilitation, but not nearly as efficient as those built since 1975. Their aging will allow for careful retrofit in combination with other modifications.

The evolution of future energy use intensities is uncertain because of their dependence on the price of energy relative to the cost of conservation investments. For example, the new ELAK code in Sweden (for homes heated by direct acting electric resistance panels), was evaluated for its payback by the Board of Planning in 1982 (85), when electricity prices were seen rising at 1% per year in real terms. Since that time they have fallen and there appears little pressure for increases in the foreseeable future. This has led to serious thought as to the private economic value of the marginal investments called for in ELAK. Prices influence consumer, builder, and manufacturer decision-making in unpredictable ways, but lower prices rarely inspire increased investment in conservation, even if today's level appears suboptimal. The rate at which consumers and producers will opt for more energy-efficient equipment is therefore a major uncertainty.

4.2   New Uses for Energy?

Implicit in the above discussion of the diminishing effect of structural increases on energy consumption is the assumption that existing components of structure will continue to dominate energy consumption. Will new energy (or electricity) uses emerge with increased incomes?

Air-conditioning is one high-consumption end use that now has high penetration only in the United States ($>60\%$) and Japan (50%). Penetration of this appliance is closely linked with income (27) and is therefore likely to continue to rise. Furthermore, in the United States, central units are replacing window units, driving up average consumption. Two additional considerations should mitigate the potentially large increase in consumption from air conditioners. Penetration has remained extremely low in most European countries in spite of income increases, which suggests that they may never be as popular as in the United States, perhaps because traditional building practices in the warm parts of Europe reduce cooling needs. New air-conditioner efficiencies are also much higher than stock-average efficiencies in both the United States and Japan.

The potential added consumption of new kitchen electric devices is less than that of air-conditioners because they often replace less efficient methods of performing the same task. Coffee machines, for example, eliminate inefficient cooking of water on the stove, egg cookers reduce the amount of hot water needed to boil an egg, toaster ovens reduce the volume heated to bake a potato. Similarly, many appliance improvements lead to efficiency improvements: self-cleaning ovens need more insulation to stand higher temperatures; washers that are equipped with small computers for sophisticated programming also feature energy saving routines; and security lighting timers can be used to turn lights off as well as on.

Other possible new uses include outdoor space heating, which is probably unlikely in view of costs. Garage heating in cold climates is also potentially important. In Sweden this was widespread, but it has yielded to electric heating of engine blocks and car seats (in 11% of homes in 1980) controlled to start just before the car is required, avoiding heating the entire garage space. Greatly increased security lighting will be a minor energy use because lighting systems designed for long hours of high illumination have become very efficient (86).

Electronics and home computers in the United States and Sweden are growing rapidly in popularity. Video systems had reached 20% of all US households in 1984 (87), 9% in Germany (60), and an even higher level in Sweden. We would expect this growth to occur soon in other countries.

Nevertheless, the electricity needs from this growth will be small, if not insignificant. First, virtually all electronic equipment for home use operates at low wattage, since high wattage increases the cost of removing heat with fans or expensive heat radiation surfaces. Electricity use is therefore low. Most important, no single new device is used more than a few dozen hours per year; the reports from Sweden (88) and Germany (22) tend to dismiss these new uses because of this fact, and the Swedish report assigns only 85 kWh per year to the household of 1990 to cover all of them. Thus, our comparison suggests that neither small appliances nor other new uses of electricity will be significant in future electricity use.

The international comparison points to one important use that we have not treated, however: energy use in second residences or summer homes, which are most common in Sweden (89), Italy (90), and France (91). Nevertheless, even in Sweden, with the most space heating, such use still represents less than 5% of use in principal residences (89).

What could be the most significant change, however, is a trend toward great increases in home offices. Employed persons working at home would effectively transfer electricity use from the workplace to the home and raise daytime space conditioning needs as well. Clearly major life-style changes affect household energy use (92), but increases are not necessarily more likely than decreases. Although a full study of these possibilities would be important, there do not appear to be any truly new uses of energy in homes that would have significant impacts on use, compared with space conditioning and hot water. We conclude that no new uses in the foreseeable future will increase household energy use at pre-1973 rates.

4.3   Prospects for Individual Fuels

Along with energy intensity, fuel choice is an important planning uncertainty. While use per household should hold steady (for the 11 countries considered here in some detail), fuel shares will change significantly.

4.3.1   OIL   Oil use is falling and will fall for some time in the major countries studied in (6). In the three largest oil users not included there— Belgium, Italy, and Japan—natural gas (LNG or LPG in Japan) is slowly but surely replacing oil in many homes.

4.3.2   NATURAL GAS   Natural gas is the most important substitute for oil in every country except Sweden and Norway, and has recently gained market share from electricity in new homes in Germany and the United States as well. The great transition from city gas, used principally for cooking, to natural gas, for all uses, is virtually complete in France, Germany, and the United Kingdom, so that growth in use will come with new hookups and conversions but will be held down by conservation by all current users. Gas industry experts in Italy, Germany, Belgium, Holland, and Denmark have indicated to us that current and future gas use was overestimated and that there is a surplus of gas relative to pipeline capacity or contracted quantities. Since new homes and appliances are relatively efficient, growth will be slower than many imagined when gas system expansion was contemplated in the 1970s.

4.3.3 ELECTRICITY   Electricity use is increasing but the increase is slowing relative to income except where electric heating is increasing in new or existing dwellings (in Canada, the United States, France, Sweden, and Norway). In the United States the penetration of cooling has passed 60% (in 1984), and in Japan it is nearing 50%; prospects for other countries are uncertain. Growth in other appliance uses will probably slow significantly except in the lower-income OECD countries (Italy, Japan, and others not covered here).

The potential market for electric space heating is the largest uncertainty. The price of electricity has fallen, relative to those of gas and (especially) oil, pushing up the share of electric heating in many countries. In the United States, Canada, France, and Denmark, the share of electricity in new homes lies between 25 and 50%, in Germany and the United Kingdom considerably lower. In Sweden and Norway electricity heats more than 80% of new SFD, and there have been wholesale conversions from oil to electricity in existing SFD; fully 25% of the stock of oil-heated SFD in Sweden in 1978 had converted to electricity by 1983 (93). In other countries there is much less evidence of conversion to electricity in homes with central heating based on gas or oil. This observation suggests that the price differential between electricity and fuels must be very large (i.e. electricity must have significantly lower running costs, as it presently does only in Norway and Sweden) before consumers will defect from fuels in large numbers. On the other hand, electric heat pumps and zoned resistance heating in very tightly built homes can be competitive with fuels, particularly if builders can install resistance heating at a lower initial cost than that of a fuel boiler or furnace. Electric water heating should replace oil in homes that continue to use oil for heating, as in the United States, Sweden, and Germany. What could limit electricity growth would be widespread use of more efficient gas and oil-fired heating and water heating systems, such as condensing or pulse combustion gas furnaces, gas driven heat pumps, or storage systems that allow the furnace to run only at full-rated output. These systems would reduce the cost of obtaining heat and hot water from fueled systems, in analogy with electric heat pumps. Growth in electricity for appliances as we argued above (cf Sections 2.4 and 4.1), will be lower in relation to income growth than in the past because of saturation and efficiency effects. Thus, while electricity will likely grow more rapidly than the other energy carriers, its growth should fall far behind the 7–8% growth rates in the residential sector of the 1960s and early 1970s.

4.3.4 OTHER FUELS   Wood consumption has nearly doubled from the mid-1970s to the early 1980s in the United States, Norway, and Sweden to more than 10% of the heating market in these countries and an undetermined but rising share in Canada. Since wood is used most often as a substitute or complement for oil or electricity, stable or falling prices in these markets will probably limit short-term growth. Free wood still dominates consumption, representing a source of income that is not taxed. Coal and coke staged a slight comeback in Denmark and Sweden, but the return was not as sharp as the rise in wood, and in Germany the long-term downward trend in coal use has simply slowed, so that coal and coke accounted for only 6% of delivered household energy in 1983. The two oil price shocks of the 1970s caused only temporary, slight increases in use and share. District heating grew dramatically in both Sweden and Denmark; now more than 35% of all homes are heated this way, but the growth is slowing because of new natural gas in Denmark and electricity in Sweden. In France and Germany growth was marked during the period to 1980, but natural gas appears to have taken much of the MFD market from district heating.

5. CLOSING REMARKS

International bottom-up comparisons are useful in explaining the differences in residential energy use among countries, and in understanding past changes in use within single countries. These comparisons are also helpful in predicting future use. When the differences among countries have been isolated, history in one country—in incomes, price levels, building characteristics, equipment efficiencies, or technologies—may point the way to the possible future in another. In this way, a careful bottom-up understanding of international patterns of demand can be transformed from an interesting academic exercise to a forecasting tool, although forecasting has not been the focus of the work reported here.

Our review has touched on several issues that remain unresolved. One is the role of new technologies. How much less (or more) energy will they require? Will the recent slack in world oil prices lead to abandonment of research on more efficient heating equipment and appliances? How much more efficient will equipment become even if prices remain constant? These questions cannot be answered by statistical means, but only by comparing differences in technologies among countries or between existing and new technologies in single countries, and estimating how far and quickly the newer, more efficient ones will penetrate.

A related issue is life-style. Although its impact is hard to measure, several qualitative features are worth noting. Gas cooking is so popular in France, Italy, and Japan, the three countries with the most renowned cuisine in our study, that consumers not living near gas mains will pay a premium for LPG. Is there a connection here between culture and energy choice? Second homes are more popular in some countries [France (5), and Sweden (4)] than elsewhere. Where do others spend their free time? Is the high ownership of second homes in Sweden related to the high share of apartments in the dwelling stock?

What about the difference between high indoor temperatures in Scandinavia, the moderate temperatures reported in the United States, Canada, France, and Germany (17–19°C), and the considerably lower temperatures reported in Holland, Norway, and the United Kingdom (13–17°C), as well as the apparent differences in hot water habits? Do these represent true cultural preferences in indoor environment and hygiene, i.e. the perception of benefits, or are they simply products of prices and marginal utilization costs?

Another issue is often called the boomerang effect (94). The elasticity of energy use with respect to energy price depends on the cost of obtaining a given energy service, not just the cost of energy. If houses are insulated, the cost of raising indoor temperatures decreases; the Swedes have both the highest insulation levels and the highest temperatures. Hirst et al (95) found that 10% of the expected savings from energy-efficient retrofit was offset by a 1°F increase in indoor temperature. A major uncertainty for planning is the degree to which technology improvements will offset behavioral changes that to date have caused significant energy savings. Another is the extent to which adaptation to lower levels of indoor comfort and hot water use will seriously reduce the attractiveness of energy efficient technologies.

The process by which households make energy-related decisions, particularly those related to equipment purchases, is poorly understood. Ruderman et al (80) found that in the United States consumers appear to demand rates of return of 30–200% in order to make incremental energy-saving investments when buying insulation or new equipment. Schipper et al (50), by contrast, indicated that Swedish consumers, using a unique housing loan system that effectively finances all increased efficiency investment costs, demanded only a 5% real rate of return. There appear to be fundamental differences in the way families (and decision makers such as builders) view energy saving investments. We believe that a great deal more cross cultural research is needed to see whether conservation investment patterns are amenable to manipulation by broad government programs (as the Swedish example shows), or whether there are some cultures where such investments will be resisted no matter how sweet the incentive.

Other issues have been touched on briefly, including the impact of an aging population on housing, life-style, and energy use patterns. The spreading of the population into smaller household units will tend to increase total energy use, but by how much? Another is the use of secondary fuels. Gathered wood, for example, represents an untaxed source of income, which probably explains its popularity in high income strata of the United States (81) and Sweden. Already in Sweden, however, there is great concern over the outdoor (and indoor) air pollution effects of widespread use of wood in built up areas. This problem may limit growth in wood use.

The final issue we raise is that of programs, which we have discussed elsewhere (3, 6, 7, 8, 10). How effective are conservation programs in changing energy use? How much extra conservation occurs because of public investment support? That is, are programs sufficient to cause conservation? Are they necessary, or do they only accelerate investments that will be made sooner or later? An international comparison (96) is still under way, but preliminary results are mixed, in part because data from many countries participating are too aggregated to permit isolation of factors which may be influenced by programs. Other international reports present virtually no residential energy use data, making it difficult to draw any conclusions (97, 98). We believe that it is important to combine available data on residential energy use with information about program participation in order to see what levels of public involvement bring the greatest benefits. An important series of studies by Hirst and colleagues (99) has shown that it is possible to both measure changes in energy use and estimate the changes attributable to conservation programs. The material we present herein suggests that most of the changes heretofore—except in Sweden—arose without any great investment, and it is difficult to see how national conservation programs contributed directly, except through important avenues of information and demonstration. At the same time, programs could ensure the orderly retrofit of most of the housing stock in the countries we have studied during the rest of this decade. Here only the bottom-up approach can show the possible impacts of conservation programs in the future.

ACKNOWLEDGMENTS

This work was supported by the Assistant Secretary for Conservation and Renewable Energy, Office of Building Energy Research and Development, Buildings Equipment Division of the US Department of Energy under contract No. DE-AC03-76F00098. We acknowledge the helpful comments of Michael Rothkopf and the assistance provided by Peter Goering and Steve Meyers. Chris Pignone prepared several of the charts. We also acknowledge the assistance of many experts in the energy industries, ministries, and in universities in the countries we studied. Adam Kahane was supported by a Natural Science and Engineering Research Council (Canada) fellowship, and Lee Schipper acknowledges the hospitality of VVS Tekniska Foereningen, Stockholm, and prior support of the Swedish Council for Building Research for the preliminary version of this review.

Literature Cited

1. Meyers, S. 1981. *Residential Energy Use in the United States.* LBL 14932. Berkeley: Lawrence Berkeley Lab. Available from Natl. Tech. Inf. Serv.
2. Chern, W., Ketoff, A., Schipper, L., Rosse, S. 1983. *Residential Demand for Energy. A Time-Series and Cross Sectional Analysis for the OECD.* LBL-14251. Berkeley: Lawrence Berkeley Lab., Natl. Tech. Inf. Serv.
3. Ketoff, A. 1985. *Residential Energy Use and Conservation in France.* Berkeley: Lawrence Berkeley Lab. In preparation
4. Schipper, L. 1985. Residential energy use in the OECD. The bottom up approach. *Energy Econ.* In press
5. Ketoff, A. 1984. Facts and prospects of the Italian end-use energy structure. In *Global Workshop on End-Use Energy Strategies,* ed. J. Goldenberg, T. Johansson, A. Reddy, R. Williams. In preparation
6. Schipper, L., Ketoff, A. 1984. *Oil Conservation: Permanent or Reversible? The Example of Homes in the OECD.* LBL-17785. Berkeley: Lawrence Berkeley Lab. *Science.* Submitted
7. Schipper, L. 1984. Residential energy use and conservation in Sweden. *Energy and Buildings,* 6:11
8. Schipper, L. 1984. *Internationella Jämförelse av Bostädernas Energiförbrukning.* (Int. Comparison of Residential Energy Use.) R131:84. Stockholm: Swed. Counc. Build. Res. (In Swedish)
9. Meyers, S., Schipper, L. 1984. Energy in American homes: changes and prospects. *Energy* 9:6
10. Schipper, L. 1983. Residential energy use and conservation in Denmark. *Energy Policy* (Dec.)
11. Schipper, L., Ketoff, A. 1983. Home energy use in the OECD: how much have we saved? *Energy Policy* (June)
12. Schipper, L., Ketoff, A., Meyers, S. 1982. Indicators of residential energy use and conservation. *Energy* 7 (2)
13. Schipper, L., Ketoff, A., Meyers, S. 1981. *International Residential Energy Use: Indicators of End Use and Conservation.* LBL-11703. Berkeley: Lawrence Berkeley Lab., Natl. Tech. Inf. Serv.
14. Schipper, L., Ketoff, A. 1980. International comparison of residential gas use and conservation. Published in *Proc. the Workshop on Gas Use and Conservation.* Chicago: Inst. Gas Technol.
15. Schipper, L. 1980. *International Residential Energy Use and Conservation: Demographic and Economic Data Base.* LBL-10690. Berkeley: Lawrence Berkeley Lab., Natl. Tech. Inf. Serv.
16. Thompson, W. 1984. *Residential Energy Consumption Survey: Housing Characteristics 1982* (and previous volumes) DOE/EIA 0314(82). Washington, DC: Energy Inf. Admin.
17. Björck, G., 1977–1984. *Energistatistik för småhus 1976–83.* (Energy Statistics for Single-Family Dwellings); Pettersson, K., 1977–84 *Energistatistik för flerfamiljhus 1976–83.* (Energy Statistics for Multi-Family Dwellings). Oerebro: Statstistics Sweden (In Swedish)
18. Ljones, A., Flagstad, K. 1983. *Energiundersøkkelsen 1983.* Preliminary version in SU Nr. 46, 1983. Oslo: Central Bur. Stat.
19. Hem, K. G. 1983. *Energiundersøkkelsen 1980.* Rapport 83/12. Oslo: Central Bur. Stat. (In Norwegian)
20. Haehnel, I. 1982. CEREN's experience in evaluating the energy demand for the residential sector in France. In *Residential Energy Use. Proc. Workship at the Joint Research Centre,* ed. C. Zanantoni. Ispra (Italy): Joint Res. Centre
21. Holm, K. F., Wolke, G. 1968. Der Verbrauch von leichtem Heizöl um

Haushalt. *GFM Mitteilungen fuer Markt- und Absaztforschung.* 14(2):55–61. Private communications from K. F. Holm, A. G. Esso. Hamburg (In German)

22. VDEW. 1981. *Ergebnisse der Haushaltkundenbefragung* (Results of the Residential Survey). Frankfurt: Verein der Deutscher Elektrizitäts Werke (In German)

23. Lemmens, T., Sypkens-Smit, A., Vellema, A. 1982. *Electriciteitsverbruik in gezinshuishoudens* (Electricity Use in Private Households). Arnhem: Vereninging van Directeuren van Elektriciteitsbedrijven in Nederland (In Dutch)

24. Sweden State Power Board and Central Board of Management. 1964, 1969, and 1971 Customer Surveys. Stockholm: Swed. State Power Board (In Swedish)

25. Gasunie. 1973. 1980–1983. *Basisonderzoek Aardgas Kleinverbruik (BAK).* Groningen: Gasunie. Published in summary form in Naber, W., Zwetsloot, M. 1979. *Gas* (Dec.) p. 590; Boonstra, A., Zwetsloot, M. 1981. *Gas* 6:301 (June); Boonstra, A., Zwetsloot, M., Mengelberg, W. 1982. *Gas* 7 (July); Minderhoud, P., Zwetsloot, M. 1983. *Gas* (Jan.)

26. Italgas, Servizio Tecnico. 1980. *Risparmio di energia negli impianti per la produzione di acqua calda.* (Energy conservation in hot water equipment) Torino: Italgas (In Italian)

27. Latta, R. 1983. *Regression Analysis of Energy Consumption by End Use* (Including unpublished cross tabs.) DOE/EIA 0431. Washington, DC: Energy Inf. Admin.

28. VDEW. 1977. *Ueberlegung zur kuenftigen Entwicklung des Stromverbrauchs privater Haushalte in der BRD* (Considerations of the future development of residential electricity use in the FRG). Frankfurt: Verein der Deutscher Elektrizitätswerke (In German)

29. Fels, M., Goldberg, M. 1983. With just billing and weather data, can one separate lower thermostat settings from extra insulation? In *Families and Energy: Coping with Uncertainty*, ed. B. M. Morrison, W. Kempton. Proceeds of a Conference. East Lansing, Michigan: Inst. Family and Child Study

30. Energiwirtschaftsinstitute. 1966. *Der Energieverbrauch der Haushalte in der Bundesrepublik Deutschland* (Energy use in households in the FRG). Köln: Energiewirtschaftsinstitut and Brussels, Comm. Eur. Community (In German)

31. Suding, P. 1982. *Detaillierung der Endenergieverbrauch der Bundesrepublik.*

Köln: Energiewirtschaftsinstitut (In German)

32. Schipper, L. 1985. Reconstruction and analysis of residential energy use in Sweden 1960–1982. In *Internationell Jämförelse av Bostädernas Energiförbrukning.* R:131. Stockholm: Swed. Counc. Build. Res.

33. Carlsson, L. G. 1983. *Energiförbrukning för uppvärmnings ändemål i övrigsektorn.* (Energy use for heating in the "other sector.") Stockholm: Statens Energiverk (In Swedish)

34. Schiffer, H. W. 1984. Der oelmarkt ist besser als seine statistik (The oil market is better than its statistics). *Brennstoffspiegel* 4, June (In German)

35. Energy Inf. Admin. 1984. *Estimate of U.S. Wood Energy Consumption 1980–1983.* DOE/EIA 0341(83). Washington, DC: Energy Inf. Admin. Also the first edition, covering 1949–1981, DOE/EIA 0341

36. Rosland, A. 1982. *Forbruk av fast brensel i husholdninger 1960–1980.* (Use of solid fuels in households 1960–1980) Report 81/11. Oslo: Central Bur. Stat. (In Norwegian)

37. Statens Industriverk. 1983. *Oekad Eldning med Skogsråvara.* (Increased firing of forest products) Stockholm: Board of Forestry and Board of Industry. See also *Skogsindustri, Skogsenergi.* SIND 1983:2. Stockholm: Liber Förlag (In Swedish)

38. Leach, G. 1981. *Energy Related Statistics for UK Dwellings.* London: Int. Inst. Environ. and Dev.

39. Elmroth, A., Levin, P., eds. 1983. *Air Infiltration Control in Housing: A Guide to International Practice.* Stockholm: Swed. Counc. Build. Res.

40. Scott, A. 1980. The economics of space heating. *Energy Policy* Dec.

41. Parikh, S. C., Rothkopf, M. 1980. Long-run elasticity of US energy demand. *Energy Econ.* Jan.

42. Sjölund, J. 1979. *Värmeisoleringsekonomi.* (Insulation Economics) R8:1979. Stockholm: Swed. Counc. Build. Res. (In Swedish)

43. Rota, G. 1983. *Confronto sperimentale fra riscaldamento autonomo e centralizzato.* Milano: Snam, Pianificazione e sviluppo attività gas n. 2 (In Italian)

44. Giorgetti, G., Lanaro, L., Oggioni, A. 1982. Industrializzazione degli usi razionali dell'energia nel settore del riscaldamento: problemi, limiti e prospettive dei servizi energetici. *Economia delle Fonti di Energia*, anno XXV, n. 18 (In Italian)

45. AgipPetroli. 1982. *Indagine sul riscaldamento domestico, stagione 1980–1981.* Roma (In Italian)
46. Carlsson, L. G. 1984. *Energianvändning i Bostäder och Lokaler 1970–1983.* R : 132. Stockholm : Swed. Counc. Build. Res. (In Swedish)
47. Klingberg, T., ed. 1984. *Energy Conservation in Rented Buildings.* M84 : 11. Gävle : Natl. Swed. Inst. Build. Res.
48. See Ref. (13)
49. Schipper, L., Brunsell, J. 1985. *Energy Use in Norwegian Homes.* In preparation
50. Shoda, T., Murikami, S., Yoshino, H. 1979. *A Study of the Performance of Household Equipment.* Rep. 165. Tokyo. Inst. Industrial Sci., Univ. Tokyo
51. Schipper, L., Meyers, S., Kelly, H. 1985. *Coming in From the Cold: Swedish Energy Efficient Housing from the US.* Washington, DC : Seven Locks. In press
52. Geller, H. 1983. *Energy Efficient Appliances.* Washington, DC : Am. Counc. Energy Efficient Econ.
53. Pagani, R., Pavoni, G., Pavarotti, L., Scotto, N., Ketoff, A. 1983. La pianificazione del risparmio energetico nel settore residenziale. In *Edilizia Popolare,* Milano, 4 (In Italian)
54. Tolstoy, N., Sjöström, C., Waller, T., Björling, J. 1984. *Bostäder och lokaler från energisynpunkt.* (Homes and non-residential buildings from an energy point of view). M84 : 8. Gävle : Statens Inst. för Byggnadsforskning (In Swedish)
55. Christensen, F., Ljungmark, P., Melchior, L. 1981. *Existerende Bygningers Energimaessig Tilstand.* Fase 1. Tåstrup : Teknol. Inst. (In Danish)
56. Hofer, P., Masuhr, K., Furtwängler, W. 1984. *Die Entwicklung des Energieverbrauchs in der Bundesrepublik Deutschland und Seine Deckung bis zum Jahr 2000.* Basel, Switzerland : Prognos AG (In German)
57. These include Shell Canada (unpublished model estimates). See Refs. 20, 21, 31, 38, 50, 55
58. Energiberedskapsutredning. 1975. *Energiberedskap för Kristid.* (Energy preparedness during crises). SOU-1975 : 60/61. Stockholm : Liberförlag (In Swedish)
59. Munther, K. 1979. *Oljeförbrukning i Småhus 1973–1979* (Oil use in single-family dwellings). Stockholm : Swed. Counc. Build. Res. (In Swedish)
60. HEA. 1984. *Statistisches Faltblatt 1983.* (Stat. tables) Frankfurt : Hauptberatungsstelle fuer Elektrizitätsanwending E.V. (In German)
61. US Dept. Commerce. 1984. *Statistical Abstract of the United States.* Washington, DC : USGPO. 104th ed.
62. Møller, J. 1981. *Danmarks elforbrug frem mod 1990* (Denmark's electricity consumption towards 1990). Lyngby : Danmarks Elvaerkersforening Udredningsinstitut (In Danish)
63. ZVEI. 1981. *Energiebericht der Elektroindustrie* (Energy report of the electric equipment industry). Frankfurt. Zentralverband der Elektroteknischen Industrie (In German)
64. Konsumetverket. 1982. *Tvättmaskiner (Washing Machines).* Råd och Rön faktabok. Vällingby, Sweden : Konsumentverket (In Swedish)
65. Energy Conservation Center. 1983. *Energy Conservation in Japan.* Tokyo : The Energy Conservation Center. 71 pp.
66. Goldstein, D., Rosenfeld, A. 1978. *Energy Conservation in Home Appliances through Comparison Shopping: Facts and Fact Sheets.* LBL-5910. Berkeley : Lawrence Berkeley Lab.
67. Diamond, R. 1984. *Energy and the Low-Income Elderly.* Berkeley : Lawrence Berkeley Lab. and University Energy Res. Group. See also Diamond, R. 1983. *Energy and Housing for the Elderly: Preliminary Observations.* In (29)
68. Lundström, E. 1982. *Boendevanornas inverkan på energiförbrukningen i småhus* (Occupant behavior influence on energy use in single-family dwellings). T46 : 1982. Stockholm : Swed. Counc. for Build. Res. (In Swedish)
69. Gaunt, L., Berggren, A. M. 1982. *Brukarvanor och Energiförbrukning* (Occupant habits and energy use). Gävle : Statens Institut för Byggnadsforskning. See also Ref. 47 (In Swedish)
70. Socolow, R., Sonderegger, R. 1976. *The Twin Rivers Program: A Four Year Summary Report.* Princeton : Center for Energy and Environmental Studies
71. Claxton, J. D., Anderson, C. D., Ritchie, J. R. B., McDougall, G. H. G., eds. 1981. *Consumers and Energy Conservation.* New York : Praeger. See also Ester, P., Gaskell, G., Joerges, B., Midden, C. J. H., van Raaij, W. F., deVries, T. eds. 1984. *Consumer Behavior and Energy Policy.* Amsterdam : Elsevier
72. Bygestyrelsen. 1982. *Building Regulations* Copenhagen : Ministry of Housing
73. Bundesministerium fuer Wirtschaft. 1982. *Wärmeschutz bei Gebaueden.* Bonn : Fed. Ministry Econ. (In German)
74. Olive, G. 1984. Industrial policy for saving energy. in Blumstein, C. 1984. *Doing Better: Setting an Agenda for the Second Decade.* Volume J. Washington : Am. Counc. for an Energy Efficient Econ.
75. Board of Planning. 1982. *Direktelvärme i småhus, lågtemperaturuppvärmning av*

byggnader mm. (Direct electric heating ...) PFS 1982:3. Stockholm; Natl. Board of Physical Planning. See also the regular building code, SBN-80, also published by the Board of Physical Planning (In Swedish)

76. Statens Husholdningsråd. 1980. Unpublished material submitted to the Ministry of Energy for "Energiplan 1981." Copenhagen: Ministry of Energy

77. Nørgaard, J. 1980. Improved efficiency in domestic energy use. *Energy Policy* 7, 1.

78. ZVEI. 1983. *Energieeinsparpotential von Elektrohaushaltsgeräten.* also ZVEI. 1983. Energiesparbericht der Elektroindustrie. Frankfurt: Zentralverband der Elektroindustrie (In Germany)

79. Horovitz, A. 1984. Unpublished compilation of all refrigerators and freezers sold in Sweden. Vällingby: Konsumentverket

80. Ruderman, H., Levine, M., McMahon, J. 1984. *The Behavior of the Market for Efficiency in the Purchase of Appliances and Home Heating Equipment.* LBL-15304. Berkeley: Lawrence Berkeley Lab.

81. Thompson, W. 1984. *The Residential Energy Consumption Survey (RECS): Consumption and Expenditures April 1982 through March 1983.* DOE/EIA-0321/1(82). Washington, DC: US Dept. of Energy

82. Ekelund, T. 1983. *Stabilisering av Energiforbruket i Norge. En analyse av det primaere energiforbruket i husholdingene i perioden 1965–1980* (Stabilization of energy use in Norway: an analysis of primary energy use in households). GRS 478. Oslo: Gruppen for Resursstudier (In Norwegian)

83. Klingberg, T., Wickmann, K. 1984. *Energy Trends and Policy Impacts in Seven Countries.* Bulletin M84:12. Gävle: Natl. Swed. Inst. for Build. Res.

84. EHUS-85. 1984. *Energi-85. Energianvänding i Bebyggelse* (Energy 1985, Energy Use in Buildings). G26:1984. Stockholm: Swed. Counc. for Build. Res. (In Swedish)

85. Abramsson, L. 1982. *Ekonomiska Konsekvenser av Planverkets Bestämmelser.* (Economic consequences of the Board of Planning's Code). Stockholm: Statens Planverk (In Swedish)

86. Verderber, R., Rubenstein, R. 1984. Comparison of technologies for new energy-efficient lamps. *IEEE Transactions on Industry Applications.* Vol. IA-20(5) (Sept.)

87. *San Francisco Chronicle,* Nov. 13, 1984

88. Central Board of Management. 1981. *Elkonsumtionen i Sverige 1978–1990* (Electricity use in Sweden 1978–1990). Stockholm: Central Board of Management (now Kraftsam) (In Swedish)

89. Fredbäck, K. 1979. *Energianvänding i Fritidsbegyggelse* (Energy use in vacation homes). R37:1979. Stockholm: Swed. Counc. Build. Res. (In Swedish)

90. Mostacci, R. 1982. *Mercato abitativo e risparmio energetico* (Residential market and energy conservation). Rome: Centro Ricerche Economiche Sociologiche e di Mercato nell' Edilizia (In Italian)

91. Observatoire de l'Energie. 1984. *L'Energie dans les secteurs economiques, Edition 1984.* Paris: Ministry of Research and Industry (In French)

92. Nader, L. ed. 1978. *Energy Choices in a Democratic Society.* (Report of the Consumption, Location, and Occupation Panel) Committee on Nuclear and Alternative Energy Systems. Washington: Natl. Res. Council

93. Hedenström, C. 1984. *Småhusens Uppvärmning (Heating of SFD).* Pub. 3. Stockholm: Kraftsam. (See also the 1983 report by the same author.) (In Swedish)

94. Levine, M. D., Chan, P., McMahon, J., Ruderman, H. 1984. *Analysis of Federal Appliance Energy Efficiency Standards.* LBL-17903. Berkeley: Lawrence Berkeley Lab.

95. Hirst, E., White, D., Goeltz, R. 1984. *Indoor Temperature Changes in Retrofit Homes.* Oak Ridge, Tenn: Oak Ridge Natl. Lab.

96. Mueller, H. 1984. *Consumer Energy Conservation Policies and Programs: A Comparative Analysis of Program Design and Implementation in Eight Western Countries.* Berlin: Int. Inst. for Environment and Society. See also the series from the IIEG, edited by B. Joerges, covering each individual country

97. Commission of the European Community. 1984. *Comparison of Energy Saving Programmes of EC Member States.* COM (84) 36. Brussels: Comm. of the Eur. Community

98. Int. Energy Agency. 1984. *Energy Conservation Programs of IEA Member Countries.* Paris: OECD

99. Soderstrom, E. J., Berry, L. G., Hirst, E., Bronfman, B. H. 1981. *Evaluation of Conservation Programs: A Primer.* ORNL/CON-76. Oak Ridge: Oak Ridge Natl. Lab.

Ann. Rev. Energy. 1985. 10:407–29
Copyright © 1985 by Annual Reviews Inc. All rights reserved

FUELWOOD AND CHARCOAL USE IN DEVELOPING COUNTRIES

Timothy S. Wood and Sam Baldwin

Volunteers in Technical Assitance, Arlington, Virginia 22209

INTRODUCTION

Nearly two million metric tons of fuelwood and charcoal are consumed daily in the developing countries, about one kilogram each day for every man, woman, and child (1). Some of the wood is converted into charcoal, but most is burned directly. Although the energy obtained represents only about 10% of the energy consumed worldwide, nearly half of the world's people absolutely depend on it to cook their food, heat their homes and water, and produce marketable goods (2).

Fuelwood and charcoal derived from wood, along with animal dung and agricultural residues, provide over half of the total energy consumed in some 60–70 developing nations (3). This fuel supplies as much as 95% of the domestic energy in these countries, as well as making a significant contribution to commercial and industrial needs (4).

While technically a renewable resource, the world's forests are disappearing faster than they are being replaced. The United Nations Food and Agricultural Organization (FAO) estimates that 11.3 million hectares of forests are being lost annually to agriculture, grazing, commercial timbering, uncontrolled burning, fuelwood consumption, and other factors, with 90% of cleared land never replanted (5). The problem is especially acute in the Caribbean, in drought-prone regions of Africa, and on the Indian subcontinent. The rapid clearing of rain forests in Central and South America and in the Far East have also raised concern (6–8).

The destruction of forests is not new in human history. Since the days of Plato, deforestation, fuelwood shortages, and environmental degradation have been repeatedly noted with alarm (9). Even in the United States,

407

0362–1626/85/1022–0407$02.00

deforestation is said to have reached rates as high as 2.2 million hectares per year during the 1800s (10). However, today's pressure on forest resources is unprecedented. In arid regions, the loss of trees and ground cover, overcultivation, and the breaking of essential nutrient cycles can eventually lead to desertification (11–13). For man to reverse this trend will require intensive resource management, innovative technology, and profound social adjustment.

This paper examines the use of fuelwood and charcoal in terms of their supply, demand, economics, and environmental impacts. It then explores various policy options to ensure adequate supplies of fuelwood for those dependent on it and to help ease the pressure on existing forests.

GENERAL REMARKS

Supply of Wood Fuels

The total global annual growth of forest biomass has been variously estimated to be from 5 (Ref. 3) to 100 (Ref. 14) times the current annual consumption. Country and regional estimates have also been made (3, 15–19, 20a). Representative listings by region are presented in Tables 1 and 2. Techniques for making such estimates are discussed in (20b).

Demand for Wood Fuels

Despite the large supply, there are acute and growing shortages of fuelwood locally and regionally. Estimates of the number of people affected by these shortages are listed in Table 3. Of course, such estimates are very imprecise. Numerous errors and inconsistencies occur in measuring fuelwood, and very few reliable quantitative studies of domestic and commercial fuelwood uses have been conducted.

Table 1   Potential annual renewable biomass energy resources in developing countries

Region	Forest growth 10^{18} J	Animal waste 10^{18} J	Crop residues 10^{18} J	Total 10^{18} J	Estimated 1976–1977 firewood and charcoal 10^{18} J
Africa	8–80	3.0	0.9	12–84	3.0
Latin America	8–80	3.0	3.0	16–88	2.5
Asia	5–51	11.0	12.0	28–74	5.6

Source: Derived from Ref. (3).

Table 2   Forest resources in 1978 per capita by geographic region

Region	Closed forest area (ha/cap)	Open forest (ha/cap)	Growing stock (m³/cap)
North America	2.0	0.7	179
Central America	0.5	0.02	50
South America	2.4	0.7	428
Africa	0.4	1.3	92
Europe	0.3	0.1	27
USSR	3.0	0.4	310
Asia	0.2	0.3	17
Pacific	3.6	4.8	340
WORLD	0.7	0.3	80
More industrialized	1.3	0.4	128
Less industrialized	0.4	0.3	61

Source: (79).

Quantifying Wood Fuels

Study of the world fuelwood situation is complicated by the difficulty of quantification. In the timber industry, wood harvested for lumber is measured volumetrically, usually in solid cubic meters. However, these terms are not easily applied to most fuelwood, which comes in sticks of varying size, shape, and length, from split logs to fragments more resembling forest litter.

Table 3   Number of people (in millions) facing current and projected fuelwood shortages in developing countries

	Year 1980				Year 2000			
	Acute scarcity		Deficit		Acute scarcity		Deficit	
	Total	Rural	Total	Rural	Total	Rural	Total	Rural
Africa	55	49	146	131	88	74	447	390
Near East and North Africa	—	—	104	69	—	—	268	158
Asia, Pacific	31	31	645	551	238	53	1532	1441
Latin America	15	9	104	82	30	13	523	236
Total	101	89	999	833	356	140	2770	2225

Source: Pasca, T. M. 1981. *Unasylva* 33(131): 2–3, citing data from UN Food & Agric. Organ.

For stacked wood, the *cord* and the *stere* (stacked cubic meter) are reasonable volumetric measures, but for fuelwood they cannot be standardized. Thin, crooked pieces, often including roots, make a very loose stack compared with straight, uniform stemwood. Moreover, the fuelwood literature frequently neglects to distinguish "stacked cubic meter" from "solid cubic meter" resulting in considerable confusion.

For field measurements of small quantities of wood, the use of spring balances to determine bulk weight is most convenient. However, there is the potential for major error if the moisture content is not also measured. Even air-dried wood contains some moisture as a function of the ambient humidity, typically estimated as: Water mass = 0.2 (relative humidity) × (mass of dry wood) (Ref. 21). In wood freshly cut or exposed to rain, the wet weight can be nearly double the dry weight.

In estimating the calorific value of wood, it is important to specify whether the weight of the wood is on an oven-dry or wet basis. The heat value of oven-dry wood is typically about 18 MJ/kg (22). For wet wood, both the mass of water and additionally the energy to evaporate it, 2260 kJ/kg, must first be subtracted to determine the calorific value. Wet fuelwood may also deliver less than its calculated calorific value due to less complete combustion.

Local people are often best able to express the quantity of wood they burn in terms of cartloads, headloads, donkeyloads, or purchase price. These may be precise enough to serve as valid units, but they must be used with caution. References (23) and (24) discuss survey techniques in detail.

FUELWOOD

Domestic Uses of Fuelwood

Estimates of domestic fuelwood consumption in developing countries are necessarily imprecise. Little of this wood passes through official channels, so there is little reliable documentation of the quantities collected, transported, or burned. Instead, estimates are generally extrapolated from individual household surveys, based on actual weighings of fuelwood consumed or on personal interviews. In published studies, typical estimates of daily fuelwood consumption range from around 1.0 to 2.0 kg per person, clustering around the 1.5 kg level. Table 4 presents typical values found in some surveys. References (18) and (25) have additional values.

With the exception of certain fruits in season, almost all food consumed in poor countries must first be cooked. Staple foods are generally starchy items, like grains, beans, and tubers, that can be stored over long periods but without cooking are practically indigestible. Cooking also provides important protection from pathogens and parasites that are otherwise spread by contaminated food and water (26).

In addition, fuelwood serves other needs besides cooking. It provides warmth and light in the cooking area, yields charcoal for ironing or making tea, heats bath water after the cooking pot has been removed, provides a social focus, and so forth. Fuelwood may also be used to prepare special products to be sold, such as shea butter or roasted peanuts. Finally, smoke from a kitchen fire may make thatched roofs more insect- and water-repellent while helping preserve food stored in the rafters.

The efficiency with which fuelwood is used for cooking in developing countries is quite low, and leads to energy requirements several times higher than in developed countries (28). The reasons for this low efficiency include

Table 4 Domestic consumption of fuelwood

Country	Fuelwood consumption (GJ/cap/year)	Reference
Asia		
Bangladesh		
(Dhanishwar)	5.6	Montreal Eng. Co. et al 1956[a]
(Country average)	2.8	Islam, 1980[b]
India	7.5	Vergara & Pimentel, 1978[a]
(Six villages in		
Karnataka)	10.4	ASTRA, 1980[b]
(Country average)	6.2	Desai, 1978[b]
Thailand	14.1	Openshaw, 1976[a]
	17.9	Arnold & Jongma, 1978[a]
Nepal	6.6	Earl, 1975[a]
Africa		
Burkina Faso	7.5–10.3	Ernst, 1977[a]
	6.6	Floor, 1977[a]
Gambia	11.3	Floor, 1977[a]
Liberia (Monrovia)	12.2	Roitto, 1970[a]
Nigeria (Batagawara)	13.2	Makhijani & Poole, 1975[a]
(Ibadan)	9.4	Ay, 1980[a]
Kenya	10.3	Marquand & Githinji, 1978[a]
	9.4	Murchiri, 1978[a]
	9.4–18.8	Western & Ssemakula, 1978[a]
	9.6	O'Keefe et al 1984 (Ref. 15)
Tanzania	19.7	Openshaw, 1978[a]
Sudan (Bara)	41.4	Digernes, 1980 (Ref. 42)

[a] Originally compiled by (110).
[b] Originally compiled by (28).

long cooking times for the food, inefficient stoves and pots, and poor control of fires.

Cooking starchy foods is a process that often requires several hours. Even when soaked overnight, beans may be boiled for two to three hours before they are considered ready to eat (29). In much of Latin America and Africa the presoaking of beans is said to alter their taste, so boiling begins with dry beans, adding as much as 40 minutes to the cooking time. Baking bread and preparing porridgelike dishes also require large amounts of heat energy.

Traditional stoves typically have efficiencies of less than 20%.[1] In West Africa, measurements on traditional three-stone stoves revealed an average efficiency of about 17% with a strong correlation to the degree of protection from air currents (30). Measurements on traditional metal stoves found efficiencies of about 18% (31). A compilation of values found in the literature, most undocumented, is given in Ref. (32).

Although protection from the wind dramatically increases the efficiency of the open fire, in the extreme case of cooking in an enclosed kitchen it constitutes a severe health hazard. Measurements from India showed that cooks were inhaling as much carcinogenic benzo(a)pyrene as they would smoking 20 packs of cigarettes per day (33). Exposure to total suspended particulates was more than 100 times the recommended standards of the World Health Organization.

In addition to the influence of air currents, the efficiency of open stoves depends heavily on the distance between the pot and fire. For wood-burning fires without grates, the optimal ratio of pot height to fire diameter is roughly $1:2$ (34), and this ratio is maintained surprisingly well in traditional West African metal stoves (35).

Cooking efficiency is also affected by the type and size of the pot. Clay pots require about 45% more fuelwood than aluminum ones, because of their lower thermal conductivity and higher porosity (36, 37). Larger pots have larger surface areas, and thus better heat gain, than smaller pots. This reduces per capita fuel consumption when cooking is done for large groups, as shown from field surveys of fuel consumption. Studies from Niger, for instance, indicate that per capita fuel consumption can decrease by as much as one third when average family size is increased from five to ten (38).

Finally, much fuelwood is apparently wasted by the common failure of the cook to pay close attention to her fire while performing her myriad other tasks. Sensitizing cooks to fuel consumption through daily wood weighings alone has been shown to reduce fuel use by as much as 25% (39).

[1] Cookstove efficiency is commonly calculated from the heat output (to bring to boil and to simmer a known quantity of water) divided by the estimated energy input (calorific value of fuel consumed).

Commercial Uses of Fuelwood

In countries that depend heavily on energy from biomass, a significant part of the fuel is used for institutional, commercial, and industrial needs. Studies from Kenya, Sudan, and the Philippines suggest that overall, nondomestic sectors may consume 8–15% of the total wood energy budget (40). This amount is increasing both in absolute terms and as a proportion of all fuelwood consumed (41).

At the local level, the impact of this demand takes on special significance. In a town in central Sudan, for example, fully 38% of all fuelwood is used by bakeries, restaurants, schools, and the local prison and hospital (42). In Ouagadougou, the making of traditional millet beer alone consumes 15% of that city's daily wood supply (43). A brick-making plant in Niger is reported to require 30 steres of wood every day, roughly equivalent to the domestic energy needs of 6000 people (44).

In tobacco-growing regions, the curing of tobacco leaves consumes an extraordinary amount of wood. In Tanzania it is estimated that an entire hectare of woodland is used to cure one hectare of tobacco, or 1 stacked cubic meter of wood for every 7.5 kg of tobacco leaf (45). Additional wood is burned by tobacco farmers when they clear land for new fields in their pattern of shifting agriculture. On Ilocos Norte, the Philippines, the curing of tobacco uses 11% of all fuelwood, and represents over 90% of the combined commercial and institutional demand (46). In Malawi this same enterprise takes 17% of the national energy budget, annually consuming one million cubic meters of wood (47).

Elsewhere, tea, coffee, or copra replaces tobacco as the major fuelwood-consuming crop. In Kenya, tea processing requires an average of 275,000 tons of wood per year (15). In some areas of rural Fiji farmers use wood-fired driers to prepare copra, consuming as much as three times the wood used for all home cooking (49).

The smoking and drying of fish are also fuelwood-intensive enterprises with a heavy local impact. They are essential in villages where fish are preserved for inland markets. Around Lake Victoria, where 90% of the fish catch is smoked, 3 kg of wood will process 12 kg of fish (45). In this region alone, over 120,000 tons of fish per year are smoked using fuelwood.

The list of other major commercial users of fuelwood includes bakeries, restaurants, and butcheries (where meat is prepared). Institutional kitchens for schools, hospitals, and military camps often consume wood by the truckload, differing only in scale from their domestic counterparts.

Many kinds of fuel are used to bake tiles, bricks, and pottery, depending partly on the items to be fired and on the kiln design. For large brickworks, however, wood is commonly used in huge quantities. In Tanzania nearly

Table 5 Energy consumption in Kenya, in percent by category of use

Energy use	Nontraditional fuel	Wood	Charcoal	Biomass
Urban household				
Cooking/heating	0.8	1.0	3.3	—
Lighting	0.6	—	—	—
Other	0.2	—	0.5	—
Rural household				
Cooking/heating	0.2	45.3	2.8	2.7
Lighting	1.1	—	—	—
Industry				
Large	8.6	5.3	0.3	—
Informal urban	—	0.1	0.6	—
Informal rural	—	9.1	0.1	—
Commerce	0.6	0.5	0.1	—
Transportation	13.7	—	—	—
Agriculture	2.5	—	—	—
TOTAL	28.4	61.3	7.6	2.7

Source: Ref. (15).

400,000 stacked cubic meters are estimated to have been used for this purpose in 1980–1981 (45).

Table 5 represents an overall breakdown of the total 332 million GJ of energy used in Kenya in 1980. Tables 6 and 7 further break down the major portions of industrial consumption by end use. Other studies of this nature are given by Refs. (52) and (53).

CHARCOAL

Since charcoal and wood are derived from the same living resource, they are generally considered together. In many respects, however, they are distinct fuels, each with its own market and applications.

Unlike fuelwood, charcoal is always manufactured, transported, and sold commercially. People do not forage for charcoal as they might for wood. Nor is charcoal likely to be widely used where firewood can be readily obtained free of charge. As a consumer good, charcoal competes with other commercially available fuels, such as kerosene, propane gas, electricity, and marketed fuelwood. Thus the domestic consumption of charcoal depends largely on its price relative to these (1).

Table 6 Annual consumption of fuelwood and charcoal in Kenya by rural cottage industries

Activity	Annual consumption, GJ/capita	
	Fuelwood	Charcoal
Brewing	1.07	—
Clay brick firing	0.06	—
Blacksmithy	—	0.06
Crop drying	0.04	—
Fish curing	0.02	—
Tobacco curing	0.04	—
Butchery	0.24	0.06
Baking	0.13	—
Restaurants	0.17	0.04
Construction timber	0.50	—
Total	2.27	0.15

Source: Ref. (15).

Traditional Charcoal-Making

"Charcoal" is the general term for a range of carbonized materials, with varying combustion properties. It is made by raising the temperature of wood beyond the point at which many of its organic components become chemically unstable and begin to break down. The details of this process, called pyrolysis, are still incompletely understood. Most of the newly

Table 7 Consumption of fuelwood in Kenya by large industries

Activity	Fuelwood consumption (GJ/capita/year)
Tea (average)	0.28
Tobacco	0.08
Sugar	0.05
Wood processing	0.30
Wattle	0.04
Clay brick	0.03
Baking	0.30
Total	1.08

Source: Ref. (15).

416   WOOD & BALDWIN

formed materials are vaporized. The material left behind is a black, porous charcoal that retains the original form of the wood but has just one fifth the weight, one half the volume, and about one third the original energy content (55, 56). Despite the large net energy loss, however, charcoal typically has a calorific value of 32–33 MJ/kg, compared to the original wood's value of 18–19 MJ/kg (55, 104).

To prevent most of the wood from igniting during pyrolysis, charcoal must be made in an environment of restricted air flow. Most of the developing world's charcoal makers build temporary earthen kilns for each batch. The wood is stacked compactly in a pit or on the ground. The stack is covered with straw or other vegetation, then buried under a layer of soil. It is ignited with burning embers introduced at one or more points at the bottom of the stack. The task of the charcoal maker throughout the ensuing "burn" is to open and close a succession of vent holes in the soil layer to draw the fire evenly around the wood stack, heating the wood while burning as little of it as possible.

In charcoal-making operations of the Ethiopian government, the kilns contain as much as 200 steres of wood and are 10 meters in diameter and 5 meters high (57). Elsewhere in the world, kilns of 20–100 steres are not uncommon, and individual part-time workers often keep kilns under 6 steres. The actual burn normally takes from three day to three months, depending on the kiln's size and the moisture content of the wood.

Reported yields from these simple kilns are extremely variable. One report from Kenya indicates a loss of 76% of the wood's energy, with a further loss of 5% of the charcoal itself during distribution (14). In Senegal, where charcoal-making techniques are similar, calculations on data by Karch (59) indicate net energy losses of only 71–75%. In Ethiopia, careful weighing of wood and the charcoal it produced from a small earthen kiln revealed an energy loss of 71% (60). The differences among all these results may reflect a normal variation in charcoal making. Yields are strongly affected by the rate of heating, the sizes and moisture contents of the wood, and the range of temperatures achieved during the burn (61). There are also the inevitable differences in estimating fuel volume, moisture content, and energy value.

Although traditional charcoal production makes excellent use of the resources available (wood and soil) it does not approach the high yields possible with more sophisticated techniques. Improved kilns often circulate the hot, volatile gases through the wood for preheating and drying. In Senegal this is achieved by using a novel stacking pattern and a metal chimney, increasing the charcoal yield by over 45% (59). Similar improvements in the traditional technology have been described from the Caribbean (63). Well-designed steel kilns can increase yields, although they are beyond the means of most local charcoal makers (64). Often it is even possible to condense the volatile products of pyrolysis and collect phenolic tars and other marketable organic products.

Despite these possibilities, there is no avoiding the fact that charcoal is a residual part of the original wood, and even the most elaborate charcoal-making operation must sustain a major energy loss in the final product.

Transporting Charcoal

Contrary to the popular view, it is as expensive to transport charcoal energy as wood energy.[2] This was not the opinion of Earl (65) when he assumed that transport costs were a function of weight alone. At US $0.10 per metric ton-km, he concluded that charcoal energy would cost less than wood energy for hauling distances greater than 82 km. Chauvin similarly used a fixed cost per ton-km in his analysis of the economics of transporting charcoal from the Ivory Coast to Upper Volta by rail (36).

Expressing transport costs in terms of cost per ton-km is standard practice in aggregated transportation statistics, but it does not describe the actual situation. Most of the transport energy is used to move the vehicle itself—to overcome wind resistance, internal friction, and so forth. Thus, an empty truck uses nearly as much energy as one that is full. A linear regression on data presented in (67) shows that the energy intensity of transport by tractor-trailers in the United States is related approximately to the payload for the range of 8–25 metric tons by the equation:

$$E = 23.6/M + 0.476$$

where E is the energy intensity in MJ per metric ton-km the load is moved, and M is the mass of the load in metric tons.

Transport is more often limited by volume than by weight, and this is particularly true in the developing world, where vehicles are often filled to overflowing. In this case of volume-limited transport, as shown in Table 8, 13% more energy can be transported per truckload of wood than of charcoal at the cost of a 21% increase in fuel use.

However, fuel costs are only a small part of the total transport cost, and do not substantially increase even on unimproved roads (68). Maintenance and repair of vehicles operating on unimproved roads are large factors (68), and vehicle depreciation and labor are even larger (70). When these costs are considered, as in Table 9, the costs of hauling energy in the forms of wood and charcoal are virtually the same. When the cost of producing

---

[2] The arguments presented in this section are more fully elaborated by one of the authors (Baldwin) in *Engineering Heat Transfer, Rural Energy Technologies, and Development*, to be published in 1985 by Volunteers in Technical Assistance, Arlington, Va.

Table 8 Energy required to transport wood and charcoal

Factor	Fuelwood	Charcoal	Notes
Assumed specific/volumetric gravity	0.7	0.33	a
Assumed packing density	0.7	0.7	b
Effective volumetric gravity	0.49	0.23	
Energy content per truckload	390 GJ	345 GJ	c
Weight per truckload	24.5 MT	11.5 MT	d
Transport energy per km/energy content	91×10^{-6}	84×10^{-6}	

[a] Based on (55). $y = 0.575x - 0.069$, where x is the specific gravity of the wood and y is the volumetric gravity of the charcoal.

[b] For wood based on (65). Charcoal may have a higher or lower packing density depending on its size and whether or not it is bagged for transport.

[c] Calorific value for wood, 16 MJ/kg; for charcoal, 30 MJ/kg; both including moisture.

[d] Based on a payload volume of 50 m^3. This is less than a standard tractor trailer, but is chosen to remain within the limits of the correlation of weight to transport energy.

charcoal is considered (65), charcoal nets a financial loss. These costs are reflected in the relative prices found for fuelwood and charcoal: the price per GJ of charcoal is usually about twice that of fuelwood (71).

Despite its higher price, charcoal is a very popular fuel, particularly in urban areas where people have cash incomes. There are two primary reasons for this. First, charcoal is nearly smokeless. Cooking can be done indoors in relative comfort without blackening the walls with soot. Metal pots stay clean, and there is no smoke irritation to eyes or lungs. Although there is a high output of dangerous carbon monoxide, which is a health

Table 9 Transport costs of wood and charcoal[a]

| Factor | Percent of total | | Notes |
	Wood	Charcoal	
Labor and management	12	12	
Fuel	18	15	b
Maintenance and repair	40	30	c
Licenses and tolls	1	1	
Vehicle depreciation	42	42	
Total costs	113	100	
Energy hauled	113	100	b

[a] From reference (70) using charcoal as the baseline.

[b] From Table 8.

[c] Estimated from Ref. (70), tire depreciation and vehicle repair charges, assuming that these costs increase proportionately to the total vehicle weight.

hazard in poorly ventilated kitchens, this does not cause obvious discomfort to the user. Secondly, once it is lit, a charcoal fire needs little further attention from the cook, while a wood fire requires frequent adjusting of the fuel.

The high value placed on charcoal fuel clearly demonstrates the importance of stoves that offer more than just fuel efficiency. This should encourage designers of improved wood-burning stoves to continue seeking ways to minimize smoke, ease the drudgery of cooking, and further reduce fuel costs.

Domestic Use of Charcoal

For domestic cooking, charcoal is most often burned in a shallow metal or ceramic stove. A grate under the charcoal allows ashes to drop through. If the cooking pot is metal, it is usually placed directly on the burning charcoal. In standard water-boiling tests of widely used stoves, the energy doing measurable work is shown to be around 15–25% of the amount released from the burning fuel (72a).[3]

Unfortunately, the frequently superior efficiency of traditional charcoal stoves compared to wood does not compensate for the heavy energy loss during charcoal manufacture. When a household switches its cooking fuel from wood to charcoal, all other conditions being constant, its demand on wood resources can triple.

It is not uncommon for both wood and charcoal to be used in the same household for different heating or cooking tasks. In the Machakos District of Kenya, for example, all rural households surveyed used wood for heating and cooking, while over one third also used some charcoal (72b). Similar cases of mixed fuels are widely reported elsewhere. In Ethiopia, urban families that simmer their sauces over charcoal will prefer wood or even electric hotplates to cook the large "injera" pancakes (37). In Kenya, foods such as "chapati" are invariably cooked over charcoal (72b).

In urban regions, the preference for charcoal over wood can be very high. According to a 1970 report from Thailand, 90% of the wood cut for urban markets is converted into charcoal (75). In Tanzania that figure is 76%, although overall consumption of biomass fuels is substantially higher there than in Thailand (76). Ninety percent of the households in Dakar cook with charcoal, consuming annually over 140 million metric tons (77).

Commercial Uses of Charcoal

Probably the single greatest nondomestic use of charcoal is for the smelting of pig iron. In China, during the 1950s, forests were cut around tens of

[3] Measured energy efficiencies of charcoal stoves should be regarded with caution owing to the difficulty in quantifying changes in volatile content of the fuel during burning (72a).

thousands of villages to provide charcoal for small furnaces producing pig iron (78). In Brazil, the world's largest charcoal producer, 45% is used in the iron smelting industry. This amounted to 3.6 million tons in 1978, and consumption was growing by 10% per year (79).

Other charcoal-consuming activities include metal casting (mostly bronze and aluminum), blacksmithing, and cooking by street vendors, restaurants, and butcheries.

OTHER BIOMASS FUELS

The scarcity of fuelwood drives people to seek alternative sources of energy for cooking their food. Agricultural residues are frequently used when they are locally and seasonally available—materials such as millet stalks, rice straw, leaves, tea bush prunings, bagasse, and coconut or peanut shells. In Burundi and Senegal small peat resources are being developed as a possible domestic fuel (80). In some areas even sawdust represents a possible source of heat energy. All of these are awkward and inconvenient fuels, used only when wood is too scarce or expensive, or when it is considered more valuable as a marketable good than as a domestic fuel. Small-scale attempts to distribute agricultural residues as compressed briquettes have so far met with little success.

For an estimated 4 million households worldwide, animal dung is the primary fuel. When occasional users are included, as much as 400 million tons of cow dung is burned each year in Africa and Asia alone (20a). But like other biomass alternatives, dung is a difficult fuel. It burns with a slow, even flame, releasing only three fourths of the energy of an equivalent weight in dry wood. Without constant ventilation, the heat of combustion is often insufficient to sustain the flame, causing the dung to smolder with a heavy smoke. Compared to a wood fire, dung requires more constant attention and frequent fanning.

Obviously, the burning of dung precludes its use as a fertilizer or soil conditioner. By one estimate, potential grain production drops by about 50 kg for every ton of dung not applied to the soil (82).

FUEL ECONOMICS

There are basically two distinct types of fuel consumers in the Third World: the rural subsistence farmer who gathers fuel, and the urban dweller who buys it.

For the rural dweller, the increasing scarcity of fuelwood supplies means longer foraging time. On this point the literature is filled with anecdotal references. In central Burkina Faso, for example, it is said to take four to five hours to collect a headload of wood weighing only 27 kilograms (83). For many women in Sri Lanka this task may now consume 25% of the working day (84). A rough correlation has been proposed relating the population size to the collection distance (85a, 85b). Such correlations are developed rather simply by equating the population's average consumption to the area required to provide a sustained yield.

In urban areas where fuel is purchased there is a choice between wood-derived fuels and fossil fuels. Oil price hikes and the increasing wood shortages have combined to raise fuel costs dramatically for Third World urban dwellers. During the 1970s, the cost of fuelwood and charcoal increased at an annual rate of 1.5–2.0% relative to the costs of other goods (71). It is not unusual for a large urban family in West Africa to spend one third of its income on wood, often exceeding the amount spent on food (86). Oil costs, of course, have increased even more rapidly, reducing the use of kerosene or liquefied petroleum gas as alternatives.

The economic performance of different fuels can be compared on the basis of the cost of delivered energy. As seen in Table 10, fuelwood is slightly cheaper than the others, particularly when used in a higher-efficiency stove.

Price ratios such as these are not consistent. A survey in Addis Ababa, for example, showed that for equivalent energy, the official price of kerosene is nearly three times that of fuelwood. However, the higher efficiency of kerosene stoves compared to open wood fires makes kerosene a better buy. Disregarding the wood used to prepare "injera," wood costs a large urban family around $8 per week, compared to $6 per week for a family using kerosene (60).

In urban areas of India kerosene is a common cooking fuel, burned in stoves costing $3–7 (G. S. Dutt, personal communication). However, in

Table 10   Relative cooking costs for different fuels in Ouagadougou, Burkina Faso

Fuel	Stove	Cost (CFA/kg)	Calorific value (MJ/kg)	Effective end use (Percent)	Cost of delivered energy (CFA/MJ)
Wood	3-stone	15.0	16	17	5.5
	Improved	15.0	16	30	(3.1)
Charcoal	Traditional	50.0	29	25	6.9
	Improved	50.0	29	35	(4.9)
Kerosene		137.5	43	50	6.4
Butane		256.0	45	55	10.3

Sources: Cecelski, E. 1983. Energy needs, tasks, and resources in the Sahel: relevance to woodstoves programmes. *Geojournal* 7(1): 15–23; Reference: 30, 35. 400 CFA = $1.

most of East and West Africa, cooking with kerosene is rare, possibly because the imported stoves cost at least $25. Poor families appear unable or unwilling to invest so much in a flimsy kerosene stove when their three-stone fireplace is free, versatile, and effective, requires no maintenance, lasts almost forever, is unlikely to be stolen, and is backed by many respected generations of tradition.

ENVIRONMENTAL IMPACT OF FUELWOOD CONSUMPTION

Wood is an extremely versatile material with a large worldwide market. Poles and sawn lumber are major uses. The paper industry consumes about 10% of commercially harvested timber. Despite this, more wood is now being cut for fuel than for all other uses combined, even without including the wood scraps that provide energy to various wood processing industries (1). Altogether, the net annual loss of forests from all of these demands exceeds 10 million hectares (90).

Given the importance of forest resources to most developing countries, one might expect to find a heavy emphasis on their careful management, to ensure reliable future supplies. In most places, however, deforestation is now out of control, driven by growing population pressure and economic priorities that recognize no limit to the resource. Shifting agriculture alone destroys 125,000 square kilometers of forest annually in tropical Africa and Southeast Asia (6). Globally, the timber and wood pulp industries consume over 1300 million cubic meters annually, and this figure is expected to double in the next 15 years. In West Malaysia and the Philippines, all accessible virgin forest will have been logged within a few years (92). In Central America, forests are being cleared at over 4000 square kilometers per year, replaced mostly by cattle ranches to produce beef for export to the North American fast food market (93).

In the African Sahel, where fuel consumption is very high and biomass production low, the loss of many trees is attributed to the continuing drought. However, some have argued that effects of the drought would be less severe had the forest and ground cover not already been so extensively damaged by overgrazing, overcultivation, brushfires, and urban fuelwood demand (11).

Firewood is a relatively heavy and bulky fuel to transport, and so most is consumed relatively close to its point of origin. A natural result is the creation of badly deforested zones around major population centers. Viewed from the air, these treeless zones can be quite impressive. The town of Bara, Sudan, with about 11,000 people, is surrounded by a "desertified" area at least 4 kilometers wide (42). Traveling along a main road in southern

Mauritania, where the near-barren landscape includes many drought-killed trees, the sudden disappearance of all wood down to the last twig is a reliable indication of a small town ahead.

Most damaging of the fuelwood gatherers are not the rural peasants, who normally take only sticks and twigs from fallow land in their own vicinities. It is commercial fuelwood and charcoal operations, serving urban areas, that most often cut whole trees and clear significant sections of forested land.

The environmental consequences of deforestation—from whatever cause—have long been surmised (9). Recent studies have linked extensive deforestation with major climatological and hydrological change (97). When natural vegetation is cleared from the land, the soil is altered in profound ways. Exposed to direct sunlight, the soil rises in temperature, and any moisture is evaporated more quickly. Without the protection of ground cover, the soil receives the full force of pounding raindrops, bringing the smallest clay particles to the surface to form a nearly impervious layer. The absence of continuous organic input from roots and leaves reduces the soil structure and makes it more vulnerable to wind and water erosion. With the decline in humus content, the soil loses its fertility and its capacity to absorb and hold moisture (11, 13).

With less rainwater penetrating the ground there is more water to flow across the soil surface; with little vegetation to stabilize the soil, huge amounts are carried away. Studies in Tanzania on a 3.5-degree slope found that no soil and only 0.4% of the rainwater ran off an ungrazed thicket. During the same two-year period on an adjacent 150-hectare plot of bare fallow land, half the rainwater ran off, carrying nearly 150 tons of soil with it (100).

In the Himalayas of Nepal and northern India the threat of mudslides from deforested slopes has become a major concern during monsoon rains (53). As riverbeds fill with upland sediment they become wider and shallower, resulting in dry streambeds for much of the year, and increasing the severity of seasonal monsoon flooding.

FUTURE OPTIONS

In the developing countries it is increasingly evident that the demands of rapidly growing populations and a lack of resource management are rapidly diminishing the ability of the land to sustain life. The increasing need for biomass fuels stands out as both an important cause and a consequence of this syndrome.

What is not so clear is the most effective way to deal with the problem. Its economic, social, and technical dimensions are so vast that it seems to

424 WOOD & BALDWIN

require corrective action simultaneously along many fronts. The following discussion assesses the attempts of the past decade to deal with the fuelwood situation.

Increasing the Resource Base

Planting trees is a natural response to the growing fuelwood scarcity. There is still unused land available, for shelterbelts and even large plantations, in and around villages and along roadsides. In addition to providing wood, selected fast-growing trees can often supply forage for livestock, nectar for honeybees, shade, windbreaks, and other benefits.

International assistance agencies have responded with a variety of tree-planting schemes, spending $100 million per year (102). Since 1975, over 300 square kilometers of new forests have been planted in the African Sahel alone (103). On the island of Java, Indonesia, a fast-growing ornamental shrub from Guatemala is now widely planted by farmers who make more from selling wood than from food crops (104). Around Addis Ababa a growing forest of *Eucalyptus*, derived from trees initially imported from Tasmania, is managed by individuals and neighborhood associations to provide wood for the urban market. The increasing importance of fuelwood plantations has been underscored by the National Academy of Sciences, which has published a comprehensive survey of fast-growing tree species as potential crops for firewood production (105). In semi-arid regions, imported fast-growing trees can produce wood two to five times faster than indigenous species. Under the best conditions they yield up to 15 cubic meters per hectare per year, but less than 5 cubic meters where annual rainfall is below 900 mm (77).

While all this may sound impressive, reforestation programs have so far had no significant impact. The World Bank estimates the need for 600,000 square kilometers of new forests, producing at least 10 cubic meters of wood per hectare per year (107). Unfortunately, the results of plantation efforts to date have been almost uniformly discouraging. The costs have been high— over $1000 per hectare in some places—and the yields often far below expectations (42).

Problems commonly encountered in plantation efforts include the following (103, 108):

1. Unrealistic expectations by project planners, given the known variations in soils and climate.
2. Failure to attract the interest and active involvement of potential fuelwood users in project planning and implementation, resulting in local hostility, suspicion or simply indifference.
3. Failure to arrange adequate protection of saplings from poachers, brushfires, and grazing animals.

4. Inadequate or inconsistent strategies at the national level for the production, transformation, transport, and use of the wood produced, including the appropriate legal framework to promote reforestation.

Despite recent setbacks, independent efforts are continuing to establish local, highly productive centers of fuelwood production. Besides the creation of new plantations, projects are under way to manage effectively natural stands of vegetation for maximum productivity, especially in open forest and semi-arid savanna regions. The concept of "social forestry" developed in India is beginning to take hold in other countries, leading to the more effective establishment of community woodlots. Still, however, reforestation efforts have so far been far less than the assessed needs.

Reducing Domestic Consumption

There is little question that in developing countries the traditional wood-burning stove is appropriate to the needs and technical levels of most households. However, it is also unhealthy, dangerous, and especially wasteful of energy. There seems little sense in producing more wood if most of its energy is lost in the stove. An obvious solution would be to promote domestic fuel conservation practices, including the use of more efficient cookstoves. Details of improved stove programs are reviewed in the 1984 *Annual Review of Energy* by Manibog (109). The recent and promising developments in high-efficiency portable stoves have been reviewed by Baldwin (35).

CONCLUSION

Fuelwood and charcoal will continue to be essential components of Third World energy supplies for the foreseeable future. However, deforestation, in part due to fuelwood use, seriously threatens supplies of fuelwood and charcoal in many areas, undermining local economies and damaging the environment. Fossil fuels, with their current high prices and expected longer-term shortages, do not offer a real alternative, especially for the rural poor. Although reforestation and improved stoves have so far shown little success, the technical, social, and managerial experience gained to date promise improved project performance in the future. Only through such efforts does there appear to be a potential for easing the increasing scarcity of energy suffered by the world's poor.

Acknowledgments

The authors are pleased to acknowledge the valuable contributions of Dr. Gautam S. Dutt, Dr. John W. Tatom, and Dr. Hans H. Gregersen, who reviewed an early draft of this paper.

426   WOOD & BALDWIN

Literature Cited

1. De Montalembert, M. R., Clement, J. 1983. *Fuelwood supplies in the developing countries.* Rome, FAO. 123 pp.
2. Hughart, D. 1979. *Prospects for traditional and nonconventional energy sources in developing countries.* Washington, DC, World Bank. 132 pp.
3. Dunkerley, J., Ramsay, W., Gordon, L., Cecelski, E. 1981. *Energy strategies for developing nations.* Baltimore, Johns Hopkins Univ. Press/Resources for the Future
4. United Nations. 1981. *Fuelwood and Charcoal.* Rep. Tech. Panel, Second Session, February 21. (A/CONF. 100/PC/34)
5. UN Food & Agric. Org. 1982. *Tropical forest resources.* Forestry Paper No. 30. Rome. 106 pp.
6. Myers, N. 1980. *Conversion of Moist Tropical Forests: Report for the Committee on Research Priorities in Tropical Biology of the National Research Council.* Washington, DC Natl. Acad. Sci.
7. Salati, E., Vose, P. B. 1983. Depletion of tropical rain forests. *Ambio* 12(2): 67–71
8. Nations, J. D., Komer, D. I. 1983. Central America's tropical rainforests: Positive steps for survival. *Ambio* 12(5): 232–38
9. Eckholm, E. 1976. *Losing ground: Environmental stress and world food prospects.* New York. Norton/Worldwatch Inst. 223 pp.
10. Perlin, J., Jordan, B. 1983. Running out: 4200 years of wood shortages. *The CoEvolution Q.* Spring, 18–25
11. Grainger, A. 1982. *Desertification: how people make deserts, how people can stop, and why they don't.* London. Int. Inst. for Environ. and Develop.
12. BOSTID. 1983. *Agroforestry in the West African Sahel.* Washington, DC Natl Acad. Sci.
13. BOSTID. 1984. *Environmental change in the West African Sahel.* Washington, DC Natl. Acad. Sci.
14. Zsuffa, L. 1982. The production of wood for energy. In *Energy from forest biomass,* ed. W. R. Smith. New York: Academic
15. O'Keefe, P., Raskin, P., Bernow, S. eds. 1984. *Energy and development in Kenya: Opportunities and constraints.* Stockholm: The Beijer Inst. 187 pp.
16. Keita, M. N. 1982. *Les disponibilites de bois de feu en region sahelienne de l'Afrique Occidentale: situation et perspectives.* Rome. UN Food & Agric. Org. 79 pp. (In French)
17. Comité Permanent Interétat de Lutte Contre la Secheresse dans le Sahel (CILSS). 1982. *Quantification des besoins en bois des pays saheliens: une analyse des bilans/programmes.* Presented at CILSS meeting in Banjul, October 18–22 (In French)
18. Clement, J. 1982. *Estimation des volumes et de la productivité des formations mixtes forestières et graminées tropicales.* Nogent-sur-Marnes, France, Centre Technique Forestier Tropical. (In French)
19. Hall, D. O., Barnard, G. W., Moss, P. A. 1982. *Biomass for energy in the developing countries.* New York. Pergamon. 205 pp.
20a. Council on Environmental Quality and the Dept. of State. 1980. *The global 2000 report to the President of the United States. Volume 2: The technical report.* Washington, DC. 766 pp.
20b. Wegner, K. F., ed. 1984. *Forestry Handbook.* New York: Wiley. 1335 pp.
21. Volunteers in Technical Assistance. 1982. *Testing the efficiency of woodburning cookstoves.* Arlington, Va. Volunteers in Technical Assistance. 76 pp.
22. Graboski, M., Bain, R. 1981. Properties of biomass relevant to gasification. In *Biomass gasification,* ed. T. Reed. Park Ridge, NJ: Noyes Data Corp. 401 pp.
23. Openshaw, K. 1980. *Woodfuel surveys: Measurement problems and solutions to these problems*
24. Food and Agric. Org. 1983. *Woodfuel surveys.* Rome, FAO. GCP/INT/365/SWE. 202 pp.
25. Bhagavan, M. K. 1984. The woodfuel crisis in the SADCC countries. *Ambio* 12(1): 25–27
26. Obeng, Y. L. E. 1983. The control of pathogens from human wastes and their aquatic vectors. *Ambio* 12(2): 106–8
27. Deleted in proof
28. Krishna Prasad, K. 1982. *Cooking energy.* Workshop on end-use focussed global energy strategy, Princeton Univ., Princeton, NJ. April 21–24. 51 pp.
29. Smale, M. 1984. *Wood fuel consumption and cooking practices in selected sites of Lower Shabelle, Banaadir, and Gedo regions of Somalia.* Arlington, Va. Volunteers in Technical Assistance. 151 pp.
30. Yameogo, G., Bussman, P., Simonis, P., Baldwin, S. 1983. *Comparison of improved stoves: lad. controlled cooking, and family compound tests.* Arlington, Va. Volunteers in Technical Assistance. 67 pp.

31. Ouedraogo, I., Yameogo, G., Baldwin, S. 1983. *Lab tests of fired clay and metal one-pot chimneyless stoves.* Arlington, Va. Volunteers in Technical Assistance. 38 pp.
32. Gill, J. 1981. *An unpublished study on wood stoves.* Open University, Milton Keynes. Cited in: Prasad, K. K. 1983. *Woodburning stoves: Their technology, economics and deployment.* Geneva. Int. Labor Org. 168 pp.
33a. Smith, K. R., Ramakrishna, J., Menon, P. 1981. *Air pollution from the combustion of traditional fuels.* Conference on Air Quality Management and Energy Policies, Barodu and Bombay, Feb. 16–25. Honolulu, Resource Systems Inst., East-West Center
33b. Smith, K. R., Aggarwal, A. L., Dave, R. M. 1982. *Air pollution and rural fuels: a pilot study in India.* Working paper WP-82-17. Honolulu, Resource Systems Inst., East-West Center
33c. Smith, K. R., Aggarwal, A. L., Dave, R. M. 1982. *Air pollution and rural fuels: implications for policy and research.* Working paper WP-83-2. Honolulu, Resource Systems Inst., East-West Center
34. De Lepeleire, G., Krishna Prasad, K., Verhaart, P., Visser, P. 1981. *A woodstove compendium.* Eindhoven, The Netherlands. 379 pp.
35. Baldwin, S. 1984. New directions in woodstove development. *VITA News.* January, 3–23
36. Geller, H. S. 1982. *Fuel efficiencies and performance of traditional and innovative cookstoves.* Bangalore, India. ASTRA. 34 pp.
37. Wood, T. S. *Report on domestic energy use for cooking.* (*Energy Assessment Mission, Ethiopia.*) Washington, DC: World Bank. 33 pp.
38. Boureima, I., Deschambre, G. 1982. *Rapport sur l'evaluation du programme foyers ameliores.* Niamey, Niger. Assoc. des Femmes du Niger/Church World Service. 23 pp. (In French)
39. Wood, T. S. 1981. *Laboratory and field testing of improved stoves in Upper Volta.* Washington, DC. Natl. Acad. Sci. (BOSTID). 23 pp.
40. Volunteers in Technical Assistance. 1983. *Commercial and institutional stoves: potential for integration into improved stove programs.* Presented at Int. Workshop on Woodstove Dissemination, Wolfheze, The Netherlands
41. Deleted in proof
42. Digernes, T. H. 1977. *Wood for fuels: energy crisis implying desertification The case of Bara, the Sudan.* Bergen,

Norway. Geografisk Institutt. Thesis. 128 pp.
43. Chauvin, H. 1981. When an African city runs out of fuel. *Unasylva.* 33(133):11–20
44. Ide, B., Kanta, R. 1980. *Rapport des representants du Niger.* Seminaire sur le developpement des combustibles et de l'energie pour les femmes africaines en zone rurale. CEA/FAO, Bamako, Mali. (In French)
45. Mnzava, E. M. 1981. Village industries, savanna forests. *Unasylva* 33(131):24–29
46. Hyman, E. L. 1982. *The demand for woodfuels by cottage industries in the Province of Ilocos Norte, Philippines.* Rome. UN Food & Agric. Organ.
47. Arnold, J. E. M. 1979. Wood energy and rural communities. *Natural Res. Forum* 3:229–52
48. Deleted in proof
49. Siwatibau, S. 1981. *Rural energy in Fiji: A survey of domestic rural energy use and potential.* Ottawa. Int. Develop. Res. Center
50. Deleted in proof
51. Deleted in proof
52. Nkonoki, S. R., Sorensen, B. 1984. A rural energy study in Tanzania: the case of Bundilya village. *Nat. Res. Forum* 8(1):51–62
53. Singh, J. S., Pandey, U., Tiwari, A. K. 1984. Man and forests: A central Himalayan case study. *Ambio* 13(2):80–87
54. Deleted in proof
55. Abe, F. 1982. Manufacture of charccal from fast-grown trees. In *Energy from forest biomass,* ed. W. R. Smith. New York: Academic
56. Deleted in proof
57. Uhart, E. 1975. *Preliminary charcoal survey in Ethiopia.* Doc. M75-1122. UN Econ. Commission for Africa; FAO Forest Industries Advisory Group for Africa. 30 pp.
58. Deleted in proof
59. Karch, G. E. 1980. *Carbonization: Final technical report of forest energy specialist,* UN FAO Document SEN/78/002
60. Wood, T. S. 1983. *Report on domestic energy use for cooking* (for Energy Assessment Mission in Ethiopia). Washington DC: World Bank. 33 pp.
61. Zaror, C. A., Pyle, D. L. 1983. The pyrolysis of biomass: a general review. In *Wood heat for cooking,* ed. K. Krishna Prasad, P. Verhaart, pp. 15–31. Bangalore: Indian Acad. Sci.
62. Deleted in proof
63. Wartluft, J., White, S. 1984. *Comparing simple charcoal production technologies*

for the Caribbean. Arlington, Va. Volunteers in Technical Assistance. 26 pp.

64. Wartluft, J. 1983. Team compares charcoal production methods. *VITA News,* October. pp. 8–11

65. Earl, D. E. 1975. *Forest energy and economic development.* Oxford: Clarendon

66. Deleted in proof

67. Rose, A. B. 1979. *Energy-intensity and related parameters of selected transportation modes: Freight movements.* Paper No. 5554. Oak Ridge, TN: Oak Ridge Natl. Lab., Paper No. 55545554

68. Bonney, R. S. P., Stevens, N. F. 1967. *Vehicle operating costs on bituminous, gravel, and earth roads in East and Central Africa.* Road Research Lab., Ministry of Transport

69. Deleted in proof

70. Ministry of National Planning, Transport and Communications Section. 1978. *Truck operating characteristics in the Sudan.* Khartoum

71. Wardle, P., Palmieri, M. 1981. What does fuelwood really cost? *Unasylva* 33(191): 20–23

72a. Baldwin, S. 1983. *Technical notes for the Senegalese "Ban ak suuf" (improved stove) program.* Arlington, Va. Volunteers in Technical Assistance. 80 pp.

72b. Mung'ala, P. M. 1979. *Estimation of present and likely future demand for fuelwood and charcoal in Machakos District.* PhD thesis. Univ. of Dar es Salaam, Morogoro

73. Deleted in proof

74. Deleted in proof

75. Arnold, J. E. M. 1979. Wood energy and rural communities. Natl. Res. Forum 3: 229–52

76. Arnold, J. E. M. 1978. Fuelwood and charcoal in developing countries. *Unasylva* 29(118): 2–9

77. Weber, F. 1977. *Economic and ecologic criteria of forestry/conservation projects in the Sahel.* Boise. Int. Res. Develop. & Conservation Services

78. Smil, V. 1983. Deforestation in China. *Unasylva* 12(5): 226–31

79. Brown, L., Chandler, W., Flavin, C., Postel, S., Storke, L., Wolf, E. 1984. *State of the World 1984.* Worldwatch Inst. New York: Norton

80. Hinrichsen, D. 1981. Peat power: Back to the bogs. *Ambio* 10(5): 240–42

81. Deleted in proof

82. Spears, J. 1978. *Wood as an energy source: The situation in the developing world.* Paper presented at the 103rd Annual Meeting of the American Forestry Association. Washington, DC. World Bank

83. Ernst, E. 1978. *Fuel consumption among rural families in Upper Volta, West Africa.* Presented at Eighth World Forestry Congress, Jakarta. Rome. UN Food & Agric. Organ.

84. Navaratna, H. 1983. *Sri Lanka case study on the Sarvodaya Stove Project.* Presented at Int. Workshop on Woodstove Dissemination, Wolfheze, The Netherlands. 30 pp.

85a. Krishna Prasad, K. 1983. *Woodburning stoves: Their technology, economics and deployment.* Working Paper. Geneva. Int. Labor Organisation. 168 pp.

85b. FRIDA. 1980. *Domestic energy in Sub-Saharan Africa: The impending crisis, its measurement and the framework for practical solutions.* London. (Cited in Ref. 85a)

86. Ki-Zerbo, J. 1981. Women and the energy crisis in the Sahel. *Unasylva* 33(133): 5–10

87. Deleted in proof

88. Deleted in proof

89. Deleted in proof

90. UN Food & Agric. Organ. 1982. *Tropical forest resources.* Forestry Paper No. 20. Rome, FAO

91. Deleted in proof

92. Soepadmo, E., Singh, K. G., eds. 1972. Proceedings of the symposium on biological resources and national development, 5–7 May, University of Malaya. Special issue, *Malay. Natl. J.* Cited in: Myers, N. 1976. An expanded approach to the problem of disappearing species. *Science* 193(4249): 198–201

93. Myers, N. 1981. The hamburger connection: How Central America's forests become North America's hamburgers. *Ambio* 10(1): 3–8

94. Deleted in proof

95. Deleted in proof

96. Deleted in proof

97. Sagan, C., Toon, O., Pollack, J. 1979. Anthropogenic albedo changes and the earth's climate. *Science* 206(4425): 1363–68

98. Deleted in proof

99. Deleted in proof

100. O'Keefe, P. 1983. The causes, consequences, and remedies of soil erosion in Kenya. *Ambio* 12(6): 302–5

101. Deleted in proof

102. Eckholm, E., Foley, G., Barnard, G., Timberlake, L. 1984. *Fuelwood: The energy crisis that won't go away.* London. Int. Inst. for Environ. & Devel./Earthscan

103. Organisation for Economic Cooperation and Development/Comite Interetat Permanent de Lutte contre la Secheresse. 1981. *Drought control and development in the Sahel: situation at the start of the 1980's, overview and prospects.* Presented at Fifth Conference of Club du Sahel, Paris

104. Natl. Acad. Sci. 1980. *Firewood crops.* Washington, DC. 237 pp.

105. Natl. Acad. Sci. 1983. *Firewood crops. Volume 2.* Washington, DC

106. Deleted in proof

107. Spears, J. S. 1978. *The changing emphasis in World Bank forestry lending.* Presented at Eighth World Forestry Congress, Jakarta. Rome, UN Food & Agric. Organ.

108. Noronha, R. 1981. Why is it so difficult to grow fuelwood? *Unasylva* 33(131): 4–12

109. Manibog, F. R. 1984. Improved cooking stoves in developing countries: Problems and opportunities. *Ann. Rev. Energy* 9:199–227

110. Moss, R. P., Morgan, W. B. 1981. Fuelwood and rural energy production and supply in the humid tropics. *Report for the United Nations Univ. with Special Reference to Tropical Africa and Southeast Asia.* Dublin, Ireland: Tycoly International

Ann. Rev. Energy. 1985. 10:431–62

INHERENTLY SAFE REACTORS[1,2]

Irving Spiewak and Alvin M. Weinberg

Institute for Energy Analysis, Oak Ridge Associated Universities, Post Office Box 117, Oak Ridge, Tennessee 37830

INTRODUCTION

No power reactor has been ordered in the United States, and 70 reactors have been cancelled, since 1978; one must therefore wonder whether nuclear energy will survive here. This same question is being asked in many European countries, for example in Sweden, where a 1980 referendum calls for shutting down all of Sweden's 12 reactors by 2010—even though they now produce more than half of Sweden's electricity. David Lilienthal, the first Chairman of the Atomic Energy Commission, in his most recent book, *Atomic Energy: A New Start,* published in 1980 (1), recognized the pre-cariousness of the nuclear option. He called on nuclear engineers to come up with a technical fix: a reactor that both friends and foes of nuclear energy would agree could not under any circumstances suffer the fate of the Three Mile Island-2 reactor—in short, a reactor that was transparently and patently immune from a core melt. This he regarded as the key to a rebirth of nuclear energy.

Lilienthal's views had been anticipated by several technologists and energy analysts, particularly in Sweden and Germany, and in the United States. Soon after Lilienthal's book appeared, the Institute for Energy Analysis convened a dozen old-timers who had been responsible for setting nuclear energy on its main technical paths, to consider Lilienthal's

---

[1] The US Government has the right to retain a nonexclusive, royalty-free license in and to any copyright covering this paper.

[2] Abbreviations used: LWR, light water reactor; PRA, probabilistic risk assessment; CMP, core melt probability; RY, reactor year; HTGR, high-temperature gas-cooled reactor; PWR, pressurized water reactor; BWR, boiling water reactor; APWR, advanced pressurized water reactor; PIUS, process inherent ultimately safe; LOCA, loss-of-coolant accident.

challenge: Could inherently safe reactors be designed? (2) If inherently safe reactors could be built at competitive costs, would their use restore the public's confidence in nuclear energy? The old-timers' workshop concluded that such a reexamination of the technical options was appropriate; and the Mellon Foundation supported the Institute for Energy Analysis in an examination of a "Second Nuclear Era" that would be based on inherently safe reactors. About the same time, studies by the Office of Technology Assessment (3) and by the Massachusetts Institute of Technology (4) addressed many of the same questions.

All three studies concluded that inherently safe reactors were feasible. The three studies, however, differed in emphasis; IEA focused strongly on inherently safe reactors as technical fixes, whereas OTA and MIT considered institutional improvements as well as improved light water reactor technology as the probable route to a revival of nuclear energy in the United States. Here we shall review mainly the technology of inherently safe reactors.

ACTIVELY SAFE AND PASSIVELY SAFE REACTORS

When nuclear reactor design began in 1942, safety was always a strong concern, but the idea of a reactor whose core could not melt never occurred to the early designers. Commercial power reactors in the United States grew out of pressurized water reactors, which were originally designed for ships. Compactness and simplicity, not inherent safety, were their primary design criteria. Thus the low thermal inertia of the light water reactors (LWRs), which meant that if uncooled, the temperature of the water in a 1000-MW(e) reactor would rise 30°C/min even after a scram, was a design constraint that had to be lived with. As commercial reactors grew in size, the possibility of a core melt became an increasingly dominant challenge for reactor designers. Since the LWR leaves so little time for remedial action if something goes awry, designers have festooned LWRs with many safety systems—e.g. the High Pressure Safety Injection System, Fast Acting Scram Rods, and Core Sprays. In consequence, modern LWRs are immensely complicated. They depend for their safety not on intrinsic properties of the reactor itself or on passively activated systems, but rather on the active intervention of electromechanical devices, such as valves, scram rods, emergency pumps, and backup diesels. What is true of LWRs is to some degree true of all other commercial reactors—their safety depends on active interventions, sometimes including action by the operators. We shall call such reactors "actively safe."

To estimate the likelihood of mishap in an actively safe reactor, one must resort to probabilistic risk assessment (PRA)—i.e. one tries to imagine all events that can lead to core damage; one then assigns a probability to each of these events. This procedure was used by Rasmussen in his famous WASH-1400 (5): he concluded that the median core melt probability (CMP) in light water reactors was 5×10^{-5} per reactor year. He assigned an uncertainty factor of around 5 to this estimate, but others (as well as Rasmussen, on further reflection) have put the uncertainty higher (6, 7).

Regardless of one's degree of confidence in PRA as a means of estimating the likelihood of a core damaging event, all must agree that one can reduce that likelihood by adding redundancy. For example, a PRA for the Calvert Cliffs reactor revealed the most important contributor to its rather high core melt probability (2×10^{-3}) was the failure of its single auxiliary feedwater train. A second train was added; this reduced the estimated core melt frequency some fivefold (8). Thus one approach to achieving Lilienthal's safe reactor would be to add more active safety systems. This "incremental" approach to safety builds entirely on existing technology; it suffers, however, from its dependence on active systems that may fail. Though enough redundancy can reduce the probability of failure to arbitrarily small values, skeptics can always claim that not all events leading to accidents can be imagined, that the probabilities used in PRAs are faulty, or that acts of war or sabotage or earthquake (so-called external events) can nullify the active safety systems.

The alternative way of meeting Lilienthal's call for a "safe" reactor is through passive systems that rely on inherent characteristics of the reactor. Such reactors we shall call passively or inherently safe. For example, if a reactor has large thermal inertia, it is obviously less likely to melt than if it has low thermal inertia. Or if it is encased in a totally inaccessible concrete structure, it is probably immune from sabotage or acts of conventional war. Thus, inherently safe reactors are safe not because of the intervention of active systems, which always have some probability of failure, but because of the workings of immutable laws of thermomechanics, of gravity, and of nuclear physics. The trick is to choose reactor configurations that embody such immutable principles.

In some sense there is no such thing as a totally safe reactor. Some events, with probabilities of, say, 10^{-9} per reactor year (RY), that could damage even the most inherently safe reactor can probably always be conceived. One can argue that, ultimately, one relies on a PRA, albeit a very far-fetched one, for ensuring safety. Thus one can hardly avoid answering the question, "How safe is safe enough?", even with inherently safe reactors. However, as we shall see, some of the advanced actively safe reactors yield PRA estimates of core melt in the range of 10^{-7}/RY, and passively safe reactors yield PRA estimates of 10^{-8}/RY—1000–10,000 times lower than the safety goals promulgated by the NRC. Reactors with such low CMPs ought to be regarded as meeting Lilienthal's call for a safe reactor.

A reactor that puts out little heat, and that by design is resistant to large overpower transients, can hardly cause much damage to itself or to its environs. This had been realized at the time of the Ergen task force on the China Syndrome (9). Ergen, in private discussion at the time, asked at what maximum power would a reactor, after shutdown, be able to dissipate its heat to the environment by natural convection and conduction. His answer, around 30 MW, was not considered of practical importance; but it does illustrate that smaller, by and large, is safer.

This perception of "smaller is safer" fits nicely with the current outlook of almost all American utilities. Having been stung by brutal cost overruns in 1000-MW nuclear plants, and faced with all but indeterminate projections of future load growth, utilities are interested in modular plants, whether nuclear or fossil-fueled. The modules might be as small as 100 MW, could be built quickly, and could be grouped to make large plants. Modular plants are being examined by several manufacturers.

Though modular plants fit better into most utilities' plans for expansion than do large plants, they would appear to sacrifice economy of scale for greater intrinsic safety. Does this imply that utilities choosing modular plants might be committing to higher cost electricity than they would were they to buy a small share of a larger plant? Perhaps, but large plants might also be made intrinsically safe, and modular plants might be made economical through factory construction and very rapid deployment at the utility's site. Of course, these considerations are rather speculative since we have no firm data on the economics of small modular plants, nor on the intrinsic safety of large, advanced ones.

THE TECHNOLOGY OF SAFER
NUCLEAR PLANTS

The terms "intrinsically safe" and "forgiving" for reactors were introduced around 1980 by Fortescue, Hannerz, and O'Farelly (10–12). In the intervening years the idea has caught on remarkably; intrinsic or passive safety features have been incorporated in newly developed actively safe reactors; and several truly inherently or passively safe reactors are now under consideration (13). We now describe these proposals for both actively and passively safe reactors as of 1984.

Actively Safe Reactors

The world's power reactors are classed as actively safe reactors, though they incorporate passive features to varying degrees. Here we discuss only those that might be considered for future deployment in the United States, the light water reactor, the CANDU pressurized heavy water reactor, and the high temperature gas-cooled reactor (HTGR).

Light water reactors account for 85% of the world's commercial power reactor capacity. In addition, all naval reactors and most research reactors are moderated and cooled by light water. Because this type is so dominant, and perhaps also because TMI-2 was a light water reactor (of the pressurized water type), most of the actual development effort has gone toward improving the safety of LWRs, both by incorporating passive safety features and by strengthening active safety features.

After the TMI-2 incident, the Nuclear Regulatory Commission mandated a number of backfits on existing LWRs to correct deficiencies revealed by the incident. The Rasmussen PRA had estimated a median core melt probability of the Surry pressurized water reactor (PWR) and the Peach Bottom boiling water reactor (BWR) to be 6×10^{-5}/RY and 3×10^{-5}/RY respectively, for an average probability of 5×10^{-5}/RY. This number has been widely quoted both in the literature and in the nuclear debate. An analysis by Minarick & Kukielka (14) of actual precursors to potentially serious events in operating reactors between 1969 and 1979 suggests that the actual core melt probability of the LWRs operating during this period was closer to 10^{-3}/RY. This number agrees with Okrent's pessimistic estimate (7). An analysis by these authors of precursors during 1980 and 1981 suggests that the core melt probability in the post-TMI period had fallen to around 15×10^{-5}/RY—within a factor of 3 of the core melt probability estimated in WASH-1400 (15). Following TMI-2, every LWR in the United States has been required to strengthen its equipment and operations and many have conducted PRAs, both with and without the NRC-mandated backfits. D. Phung has concluded from these analyses that the median core melt probability of US reactors with the TMI-2 backfits is now comfortably within the range estimated by WASH-1400 (16). This means that reactors today are perhaps six times safer than they were before 1979, when Minarick & Kukielka's semiempirical analysis found them to be considerably less safe than indicated by the WASH-1400 median estimate. The backfits responsible for these improvements in PRA estimates include direct measurement of the water level in the reactor—(recall that ambiguity in measurement of this water level at TMI-2 prompted the operators to shut off the emergency core cooling system, with disastrous results)—and direct measurement of the state of the pilot-operated relief valve (whose failure to close, and incorrect indication of this state, led directly to the accident). Altogether Phung estimates that the backfits mandated in response to the TMI-2 accident will cost about $25/kW(e) (17). Probably more significant than the plant changes have been changes in operator training and procedures. One must therefore concede that, insofar as PRA is a believable estimator of core melt probability—or better, PRA backed up by the semiempirical evidence of Minarick & Kukielka—today's LWRs are about as safe as Rasmussen claimed (CMP = 0.5×10^{-4}/RY). This is well within NRC's promulgated safety goal of 10^{-4}/RY.

IMPROVED PWRs   From PRA one finds that small break loss-of-coolant acidents (such as occurred at TMI) are the largest contributors to the estimated core melt probabilities for PWRs (Figure 1) (18). To reduce the CMP, designers of PWRs have therefore focused strongly, though not exclusively, on measures to avoid or mitigate loss-of-coolant accidents. Such design precepts are exemplified in the Sizewell B PWR (19), the Advanced PWR (APWR) (21), and the Combustion System-80 PWR (20).

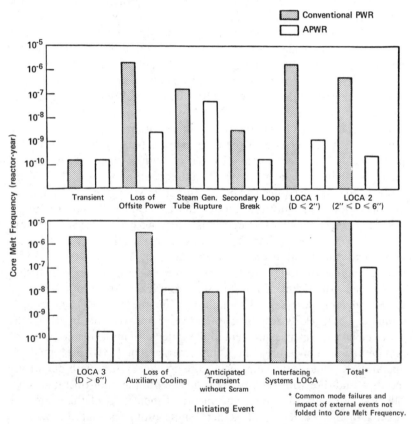

Figure 1   Initiating events contributing to the probability of core melt in pressurized water reactors. Source: (18).

The Sizewell-B PWR   The United Kingdom, in collaboration with Westinghouse, has incorporated additional redundancy and diversity in the latter's Callaway Standardized Nuclear Unit Power Plant System (SNUPPS) design for its proposed 1300-MW(e) Sizewell-B PWR. Sizewell-B achieves a lower core melt probability through additional safety equipment, e.g.

1. Four high-pressure safety injection pumps, each with heads lower than 2000 pounds per square inch (psi) and higher flows than Callaway's.
2. Four accumulators, any two of which are sufficient for core cooling at the 600-psi range (instead of the required three at Callaway).
3. A steam-driven auxiliary feed pump, in addition to the two electric pumps already in SNUPPS.
4. Four diesel generators (instead of two) to provide emergency power.
5. An emergency boration system as a backup reactor trip system to cope with anticipated transients without scram.

These exemplify the dozen or so redundancies added to Sizewell-B. The United Kingdom Atomic Energy Authority and Westinghouse have estimated, by means of PRA, the median core melt probability of the Sizewell-B reactor to be only 1.1×10^{-6}/RY—almost 50 times lower than the WASH-1400 value—and the estimated probability of a large release of radioactivity to be only 3×10^{-8}/RY. The various contributors to this core melt probability are summarized in Table 1. The measures added to Sizewell-B to achieve its lower estimated core melt probability add about 20% to the cost of the reactor.

The APWR   Westinghouse and Mitsubishi, together with five Japanese utilities and the Japanese government, are spending $\$150 \times 10^6$ to design a 1350-MW(e) Advanced Pressurized Water Reactor for deployment in Japan. The total CMP for APWR is estimated to be only 2×10^{-7} per RY—some 300 times lower than the CMP for the PWR analyzed by Rasmussen (Figure 1). The APWR contains a number of innovations compared to earlier Westinghouse four-loop designs. Here we dwell on those improvements that lead to the much lower estimated core melt probabilities for the APWR. These include:

1. The increased volume of primary coolant in the reactor vessel above the core increases the time available to deal with a loss of coolant.
2. A lower core power density increases safety margins.
3. There are four complete safeguard trains of mechanical equipment in the safety system.
4. A large emergency water storage tank is provided inside the containment as the water source for four safety injection pumps. This storage

Table 1   PRA for the Sizewell-B reactor core melt by initiating event (23)

Initiating event	Core melt frequency	Percentage of total core melt frequency
Large LOCA[a]	1.83×10^{-7}	15.8
Medium LOCA	2.58×10^{-7}	22.2
Small LOCA	3.83×10^{-7}	33.0
Steam generator tube rupture	1.91×10^{-8}	1.6
Secondary side break inside containment	2.32×10^{-8}	2.0
Secondary side break outside containment	3.54×10^{-8}	3.0
Loss of main feedwater	1.58×10^{-8}	1.4
Closure on one MSIV[b]	5.71×10^{-11}	<0.01
Loss of RCS[c] flow	8.11×10^{-11}	<0.01
Core power excursion	5.11×10^{-12}	<0.01
Turbine trip	8.36×10^{-10}	0.07
Spurious safety injection	1.44×10^{-10}	0.01
Reactor trip	8.54×10^{-10}	0.07
ATWS[d]	1.37×10^{-7}	11.8
Loss of off-site power/turbine trip	6.03×10^{-9}	0.5
Interfacing systems LOCA	2.37×10^{-9}	0.2
LOCA beyond capacity of ECCS[e]	1.00×10^{-7}	8.6
Total	1.16×10^{-6}	100.0

[a] LOCA = loss-of-coolant accident.
[b] MSIV = main steam isolation valve.
[c] RCS = reactor cooling system.
[d] ATWS = anticipated transients without scram.
[e] ECCS = emergency core cooling system.

tank automatically receives water flows from ruptures in steam generator tubes.

5. Containment sumps are kept filled with water to increase the available heat capacity.
6. Safety and control systems are separated to increase reliability and reduce common mode failures.
7. Four separate and hardened compartments are provided to house high- and low-pressure safety injection pumps. This feature reduces the likelihood of radioactivity release to the atmosphere and makes sabotage of the safety systems very difficult.
8. The control room is improved, with improved diagnostic capabilities.
9. The larger pressurizer and core provide for improved response to transients.
10. The large dry containment vessel is conservatively designed.
11. The steam generator secondary-side water inventory is controlled automatically.

12. There is injection of pressurized water to reactor coolant pump seals.
13. The reactor vessel neutron fluence is reduced.
14. The overall improvements in plant availability, reliability, and maintainability translate into improved safety.

Westinghouse has performed a comparative PRA of the APWR and a conventional PWR for internal events. The results are shown in Figure 1. The internal risk in the APWR appears to be dominated by the steam generator tube break accident, itself at the very low level of 10^{-7} core melts per reactor year. The reported total risk from internal events of less than $2 \times 10^{-7}/RY$ is well below Westinghouse's target of $1 \times 10^{-6}/RY$ overall risk from the APWR. External event analysis, which is site-specific, will be carried out later.

Combustion System-80 Combustion Engineering, Inc. (CE) currently offers a standard nuclear steam supply system, System-80, rated at 3800 MW(t) (1270 MW(e)). The first System-80 plant to be completed will be the Palo Verde plant of Arizona Public Service Company.

CE believes that the System-80 design has pioneered many features contributing to the safety of the plant. These features are now being offered by other vendors as well, either as backfits or incorporated into new designs. These features include:

1. Greater core thermal margin.
2. Large pressurizer volume to absorb loss of electrical load.
3. Improved secondary-side steam generator materials (stainless steels).
4. Use of two-of-four concurrent measurements of the same parameter to actuate safety systems, with one channel available for off-line testing.
5. Use of a core-monitoring computer to monitor core thermal-hydraulic parameters continually.
6. Advanced control room.

CE's efforts on future plants are concentrated on upgrading certain aspects of the System-80 design to increase reliability, decrease costs, and provide greater assurance of safety. To a certain extent, modifications may be required to satisfy new NRC regulations, but the primary driving force for these changes would be innovation based on construction and operating experience.

The further evolution of System-80 is expected to include the following features:

1. Simplification, to the extent possible.
2. Emphasis on reliability and maintainability.
3. High-quality steam generators and improved steam generator materials.

4. High-quality heat exchangers and condensers to avoid ingress of contaminants to the steam generator.
5. Upgraded control room and instrumentation.
6. Fewer pipe supports (seismic and pipe-whip criteria).
7. Optional full-pressure decay heat removal system.
8. Reactor pressure vessel design that greatly reduces the impact of neutron fluence.
9. Improved feedwater systems and control systems.
10. Design to avoid spurious trips.
11. Fully replaceable major equipment.

No PRA has been reported thus far for the System-80. It is anticipated that the design could achieve core melt frequencies of the magnitude reported earlier for the Sizewell-B plant.

BOILING WATER REACTORS   The relative contribution to core melt probabilities of various initiating events in BWRs are shown in Figure 2 (22). In PWRs a pipe break followed by a loss-of-coolant accident poses the dominant threat of a core melt; in BWRs, transients—in particular loss of off-site power—are the dominant threat. Thus the main strategy for improving the safety of BWRs has to do with providing additional protection against loss of off-site power and other transients, and improving the on-site emergency power supplies.

The BWR/4, which represents the kind of reactor analyzed in WASH-1400, yields an estimated CMP of around 3×10^{-5}/RY. The standard General Electric (GE) offered, the BWR/6 GESSAR, has a reported core melt probability from internal sources of 4.7×10^{-6}—about 8 times lower than the CMP for the BWR/4. Most of this improvement results from better instrumentation and automatic depressurization of the reactor to make it easier to supply emergency feedwater. By contrast, the loss of off-site power remains the dominant contributor to CMP, at approximately 6×10^{-6}, compared to 8×10^{-6} for the BWR/4.

GE, Hitachi, and Toshiba are developing a 1350-MW(e) Advanced Boiling Water Reactor (ABWR) for which the CMP is claimed to be substantially below that of the BWR/6 GESSAR. The key changes in going from BWR/6 GESSAR to ABWR are summarized in Table 2.

GE has described a small BWR concept (23) with modest innovations to simplify safety functions (Figure 3). Gravity-driven shutdown rods and liquid poison injection provide assurance of shutdown. An isolation condenser increases assurance of decay heat removal. Water from the elevated pressure suppression pool will automatically flood the reactor vessel should it become depressurized. At the low end of the projected capacity range (200 MW(e)) the power is normally removed without

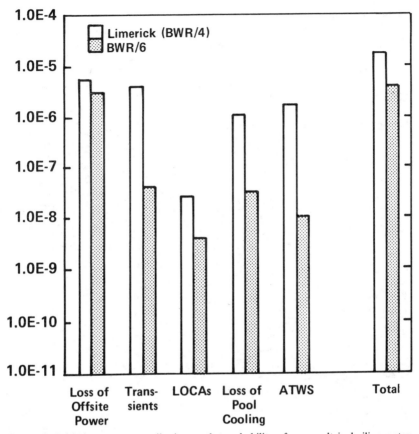

Figure 2   Initiating events contributing to the probability of core melt in boiling water reactors. Source: (22).

mechanical recirculation pumps; at the high end of the capacity range (600 MW(e)) internal circulation pumps are used, as in the Advanced Boiling Water Reactor.

THE CANDU-PRESSURIZED HEAVY WATER REACTOR   The CANDU reactor has operated successfully since 1968, and a total of 16 CANDU reactors are in commercial operation in four countries.

The core consists of an array of horizontal pressure tubes surrounded by a low-pressure calandria containing the relatively cool moderator. Zircaloy-clad natural uranium fuel bundles are loaded and removed from the pressure tubes during operation by fueling machines operating at the two faces of the core. Heat is transferred from the heavy water primary

Table 2   Key differences between ABWR and GESSAR designs

Plant feature	GESSAR	ABWR
Recirculation system	External pumps	Internal pumps
	Flow control valve	Solid state power supply
Control rod drives	Hydraulic	Electric/hydraulic
		fine motion
Emergency core cooling	Three divisions:	Three completely
	1 high-pressure spray	separate divisions:
	1 low-pressure spray	2 high-pressure sprays
	3 low-pressure flooders	1 steam-driven reactor
		core isolation
		cooling system
		3 low-pressure flooders
Core spray sparger	Side	Overhead
Decay heat removal	2 steam-condensing	2 modulating valves
	heat exchangers	3 wetwell/drywell heat
		exchangers
Control of reactor flow,	Analog	Digital
feedwater, and pressure		
Transmission of control	Wires	Multiplexed
and safety signals		
Containment	Horizontal vents	Vertical vents
	Steel	Concrete
	Open pool	Covered pool
	Air	Inerted
Steam bypass capacity	35%	100%
Fuel transfer	Inclined tube	Cask lift

Source: (20).

system to a light water secondary system in vertical U-tube steam generators.

Control of the CANDU reactor is maintained primarily through on-stream refueling. Computers control the routine plant operation.

The CANDU reactor is subject to many of the same potential accident initiators as a PWR. However, transients would generally occur more slowly because of the large thermal inertia of the moderator and pressure tubes. No PRA of a CANDU reactor has yet been reported, but preliminary estimates indicate a CMP of 10^{-5}/RY, in the same range as that of an LWR (24).

Studies of CANDU reactors for US construction (25, 26) have indicated that some design and licensing decisions could prove difficult. The pressure tubes, which contain the full primary system pressure, do not currently

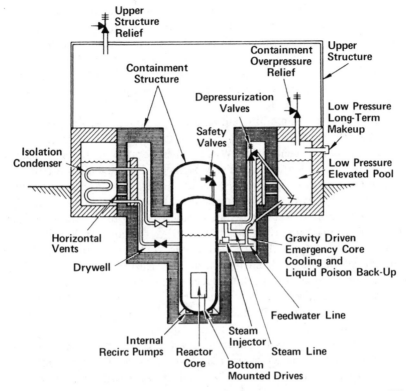

Figure 3   Small boiling water reactor concept with innovative safety features. Source: (23).

conform to the American Society of Mechanical Engineers Pressure Vessel Code. If the tubes were required to have thicker walls, then slightly enriched fuel would be preferred to natural uranium. Other features that would be novel to the Nuclear Regulatory Commission include the seismic analysis of the core, the use of on-stream refueling (and the security problems related to continuous fuel handling), and computer control of the reactor. It is our understanding that some Canadian studies of advanced CANDUs suited for Japanese or perhaps US application are under way but have not been publicly disclosed.

THE HIGH-TEMPERATURE GAS-COOLED REACTOR (HTGR)   A prototype HTGR has been in operation at Fort St. Vrain in Colorado since 1979. A related high-temperature reactor, but with pebble bed instead of prismatic fuel, is in a startup process in West Germany. The HTGR is a helium-cooled, graphite-moderated reactor assembled in a prestressed concrete pressure vessel. The primary system pumps and steam generators are located in cavities in the vessel. Fuel is in the form of graphite-coated uranium oxide or uranium carbide particles in graphite blocks dispersed in a large stack of graphite moderator blocks. The superior high-temperature characteristics of the fuel and moderator allow the HTGR to generate steam at temperature and pressure conditions approximating those of modern fossil-fueled boilers.

The HTGR has some inherent safety advantages (27). Probably most important is the relatively low power density of HTGRs—between 5 and 10% of that of a conventional PWR. Further, because the fuel is dispersed throughout the moderator, the heat capacity closely associated with the fuel is over 100 times that of an LWR. Thus, in the event of an accident that interrupts the flow of coolant to the core, the elapsed time between the cessation of coolant flow and severe damage to the core (if no automatic or operator action is taken) is of the order of 10 hours in an HTGR, rather than tens of minutes for a PWR—a difference of more than an order of magnitude. One factor contributing to this feature is that lateral heat conduction through the graphite blocks in the HTGR core is sufficient to remove a substantial fraction of the afterheat from the core and carry it out through the reflector, from which the heat is radiated to the water-cooled steel liner of the reinforced concrete pressure vessel.

The HTGR's prestressed concrete pressure vessel has redundant load-carrying steel tendons, which are readily inspectable and replaceable. The tendons keep the concrete and the vessel liner in compression. The vessel is designed to withstand 2400 psi, over twice its operating pressure. Should a crack form during a pressure transient, the resulting small gas leak would tend to be sealed when the gas pressure was reduced. Thus, catastrophic failure of the vessel is not possible under loads possibly imposed by the HTGR.

The PRA of the reference HTGR indicates a fuel damage probability of 4×10^{-5}/RY, comparable to those of PWRs. It should be recognized that the character of the fuel damage in an HTGR would be less severe than that in an LWR. The fuel and graphite structure would remain basically as before, except that some of the more volatile fission products would escape from the coated fuel particles into the gas stream. Severe accident analysis of HTGRs indicates that there would be damage to the heat transport systems in the event that the dedicated core cooling systems fail and that there would be damage to the vessel should the linear cooling system be lost. However, most of the non–rare gas fission products are retained in the reactor vessel even in the worst accident scenario that has so far been envisioned.

Inherently or Passively Safe Reactors

Both the APWR and the ABWR, with CMPs below $10^{-6}/RY$, must be regarded as extraordinarily safe, at least insofar as one is willing to trust probabilistic risk assessment. It is also plausible that advanced CANDUs and the HTGR could provide equal safety. Even if one is skeptical of PRA for absolute estimates of safety, one can hardly deny that PRA ought to be much more reliable for estimating the additional safety added by an incremental improvement on an existing reactor. One must therefore be impressed with the two and more orders of magnitude in reduction of CMP as one goes from existing PWRs and BWRs to APWRs and ABWRs, even if these estimates depend on PRA.

This additional safety incurs cost and complication. Anyone who has walked through a 1000-MW nuclear power plant must be impressed by its complexity; utility executives canvassed by the Electric Power Research Institute in 1982 (28), almost without exception, expressed dissatisfaction with the complexity of their nuclear plants.

Can this complexity be avoided, and safety possibly even improved, by designs that exploit inherent characteristics of the reactor? Can some external events, such as acts of nonnuclear war or sabotage, which are never considered in PRAs, be protected against? And can estimates of safety be made deterministic rather than probabilistic if the reactor incorporates clever enough passive safety features?

At least two well thought out proposals for reactors that meet such stringent criteria have been seriously proposed: The Secure-P or PIUS reactor, invented by K. Hannerz of ASEA/ATOM in Sweden (29); and the modular HTGR, originally proposed by G. H. Lohnert of Kraftwerk Union AG (KWU) (Interatom/GE) (30) and also by General Atomic (31). In addition, several other proposals for inherently safe reactors have appeared in the past few years.

THE SECURE-P OR PIUS (PROCESS INHERENT ULTIMATELY SAFE) REACTOR PIUS is a pressurized water reactor whose entire primary system, including the steam generator, is submerged in a large pressurized pool of cold water containing boric acid. The water containing boric acid, which is held in a huge prestressed concrete pressure vessel, is connected at top and bottom to the circulating primary system via a mechanically unblockable natural circulation circuit (see Figure 4). If the submerged pumps in the primary system are operating normally, the interfaces between pool and primary circuit are maintained. Should the primary circulation be disturbed in any way (pump failure or boiling in core, for

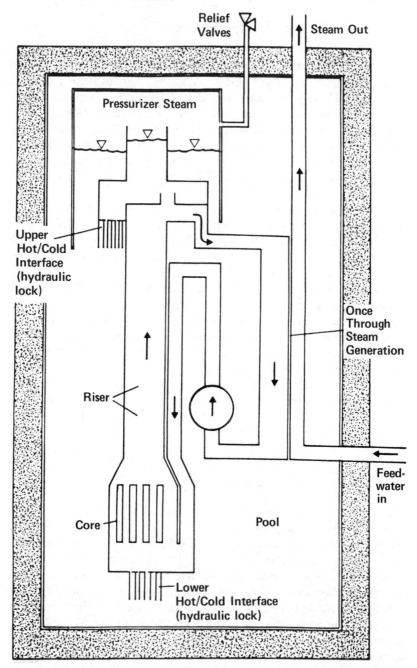

Figure 4   Schematic diagram of the PIUS primary system. Source: (32).

example) the cold boric acid is forced by gravity into the reactor, causing the reactor to shut down (because boron absorbs neutrons so strongly), and a natural convection circuit through the reactor and the pool is established. Enough water is in the pool to keep the reactor cool for at least a week, after which time a fire truck can provide additional water if necessary.

The PIUS reactor, through its very clever hydraulic locks, eliminates the low thermal inertia of the standard PWR. Because PIUS can thereby guarantee core integrity, it eliminates the risk of major releases of radioactivity by virtue of the laws of thermohydraulics and gravity alone. The redundant, diverse, spatially separated engineered safety systems now distributed throughout most of today's nuclear plants are no longer needed.

The response of PIUS to various transients has been studied by groups in Sweden and in Japan. As pointed out by Babala (32), all such transients can be divided into two groups according to the qualitative character of the final state: 1. transients that end with an autonomous reactor shutdown and 2. transients that end in a quasi–steady state at a new power level. For type 1 transients, caused, for example, by recirculation pump trips, total loss of secondary system heat sink, or depressurization of the system, the reactor automatically shuts down within 0.5–2 minutes because of the massive ingress of borated water. Figure 5 shows the results of Babala's simulation of the initial phase of a feedwater trip without scram (an event in an ordinary LWR that requires immediate intervention). The secondary side of the steam generator boils dry within 10 seconds, whereupon the primary system starts to heat up. The density of the fluid in the riser section of the loop decreases, and the event culminates in a brief void episode (between 23 and 28 seconds), which eventually overrides the pump controller that tries to keep the pool water out of the loop. The boron concentration in the core starts to rise at 28 seconds, and the reactor power is down to 10% after about 90 seconds.

Of type 2 transients, the one that often concerns critics of PIUS is inadvertent dilution of the boron in the pool. A simulation of continuous boron dilution caused by injecting 100 kg per second (twice the design maximum) of clean water into the loop leads to the following: the coolant density decreases as reactor power rises and borated water enters the loop at 340 seconds. A quasi-equilibrium is reached at 500 seconds, when the effect of the injected clean water is neutralized by a steady inflow of pool water. The 10% overpower that cannot be dissipated in the steam generators is deposited in the pool by natural circulation. Babala states that the external inventory of clean water will generally be exhausted before a dangerous increase in the temperature of pool water occurs.

The response of PIUS to transients has been studied, independently, in Japan (33), with results that generally confirm the claims made for PIUS—

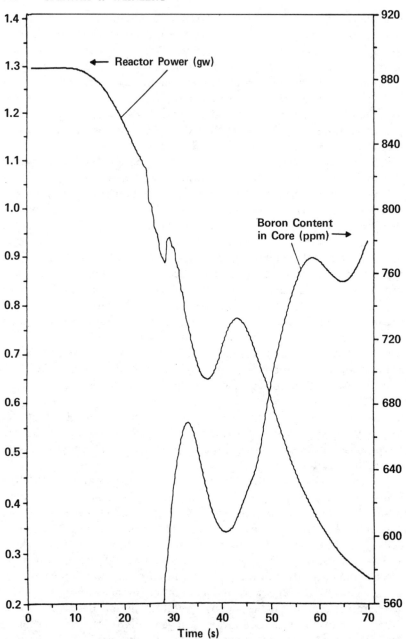

Figure 5   Simulation of a feedwater trip without scram, PIUS reactor. Source : (32).

that no transient has been identified that can lead to core damage. PIUS is also remarkably resistant to most loss-of-coolant accidents (LOCAs), since any break in the primary system immediately breaks the hydraulic lock, and borated pool water rushes in. A catastrophic failure of the steam generators, though it too would cause the reactor to be flooded with borated water, conceivably might offer a path for partial draining of the pool. For this reason, the steam generators are located above the reactor so that, at the very worst, the level of the pool would be lowered to the point of the break; enough of the pool would be left in place to provide cooling, by evaporation, for several days. Even farther fetched is a massive failure, induced by an earthquake, of the prestressed concrete vessel itself. This event is regarded as incredible by ASEA/ATOM. In any event, the vessel is built into the ground, much like an ICBM silo. Water leaking from a massive failure would accumulate in the hole around the vessel, which would in any case be lined.

PIUS is a derated PWR, and therefore requires no development of fuel elements or pumps; nevertheless skeptics have raised questions about its feasibility.

1. Will the hydraulic locks work as planned? Two experimental rigs, one a table-top model tested at the Tennessee Valley Authority in the United States (34), and the second a one-tenth scale model tested at ASEA/ATOM, have demonstrated that the locks work exactly as predicted. A 2500-kW, electrically heated mock-up is under construction at ASEA/ATOM; results from it should be forthcoming in 1985.
2. Is the steam generator feasible? The bayonet-type, once-through steam generator, located in the pool above the core, requires considerable development. Since the hot steam generator is immersed in the cold pool, it must be insulated from the pool. ASEA/ATOM proposes stainless steel sheeting to cover, and thus insulate, the entire primary system, but this approach needs testing.
3. What about thermal shock caused by ingress of the pool water? This presumably can be obviated by design, but this presumption needs to be demonstrated.
4. Will PIUS operate continuously, or will it always be shut down due to boron's inadvertent invasion of the primary cooling system? The remarkable stability of the hydraulic locks demonstrated in the ASEA/ATOM mock-up, as well as the computer simulations, should be reassuring on these scores. However, not enough is yet known about ingress of boron across the hydraulic lock.
5. Can PIUS be maintained? ASEA/ATOM claims that maintenance

through the large pool should be no more difficult than the maintenance now performed routinely during refueling of LWRs.

As presently conceived, a single large prestressed concrete vessel would house up to four PIUS modules, each rated at 200 MW(e). Whether these modules would be regarded as separate reactors by licensing authorities remains to be seen. A single module of 500 MW(e) is also being studied.

Economics   Whether or not PIUS would be economic depends on how it fares with licensing authorities. Though the large concrete vessel is very expensive, its use eliminates essentially all the safety systems needed in conventional reactors. Moreover, there is no reason to build the balance-of-plant (BOP) to nuclear standards since the PIUS is resistant to transients caused by failures in the balance-of-plant. In particular, a full-fledged containment vessel is not needed. If these contentions can be sustained before licensing authorities, the PIUS might cost no more per kilowatt, and possibly less, than a conventional LWR of 1000 MW(e).

Present status   As of this writing, PIUS is being developed only by ASEA/ATOM. A 400-MW reactor with many features of PIUS has been offered by ASEA/ATOM for district heating in Helsinki. Whether this tender will be accepted will be known within the coming year.

These writers conclude that PIUS, with its clever hydraulic locks, must be regarded as inherently safe. Moreover, in extremis, it protects not only the public, but also the owner of the reactor; and neither an act of nonnuclear war nor sabotage could cause an accident that threatens the public. K. Hannerz's PIUS reactor therefore represents a quantum leap in the quest for inherently safe reactors.

THE MODULAR HTGR   We have already seen that the thermal inertia of the HTGR is some hundred times greater than that of the LWR. Moreover, the volumetric power density, around 6 MW/m^3, is 15 times lower than in an LWR. G. H. Lohnert (35) of KWU has proposed lowering the specific power of the HTGR even further as well as reducing the power output, so that even if all systems for removing after heat were to fail, the reactor could still cool itself by heat conduction, radiation, and natural convection. KWU, GE, and General Atomic (GA) (36) have proposed modular HTGRs whose maximum power is around 200 MW(e). A 1000-MW(e) plant would consist of five or more such modules.

The KWU design uses a pebble bed; the GA design uses either prismatic or spherical fuel elements. The fuel itself, as in large HTGRs, consists of tiny uranium oxide spheres triply coated with silicon carbide and graphite; these are embedded, like raisins in a cake, in fuel elements, either graphite prisms or spheres ("pebbles"). At a maximum power density of 3 kW(t) per liter (some 30 times lower than the power density of an LWR), and with an elongated core that maximizes surface area, the highest temperature reached by any fuel sphere is around 1550°C even if all active cooling fails. At this temperature the release of fission products to the atmosphere poses no hazard to the public, although the release may contaminate the reactor itself. Of course, loss of all cooling is a very unlikely event; and in almost any situation in which the system retains pressure, natural convection keeps the fuel well below 1550°C. Nevertheless, it is remarkable that a 200-MW(e) HTGR poses no danger to the public even if the reactor loses pressure and all cooling is lost.

A schematic of the KWU/GE modular HTGR is shown in Figure 6. In Figure 7 we see how the hottest fuel element temperature rises, but remains below 1550°C in a core heat-up accident initiated by depressurization combined with a failed steam generator or blower. In Figure 8 we show how the release of ^{137}Cs in these extreme circumstances varies with the volumetric power density. Below 3 MW/m^3, the release is well below that allowed by the German safety authorities; the release rises quickly with power density, and reaches unacceptable levels at power densities above 3.5 MW/m^3.

The only event that might compromise the inherent safety of a modular HTGR would be a graphite fire. This would require a catastrophic failure in both the main inlet and outlet to the reactor, and this could be initiated, if at all, only by a saboteur, or possibly by an earthquake that exceeds design specifications. These possibilities must be regarded as extraordinarily remote. One must therefore regard the modular HTGR as inherently safe.

Can a modular HTGR be afforded? At 200 MW(e) one is sacrificing economy of scale for safety. On the other hand the reactor is now so small that much of it can be fabricated in a shop; and, as with PIUS, the BOP need not be built to nuclear standards.

One is naturally tempted to compare the modular HTGR and PIUS as to practicality and inherent safety. Though no serious comparison of the two has appeared in the literature, we would point out the following:

1. Favoring modular HTGR. (a) No new mechanical devices or physical principles are involved; several HTGRs have been built, and their forgiving nature has been demonstrated. (b) Following a loss-of-coolant accident, no intervention would be required to protect the public.
2. Against modular HTGR. (a) In the worst accident, even though the public is unharmed, the reactor and its building might be contaminated. (b) HTGR is not absolutely immune from earthquakes and graphite fires.

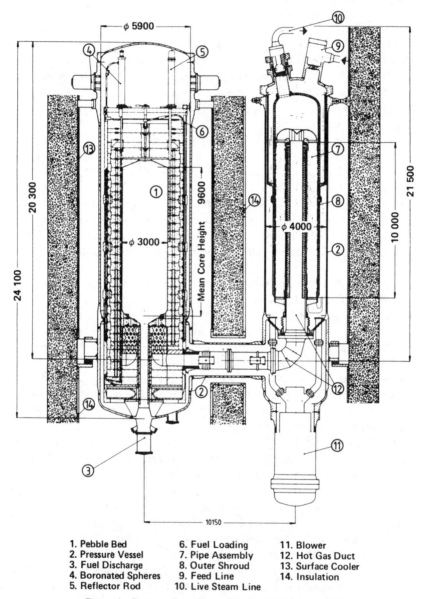

Figure 6   Cross section of a modular HTGR. Source : (17).

1. Pebble Bed
2. Pressure Vessel
3. Fuel Discharge
4. Boronated Spheres
5. Reflector Rod
6. Fuel Loading
7. Pipe Assembly
8. Outer Shroud
9. Feed Line
10. Live Steam Line
11. Blower
12. Hot Gas Duct
13. Surface Cooler
14. Insulation

3. Favoring PIUS. (*a*) Even in the worst accident, the reactor would remain intact. (*b*) No fuel element or pump development is needed. (*c*) The reactor is immune from sabotage and acts of war.
4. Against PIUS. (*a*) Intervention is required after a loss-of-coolant accident, though only after about a week. (*b*) PIUS probably is

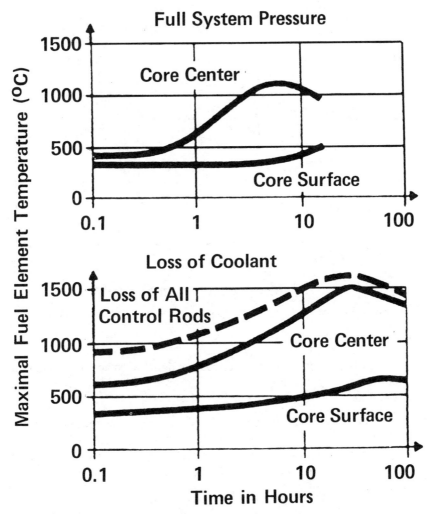

Figure 7 Time-dependent maximum fuel element temperature following loss of flow (upper figure), loss of coolant (lower figure), modular HTGR. Source: (35).

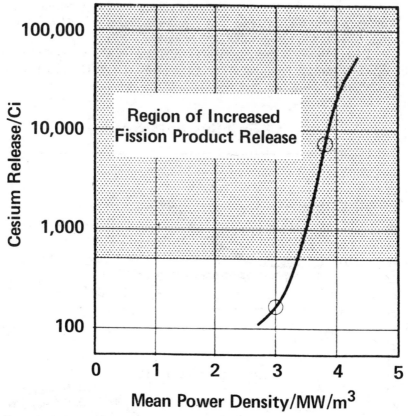

Figure 8   Cesium release during core heat-up accident vs power density, modular HTGR.
Source: (35).

sufficiently different from a conventional LWR to require a demonstration; in particular, the steam generator is novel and must be developed.

At present, the US Department of Energy is supporting the development of the modular HTGR at about $30 million in fiscal year 1985. In contrast, the PIUS reactor development is being supported by ASEA/ATOM, a private company, at a much lower level.

Other Ideas for Inherently Safe Reactors

THE SCHULTZ-EDLUNG STEAM-COOLED FAST REACTOR (37)   In 1982, M. Schultz and M. Edlund, while participating in IEA's Second Nuclear Era Study, proposed a steam-cooled fast breeder reactor that incorporated many of the safety features of PIUS. These designers noticed that a steam-cooled fast reactor could be stable against ingress of water if enough boron-10 were incorporated in the fuel. If the density of steam decreased (because the reactor power increased), neutron leakage would increase and thereby reduce power. On the other hand, if the steam condensed in the reactor, the neutron spectrum would soften. Since the capture cross section of boron increases as $1/v$, v being the neutron velocity, whereas fission cross sections do not rise as sharply as neutron energy falls, the reactivity of the reactor would tend to decrease as the result of the ingress of water. Thus the reactivity peaks at some intermediate steam quality. An example of the reactivity as a function of steam quality in a Schultz-Edlund fast reactor is shown in Figure 9. Depending on the detailed configuration of the reactor, the peak in reactivity corresponds to saturated steam pressures between 3.45 and 22.1 MPa (500–3200 psia).

With such a characteristic reactivity curve, one can configure an inherently safe reactor that embodies the hydraulic locks of PIUS, but—and this is a great advantage—the pool water need not be borated. Thus a Schultz-Edlund steam-cooled reactor would sit at the bottom of a large pool of ordinary water. As in PIUS, the steam circuit is separated from the pool water by hydraulic locks. At some steam quality, the chain reaction sets in; and this operating point should be stable. Any deviation from the

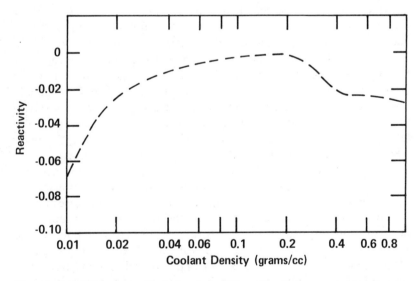

Figure 9 Reactivity vs water density for a typical inherently safe steam-cooled faster reactor. Source: (37).

operating point would shut the reactor down, with a resulting ingress of water. The reactor would continue cooling itself by natural convection.

The Schultz-Edlund reactor embodies a very ingenious idea that should be pursued further.

LIQUID METAL FAST REACTORS (LMFRs)   With the recent interest in inherently safe reactors, designers of LMFRs have incorporated passive safety features into them. This trend in design coincides with the collapse of the Clinch River Breeder Reactor project; out of these studies a new, intrinsically safe design for an LMFR may emerge. One would hope that such a reactor would not evoke the bitter antagonism that ultimately led to cancellation of CRBR.

Sodium-cooled reactors with enhanced passive safety have not received as much attention as have PIUS or the modular HTGR. Nevertheless one can discern two main threads in the attempts to design an intrinsically safe LMFR.

1. *Inherently safe oxide-fueled LMFR*   This approach is being followed by Westinghouse, Rockwell International, and GE. As described by Schmidt et al (38) of Westinghouse, conventional LMFRs are vulnerable to Loss of Flow with Failure to Scram (LOFS); to Transient Overpower; and to Loss of Normal Heat Sink. Any of these events, when coupled with failure of both primary and secondary shutdown systems, could lead to a core disruption.

To deal with all three events occurring simultaneously, Westinghouse has developed a design that limits the amount and rate of reactivity insertion to a level that does not heat the sodium to boiling. This is done in one of three ways: independently operated control rods, each of limited worth, gravity-operated shutdown rods actuated by magnetic couplings that lose their magnetic properties when sodium is heated beyond the Curie point of the magnet material, and/or injection of a liquid poison (indium) under transient conditions. There is a passive decay heat removal system that allows heat to move through the reactor vessel to the guard vessel. The latter is cooled by natural air convection, enhanced by fins on the guard vessel. Westinghouse proposes intentionally extending the primary pump coastdown by means of a flywheel or equivalent, which would keep the coolant below its boiling point even during an overpower transient.

In addition, the design of the core has been optimized to increase inherent negative reactivity feedbacks so as to limit the overpower excursion. This is accomplished through thermal expansion of the core supports, which increases neutron leakage as the core heats up. Advantage is also gained from the differential thermal expansion of the control rods and core. These negative reactivity inputs as the core heats up must offset positive reactivity inputs, such as the Doppler effect due to the fuel's cooling off from operating conditions to shutdown condition.

The Westinghouse approach is claimed to be feasible for both small and large cores, and for both pool and loop plant configurations. However, inherent safety can be gained more easily at smaller size (simpler decay heat removal) and with a pool design (more thermal inertia slowing temperature excursions).

General Electric (39) and Rockwell International (40) are proposing modular fast reactors. Such pool-type reactors would be shop fabricated. Several modules could be put together to form a large power plant. The principles for gaining inherent safety are similar to those described for the Westinghouse plant.

2. *Reemergence of the metal-fueled LMFR* (41) Argonne National Laboratory has reintroduced the metal-fueled LMFR on the grounds that such a reactor is intrinsically safer than an oxide reactor of the same size. Argonne's case rests on the better thermal conductivity of the metal fuel. Since the temperature rise in fuel on going to power is lower in metal than in oxide, the reactivity that must be compensated for because of the Doppler effect is less in a metal fuel than in an oxide fuel. Thus less reactivity must be dealt with during a failure to scram, and the Doppler coefficient can be counted on to terminate the LOFS transient without allowing the sodium to reach its boiling point. This may be seen in Figure 10; the metal-fueled system rides out the LOFS accident with a lower temperature excursion than in the oxide system.

Metal fuel elements have operated flawlessly in the EBR-II, with burnups of 100,000 MW-days per ton—i.e. about 10% of the fissile atoms are fissioned. On the other hand, this remarkable performance at EBR-II cannot be extrapolated directly to a full-scale LMFR since the fast neutron flux in EBR-II, for a given burnup, is about one third that in a commercial reactor (a consequence of the three times higher enrichment in EBR-II as compared to the full-scale reactor). The cladding in the commercial reactor, being bombarded by a three times higher fast fluence, might be expected to deteriorate faster than that in EBR-II.

Argonne proposes to reprocess its LMFR fuel on site; the pyrometallurgical process had been demonstrated on fuel elements at EBR-II. Keeping the fuel on-site would, according to proponents of the scheme, all but eliminate the likelihood of clandestine diversion of plutonium, since the fuel is never decontaminated enough to be handled directly.

Loop or pool The current reexamination of LMFRs has reopened a sensitive subject: which configuration, pool or loop, is safer? At the time

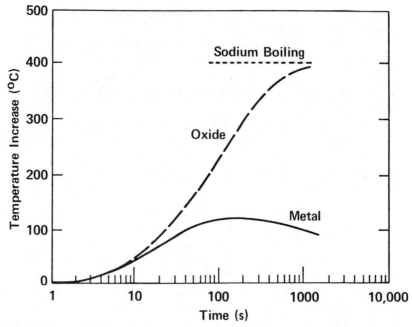

Figure 10   Increase in reactor outlet temperature during loss-of-flow transient without scram, liquid metal fast reactor. Source: (41).

when the Atomic Energy Commission (AEC) decided on the loop configuration for the fast flux test facility and for the Clinch River Breeder Reactor, the standard doctrine held that loop and pool were equally safe. This assertion says nothing about the relative inherent safety of the two configurations. The pool reactor, immersed in a large pot of sodium, would appear to be intrinsically the safer, because the pool of sodium provides a heat sink that is passive and reliable. The matter is hardly settled; during this period of reexamination the relative safety of these two systems will likely be debated in a more serious and revealing way than has previously been the case, at least in the United States.

THE MOLTEN SALT BREEDER (42)   Reactors based on molten salt fuel can hardly suffer a core melt since the fuel is already molten. In the 1970s, when molten salt reactors were still actively pursued, the late Professor T. Thompson (who became an AEC Commissioner) argued on general grounds that molten salt reactors were inherently safer than sodium-cooled solid fueled reactors. No definitive comparison of the two types was possible since the design of molten salt reactors (MSRs) had not been fully worked out. Experience drawn from the operation of the Molten Salt Reactor Experiment (MSRE), an 8-MW molten salt reactor that operated from 1965 to 1969, suggests that MSRs, though they may be less vulnerable than LMFRs to accidents that might break containment, are perhaps more vulnerable to incidents in which small amounts of radioactivity are released. The issue remains an academic one unless a molten salt reactor is built.

INTRINSICALLY SAFE FUSION REACTORS   The afterheat generated immediately after shutdown in a 1000-MW(e) fusion reactor is about 20 MW (if the first wall is made of steel cooled with Pb-Li), and about 12 MW (if the first wall is made of V-20 titanium alloy). Though these afterheats are much smaller than the 200 MW of after heat generated in a fission reactor of the same electrical output, they are sufficient to melt the first wall of the fusion reactor unless the wall is cooled vigorously. Thus, contrary to some popular belief, fusion reactors are, in principle, subject to "wall melt" accidents that could release tritium and other isotopes to the environment, as well as being awkward and expensive for the operator of the device. G. Logan (43) of Lawrence Livermore National Laboratory has therefore proposed that fusion reactors be designed with wall loadings of only 3–5 MW/m^2, rather than the 10–20 MW/m^2 assumed in many current design studies. At this lower wall loading, the first wall would cool itself by radiation, conduction, and natural convection. Logan points out that a 300-MW(e) fusion reactor with such low wall loading would therefore be inherently safe. Though the cost per kilowatt of the fusion reactor itself would thereby be increased, this increase would be compensated for by simplification of the balance-of-plant. Logan's ideas have just been presented to the fusion community; one cannot as yet estimate how much influence they will have on the design of fusion reactors.

CONCLUSIONS

The case for inherently safe reactors has been neatly summarized by K. Hannerz and P. Isberg (26):

> Nearly all the problems facing the nuclear industry can be traced back, directly or indirectly, to the nuclear safety issue. There is no doubt that the latter is causing the technology to be too complex and demanding to be attractive to the utilities, at least in the U.S.
>
> There is no agreement on how to remedy this situation. Should the technology be improved and made less demanding, or should the competence of the utilities be strengthened, licensing streamlined, etc.?
>
> The latter path seems to be the preference so far. However, the choice between "institutional" and "technological" approaches should not be based solely on a myopic

view of present problems but must also consider broader aspects of the expected world energy future. Institutional reforms, e.g. in the U.S., are obviously of little relevance for the situation in the third world when large scale application of nuclear energy there comes under way, as will inevitably happen.

This duality between technological fixes and institutional fixes is paramount. As J. Ahearne (44) has said, nuclear power is a complex, demanding technology. To deal with it safely requires sophisticated people and appropriate institutions. Obviously, the more inherent safety is built into a reactor, the fewer demands it makes on the people entrusted to run it. Inherently safe reactors might therefore find a niche where the institutions are not up to the demands conventional nuclear power has made of them— notably in the Third World, or even among less qualified utilities in the United States.

Though inherent safety has now become a watchword among US reactor designers, many in the industry seem to view inherently safe reactors as an unnecessary diversion, or even as a threat. For example, the Atomic Industrial Forum (45) recently stated:

> Increasing discussion in recent months, presumably as a result of the accident at Three Mile Island, has been directed at the rhetorical question of whether renewed utilization of the nuclear option should not be based on some system other than the light water reactor (LWR). The discussions, however, have failed to acknowledge the extensive research, development and demonstration effort that went into alternative systems in the late 50s and early 60s. They have failed to recall the deliberative reasoning that went into the selection of the LWR, not only in the U.S. but subsequently in Europe and the Far East. They have failed to recognize the improvements that have been incorporated into the LWR as a result of 25 years of design and operating experience, including the improvements made since the accident at Three Mile Island. And finally, they have failed to specify how they consider the LWR system to be flawed or why alternative systems could be expected to perform any better.
>
> None of this is by way of suggesting that research and development on the LWR as well as alternative reactor designs should not be vigorously pursued. Nuclear power is no different than the product of any other technology in that there should always be room for improvement. At this point in time, decisions as to which concepts are to be pursued should be made in the marketplace.

One must therefore ask, "Can passively safe and actively safe reactors coexist?" Or, more practically, "Would the demonstration of a reliable, passively safe reactor like PIUS or the modular HTGR bring on demands to shut down existing reactors because they are not as safe as PIUS or modular HTGR?" We cannot judge this question; but we can point out that DC-3s are not as safe as 767s, yet no one is demanding that the DC-3 be decertified. The public apparently does accept devices with which it has had long and acceptable experience. Should the next 15 years pass without a repetition of the TMI-2 accident, one might hope that the public would become less concerned about reactor safety. We can estimate the likelihood of a serious core melt over the next 15 years as follows: the 400-odd reactors now on line or under construction will amass 6000 reactor years by the year 2000. If the core melt probability for these reactors is 5×10^{-5}/RY, the probability of a core melt by then is 0.3; if improvements now being installed reduce the CMP to 10^{-5}/RY, this probability falls to 0.06. We would suggest there is a good chance that we shall reach 2000 without a repetition of TMI-2.

But in a way this misses the point. Though nuclear energy is at a low ebb in many countries, notably the United States, most analysts continue to see nuclear power as a long-term, even permanent energy source. We are just beginning the nuclear era. Many, many thousands of reactor years may well be amassed by future generations. Core melt probabilities that are adequate for nuclear power in the short run may not be adequate over the very long run. Thus there will develop, and indeed there has already developed, pressure to reduce CMPs—to 10^{-6}/RY, to 10^{-7}/RY, or even lower. And once the existence theorem has been proven, that passively safe reactors with CMPs in this range or lower are feasible, this knowledge can never disappear. It will continue to haunt the nuclear power enterprise until some inherently safe reactors are built. With the safety issue exorcised by inherently safe reactors, perhaps we can look forward to a second nuclear era no longer tormented by visions of Class IX accidents causing vast damage. Nuclear power could then be part of the solution to the problems of acid rain and the accumulation of carbon dioxide rather than a festering source of political conflict.

Literature Cited

1. Lilienthal, D. E. 1980. *Atomic Energy: A New Start.* New York: Harper & Row
2. Firebaugh, M. W., ed. 1980. *Acceptable Nuclear Futures: The Second Nuclear Era,* ORAU/IEA-80-11(P). Oak Ridge, Tenn: Oak Ridge Associated Universities, Inst. Energy Analysis
3. US Office Tech. Assessment. 1984. *Nuclear Power in an Age of Uncertainty,* OTA-E-216. Washington, DC
4. Lester, R. K., Driscoll, M. W., Lanning, D. D., Lidsky, L. M., Rasmussen, N. C., Todreas, N. E. 1983. *Nuclear Power Plant Innovation for the 1990s: A Preliminary Assessment.* MIT-NE-258. Dept. Nuclear Engineering, MIT
5. US Nuclear Regul. Commission. 1975. *An Assessment of Accident Risk in U.S. Commercial Nuclear Plants.* WASH-1400 (NUREG 75/014). Washington, DC: US Dept. Energy
6. Lewis, H. W., Budnitz, R. J., Kouts, H. J. C., Loewenstein, W. B., Rowe, W. D., et al. 1978. *Risk Assessment Review Group Report to the U.S. Nuclear Regu-latory Commission.* NUREG-CR-0400. Washington, DC
7. Okrent, D. 1981. *Nuclear Reactor Safety, On the History of the Regulatory Process,* Madison, Wisc: Univ. Wisconsin Press
8. US Nuclear Regul. Commission. 1982. *Reactor Safety Study Methodology Applications Program: Calvert Cliffs #2 PWR Power Plant, Vol. 3.* NUREG/CR-1659 (SAND80-1897, 3 of 4). Sandia National Lab.
9. Ergen, W. K., chairman. 1967. *Report of Advisory Task Force on Power Reactor Emergency Cooling.* TID-24226. Washington, DC: US Atomic Energy Commission
10. Fortescue, P. 1980. See Ref. 2, p. 16
11. Hannerz, K. 1983. *Towards Intrinsically Safe Light Water Reactors.* ORAU/IEA-83-2(M) Rev. Oak Ridge, Tenn: Oak Ridge Associated Universities, Inst. Energy Analysis. Also Hannerz, K. 1973. Applying PIUS to power generation: the SECURE-P LWR. *Nuclear Eng. Int.* 28(349):41

462   SPIEWAK & WEINBERG

12. O'Farelly, C. 1980. Towards a more forgiving reactor. Presented at *Int. Conf. on Current Nuclear Power Plant Safety Issues*. Stockholm, Sweden. IAEA-CN-39/58
13. Forgiving or Inherently Safe Reactors, I and II, *Trans. Am. Nuclear Society* 47:286–304
14. Minarick, J. W., Kukielka, C. A. 1982. *Precursors to Potential Severe Core Damage Accidents: 1969–1979, A Status Report*. ORNL/NSIC-182 and NUREG/CR-2797. Oak Ridge, Tenn: Oak Ridge Natl. Lab.
15. Cottrell, W. B., Minarick, J. W., Austin, P. N., Hagen, E. W., Harris, J. D. 1984. *Precursors to Potential Severe Core Damage Accidents: 1980–1981, A Status Report*. ORNL/NSIC-217/V1 and NUREG/CR-3591. Oak Ridge, Tenn: Oak Ridge Natl. Lab.
16. Phung, D. L. 1984. *Assessment of Light Water Reactor Safety Since the Three Mile Island Accident*. ORAU/IEA-84-3(M). Oak Ridge, Tenn: Oak Ridge Associated Universities, Inst. Energy Analysis
17. Weinberg, A. M., Spiewak, I., Barkenbus, J. N., Livingston, R. S., Phung, D. L. 1984. *The Second Nuclear Era*. ORAU/IEA-84-6(M). Oak Ridge, Tenn: Oak Ridge Associated Universities, Inst. Energy Analysis. Also Weinberg, A. M., Spiewak, I. 1984. Inherently safe reactors and a second nuclear era. *Science* 244:1398–1402
18. Paulson, C. K., Iacovino, J. Jr. 1984. *Increasing Safety Margins at Nuclear Power Plants*. Westinghouse Electric Co.
19. Weinberg, A. M. 1984. See Ref. 17, pp. 18–22
20. Spiewak, I. 1985. *Survey of Light Water Designs to be Offered in the United States*, p. 24. Oak Ridge, Tenn: Oak Ridge Natl. Lab. In press
21. Spiewak, I. 1985. See Ref. 20, p. 7
22. Hucik, S. A., General Electric Co. 1984. Personal communication
23. Duncan, J. D., Sawyer, C. D. 1984. Capitalizing on BWR simplicity at lower power ratings. ANS/ENS meeting. Nov.
24. Siddall, E. 1984. *The CANDU-PHW Reactor in Relation to a Second Nuclear Era*. ORAU/IEA-83-11(M). Oak Ridge, Tenn: Oak Ridge Associated Universities, Inst. Energy Analysis
25. Combustion Engineering, Inc. 1979. *Conceptual Design of a Large HWR for U.S. Siting*. CEND-379, Vols. 1–4. Washington, DC: US Dept. Energy
26. Electric Power Res. Inst. 1977. *Study of the Development Status and Operational Features of Heavy Water Reactors*. EPRI NP-365. Palo Alto, Calif.
27. Fischer, C., Fortescue, P., Goodjohn, A. J., Olsen, B. E., Silady, F. A. 1984. *The HTGR—An Assessment of Safety and Investment Risk*. ORAU/IEA-84-7(M). Oak Ridge, Tenn: Oak Ridge Associated Universities, Inst. Energy Analysis
28. Martel, L., Minnick, L., Levey, S. 1982. *Summary of Discussions with Utilities and Resulting Conclusions*. EPRI-RP 1585. Palo Alto, Calif: Electric Power Res. Inst.
29. Hannerz, K., Isberg, P. 1984. See Ref. 13, pp. 286–87
30. Lohnert, G. H., Pflasterer, G. R. 1984. See Ref. 13, pp. 287–89
31. Silady, F. A., McDonald, C. F., Weicht, U. H. 1984. See Ref. 13, pp. 290–91
32. Babala, D. 1984. See Ref. 13, pp. 296–97
33. Asahi, Y., Wakabayashi, H. 1984. See Ref. 13, pp. 297–98
34. Tenn. Valley Authority. 1984. *A Tabletop Flow Demonstrator for the PIUS Reactor Concept*. TVA/OP/EDT-84/17, WR28-1-900-118
35. Lohnert, G. H., Pflasterer, G. R. 1984. See Ref. 13, pp. 287–89
36. Silady, F. A., McDonald, C. F., Weicht, U. H. 1984. See Ref. 13, pp. 290–91
37. Schultz, M. A., Edlund, M. C. 1984. See Ref. 13, pp. 291–92
38. Schmidt, J. E., Doncals, R. A., Markley, R. A., Lake, J. A., Coffield, R. D., et al. 1984. See Ref. 13, pp. 298–99
39. GE, Rockwell win reactor race. October 15, 1984. *The Energy Daily* 12(199):1
40. McDonald, J., Brunings, Lancet, R. 1984. See Ref. 13, pp. 299–301
41. Chang, Y. I., Marchaterre, J. F., Sevy, R. H. 1984. See Ref. 13, pp. 293–94
42. US Dept. Energy. 1980. *Nuclear Proliferation and Civilian Nuclear Power: Report of the Nonproliferation Alternative Systems Assessment Program, Vol. IX: Reactor and Fuel Cycle Description*, DOE/NE-0001/9, pp. 227–40. Washington, DC
43. Logan, B. G., *A Rationale for Fusion Economics Based on Inherent Safety*. 1984. UCRL-91761. Livermore, Calif.: Lawrence Livermore Natl. Lab.
44. Ahearne, J. 1983. Prospects for the U.S. Nuclear Reactor Industry. In *Ann. Rev. Energy*, ed. J. Hollander, H. Brooks, 8:355–84. Palo Alto, Calif: Ann. Rev. Inc.
45. AIF Study Group, *Nuclear Power in America's Future*. 1984. Bethesda, Md: Atomic Industrial Forum, Inc.

Ann. Rev. Energy. 1985. 10 : 463–93
Copyright © 1985 by Annual Reviews Inc. All rights reserved

NATURAL GAS SUPPLY AND DEMAND PLANNING IN THE ASIA-PACIFIC REGION

Fereidun Fesharaki

Asian Energy Security Project, Resource Systems Institute, The East-West Center, Honolulu, Hawaii 96848

Wendy Schultz

Asian Energy Security Project, Resource Systems Institute, and Department of Political Science, University of Hawaii at Manoa, Honolulu, Hawaii 96822

Resilience has become a primary concern for energy planners : flexibility in the face of crises, coupled with confidence in the sturdiness of available options. Resource-weak governments promote diversification of the types of energy used as well as their sources of supply. Such policies broaden the base of national energy use, decreasing vulnerability to changes in the price or availability of any one energy source. While resource-rich nations have few supply vulnerabilities, they often rely on their exports for foreign exchange. In the rocky energy markets of the late 1980s, countries that are net energy exporters would do equally well to diversify their customer lists. This strategy offers some protection from the vagaries of shifts in demand, domestic energy policies of consumers, or international relations.

During the next decade, the natural gas industry in the Asia-Pacific region will double in terms of volumes traded and the numbers of actors involved. Two factors will contribute to this expansion : the policy emphasis on resilience and the favorable economics of shipping liquefied natural gas (LNG) within the Pacific region. Understanding this dynamic process begins with a clear picture of the current state of natural gas production, consumption, and trade in Asia.

0362–1626/85/1022–0463$02.00

REGIONAL OVERVIEW

As Table 1 shows, world reserves of natural gas at the beginning of 1984 totaled over 90 trillion cubic meters; of this, a little over 6 trillion cubic meters lie within the Asia-Pacific region. As with oil, a few countries hold a disproportionate percentage of world gas reserves; the Soviet Union, Iran, and the United States lay claim to almost 60% of the world's natural gas (1). But even reserves totaling 1% or less of the world total provide sufficient volume to support commercial production and even export, as the cases of Brunei (0.2%), Indonesia (1.1%), and Malaysia (1.5%) demonstrate. Despite the fact that the Asia-Pacific region has only 7% of world natural gas reserves, it is an active arena in the world natural gas industry. Indeed, it dominates global LNG trade.

Proven Reserves

As of January 1, 1984, Malaysia possessed by far the greatest gas reserves in the Asia-Pacific, with 1.4 trillion cubic meters in proven reserves. Indonesia, Australia, and China can each command only about two thirds that volume, with proven reserves of between 1 trillion and 800 billion cubic meters (bcm). In a yet lower category of resource availability, Pakistan,

Table 1 1982–1984 proven reserves of natural gas in the Asia-Pacific region (billion cubic meters)

	1/1/82	1/1/83	1/1/84	% of World (1984)
Afghanistan	55	60	60	Negl
Australia	880	908	945	1.0
Bangladesh	317	339	340	0.3
Brunei	212	204	210	0.2
Burma	124	145	210	0.2
China	736	765	800	0.9
India	352	465	475	0.5
Indonesia	971	960	1,000	1.1
Japan	27	20	25	Negl
Malaysia	1,274	1,385	1,400	1.5
New Zealand	159	154	152	0.2
Pakistan	462	481	510	0.6
Taiwan	24	22	23	Negl
Thailand	402	350	240	0.3
Total Asia-Pacific	5,995	6,258	6,390	7.1
Total world	83,621	87,450	90,325	100

Source: Cedigaz, 1984. *Le Gaz Naturel Dans le Monde en 1983: Statistiques*, statistical pamphlet. Paris: Cedigaz. 1 p.

India, and Bangladesh each has gas reserves assessed at about one third the volume of Malaysia's, with proven reserves of 510 bcm, 475 bcm, and 340 bcm, respectively. Each remaining Asian nation has less than 250 bcm of proven gas reserves.

From January 1982 to January 1984, Malaysia, Indonesia, Australia, China, Pakistan, India, Bangladesh, Burma, and Afghanistan all added steadily to their assessed proven reserves. Aggressive exploration programs characterize energy policy throughout the Pacific, and this aggressiveness has proven profitable. In the face of intensive development and consumption, Japan, Brunei, New Zealand, and Taiwan managed to maintain their comparatively small reserve volumes. Thailand has seen a significant drop in proven gas reserves. This drop may be attributed to the reappraisal of the giant Erawan gas field, resulting in a 60% devaluation of its estimated reserve volume (2). In total, regional natural gas reserves have steadily increased over the last three years. This trend is likely to continue for at least a decade, as explorations continue in promising basins in the Timor, Natuna, and South China Seas.

Commercial Production

Table 2 presents 1983 production statistics for the Asia-Pacific region. Ranking the countries by volume of gas produced commercially results in

Table 2   1983 natural gas production in the Asia-Pacific region (billion cubic meters)

	Gross production	Reinjection	Flaring	Other	Commercial production
Indonesia	33.59	6.78	5.46	0.52	20.83
China	21.50	—	1.70	—	19.80
Australia	11.99	—	—	1.01	10.98
Pakistan	9.72	—	—	—	9.72
Brunei	9.77	—	0.39	—	9.38
Malaysia	5.55	—	1.85	—	3.70
New Zealand	3.18	—	—	—	3.18
India	3.84	0.15	0.58	0.12	2.99
Afghanistan	2.85	—	—	—	2.85
Bangladesh	2.20	—	—	—	2.20
Japan	2.12	—	—	—	2.12
Thailand	1.61	—	—	—	1.61
Taiwan	1.45	—	—	—	1.45
Burma	0.53	—	—	—	0.53
Total Asia-Pacific	109.90	6.93	9.98	1.65	91.34
Total world	1869.85	151.96	106.37	63.78	1547.74

Source: Cedigaz. 1984. See Table 1.

four distinct groups. Indonesia and China commercially produced approximately 20 bcm each, almost twice as much as their nearest competitors. Australia, Pakistan, and Brunei, with commercial gas production of about 10 bcm each, fall into the next group. Below these five countries, volumes produced drop off rapidly, more than halving again; Malaysia, New Zealand, India, and Afghanistan each produce about 3 bcm. The final group consists of those five countries in the region that produce approximately 2 bcm or less. While resource scarcity may constrain production in Japan and Taiwan, low production in Bangladesh, Thailand, and Burma results primarily from a late start in resource development.

It is also easy to deduce from the data in Table 2 which countries are producing their natural gas in association with crude; Indonesia, China, Brunei, Malaysia, and India all flared some natural gas. By far the worst offender, Indonesia burned the equivalent of one quarter of its commercial gas production. Reinjection and flaring together account for almost one third of Indonesia's 1983 gross production volume.

Domestic Consumption

In the following pages we will distinguish between those countries that simply produce all the gas they need for domestic usage, and so are self-sufficient, and those that require either additional volumes for domestic demand, or additional markets for their commercial production. As Table 3

Table 3   1983 commercial production and consumption of natural gas in the Asia-Pacific region (billion cubic meters)

	Commercial production	Exports	Imports	Consumption
Japan	2.12	—	25.46	27.58
China	19.80	—	—	19.80
Australia	10.98	—	—	10.98
Pakistan	9.72	—	—	9.72
Indonesia	20.83	12.97	—	7.86
New Zealand	3.18	—	—	3.18
India	2.99	—	—	2.99
Brunei	9.38	7.16	—	2.22
Bangladesh	2.20	—	—	2.20
Malaysia	3.70	1.55	—	2.15
Thailand	1.61	—	—	1.61
Taiwan	1.45	—	—	1.45
Afghanistan	2.85	2.28	—	0.57
Burma	0.53	—	—	0.53
Total Asia-Pacific	91.34	23.96	25.46	92.84
Total world	1547.74	193.45	193.45	1547.74

Source: Cedigaz. 1984. See Table 1.

demonstrates, all but five of the Asian-Pacific nations are self-sufficient; imbalances exist between production and consumption in Japan, Indonesia, Brunei, Malaysia, and Afghanistan. Of these five, only Japan uses more gas than is produced domestically, importing over 25 bcm in 1983. Indonesia, Brunei, and Malaysia all export LNG; Afghanistan exports gas via pipeline.

Comparing Tables 2 and 3, Indonesia drops from first place in commercial production to fifth in consumption; Japan moves from eleventh in production to first place as the region's largest consumer of natural gas. On the other hand, China, Australia, and Pakistan, self-sufficient countries ranked second, third and fourth in commercial production, carry that rank order over to the consumption listing. With total 1983 gas consumption of more than 27 bcm, Japan uses about one third again as much gas as China. In turn, China's consumption of almost 20 bcm is twice as much as that of either Australia or Pakistan, both of which consume about 10 bcm. Of the rest of the region, only Indonesia consumes a sizable volume of gas, with 1983 demand of approximately 8 bcm; the other countries currently use less than 3 bcm of natural gas.

Natural Gas Trade

As of fall 1984, six natural gas trade partnerships existed in Asia. Five involved LNG, with Japan buying from the United States, Brunei, Indonesia, Abu Dhabi, and Malaysia. The sixth involved gas exports by pipeline from Afghanistan to the Soviet Union. Table 4 compares the import and export volumes of these trade agreements with world totals. While these transactions accounted for only 14% of total world natural gas trade, Japan's natural gas imports made up 60% of the global LNG trade. The predominance of LNG in area exports distinguishes Asian gas trade from that elsewhere. In 1983, LNG accounted for only 22% of natural gas traded worldwide; in the Asia-Pacific region, over 91% of the gas traded moved as LNG.

Gas will never be as cheap to move and store as crude oil. Compared with crude transport on a thermal equivalent basis, piping gas costs twice as much, and LNG shipments approximately five to seven times as much. Obviously, the archipelagic geography of Asia limits the possibilities of natural gas trade by pipeline among these countries. However, relatively short shipping distances in the region lower transportation costs, allowing maximum returns for an LNG export chain (an LNG chain includes the liquefaction plant, tankers, and receiving terminal with regasification facilities). Shipping distances in the three LNG chains currently on-stream in Asia range from 2400 to 3200 nautical miles (see Appendix Table A). The shipping distance from Alaska is approximately 3200 nautical miles. While LNG netbacks are more sensitive to capital costs varying on plant capacity, analysis has shown that shipping distance does affect netbacks. At any given plant capacity, netbacks will be $0.80 per MBtu (million Btu) higher shipping the LNG 3000 miles than shipping it 7000 miles (3) (an approximate comparison between the Indonesian Arun chain to Japan, and the Abu Dhabi chain to Japan).

Until 1980, that disadvantage in costs to Abu Dhabi was not reflected in price differentials. Table 5 shows that, through 1980, prices for LNG from Alaska, Brunei, and Abu Dhabi differed by only $0.15/MBtu. Indonesia negotiated much more successfully for its LNG. Beginning service the same year as Abu Dhabi, it received $0.50/MBtu more for its LNG than Abu Dhabi in 1977; $0.60 more in 1978; almost $1.15 more in 1979; and about $1.65 more in 1980.

In 1980, two events changed pricing for natural gas exports worldwide. Algeria pushed hard for gas producers to write escalation clauses into their contracts based on thermal parity with crude oil, and the price of oil doubled. By the end of 1980, LNG prices had jumped, and imports from Abu Dhabi cost Japan more than gas from any other source. That remained unchanged until 1983, when Malaysia's new LNG project came on stream; the single most expensive investment in the country's history, it is supported by a very favorable price structure for its gas sales. The March 1983 drop in

Table 4   1982–1983 natural gas trade in the Asia-Pacific region (billion cubic meters)

	1982		1983	
	Pipeline	LNG	Pipeline	LNG
Importers				
Japan	0.00	23.60	0.00	25.46
USSR	2.30	0.00	2.28	0.00
Total world imports	150.55	34.34	150.77	42.68
	184.89		193.45	
Exporters				
Abu Dhabi	0.00	2.98	0.00	2.41
Afghanistan	2.30	0.00	2.28	0.00
Brunei	0.00	6.92	0.00	7.16
Indonesia	0.00	12.41	0.00	12.97
Malaysia	0.00	0.00	0.00	1.55
United States	0.18	1.29	0.19	1.37
Total world exports	150.55	34.34	150.77	42.68
	184.89		193.45	

Source: Cedigaz. 1984. See Table 1.

the price of marker crude hit all gas exporters hard, and they are waiting just as anxiously for the oil market to tighten as the big oil exporters.

Their anxiety can be explained by the high levels of capitalization involved in the gas trade. Building an LNG export chain requires at least two decades of reliable receipts to ensure profitability. An LNG chain capable of exporting approximately 4 bcm/yr of gas costs $2.6–3.8 billion (see Appendix Table B). Many investors balk at such high capital outlays, as well as the political and economic difficulties attendant upon negotiating the 20-year trade agreements necessary to pay off projects that large. Asian natural gas exporters, however, have little choice. While Asia is geographically unsuited for trading natural gas by pipeline, the conjunction of large gas reservoirs, located in Brunei and Indonesia, with Japan's rapidly increasing electrical generation requirements, encouraged the development of gas trade by sea. The production volume of those projects has made Asia the center of world LNG trade, and their profitability has sparked a continuing interest in expanding that trade.

With the newly industrialized countries in Asia expressing interest in gas imports, Asian exporters now find themselves in an excellent position for expansion. As Indonesia is proving, additions to infrastructure in place cost much less than building an entirely new export project. Additional refrigeration facilities at Badak and Arun have cost between $300 and $400 million for units with capacities of 2.7 bcm to 4.5 bcm per year. New export

Table 5   LNG prices into Japan, 1975–1984 (US $/million BTU, CIF)[a]

Year	USA-Alaska	Brunei	Abu Dhabi	Indonesia	Malaysia	Price of marker crude (US$/B)[d]
1975	1.40	1.67	c	c	c	12.376
1976	1.70	1.76	c	c	c	12.376
1977	2.01	1.91	1.93	2.44	c	13.00
1978	2.26	2.04	2.20	2.80	c	13.66
1979	2.37	2.20	2.31	3.45	c	14.34
1980[b]	3.51–4.81	2.97–4.79	3.08–5.31	4.72–5.16	c	27.37
1981	5.97	5.84	6.53	4.96	c	33.82
1982[b]	5.75	5.77	6.42	5.94	c	32.00
1983	4.97	5.04	5.18	5.00	5.78	29.00

[a] Multiply prices listed by 35.3 × 10⁶ to approximate the price in $/bcm.
[b] Fluctuations in oil prices and alterations in contracts caused a range of prices within the year.
[c] Service unavailable during this period.
[d] Provided for comparison; division by 6 will convert to approximate price per million Btu.
Source: Macris, G., ed. 1975–1984. *Pet. Int. W.* See Ref. 2.

projects can cost ten times as much, with estimates of $3 billion for Thailand's proposed 4 bcm/yr plant and $7.56 billion for Australia's 8 bcm/yr facility (see Appendix Table B). With a small additional capital investment, Indonesia has doubled its LNG output. The combined Badak-Arun complexes produce 19 bcm/yr of LNG, putting Indonesia in the best possible position for economies of scale in plant size.

ASIA-PACIFIC SUPPLY AND DEMAND PLANNING

Self-sufficient States

The self-sufficient states discussed here will use their gas resources internally at least for the next decade. Having decided on domestic consumption, most had then to wrestle with strategic economic questions concerning the uses to which their gas resources should be put. Gas can be used directly as a fuel for residential, commercial, and industrial heat, it can power electrical generation plants, and it can be used as feedstock for producing petrochemicals or synthetic gasoline and diesel fuel. The next few paragraphs will describe recent policies for domestic natural gas use in the Asia-Pacific region, beginning with New Zealand's decision to adopt the economically unproven Mobil-M process for making synthetic gasoline from natural gas.

In early 1969 exploration determined that the Maui gas field off the west coast of New Zealand held reserves in excess of 140 million cubic meters. While a joint venture combining Shell, BP, and Todd Oil Services discovered the Maui field, in April 1973 the New Zealand government bought 50% interest in field development, at a cost of NZ$30 million. As the consortium had estimated that the New Zealand government would be the only consumer with demand sufficient to warrant exploitation of Maui, the agreement included a 30 year take-or-pay clause guaranteeing government gas purchases at an agreed-upon price.

The smaller, onshore Kapuni gas field adequately served residential needs, the New Zealand petrochemical market offered insufficient demand for a viable petrochemical plant, and there was little business interest in a third option, gas-based smelting. That apparently narrowed the field to two options: electricity generation and exporting the gas as LNG. In the wake of two oil shocks, the government was reluctant to waste New Zealand's gas resource in generating electricity. In addition to being an inefficient end use for the gas, demand for electrical power itself had dropped because of energy conservation measures and economic recession. As the government reviewed plans for an LNG export project, energy advisors also broached the possibility of a synthetic gasoline plant.

New Zealand produces only about one tenth of the oil it uses, and transportation accounts for over four fifths of its liquid fuel use. Obviously, greater self-sufficiency in motor fuel supplies would ameliorate the nation's oil bill. That argued well for the synthetic gas plant; the vision of long-term contractual ties with only one buyer, coupled with the perceived dangers of handling LNG, argued against a liquefaction plant.

In the end, the New Zealand government planned a package of end uses for Maui gas; the synthetic gasoline plant, a liquefield propane gas (LPG) facility; a compressed natural gas (CNG) plant, a methanol plant for export production; and a pipeline system for eventual distribution to residential and industrial users. The loan for the synthetic gasoline project totaled $1.7 billion; although it required a massive capital investment, the project has the advantage of a gas feedstock negotiated in 1973 prices. The price negotiations attending the take-or-pay clause the government signed in early 1973 still obtain. If the feedstock costs were based on current gas prices, the synthetic gasoline plant would be risky business indeed (4).

The gasoline plant is currently projected to be on-stream by 1985. Using a feedstock of 1.55 bcm/yr of natural gas, it will produce about 570,000 tonnes of gasoline annually. This will meet approximately one third of New Zealand's present gasoline demand. A reasonable plan on the drawing board, the project must still prove itself economically. New Zealand has chosen an adventurous strategy for using its indigenous gas reserves (5).

Other self-sufficient states in the region have chosen more prosaic paths to using their indigenous gas resources; following the rank order of their 1983 gas consumption volumes, the People's Republic of China merits consideration next. Like many other countries, China pursued aggressive development of its oil resources first, with natural gas exploration and development a by-blow of the hunt for crude oil. Natural gas production now supplements oil production to a modest degree.

Widest use occurs in Sichuan province and areas nearby. Regional reserves are said to be substantial, although the gas has a fairly high sulfur content and requires treatment. This necessity renders exploitation more difficult and costly. While residential use is widespread, the government gives priority to its use in the steel and petrochemical industries.

The north and northeast areas of China also produce small quantities of gas, and the country hopes to find additional reserves offshore in the South China Sea (6). In fall 1984, Arco China tested a gas field off Hainan Island, and estimated potential production capability at several billion cubic meters per year. Although no final decision has yet been made on utilization, the government has discussed the possibility of establishing a fertilizer-petrochemicals complex on Hainan (7).

Pakistan has chosen to use production from its Mari gas field to expand electrical power generation. The $362.2 million Guddu Combined Cycle Project will use 300 MW of gas turbines and 150 MW of steam turbines, fueled by Mari gas. The Asian Development Bank has loaned the project over $140 million (8); the International Finance Corporation loaned Pakistan about $40 million for actual field development (9).

In a concerted effort to cut flaring, India established the Gas Authority of India, to oversee processing, transportation, and marketing of the associated gas flared on the Bombay High oil fields. Bombay High produces between 2.9 and 3.3 bcm/yr of associated gas, much of which is flared due to lack of equipment to convert it. Limited gas compression facilities will be in use by 1986 (10); when Bombay High's Hazira-Jagdishpur pipeline is completed in 1989, the gas reserves in the Bombay area will be used in six gas-based fertilizer plants planned for the north. The Ministry of Energy has authorized the construction of three electric power projects along the pipeline, with a total generating capacity between 1000 and 1500 MW (11).

In June, 1985, India's first major natural gas development will come on stream; over 1.8 bcm/yr of gas from the South Bassein field in Bombay High will come ashore at Umrat and go to the Hazira fertilizer complex nearby. Liquids extracted from the Bombay High field, 75 miles to the west, will be added to the pipelines. With estimated project costs of $701.5 million, India's Oil and Natural Gas Commission has assembled a loan package including $222.3 million from the World Bank, $50 million from Kuwait, and almost $250 million in commercial loans and credits. The Commission is assuming the remaining $179 million (12).

Following an aggressive policy of substituting natural gas for imported oil, Bangladesh has seen a 20% per year increase in use of natural gas in its commercial sector since 1979. Under its revised second five-year plan, the government plans to increase the share of natural gas in commercial energy supply from the 1979–1980 level of 37% to over 62% in 1984–1985. The main obstacle to increased use of these reserves has always been their location; the onshore fields lie far to the north and east of population centers (13).

The July 1984 opening of the Bakhrabad-Chittagong pipeline was a modest attack on that problem. The 110-mile pipeline, designed to carry about 3.6 bcm/yr, terminates at the port city of Chittagong. The gas supply will fuel steel mills, generating plants, and other industries (14). Bangladesh received about $185.6 million in aid for the project, with the World Bank and OPEC the largest contributors (15).

In the last two years, possibilities for developing natural gas in Burma have blossomed. Myanma Oil, the national oil corporation, began an oil and gas exploration program in November 1982 with Japanese financial and technical assistance, including a $15 million loan. In February 1983, the group discovered a potentially commercial gas field, offshore in the Gulf of Martaban (16). Burmese authorities will apply to the World Bank for additional funds to finance further exploration to establish the extent of the Martaban reserve.

Myanma estimates that developing the reserves would cost almost $1 billion; to finance the project, they have assembled different potential partners for each phase. The Norwegian government has been contacted with a suggestion that it finance contracts for offshore production platforms, to be awarded to Norwegian suppliers. Loans from the Asian Development Bank might support pipeline construction; West Germany and Austria are discussing joint financing of an onshore petrochemicals plant, with the understanding of a 50/50 split on the contract. The latter involves a $266 million complex; Burma has no set plans, but envisions possible production and export of methanol, urea (as ammonia), and liquid fuels.

In addition, July 1984 saw a major onshore gas strike near Rangoon. Initially, the government considered using these reserves, in addition to the Martaban gas supply, as feedstock for the future petrochemicals complex. It is now more likely that the onshore gas production will go to fuel power generation for the Rangoon area (17).

Trading States

This section describes the gas export projects currently operating in the Asia-Pacific region, as well as prospective export projects. It also includes overviews of current and prospective LNG chains that originate outside Asia but add to the total volume of gas consumed in the region. Current projects will be described in order of initial start-up, to convey the development of area trade. Prospective projects will be discussed in order of probability of realization. Appendix Tables A and B summarize project specifics for the LNG chains.

NET EXPORTERS   *Current Gas Export Projects*   Japan began the Asian gas trade, with imports from the United States. The Alaska-Japan LNG chain is not only the oldest Asia-Pacific LNG trade agreement, it is also the smallest. The agreement that Phillips Petroleum and Marathon Oil signed in 1966 with Tokyo Electric and Tokyo Gas specified delivery of 1.3 bcm/yr of LNG for 15 years. As the first trade contract negotiated with Japan, it necessarily became the first trade contract Japan renewed. Due to expire in 1984, the project agreement was extended five years, through May 31, 1989; the Federal Energy Regulatory Commission granted final approval of the extension in January 1983. As of April 1980, the cost, insurance, and freight (CIF) cost of Alaskan LNG to Japan matches the existing mean cost of Japan's imported crude, on a Btu basis (18).

In exporting gas via pipeline, Afghanistan is a minor anomaly in Asian gas trade. Its Khwaja Gugerdak gas field has been on-stream since 1968; trade with the Soviet Union began in the early 1970s. Although the Soviet Union produces enough gas to meet domestic demand and export as well, the Afghani gas serves to balance regional deficits in the domestic gas grid. Exports in 1981 and 1982 were 2.3 bcm; this was expected to increase as the new Jarqodoq field came into production. Repeated rebel sabotage of the pipelines, however, prevented export of those additional volumes. Kabul announced exports of 2.4 bcm for 1983, with similar volumes planned for export in 1984 (19).

Brunei started the LNG industry in the Asia-Pacific region. The Brunei government formed a joint venture corporation with Shell and Mitsubishi, and in June 1970 this corporation, Brunei Coldgas, signed an export contract with Tokyo Electric, Tokyo Gas, and Osaka Gas. Including four years initial volume buildup, Brunei agreed to ship 5.8 bcm/yr to Japan. Before deliveries began, however, the contractual volumes were raised to almost 7 bcm/yr. This required adding extra refrigeration capacity and another tanker for the project; the first delivery actually arrived in December 1972. In the wave of 1980 price renegotiations, Brunei won a contractual amendment similar to Alaska's, allowing the CIF base price of Brunei LNG to fluctuate in thermal parity to the mean cost of Japan's imported crude (20).

Only one Persian Gulf country currently exports gas to the Asia-Pacific region. Abu Dhabi initiated the venture in 1972 with British Petroleum (BP), Compagnie Francaise de Petroles (CFP), and Mitsui as its partners. While the first delivery arrived in Sodegaura in 1977, the plant did not reach peak production until 1981 (21). Despite this slow start, the complex succeeded in exceeding design capacity in 1982; in fact, project backers have solicited $500 million in loans to expand storage.

Abu Dhabi's LNG project has suffered, however, from the last two years' uneasiness in the oil market. Profits for 1981 and 1982 were approximately $280 million (22); as of December 1982, the CIF price of Abu Dhabi's LNG to Tokyo Electric was $6.02/MBtu (23). But Abu Dhabi's LNG prices are indexed, on a percentage basis, to changes in the prices of their Murban crude oil. In March 1983, the price of Murban dropped from $34.56 to $29.56; this 14.5% decrease translates into a CIF price of $5.13/MBtu (24). In addition, the production quotas OPEC instituted reduced the associated gas supplies to the plant; to compensate, Abu Dhabi developed the nonassociated gas reservoirs in the Khuff formation, which produces approximately 2.6 bcm/yr (25).

Indonesia is the heavyweight of the Asian LNG export industry. Production began at its Badak plant in 1977, and the Arun trains came on stream in 1978; combined capacity of the original plants exceeds 10 bcm/yr. It has supplied five Japanese customers for more than six years and, until 1983, exceeded plant capacity regularly at both locations, by 30% at Badak and 10% at Arun (26). Typical yearly profits from these exports are over $2.6 billion (1982 figures). In April 1983, an explosion at Badak marred this exceptional production run. Attributed to human error, the mishap put one of the two refrigeration units out of commission (27). Although spare capacity was sufficient to meet Indonesia's export commitments, loss of one fifth of its LNG capacity cost the country approximately $1 million per day (28).

Growing demand in Japan prompted construction of two extra refrigeration trains at both Badak and Arun. The additional 4.3 bcm/yr capacity at Badak was completed in October of 1983; Arun's 4.5 bcm/yr addition was not finished until early 1984. Exports from this new capacity differ contractually from those of the original trade agreements: instead of CIF pricing, shipments from the plant additions will be priced free on board (FOB); the Japanese customers plan to purchase their own tankers, and provide transportation themselves (29).

Indonesia's LNG export capacity now totals more than 19 bcm/yr, if continued capacity overruns are included. During the first half of 1984, the value of Indonesian LNG exports equaled $838 million, which represents a 16% increase over export value for the first half of 1983 (30). In Indonesia's case, a policy of exporting gas has obviously paid off; while the initial capital investments were large, Pertamina arranged for loans from the Japanese to finance all nine of its production units (31). Considering that the Arun facility alone has earned $7 billion in its six years on-stream, loan defaults will not be a problem (32).

February 1983 marked the beginning of shipments from Malaysia's new Bintulu LNG plant to Japan. The $1.6 billion complex will produce over 8 bcm/yr from three liquefaction trains once it reaches peak production in mid-1986. In its first year, however, deliveries to Tokyo Electric and Tokyo Gas will equal only 2.5 bcm. Although construction of the plant ran to schedule, the project started slowly because of both controversy over production-sharing terms and Malaysia's inexperience in coordinating large-scale projects (33). While the Bintulu plant is the most expensive single project the nation has initiated, the government estimates income of about $30 billion from LNG exports over the next 20 years (34).

The other features of Malaysia's natural gas development policy also merit brief attention. Like New Zealand, Malaysia did not merely invest in one gas superproject; the country plans to expand gas use massively in the next 15 years. Forecasts suggest gas use will increase from its current 1% of Malaysia's total primary energy supply to 39% by 1990, replacing oil as the nation's primary energy source by 2000. At the turn of the century, when oil dependence has fallen to 40%, gas will have become the major fuel for thermal power plants and industry. In Trengganu, gas will fuel the Hicom steel plant; in Sarawak, it will feed an ammonia and urea fertilizer plant; in Sabah, it will run sponge-iron and methanol plants (35).

Prospective Gas Export Projects   The following paragraphs describe expansions planned in two existing export projects, and plans in five different countries for new LNG chains. While these developments have fairly good chances of implementation, some are experiencing more difficulties than others. The following discussion therefore moves from high-assurance plans to low-assurance plans.

Indonesia is considering two expansions. The first will respond to South Korean demand for LNG use in electrical power generation; the second, more recent, plan will send LNG to Taiwan for commercial and industrial use. The Korean Electric Power Company (KEPCO) signed a contract in August 1983 with Indonesia for 2.7 bcm/yr of gas (36); to supply this volume, Pertamina will build a sixth refrigeration train for its Arun complex. The gas feedstock will come from Mobil Oil Indonesia's North Sumatra fields, and Mobil has agreed to provide the $300 million construction costs for the additional LNG unit, which will be constructed by JGC and C. Itoh of Japan. As Indonesia funded all previous construction on its LNG complexes with loans from Japan, this arrangement is unique. The $300 million will not appear as a debt on Pertamina's balance sheets, and Mobil will be repaid from the export proceeds (37). In the case of exports to Taiwan, negotiations have just begun. A contract will necessitate construction of yet another refrigeration unit, but Pertamina has not yet decided at which complex. Nonetheless, plans call for exports to begin in 1991 (38).

Malaysia will be the second Asian gas producer to pipe gas exports; plans are progressing for a pipeline link with Singapore. Petronas officials have met with representatives of the Singapore National Oil Corporation and the Public Utilities Board to draft the sales agreement; it specifies sales beginning in 1988, with an initial supply of almost 1.6 bcm/yr of gas. Anchoring the 750-km pipeline at Teluk Kalong, in Trengganu on Malaysia's east coast, Petronas envisions running it west to Port Kelang and then south to Johore Baru, where it will cross the Johore Straits to Singapore. Nederlandse Gasunie and two Malaysian firms will finish technical feasibility studies by January 1985 (39).

Expanding Malaysian LNG exports will take a little longer. Although negotiations to supply LNG to South Korea began in 1982, the initial delivery date has been pushed back from 1985 to 1990. The proposed export volume of 2 bcm/yr will not require expansions in the Bintulu LNG plant; the Japanese export contract does not utilize the plant's full capacity. Apparently the delay is partially due to Korea's plans to build an additional receiving terminal and other land facilities (40).

The history of Australia's North West Shelf gas complex is a litany of delays. Demand for LNG in Japan has been slowing; the eight Japanese utilities slated to purchase the development's LNG have asked project managers to postpone start-up from 1986 to 1988 (41). In addition, sagging domestic requirements for industrial fuel in Western Australia resulted in a similar request with respect to the domestic half of the project: the Western Australian Energy Commission deferred taking its contracted supply until early 1985 (42).

The project will develop three offshore gas fields in the Rankin Trend, which lies approximately 135 km off the port of Dampier, in the northwest section of Western Australia. Processing facilities for the domestic phase and the liquefaction units for gas export will be built in the suburbs of Dampier (43). Although total plant capacity is over 12 bcm/yr, so far only 8 bcm/yr are reserved for sale to Japanese customers (44). The estimated worth of the contracts is about $A50 billion over the first 20 production years (45).

The question of equity shares has also caused problems. Woodside Petroleum, which originally held a 50% investment in the project, in 1983 and 1984 worked to restructure finances for the export phase to reduce its huge investment commitment.The project was divided into six equal stakes, including Woodside Petroleum Ltd., Shell Development Australia Pty. Ltd., BHP Petroleum Pty. Ltd., BP Australia Development Ltd., California Asiatic Oil Co., and a partnership between Mitsui and Mitsubishi, who will jointly own a one-sixth interest in the project. Woodside was reimbursed by BHP and Shell for its preinvestment outlays, and future investment will be carried equally by the six participants (46).

Bringing in the two Japanese trading houses raises the level of foreign investment in the project above the government guideline of 50%; special approval was requested from Canberra. Mitsui and Mitsubishi plan to establish a joint investment company in Australia, with each firm supplying 100 million yen. In addition, they are requesting further funding from a group of Japanese banks, including the Export-Import Bank of Japan. With estimated project costs of $7.65 billion, the two companies will invest approximately $1.3 billion on a 50/50 basis over the life of the project (47).

The Soviet Union and Japan have finally breathed life into their Sakhalin gas development project. In fall 1984, Japan agreed to the $3.8 billion basic technical plan the Soviets designed for the joint development of their Chaivo oil and gas deposit (48). Political tensions have repeatedly troubled the arrangement, made almost a decade ago, for LNG exports from the Soviet Union to Japan. Progress was especially slow during the US equipment embargo, mainly because of the Soviet lack of expertise in developing large offshore projects. In consequence, the Soviet Union has opened international bids for construction of the plant and facilities.

With Japan the sole customer, the complex will produce and export over 4 bcm/yr of LNG, and 20,000 barrels a day of crude oil. Plans specify four production platforms at Chaivo field northeast of Sakhalin Island, a 230-km pipeline connecting Chaivo to the eastern Siberian port of Dekastri, and two 2-bcm LNG trains with attendant storage and loading facilities (49).

For several years, Qatar has contemplated large-scale development of its massive North Dome gas reservoir. A weak international gas market contributed to delays in project development and uncertainties about optimal export volumes. When potential purchasers in Japan, South Korea, and Taiwan, as well as Europe, expressed interest, Qatar approached the project with renewed vigor. In February 1983, the Qatar General Petroleum Corporation (QGPC) announced selection of British Petroleum and Compagnie Francaise de Petroles as partners in the development. QGPC holds only 70% of the LNG venture; BP and CFP each have 7.5% equity. The final 15% was set aside for potential new participants (50).

In June 1984 QGPC signed the joint venture agreement with BP and CFP that launched the project. Planners envisage production of 20.7 bcm/yr of gas, of which 8.3 bcm/yr will be reserved for domestic consumption, with 12.4 bcm/yr processed for export as LNG. Phase I, requiring a $500 million investment, will produce gas for domestic industrial and power generation use; work began on the first phase in late fall of 1984. Phase II, construction of the LNG and condensates plants, will begin in 1986 and cost at least $6 billion. Before initiating the second phase, the three partners want firm commitments in hand from potential customers (51).

To this point, QGPC has discussed the LNG plant with a Japanese consortium comprising Mitsui, Mitsubishi, and C. Itoh. The group expressed interest in obtaining the project's remaining 15% equity, for equal division among themselves. They would then be responsible for marketing the LNG. In separate meetings, Marubeni representatives also indicated interest in the project (52). Later, Qatar in essence raised the ante; for the second round of talks with the consortium, it offered the final 15% interest if the three companies could procure firm purchase commitments for 4 bcm/yr of Qatari LNG from Japan's nine electric power companies. For Marubeni, it lowered the requirement to only 2.7 bcm/yr of purchase commitments (53).

Thailand's natural gas development plans went slightly awry when in the winter of 1983 it, in effect, lost 27 bcm of gas from its reserves, after initial estimates of the Erawan gas field were revised downward from 44.7 bcm to 17.8 bcm by DeGolyer and MacNaughton (54). The firm reaudited the field to settle a production dispute between Union Oil and the Thai government. Union had fallen behind in its contracted gas deliveries, but claimed the contract itself was based on an erroneous initial estimate of Erawan reserves.

This brouhaha had put talk of an LNG export project on the shelf for a year and a half; the idea was dusted off and reconsidered by four Japanese firms. In early spring, 1984, the Thai LNG Company, Mitsubishi, and Mitsui agreed on a joint venture for Thai LNG production. With Thai LNG retaining a 60% share, the Japanese partnership will hold the remaining 40%. The Japanese will invest initial capital of 100 million yen, with Mitsubishi and Mitsui responsible for 35% each, and Marubeni and Sumitomo for 15% each. The project will cost about $3 billion; half of that will finance construction of the liquefaction plant proper. More than 4 bcm/yr of LNG will be shipped to Japan starting in 1990 (55).

In the spring of 1982, Dome Petroleum's proposed Western LNG project, for export of Canadian gas to Japan, seemed a sure bet. Nissho-Iwai was the leading Japanese partner, backed by the interest of five potential customers. Exports were due to begin in 1986. That, however, was before Dome Petroleum accumulated a 6-billion-Canadian-dollar corporate debt. Since 1983, the other parties interested in the project have been scrambling to keep it alive.

At the start, Dome held an 80% interest in Western LNG, while Union Oil of Canada and Nissho-Iwai split the remaining 20% (56). The Japanese had initially invested 400 million Canadian dollars in the project in 1981 (57). When Dome began its financial retrenchment in early 1984, it took two immediate steps with reference to its LNG scheme: it handed over project management to Union Oil and Nissho-Iwai, and it began looking for investors to absorb its share of project finances. Meanwhile, the Japanese utilities that were potential customers demanded that Dome settle all questions concerning project financial structure and Canadian resource extraction approvals. The utilities set a deadline of October 30, 1984 (58).

In early fall, the project acquired a new name: Canada LNG Corporation. In mid-October, Dome dropped its interest in the project to 10%; one week later, three new partners had apparently joined Nissho-Iwai and Union Oil in the consortium: Trans-Canada Pipelines, Suncor, and Pan-Albert Gas. The investors were encouraging the crown corporation Petro-Canada to buy into the development (59), and Westcoast Transmission Co. was expressing interest (60). The province of British Columbia issued a certificate covering 50% of the project's gas supply requirements, but the Alberta provincial government was reluctant (61). It felt that the US market would take Alberta gas at a much higher price than that offered by the Japanese customers. The province of British Columbia has more actively supported project development, however, and the possibility exists that it would cover 100% of the plant's gas supply requirements (62).

With these signs of activity, the Japanese extended their deadline until year-end 1984. The consortium partners held several meetings in Tokyo with the potential customers, at which representatives of Petro-Canada were present (63). As of mid-November, 1984, the crown corporation had still not decided to join the consortium, although further meetings were scheduled in Canada. The clock has not yet run out on Canada LNG, but the Japanese customers have grown doubtful. Mr. Kinishiro Tsukuda, vice-president of Chubu Electric, was quoted as saying, "The situation is severe. I am not optimistic." (64)

Finally, a new pipeline project merits brief attention. In early fall 1984, the Hong Kong and China Gas Company announced plans for a gas pipeline to import gas from the People's Republic of China. It did not specify a completion date or transmission capacity. Hong Kong uses manufactured gas, and apparently rapid increases in demand have prompted the move: town gas sales rose 22% in 1983 (65).

Table 6 summarizes the current and probable contracts for gas exports to the Asia-Pacific. In 1985, one country, Japan, will be importing LNG; five countries will be supplying that LNG, totaling 38.3 bcm/yr. Afghanistan will be exporting 2.4 bcm/yr of gas via pipeline to the Soviet Union. Considering only the export agreements currently functional, in 1985 regional gas trade will equal 40.7 bcm/yr. If those contracts lapse without renewal, by 2000 only two projects, Malaysia's and Indonesia's expansion for Japan, will still be operational, shipping approximately 17 bcm/yr. That represents more than a 58% drop in volume.

By 1990, of course, export projects in planning will be coming on stream. By 1990, new projects could add over 22 bcm/yr in volumes traded regionally; beyond 1995, that figure could exceed 35 bcm/yr. Even if older contracts are not renewed, Asia-Pacific gas trade could reach as much as 65 bcm/yr by 1995, for a 61% increase over 1985. Given that the capital stock of the old projects will be paid off but still productive, a more likely scenario includes contract renewals for old export projects. In that case, regional gas trade will almost double in the next decade: total contract volumes for 1995 and beyond will exceed 76 bcm/yr, an 87% increase from

Table 6 Natural gas exports to the Asia-Pacific

Exporter	Importer	Initial delivery (duration)	Volume (bcm/yr)	1985	1990	1995	2000
Operating							
LNG							
USA-Alaska	Japan	1969 (15)					
		1984 (5)	1.3	1.3	—[d]	—[d]	—[d]
Brunei	Japan	1972 (20)	6.9	6.9	6.9	—[d]	—[d]
Abu Dhabi	Japan	1977 (20)	3.0	3.0	3.0	3.0	—[d]
Indonesia-Badak	Japan	1977 (20)	4.1	4.1	4.1	4.1	—[d]
Indonesia-Arun	Japan	1978 (20)	6.1	6.1	6.1	6.1	—[d]
Malaysia	Japan	1983 (20)	8.1	8.1	8.1	8.1	8.1
Indonesia-Badak II	Japan	1983 (20)	4.3	4.3	4.3	4.3	4.3
Indonesia-Arun II	Japan	1984 (20)	4.5	4.5	4.5	4.5	4.5
Pipeline							
Afghanistan	USSR	1968[a]	2.3	2.4	2.4	—	—
Subtotal (operating)				40.7	39.4	30.1	16.9
Prospective							
LNG							
Indonesia-Arun III	S. Korea	1987 (20)	2.7	—	2.7	2.7	2.7
Indonesia IV	Taiwan	1991[b]	0.8	—	—	0.8	0.8
Malaysia	S. Korea	1990[b]	2.0	—	2.0	2.0	2.0
Australia	Japan	1989[b]	8.1	—	8.1	8.1	8.1
USSR-Sakhalin	Japan	1995[b]	4.1	—	—	4.1	4.1
Qatar	Japan, S. Korea, Taiwan	1992[b]	8.1	—	—	8.1	8.1
Thailand	Japan	1990[b]	4.1	—	4.1	4.1	4.1
Canada	Japan	1990[b]	3.9	—	3.9	3.9	3.9
Pipeline							
Malaysia	Singapore	1988[b]	1.6	—	1.6	1.6	1.6
China	Hong Kong	c	c	c	c	c	c
Subtotal (prospective)					22.4	35.4	35.4
Grand total without contract renewals				40.7	61.8	65.5	52.3
Grand total with contract renewals				40.7	63.1	76.1	76.1

[a] Information unavailable as to duration; 25 year supply agreement assumed.
[b] Information unavailable as to duration; 15–20 year supply agreement assumed.
[c] Information unavailable.
[d] Assumes contracts are not renewed; grand total including possible contract renewal volumes listed below.
Source: Derived from Appendix Tables A and B; pipeline contracts referenced in text.

1985. The number of exporters involved in trade will double; the number of importers will triple. Asia-Pacific gas trade has a potentially strong future; the succeeding paragraphs will discuss regional import demands affecting that future.

NET IMPORTERS   *Current*   Tokyo Electric and Tokyo Gas began importing 1.3 bcm/yr of LNG in 1969. By 1978, Japanese LNG imports had increased to 15.6 bcm, and in 1983 they totaled over 25 bcm. Considering the decreased growth in Japan's primary energy consumption during 1980 and 1981, the 63% increase from 1978 to 1983 is particularly interesting. It indicates LNG's growing importance as an energy source.

The electric power sector absorbs most of the LNG imported to Japan. A 1983 analysis suggests that 76% of the gas consumed in Japan goes to generate electricity, while 20% is used as municipal fuel and 4% as industrial fuel. Government policies have encouraged substitution of natural gas for other fuels in all sectors, but the government proposes to increase LNG's share of power production from 14.7% in 1982 to 21.3% by 1990 (66).

Japan's Ministry of International Trade and Industry (MITI) has revised its LNG demand estimates three times in the last two years. In the early 1980s, MITI's demand scenario proposed a 1990 supply target of 70.7 bcm/yr (67). Its April 1982 estimates revised that to 58.1 bcm in 1990, rising to 67.5 bcm in 2000. In the light of continued international economic difficulties, and overwhelming critical opinion that even the revision reflected overoptimistic growth rates, MITI revised the demand scenario yet again. The August 1983 version envisioned demand ranging from 49.8 to 52.1 bcm in 1990, and from 56.4 to 58.6 bcm in 1995 (68). Finally, in November 1983, MITI published new energy demand figures offering the most conservative perspective to date: demand estimates of 49.3 bcm for 1990, 54 bcm for 1995, and 64–66 bcm for 2000 (69).

During this period of government adjustment, both the Institute of Energy Economics (IEE) and Shell Oil published analyses implying the extreme nature of MITI's initial estimates. In late 1982, the IEE estimated that a 1990 demand figure of 51.8 bcm represented a high-growth scenario; their base scenario put LNG demand at 55 bcm for 1990, with a low-growth scenario of 51.8 bcm (70). In June 1984, the IEE revised its own demand estimates, reaching consensus with MITI on the 1995 projections: 54 bcm (71). The Royal Dutch/Shell forecasts fell even lower than those of the IEE; in June 1983, Shell analysts suggested a range of 46.5 to 56.8 bcm for 1990, and 51.7 to 67.2 bcm for 2000 (72). MITI's November 1983 estimates, however, do fall within these more conservative limits. In fact, the Shell mid-range estimate for 1990, 51.7 bcm, is higher than the new MITI estimate, 49.3 bcm.

In 1983, Japan imported approximately 60% of the LNG traded worldwide. Japanese gas imports accounted for all the LNG traded in the Asia-Pacific region, and almost 92% of all gas traded regionally. The other 8% moved by pipeline into the Soviet Union. Because the Soviet Union uses that gas merely to balance internal supply deficits in its domestic pipeline system, we will simply assume continuation of imports at approximately the same volume for the next decade, rather than estimating total Soviet gas consumption through 2000.

Prospective   In the fall of 1983, South Korea concluded negotiations with Indonesia for LNG imports, to start in 1987 (73). Although previous gas consumption was minimal, the government decided to substitute LNG as a fuel for some of the country's thermal power plants. By 1990, government plans envisage LNG providing 11.8% of generated power. By 2000 this percentage share will decrease slightly, with LNG accounting for only 11.1% of total power generation. Estimates suggest that this will require 4.1 bcm/yr of LNG in 1990, and 5.5 bcm in 2000 (74).

Construction has begun on an LNG receiving terminal in Pyongtaek, to accommodate the Indonesian imports; it will cost the country approximately $750 million. Plans to import 2 bcm/yr of LNG from Malaysia in 1990 suggest construction of an additional terminal.

The fuel-oil bill for Singapore's Public Utilities Board totaled $390 million in 1981. This prompted the government to consider replacing 30–50% of its power station fuel needs with natural gas. Its Senoko power station is already designed to use natural gas when the price is competitive.

Extrapolating from 1980 fuel oil requirements, using a growth rate of 4% per year in fuel oil demand, Singapore's 1990 and 2000 fuel oil needs are approximately 617,400 MBtu and 921,600 MBtu, respectively. Assuming a mid-range substitution of 40%, that amounts to natural gas requirements of about 2.6 bcm in 1990, and about 3.8 bcm by 2000 (75).

To supply the gas, Singapore negotiated with Malaysia during 1984 to arrange pipeline imports. Although the countries have not settled project financing or final export price, 1988 was set as the delivery date. Initially, Malaysia will supply only 1.6 bcm/yr, to fuel electrical generation. Volumes will increase later, as Singapore plans to use natural gas to cover half its overall energy requirements (76). With these facts in mind, a more appropriate estimate for 1990 would be 2.1 bcm, accounting for a gradual volume buildup from 1988. This also suggests that the 2000 estimate is conservative.

In 1984 the Chinese Petroleum Corporation proposed that Taiwan import LNG for use as town gas (77). Taiwan produces over 1 bcm/yr of indigenous natural gas, but reserve depletion will limit production past

1995. Negotiations were opened with Indonesia in 1984 for the import of 0.6 bcm/yr from 1991, and work began on a $750 million receiving terminal at the port of Hsinta, to be completed in 1990 (78).

The Energy Committee of the Ministry of Economic Affairs released forecasts in 1984 estimating energy supply structure to 2000. Its scenario projected a total energy supply of 44 million kiloliters oil equivalent (Mkloe) in 1990, and 73 Mkloe in 2000, with natural gas accounting for 5.9% in 1990 and 8.5% in 2000 (79). This equals natural gas requirements of 2.5 bcm in 1990, and 5.95 bcm in 2000. Given current production and reserve levels, Taiwan will probably sustain gas production of 1 bcm/yr through 1990, but production will begin to drop off, and by 2000 might equal only 0.65 bcm/yr. Consequently, required imports will total approximately 1.5 bcm in 1990 and 5.3 bcm in 2000.

Hong Kong uses manufactured gas, primarily in its residential and commercial sectors. In 1971, gas consumption equaled only 24 million cubic meters; by 1982 consumption had almost quintupled, reaching 119 million cubic meters (80). The growth rate is rapidly increasing; as mentioned previously, gas demand increased by 23% in 1983 alone. Due to this burgeoning demand, Hong Kong plans to import gas from the People's Republic of China to augment the domestically manufactured supply. At this time, however, too little information is available to estimate its import requirements for the next two decades.

Table 7 summarizes the preceding demand projections. In 1982, Asia-Pacific gas imports equaled almost 28 bcm. By 1990, import demand will increase at least 112%; a low case estimate suggests import requirements of 59.4 bcm, with a high case of 61.8 bcm. In the decade from 1990 to 2000, even conservative estimates envision imports increasing by 29%; projections for demand requirements range from 76.5 bcm to 82 bcm. Although the regional totals present high and low cases for 1990 and 2000, obviously they only represent the difference between the high and low LNG import requirements estimated for Japan. For 1990, this volume equals about the production of one refrigeration unit; by 2000, the disparity between estimates almost equals the volume of the Brunei LNG plant. Japan's import decisions will determine regional gas trade balances.

REGIONAL TRADE BALANCES   Prospective exporters and importers plan gas use based on their images of future market conditions. In comparing estimated required imports with projected total export capacity, Table 8 provides a rough sketch of regional gas trade through the end of the century. The low and high cases for required imports represent differences of opinion regarding Japan's LNG demand; the low and high cases for

Table 7 Demand for natural gas imports in the Asia-Pacific to 2000 (billion cubic meters)

	Total gas consumption 1982	Total gas imports 1982	1990	2000
Japan	27.442	25.664	49.3[a]–51.7[b]	59.5[b]–65.0[a]
South Korea	0.007	—	4.1[c]	5.5[c]
Singapore	0.054	—	2.1[d]	3.8[d]
Taiwan	0.997	—	1.5	5.3
Hong Kong	0.119	—	[f]	[f]
Soviet Union	422.18	2.3	2.4	2.4
			Low High	Low High
Total	450.799	27.96	59.4—61.8	76.5—82.0

[a] MITI estimates.
[b] Royal Dutch/Shell estimates.
[c] Electric Futures of Asia Project estimates.
[d] Authors' estimates.
[e] Chinese Petroleum Corporation estimates.
[f] Information insufficient for estimate.
Sources: Goss, R. M., ed. 1982. *BP Statistical Review of World Energy 1982.* London: British Petroleum Company. 32 pp.
International Energy Agency. 1984. *Energy Balances of Developing Countries.* Paris: Organization for Economic Co-operation and Development. 346 pp.
See also Refs. 68, 69, 72, 74, 79.

projected export volumes represent current trade contracts lapsing, on the one hand, and contract renewals on the other.

Analyses of demand very nearly agree for 1990, and few extant contracts will have lapsed. The market will, if anything, be a little soft, with excess export capacity available for the imports required. If all contracts are renewed, new projects come on stream on schedule, and Japan's demand remains as depressed as MITI projects, the excess export capacity could be

Table 8 Natural gas trade balances in the Asia-Pacific to 2000 (billion cubic meters)

	1990 Low	High		2000 Low	High	
Total required imports	59.4	61.8		76.5	82.0	
Total exports	61.8	63.1		52.3	76.1	
Excess export capacity	+2.4	+1.3	+3.7[a]	−24.2	−5.9	−0.4[a]

[a] Compares low estimate for required imports with high estimate for export capacity.
Source: Derived from Tables 6 and 7.

as much as 3.7 bcm. That volume is roughly equivalent to Japan's trade with Abu Dhabi—or the output of the prospective Canadian LNG project.

By 2000, the differences are marked, but the market will definitely tighten up. The least likely scenario allows contracts to lapse, in which case importers will be looking for 25 bcm/yr of gas even given low demand. In the case of high demand coupled with contract renewals, room exists for another 6 bcm/yr of export capacity, or a moderately large export project. If, however, conservative analyses of Japan's energy growth prove correct, comparing low demand in 2000 to high export capacity produces an almost perfectly balanced regional trade scenario.

CONCLUSION

As far as prospective gas exports are concerned, the preceding paragraphs have presented a conservative perspective. During the past four years, various groups have tentatively proposed gas export projects aimed at the Asia-Pacific market. One group envisioned shipping up to 22.7 bcm/yr of Alaskan North Slope gas to Japan as LNG: almost half Japan's projected 1990 import requirements (81). Yet even though the wilder export schemes were excluded from consideration, the final scenario compels the conclusion that the natural gas industry in the Asia-Pacific will grow dramatically in the next 15 years: volumes in trade will double; the number of exporters will double; the number of importers will triple. Figure 1 graphically

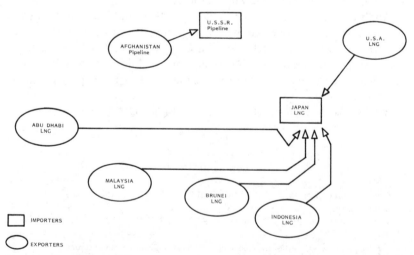

Figure 1 Natural gas trade flows in the Asia-Pacific, 1985. Diagram is not to scale: for representative purposes only.

illustrates the present regional flow of natural gas trade; Figure 2 shows the growth expected in the next decade.

In fact, this growth will present difficulties for new exporters looking for customers. By 1995, the Asia-Pacific market will face downward price pressures from two directions. First, older projects will have paid off their initial capital investments, yet still have productive facilities; they will be able to undersell the competition. Both Indonesia and Malaysia have sufficient resources to double their outputs, with only the small additional investments required to expand existing infrastructure. Essentially, the gas trade in the Pacific market will be marked by too many sellers chasing too few buyers—with Japan clearly in the dominant monopsonist position. This could mean that once the currently signed take-or-pay contracts are over, there may be an opportunity for a major decline in the price of gas due to excess resource availability. However, such a prospect will not be in the cards until after the mid 1990s.

Second, Japan will be in the position to pressure older producers actively for contractual changes by altering its future energy mix. When the Middle East export refineries start exporting in 1985, Organization of Petroleum Exporting (OPEC) nations may well exert pressure on their crude oil customers to absorb their petroleum products (82). In that case, Japan could decide that it is both economically and politically advisable to move some fuel oil back into power generation. There is ample evidence that the Asia-Pacific region, as a whole, will have a fuel oil surplus in the 1990s. This

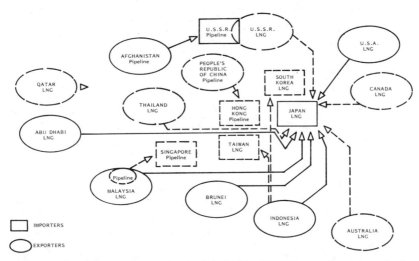

Figure 2 Natural gas trade flows in the Asia-Pacific, 1995. Diagram is not to scale: for representative purposes only. Dotted lines indicate prospective trade agreements.

will lower fuel oil prices and damage LNG's competitive position. When gas export contracts are up for renewal, the Japanese could parlay one set of pressures into another, forcing established gas producers to drop take-or-pay clauses and to break the link between the price of gas and crude oil. Established exporters could absorb these changes; newcomers could not. For the Asia-Pacific region, however, the increase in the actors involved and the gas produced means increased resource flexibility and self-sufficiency.

Finally, much has been said about government overestimation of Japan's demand for natural gas (83). Japanese colleagues have, at times, gone to great lengths to explain to us that MITI's figures are targets and not forecasts, and that they should be allowed to make mistakes—in the same way their American colleagues do. In our judgment, the most recent data made available by MITI (and IEE) represent a genuine attempt at forecasting a realistic future gas demand in Japan.

Appendix Table A LNG chains terminating in Asia

	USA	Brunei	Abu Dhabi	Indonesia		Malaysia	Indonesia	
	(Alaska)				Arun			
				Badak (#1, #2)[a]	(#1, #2, #3)[a]		Badak (#3, #4)[a]	Arun (#4, #5)[a]
Importer	Japan	Japan	Japan	Japan	Japan	Japan	Japan	Japan
Contract quantity (bcm/yr)	1.3	6.9	3.0	4.1	6.1	8.1	4.3	4.5
Buyer(s) (bcm/yr)	Tokyo Elec. (0.97) Tokyo Gas (0.33)	Tokyo Elec. (4.65) Tokyo Gas (1.43) Osaka Gas (0.85)	Tokyo Elec. (3.0)	Kansai Elec. (3.2) Chubu Elec. (2.3) Kyushu Elec. (2.1) Osaka Gas (1.8) Nippon Steel Corp. (0.8)		Tokyo Elec. (5.4) Tokyo Gas (2.7)	Kansai Elec. (1.1) Chubu Elec. (2.0) Osaka Gas (0.5) Toho Gas (0.7)	Tokyo Elec. (0.6) Tohoku Elec. (3.9)
Start-up	11/1969; 1984	12/1972	5/1977	8/1977	10/1978	2/1983	10/1983	1984
Period of contract	15 yr; 5 yr	20 yr	20 yr	20 yr	23 yr	20 yr	20 yr	20 yr
Loading port	Kenai	Lumut	Das Island	Bontang	Lho Seumawe	Bintulu	Bontang	Lho Seumawe
Vessels used[b]	71,500 m³ × 2	77,731 m³ × 2 75,000 m³ × 2	125,000 m³ (ave.) × 5 87,600 m³ × 1	125,000 m³ × 1	126,300 m³ × 7	130,000 m³ × 5	125,000 m³ × 3	125,000 m³ × 4
Approx. shipping distance in km (naut. mile)	6,000 (3,200)	4,400 (2,400)	11,900 (6,400)	4,450 (2,400)	6,000 (3,200)	4,600 (2,500)	4,450 (2,400)	6,000 (3,200)
Receiving port(s)	Negishi	Negishi Sodegaura Senboku #1	Sodegaura	Senboku #2 Tobata (Kitakyushu) Himeji Chita		Sodegaura Higashi Ogishima	Chita Himeji Senboku #2	Niigata Higashiko
Investors	Phillips 70% Marathon 30%	Brunei (33⅓%) Shell (33⅓%) Mitsubishi (33⅓%)	ADNOC (51%) Mitsui (22.05%) Mitsui Liq. Gas (2.45%) BP (16¾%) CFP (8¼%)	Pertamina (55%) Huffco (30%) Jilco (15%)[c]	Pertamina (55%) Mobil (30%) Jilco (15%)[c]	Petronas (6.5%) Shell (17.5%) Mitsubishi (17.5%)		

[a] Numbers refer to specific refrigeration trains commissioned for each contract.
[b] Capacity of LNG tankers in cm³ × number commissioned for the project.
[c] JILCO: The Japan-Indonesia LNG Corporation, a partnership of the Japanese customers.

Sources: Union Internationale de l'Industrie du Gaz. 1982. *Rapport de la Commission H, Gaz liquefies, XVe Congrès mondial du gaz, Lausanne 1982*, pp. 13–28. Switzerland: Benziger Ltd. (in French and English). 187 pp.
Morse, R. A. 1984. Japan's Liquefied Natural Gas Dilemma: Oversupply and Lower Demand. In *U.S.-Japanese Energy Relations: Cooperation and Competition*, ed. C. K. Ebinger, R. A. Morse, pp. 190–92. Boulder/London: Westview, 239 pp.
Macris, G., ed. 1983–1984. *Pet. Intel. W.* See Ref. 2.

Appendix Table B   Prospective LNG chains terminating in Asia

	Indonesia (Arun #6)[a]	Indonesia	Malaysia	Australia
Importer	South Korea	Taiwan	South Korea	Japan
Contract quantity (bcm/yr)	2.7	0.81	2.0	8.1
Buyer(s) (bcm/yr)	Korea Elec. Power Co. (2.7)	Taiwan Power Co. (0.81)	Korea Elec. Power Co. (2.0)	Tokyo Elec. (1.25) Chubu Elec. (1.25) Kansai Elec. (1.25) Chugoku Elec. (1.25) Kyushu Elec. (1.25) Tokyo Gas (0.8) Osaka Gas (0.8) Toho Gas (0.25)
Start-up	1987	1991	1989–1990	1988–1989
Period of contract	20 yr	c	c	19 yr
Loading port	Lho Seumawe	c	Bintulu	Dampiar
Vessels used[b]	c × 2	c	c	25,000 m³ × 7
Approx. shipping distance in km (naut. mile)				6,800 (3,700)
Receiving port(s)	Pyongtaek	Hsinta	Pyongtaek	d
Investors (estimated project cost)	Mobil Oil Indonesia (29%) Korea Elec. Power Co. (71%) ($1.05 billion)			($7.65 billion)

Appendix Table B (*continued*)

	USSR Sakhalin	Qatar	Thailand	Canada
Importer	Japan	Japan, South Korea Taiwan	Japan	Japan
Contract quantity (bcm/yr)	4.1	8.1	4.1	3.9
Buyer(s) (bcm/yr)	c	c	c	Chubu Elec. (2.2) Kyushu Elec. (0.4) Chugoku Elec. (0.4) Osaka Gas (0.7) Toho Gas (0.2)
Start-up	1990s	1992	1990	1990
Period of contract	20 yr	c	c	20 yr
Loading port	Sakhalin	Umm Said	c	Grassy Point
Vessels used[b]	c	c	c	125,000 m³ × 5
Approx. shipping distance in km (naut. mile)				6,600 (3,600)
Receiving port(s)	c	c	c	c
Investors (estimated project cost)	USSR Sakhalin Oil Development Cooperation Co. (Japan) ($3.8 billion)	Qatar General Petroleum Corp. (70) BP (7.5%) CFP (7.5%) Tentative: Mistui (5%) Mitsubishi (5%) C. Itoh (5%) ($6 billion)	Thai LNG Co. (60%) Japan (40%): Mitsubishi (35%) Mitsui (35%) Marubeni (15%) Sumitomo (15%) ($3 billion)	Nissho-Iwai Corp. Resources Union Oil of Canada TransCanada Pipelines Pan-Alberta Gas Suncor Dome Petroleum (10%) ($2.6 billion)

[a] Number refers to specific refrigeration train commissed for the contract.

[b] Capacity of LNG tankers in cm³ × number commissioned for the project.

[c] Information unavailable.

[d] A seven-party consortium: Woodside Petroleum Ltd., ($16\frac{2}{3}$%); Shell Development Australia Pty Ltd. ($16\frac{2}{3}$%); BHP Petroleum Pty Ltd. ($16\frac{2}{3}$%); BP Australia Development Ltd. ($16\frac{2}{3}$%); California Asiatic Oil Co. ($16\frac{2}{3}$%); Mitsui-Mitsubishi ($16\frac{2}{3}$%).

Sources: Union Internationale de l'Industrie du Gaz. 1982. *Rapport de la Commission H, Gaz liquefies, XVe Congrès mondial du gaz, Lausanne 1982*, pp. 13–28. Switzerland: Benziger Ltd. (in French and English), 187 pp.

Morse, R. A. 1984. Japan's Liquefied Natural Gas Dilemma: Oversupply and Lower Demand. In *U.S.-Japanese Energy Relations: Cooperation and Competition*, ed. C. K. Ebinger, R. A. Morse, pp. 190–92. Boulder/London: Westview, 239 pp.

Marcis, G., ed. 1983–1984. *Pet. Intel. W.* See Ref. 2.

Literature Cited

1. Fesharaki, F., Hoffman, S., Schultz, W. 1985. Oil and gas in Asia: trade potentials and the refining outlook. In *Asian Energy Supplies and Requirements*, ed. R. el Mallakh. Boulder/London: Westview. In press
2. Macris, G., ed. 1984. *Pet. Intel. W.* XXIII(24): 8
3. Isaak, D., Totto, L. 1982. *Petrochemicals vs. LNG: An Examination of Alternative Natural Gas Strategies for Developing Countries.* Presented at the Ann. North Am. Meeting of the Int. Assoc. of Energy Economists, 4th, Denver. 51 pp.
4. Wintringham, M. C. 1983. *Natural Gas to Motor Fuel: The New Zealand Decision.* Presented at CSIS: Strategic Factors in International Gas Decision-making, Washington, DC. 20 pp.
5. Boshier, J. F. 1984. *New Zealand Country Paper.* Presented at Asia-Pacific Energy Studies Consultative Group Conference, VIIth, The East-West Center, Honolulu, Hawaii. 10 pp.
6. Rahmer, B. A. 1983. China: Industrial plans on target. *Pet. Econ.* L(1): 12–14
7. Marashian, O., ed. 1984. *Platt's Oil. News.* 62(158): 1
8. Cooper, B., ed. 1984. *Pet. Econ.* LI(1): 41
9. Marashian, O., ed. 1984. *Platt's Oil. News.* 62(182): 4
10. Marashian, O., ed. 1984. *Platt's Oil. News.* 62(163): 1
11. Marashian, O., ed. 1984. *Platt's Oil. News.* 62(182): 4
12. Macris, G., ed. 1983. *Pet. Intel. W.* XXII(13): 11
13. Cooper, B., ed. 1983. *Pet. Econ.* L(9): 358
14. Cooper, B., ed. 1984. *Pet. Econ.* LI(7): 272
15. Cooper, B., ed. 1983. *Pet. Econ.* L(5): 192
16. Macris, G., ed. 1983. *Pet. Intel. W.* XXII(9): 10–11
17. Marashian, O., ed. 1984. *Platt's Oil. News.* 62(180): 1–2
18. Fesharaki, F., Schultz, W. 1984. Oil and gas trade in the Pacific basin. In *U.S.-Japanese Energy Relations: Cooperation and Competition*, eds. C. K. Ebinger, R. A. Morse, pp. 39–81. Boulder/London: Westview, 239 pp.
19. Cooper, B., ed. 1984. *Pet. Econ.* LI(4): 156
20. Fesharaki, F., Schultz, W. 1984. See Ref. 18, p. 69
21. Fesharaki, F., Schultz, W. 1984. See Ref. 18, pp. 71–72
22. Macris, G., ed. 1983. *Pet. Intel. W.* XXII(30): 7
23. Macris, G., ed. 1983. *Pet. Intel. W.* XXII(3): 9
24. Itayim, F. W., ed. 1983. *Mid. East Econ. Sur.* XXVI(27): A8
25. Itayim, F. W., ed. 1983. *Mid. East Econ. Sur.* XXVI(30): A4
26. Fesharaki, F., Schultz, W. 1984. See Ref. 18, pp. 69–70
27. Macris, G., ed. 1983. *Pet. Intel W.* XXII(17): 6–7
28. Cooper, B., ed. 1983. *Pet. Econ.* L(5): 192
29. Fesharaki, F., Schultz, W. 1984. See Ref. 18, p. 72
30. Cooper, B., ed. 1984. *Pet. Econ.* LI(8): 309
31. Marashian, O., ed. 1984. *Platt's Oil. News.* 62(150): 3
32. Hough, G. V. 1984. LNG Market: Sales increase as prices weaken. *Pet. Econ.* LI(12): 439–41
33. Hough, G. V. 1983. Malaysia: Gas opens a new era. *Pet. Econ.* L(2): 47–50
34. Hough, G. V. 1984. See Ref. 32
35. Hough, G. V. 1983. See Ref. 33
36. Cooper, B., ed. 1984. *Pet. Econ.* LI(3): 110
37. Marashian, O., ed. 1984. *Platt's Oil. News.* 62(118): 2
38. Cooper, B., ed. 1984. *Pet. Econ.* LI(5): 197
39. Marashian, O., ed. 1984. *Platt's Oil. News.* 62(122): 1
40. Marashian, O., ed. 1984. *Platt's Oil. News.* 62(159): 1
41. Macris, G., ed. 1983. *Pet. Intel. W.* XXII(21): 6
42. Macris, G., ed. 1983. *Pet. Intel. W.* XXII(12): 7
43. Speakman, D., ed. 1984. *Pet. News.* 15(5): 29–30
44. Macris, G., ed. 1984. *Pet. Intel. W.* XXIII(27): 6
45. Cooper, B., ed. 1983. *Pet. Econ.* L(5): 191
46. Cooper, B., ed. 1984. *Pet. Econ.* LI(1): 37–38
47. Speakman, D., ed. 1984. *Pet. News.* 15(5): 29–30
48. Marashian, O., ed. *Platt's Oil. News.* 62(156): 1
49. Macris, G., ed. 1984. *Pet. Intel. W.* XXIII(34): 6–7
50. Itayim, F. W., ed. 1983. *Mid. East Econ. Sur.* XXVI(36): A2
51. Marashian, O., ed. 1984. *Platt's Oil. News.* 62(124): 1
52. Marashian, O., ed. 1984. *Platt's Oil. News.* 62(165): 2–3
53. Marashian, O., ed. 1984. *Platt's Oil. News.* 62(187): 3
54. Cooper, B., ed. *Pet. Econ.* LI(1): 41
55. Speakman, D., ed. 1984. *Pet. News.* 15(5): 31

56. Marashian, O., ed. 1984. *Platt's Oil. News.* 62(109):3
57. Cooper, B., ed. 1984. *Pet. Econ.* LI(4): 157
58. Marashian, O., ed. 1984. *Platt's Oil. News.* 62(109):3
59. Marashian, O., ed. 1984. *Platt's Oil. News.* 62(206):1
60. Marashian, O., ed. 1984. *Platt's Oil. News.* 62(207):2
61. Marashian, O., ed. 1984. *Platt's Oil. News.* 62(195):3
62. Marashian, O., ed. 1984. *Platt's Oil. News.* 62(207):2
63. Marashian, O., ed. 1984. *Platt's Oil. News.* 62(223):2
64. Marashian, O., ed. 1984. *Platt's Oil. News.* 62(224):1–2
65. Cooper, B., ed. 1984. *Pet. Econ.* LI(9): 358
66. Nemetz, P. N., Vertinsky, I. B. 1984. Japan and the International Market for LNG. In *The Columbia Journal of World Business.* XIX(1):1–7
67. Fesharaki, F., Schultz, W. 1984. See Ref. 18, p. 61
68. Nemetz, P. N., Vertinsky, I. B. 1984. See Ref. 66
69. Seiki, K. 1984. *The Future of the World Oil Market (Country Paper: Japan).* Presented at the Asia-Pacific Energy Studies Consultative Group Conference, VIIth, The East-West Center, Honolulu, Hawaii. 8 pp.
70. Institute for Energy Economics (IEE). 1983. *Prospects for Japan's Economy and Energy Supply/Demand in the 1980s.* Presented at the Energy Symposium, 15th, Institute for Energy Economics, Tokyo
71. Institute for Energy Economics. 1984. *Japan's Long-term Energy Supply/ Demand Forecast.* IEE Working Paper. 28 pp.
72. Macris, G., ed. *Pet. Intel. W.* XXII(23):6–7
73. Macris, G., ed. *Pet. Intel. W.* XXII(32):8
74. Kim, Y. H. 1983. *Major Issues and Their Policy Implications in the Development of Electric Power Systems: The Cases of China, Japan, and Korea.* EFA Working Paper. The East-West Center, Honolulu, Hawaii. 86 pp.
75. Fesharaki, F., Schultz, W. 1984. See Ref. 18, p. 62
76. Macris, G., ed. 1984. *Pet. Intel. W.* XXIII(28):8
77. Cooper, B., ed. 1984. *Pet. Econ.* LI(2):77
78. Cooper, B., ed. 1984. *Pet. Econ.* LI(5): 197
79. Lee, Y. K. 1984. *The Energy and Oil Market of Taiwan, ROC—Its Today and Tomorrow.* Presented at the Asia-Pacific Energy Studies Consultative Group Conference, VIIth, The East-West Center, Honolulu, Hawaii.
80. International Energy Agency (IEA). 1984. *Energy Balances of Developing Countries, 1971/1982.* Paris: OECD. 347 pp.
81. Cooper, B., ed. 1983. *Pet. Econ.* L(3):99
82. Fesharaki, F., Isaak, D. 1984. *OPEC and Asia: Factors Affecting the Emerging Product Trade.* Presented at the Asia-Pacific Energy Studies Consultative Group Conference, VIIth, The East-West Center, Honolulu, Hawaii. 36 pp.
83. Fesharaki, F., Schultz, W. 1984. See Ref. 18, p. 67

Ann. Rev. Energy. 1985. 10:495–514
Copyright © 1985 by Annual Reviews Inc. All rights reserved

MILITARY SABOTAGE OF NUCLEAR FACILITIES: THE IMPLICATIONS

Bennett Ramberg

Center for International and Strategic Affairs, University of California, Los Angeles, California 90024

In June 1981, a squadron of Israeli military aircraft descended out of a setting sun to drop sixteen 2000-pound bombs on an Iraqi research reactor under construction in the vicinity of Baghdad. The successful attack created a worldwide media sensation. Congressional hearings were held within weeks to establish the motivations and lessons. Attention focused on proliferation. Jerusalem rationalized its action on the grounds that Iraq intended to use the facility to acquire the feedstock for a nuclear weapons program. The potential radiological implications of an attack on an operating reactor were neglected, though not entirely ignored.

Stimulated by Prime Minister Begin's assertion that the attack was timed before the reactor turned "hot," which "would have caused a huge wave of radioactivity over the city of Baghdad and its innocent civilians," (1) the Congressional Research Service was asked for an evaluation. It reported that "it would be most unlikely for the attack with conventional bombs upon the reactor to have caused lethal exposures in Baghdad, although some people at the reactor site might receive some exposure" (2). Regrettably, the report, which may have steered Congress away from examining the issue more closely, may have underestimated the consequences. Some adverse public health consequences might have resulted in Baghdad and its environs. For a review of the debate over whether destruction of the Iraqi reactor could have posed a hazard to Baghdad and its environs, see (3).

In some respects, this fact is beside the point. Given the small size of the

495

0326–1626/85/1022–0495$02.00

facility, casualties probably would not have been severe.[1] However, users of the Congressional Research Service findings failed to appreciate the severe consequences that would have resulted had the Baghdad facility been standard-sized, or 60 times larger than it was (the size of a facility Baghdad is planning). Thousands of square miles might have been contaminated. Regrettably, the strategic implications of such acts have received little attention in the global energy debate.[2] This article reviews the problem generically, examines the strategic implications in two cases, the Middle East and Europe, and suggests remedies to conventional weapons attacks.

VULNERABILITY OF NUCLEAR FACILITIES

Approximately 260 nuclear power plants are now in operation in 22 countries with a like number under construction, and by the end of this century more than 40 nations may be producing electricity with atomic power. These large commercial reactors contain significant amounts of radioactive material. As long as there are no major coolant pipe breaks, reactor vessel ruptures, or mismatches of power and coolant due to excessive fission or undercooling, the most common atomic plant—the light water reactor—operates without posing a hazard to public health. To prevent accidents, manufacturers rely on equipment of high integrity, and American and European stations (but not most Soviet plants) have

---

[1] The Iraqi plant was designed to provide 40 megawatts of heat, distinguishing it from its French twin, which produces 70 megawatts. See Richard Wilson (4). There has been some debate over how extensive the contamination would have been had the reactor been in operation. The Congressional Research Service concluded that contamination would in the main be limited to the reactor site. Save for pieces of the core, the problem would not be significant. See (5). Frank Von Hipple and Jan Beyea, nuclear physicists who have written several reactor accident consequence studies under the auspices of Princeton University's Center for Energy and Environmental Studies, reviewed the Congressional report and concluded that Donnelly's estimates were inaccurate because he analogized the effect of radium for fission products (telephone conversation with Jan Beyea on April 30, 1982). Steve Ramos, a Nuclear Regulatory Commission official who formerly directed the safety program for US research reactors, contended that "significant" radioactive fallout from such facilities would not extend beyond two miles. See (6). Herbert Kouts of the Brookhaven National Laboratory estimated that the radiation danger would be insignificant beyond 1000 feet around the plant, while Herbert Goldstein of Columbia suggested that, based on International Atomic Energy Agency calculations, some radiation would reach the Iraqi capital, whereas lethal doses might prevail near the reactor. See (7).

[2] This fact is underscored by the failure of even the most comprehensive analyses of nuclear energy policy to mention the subject. Among such studies are (8–10). The implications of nuclear weapons destruction in the United States have, however, been reviewed periodically at the Oak Ridge National Laboratory. The most recently published work is (11). See also (12) and (13). Assessments of conventional weapons destruction will be found in (14–16).

emergency core cooling systems, redundant pumps, emergency external power diesel generators, and reinforced concrete containment buildings.

Notwithstanding the care with which most reactors are built, US government studies document their vulnerability to willful destruction through disruption of coolant mechanisms both inside and outside the containment building (17–19). Israel's successful attack demonstrated the destructibility of a large research reactor with some safety characteristics, namely, several feet of concrete biological shielding immediately surrounding the reactor, similar to that of a commercial plant. Keeping in mind that the Israeli military capability is not unique, reactor vulnerability is likely to increase in the decade ahead as precision-guided munitions—some fused with shaped charges capable of destroying the hardest containments—are introduced into the arsenals of many countries.[3]

What would be the consequences of a successful conventional weapon's attack? Conceivably, they could equal those of the worst accidental meltdown in terms of radiation discharged.[4] As distinct from conventional or nuclear weapons, the effects of which can be calculated reliably, the emission of radioactivity from reactors is subject to a number of variables including (but not limited to) the quantity, composition, and rate of deposition of materials. Assuming that recent (June 1981) Nuclear Regulatory Commission accident studies are well founded, military destruction can result in moderate to major releases of radioactivity into the environment.[5] A moderate release from a reactor the size of Three Mile

[3] The United States has in its arsenal conventional munitions capable of penetrating 10 meters of concrete. See (20). For the effectiveness of other conventional weapons against hard targets, see (21). A discussion of the proliferation of precision guided munitions will be found in (22).

[4] Telephone conversation with a staff member of Sandia Laboratories who conducted a technical study on conventional weapons destruction of nuclear facilities. According to this source, conventional weapons could not release more radiation than the worst accident. See also Kaul & Sachs, *Adversary Actions*, Tables C1–C4. Note that if the core is broken apart and scattered by conventional weapons, releases of radioactivity into the environment may not be as significant as those from the meltdown of an intact core. Jan Beyea alerted me to this in an August 18, 1981, telephone conversation.

[5] There is controversy in the technical literature over whether major releases of radioactivity are possible in accident scenarios, which military destruction could induce. The nuclear industry argues that the Three Mile Island accident (along with experiments in small test reactors where a large fraction of the nuclear core was destroyed) suggests that when water and steam are present, a significant fraction of the most dangerous radioactivity is contained therein. This could reduce by a factor of 10 radiation exposure estimates forming the basis for emergency planning in the event of a nuclear accident. At the same time, the industry acknowledges that the absence of water—a dry containment—allows the release of significant radiation if the containment is breached. For elaboration, see (23–25). The Nuclear Regulatory Commission (NRC) recently examined these contentions and concluded that the con-

Island—880 megawatts of electricity (MWe)—that has been operating for more than three months could contaminate 500 square miles; a major release might affect 3000–5000 square miles, requiring occupation restrictions over a significant fraction that could last decades, as the effectiveness of decontamination is very uncertain.[6] Since nations often cluster several reactors, the problems would be compounded if the contents of more than one were discharged as is conceivable in military scenarios. Additional problems would arise through release of the inventories of spent fuel customarily located at reactor sites.

Table 1 illustrates possible consequences from another set of calculations. The third column shows land contamination resulting from a meltdown induced by a conventional weapon's attack on a large power reactor, which is analogous to a major accident. Note that one week after the release 2200 square miles could be contaminated. (A more conservative calculation to exposure of 10 rem over a 30-year period extends the zone of concern to 5300 square miles.) Two years after the meltdown the reactor

---

sequences from small reactor incidents may be "of very limited value." The NRC acknowledged that the *Reactor Safety Study* may have overpredicted "certain" accident sequences (for example, a steam explosion), but that, in general, its estimates are reliable (26). The NRC conclusions are adhered to in this article.

As this article goes to press, the source term issue, i.e. the quantity of radioactivity that could be released, is muddled by recent research. The most authoritative study was published by the American Physical Society (*Report to the American Physical Society of the Study Group on Radionuclide Release from Severe Accidents at Nuclear Power Plants*, draft, February 1985). The report concluded that the source term in accidents could be substantially smaller than those reported in the *Reactor Safety Study* for volatile radionuclides assuming late containment failure. However, in war, the Society's conclusion may be inapplicable assuming adversaries use hard munitions capable of blowing a hole through the containment. Under these circumstances the *Reactor Safety Study* calculations may still apply. Adding to the source term may be lanthanides and some tranuranics such as plutonium as a result of concrete-core interaction which the study group urged to be researched more thoroughly in accident scenarios. Given the importance of late containment failure in reducing the source term and its inaptitude in war, I believe that for the present the calculations used in this article are reliable for policy planning. Reinforcing my conclusion are telephone interviews I conducted with several technical experts. Although Richard Wilson and Fred Finlayson of the American Physical Society study group could not make a judgment of the implications of military attacks, Jan Beyea of the Audobon Society and Gordon Thompson and Stephen Sholly of the Union of Concerned Scientists concluded that releases approaching a PWR 2 were possible. Richard Denning of Battelle suggested that releases might be limited to a PWR 3 which would be two to three orders of magnitude below a PWR 2. However, even assuming a PWR 3—and Denning as well as all others interviewed acknowledged great uncertainties—hundreds of square miles still could be contaminated. The problem would be made worse in war because of the likely demolition of all co-located reactors.

[6] See Beyea (27). In a telephone conversation on August 18, 1981, Beyea reconfirmed the relevance of his estimates to conventional weapon scenarios.

can contaminate more land than a one-megaton nuclear weapon (680 versus 150 square miles), reflecting the longer-lived radiation from the material contained by the reactor. The problem is compounded when the contents of several reactors per site are destroyed, which is conceivable in wartime scenarios. Such releases can be induced by conventional weapons designed to destroy hard targets such as reactor containment buildings. Alternatively, an attacker can take advantage of critical soft spots outside the containment structure, including the off-site and emergency diesel generators required to keep the plant in operation, coolant feedlines, and the control room, to induce a meltdown. For elaboration, see (28).

Table 1 also illustrates the consequences of nuclear weapons bombardment. The fourth column shows the contamination estimated to be released by an attack with a one-megaton nuclear weapon on a reactor. Reflecting the longer-lived radiation of the nuclear facility, the addition of the reactor's core to the weapon threatens 79,000 square miles, compared with the 31,000 square miles threatened by the weapon alone, one week after the attack.

Table 1 Area in square miles that must remain uninhabited for a given time for different nuclear energy facilities release scenarios[a]

Time uninhabited	1-megaton weapon	Reactor meltdown	(1000 MWe)	Weapon on reactor	Weapon on waste storage facility
1 week	31,000	2,200	(5,300)[b]	79,000	113,000
2 weeks	26,000	2,000		72,000	110,000
1 month	21,000	1,800		64,000	103,000
2 months	17,000	1,600		54,000	100,000
6 months	5,000	1,200		33,000	83,000
1 year	1,200	900		25,000	67,000
2 years	150	680		17,000	49,000
5 years	11	320		10,000	35,000
10 years	2	140	(550–4300)	6,000	30,000
20 years	1	68		3,200	25,000
50 years		50	(240–3300)	1,200	14,000
100 years		20		180	2,400

Source: Steve Fetter and Kosta Tsipis, "Catastrophic Nuclear Radiation Releases" (Program in Science and Technology for International Security, Department of Physics, MIT, Cambridge, Massachusetts, Report No. 5 September 1980), Tables 2, 3, 6, 8. Figures in parentheses are from (27, p. B-13).

[a] Fetter & Tsipis assume that an area becomes uninhabitable for a given time assuming the maximum allowable dose is 2 rem per year.

[b] The figures in parentheses are drawn from (27). They "assume that occupation would be restricted if the resident population would otherwise receive more than a 10 rem whole body radiation dose over 30 years. This corresponds to about a threefold increase over the natural background dose in the same period. A 10-rem whole body dose has associated with it a risk of .05 to .5 percent chance of cancer death."

Within one year, 20 times more land is contaminated by the weapon and reactor combined than by the weapon alone. The fifth column illustrates that the problem is significantly worse when a nuclear weapon destroys a large waste storage facility.

The medical effects of atmospheric emissions of radioactivity are the result of exposures through inhalation, radiation deposited on the ground, and consumption of contaminated food. The early lethality of irradiation (death within 60 days) would depend on the effectiveness of prophylactic measures like evacuation, sheltering, and medical care for the exposed. If these are are instituted promptly, casualties will be minimal; delay increases the danger. For example, 24-hour residence in the contaminated region could result in early fatalities as far as 20 miles out in a wedge-shaped zone that could be several miles wide at its outermost reaches.[7] Generally, the closer one is to the source of emission, the greater the likelihood of exposure to unacceptable radiation. The possibility of early fatalities would be diminished at all distances in unstable weather; winds, for example, dilute airborne radioactivity. Unfortunately, except in a few regions, climate cannot be relied on to ameliorate radiation.

Late cancers would likely dominate the medical effects observed in the hundreds to thousands of square miles affected by low-level radiation beyond the zone defined above. Failure to impose habitation restrictions— a difficult undertaking given the extent of contamination—could result in an increase in cancers of from less than one to several percent depending on the intensity of irradiation. The number of deaths would depend on the size of the exposed population. In densely inhabited regions, late cancer fatalities could range in the tens of thousands, assuming that the much debated cancer induction models, notably the linear hypothesis, are reliable (27, 32). A like number could bear genetic effects in future generations.

In sum, nuclear energy facilities are vulnerable to destruction in time of war. However, destruction would not be easy. Facilities are usually housed in massive reinforced concrete structures, built to rigorous standards, and have a number of back-up systems to compensate for primary system failure and to minimize the consequences of accidents. The exact standards vary depending on the manufacturer; German facilities often have the largest number of compensatory systems and the Soviet Union's facilities the least. But given current military technology, the standards are not sufficient to prevent the release of nuclear products into the environment as

---

[7] This 20-mile calculation assumes a release in very stable weather, that is, little wind. Supportive medical attention will reduce the zone of concern to under 15 miles. For calculations that take these and other variables into account, see (29–31).

the result of a nuclear weapons attack, a concerted conventional weapons bombardment, or the efforts of sophisticated saboteurs. The problem is minimized today because nuclear weapons are limited to a few countries, and precision-guided munitions capable of destroying facilities still are not widespread. However, this situation will change as increasingly lethal conventional munitions are introduced into the arsenals of many countries.

STRATEGIC IMPLICATIONS

The consequences of nuclear facility destruction demand a rationale, and certainly Israel provided one: the need to preempt an adversary's acquisition of atomic weapons, for Israel believed that Iraq's plant was a guise for a weapons program. There are other rationales as well. To cripple an antagonist's industrial capabilities for waging war, combatants have targeted enemy energy sources, as was the case in Israel's 1982 incursion into Lebanon, in the Iran-Iraq conflict, in the 1973 Arab-Israeli war, in Vietnam, in Korea, and in the European and Asian theaters in World War II. Combatants also have destroyed the environment for military purposes. The Dutch destroyed their dikes during the Second World War to hamper the Germans; the Chinese did likewise in the late 1930s to impede the Japanese. The Soviet Union practices a scorched earth policy for the same reason. During the Vietnam War, the United States used herbicides to destroy enemy defensive cover and to improve target identification. In addition, combatants might destroy nuclear power stations because they represent one of the greatest concentrations of capital investment a country is likely to possess. Then, too, an aggressor may experience failures in command and control and execute accidental bombings, or parties with a stake in an ongoing conflict might consider sabotaging a facility to escalate the conflict.[8]

Still another rationale is coercion. Large populations in many countries have become acutely concerned about possible radioactive releases from nuclear power plant accidents, and a belligerent could capitalize on this fear for coercive purposes. Chester Cooper, assistant director of the Oak Ridge National Laboratory's Institute for Energy Analysis, speculated that in certain Third World situations, this coercion could have the "positive"

[8] Some of this rationale was brought to my attention by Theodore Taylor and crystallized in his unpublished manuscript, "Reactor Safety Considerations Related to Sabotage and Wartime Bombardment of Nuclear Power Plants" (1968). Note in the Iran-Iraq war both sides have bombed each other's reactors; no radiation was emitted because the plants were under construction. We do not know the rationale behind the attacks.

effect of containing belligerence by allowing weak states to threaten strong ones with unacceptable damage (14). Recent events in the Middle East, however, suggest that the presence of nuclear energy facilities among opponents may exacerbate rather than enhance stability. Certainly the costs of war would increase should plants be destroyed.

The success of such coercion lies in the target nation's psychological sensitivity to nuclear contamination. Sensitivity is a product of different cognitive factors, including estimates of the physical consequences, and of faith in countermeasures. Leaders will approach this matter differently. Some may not consider the problem significant and may optimistically argue that numerous variables act on facility radiation, thereby reducing its effectiveness as a weapon. It may be argued, for example, that population centers and valued land are not in the vicinity of the facilities, or that installations are downwind from valued locations. Meteorological conditions are likely to promote rapid wind dilution or washing out of radiation via rainfall. The numerous safety features of the threatened plants could be stressed, along with the availablity of shelters and relocation plans. Regard for the adversary's military capability may be low. Finally, contamination estimates may show few immediate effects and long-term consequences that can be dealt with reasonably. The absence of the blast, heat, and intense radiation that are characteristic of atomic bombs may further reinforce these sentiments.

On the other hand, some leaders and their compatriots may be hypersensitive about the prospect of contamination from nuclear facilities. Those nations most concerned would be the most easily manipulated and possibly the most disposed to take drastic measures to reduce their vulnerability. Threats of retaliation in kind are one option. Another option would be nuclear weapons. Case studies can illuminate these points.

THE MIDDLE EAST

Nuclear power in the Middle East is currently limited to five small research reactors in Egypt, Iran, Iraq, and Israel. None of these reactors—including the largest, Israel's Dimona installation, producing 25 megawatts of heat— poses a significant contamination hazard given its size. There are plans, however, to introduce 600–900-megawatt (electric) power stations into the region. Cairo currently is being courted by salesmen of nuclear firms in France, West Germany, Canada, and the United States. Libya has expressed interest and has negotiated with the Soviet Union to purchase a facility. Colonel Moammar Qaddafi is being wooed by the French, who have offered a power reactor to Iraq as well. The future of Israel's program is uncertain. Jerusalem's failure to sign the nuclear weapons nonproliferation treaty resulted in American refusal to sell reactors despite promises made in 1974. Israel is considering buying from other vendors, building a reactor of its own, or turning to other energy sources. Kuwait and Syria are reviewing the nuclear option. Iran, the only country to have begun construction of two atomic power stations, suspended its program with the fall of the Shah but is contemplating completion of one plant.[9]

While concern has focused on the manipulation of Middle Eastern atomic plants for nuclear weapons,[10] little attention was given to their military destruction until the Israeli bombardment. As other states in the region modernize their armed forces during the decade ahead, a number may acquire the ability to destroy atomic stations.

The consequences of military destruction could be grave. Because Israel is such a small country, a large fraction of it would be put at risk wherever future construction takes place. (A Mediterranean site 20 miles south of Tel Aviv was once under consideration: a more recent plan suggests a Negev location along the proposed Dead Sea Canal, perhaps underground.) Egypt confronts a potential hazard that is less acute but nonetheless could be significant. Safety concerns compelled the Egyptian government in 1981 to abandon a site at Sidi Kreir, 36 miles west of Alexandria. Plans now call for construction of up to nine plants by the year 2000 to be built about 80 miles west of Egypt's second largest city. Even at this distance, Alexandria and the densely populated Nile Delta could possibly be contaminated, although variable winds could carry fallout to largely uninhabited regions in the south.

In these and other instances, the timing of nuclear energy generation may also have strategic implications. Some Israelis advocate placing nuclear power on line before the end of the 1990s. Should such a plant be the first or only large station in the region, and assuming the Arabs, including the Palestinians, are able to release its contents, widespread damage could be inflicted with a few well-placed bombs. Thus, Israel might be vulnerable to a new mode of Arab intimidation. If radioactivity were released during conflict, Israel may have difficulty identifying the attacker for purposes of retaliation since Jerusalem has numerous adversaries. Further, should the antagonist be uncovered, retaliation in kind would be impossible unless the belligerent had a nuclear energy facility. In this case, Israel might instead turn to nuclear weapons.

Should Egypt be the first to acquire a nuclear power plant, it would expose itself to manipulation not only by Israel, but perhaps by Arab antagonists—notably Libya—as well. Likewise, development of nuclear

---

[9] For reviews of nuclear energy plans in the Middle East, see (34–36).
[10] For reviews of the proliferation implications of the attack, see (37–40).

energy by Iraq, Iran, or Libya would expose each to intimidation by its adversaries. If the antagonists acquired equally vulnerable installations simultaneously, stability might result if each were equally sensitive to the problem and there were no other significant asymmetries diminishing vulnerability and consequences.

The implications of these Middle Eastern considerations are important to countries outside the region as well. Should Israel acquire a nuclear power station, its vulnerability might force the United States to play a more active deterrent role to reassure Jerusalem. Iraq, a sometime Soviet client, might seek Soviet support against either Iranian or Israeli intimidation. Greater regional involvement by either superpower raises the risk of future confrontation between them.

EUROPE AND THE SOVIET UNION

Unlike countries in the Middle East, antagonists in Europe already have the capacity to inflict massive damage using nuclear weapons. It might appear at first glance that the strategic consequences of military actions against the 200 or more nuclear installations existing in or planned for the continent would be marginal. Closer scrutiny of the issue suggests otherwise.

At present, North Atlantic Treaty Organization (NATO) forces are positioned to confront a conventional Soviet assault in the belief that the USSR, at the outset of a European war, might forego the use of nuclear weapons and try to overwhelm the West with conventional weapons alone (41). In maintaining a strong conventional posture, NATO hopes it will be able to repel Warsaw Pact forces without having to resort to nuclear weapons.

The focal point of any such conflict is likely to be West Germany, which has more than 30 nuclear plants and support facilities operating or planned. Their destruction for any of the reasons discussed earlier could have significant implications for the course of war. Many of these plants are located along invasion routes; their successful bombardment by Soviet conventional forces would introduce radiation into the conflict. To be sure, the Soviet Union might be self-deterred, given prevailing westerly winds that would carry the fallout over Eastern Europe, although the Soviets might not be overly concerned about the fate of their allies. (Indeed, the West Germans could threaten destruction of their own facilities along their eastern frontier in an attempt to deter war. However, activation of that threat runs the risk of self-contamination and escalation.) Nevertheless, should destruction occur, the complexion of the war would change. NATO would have to decide to respond in kind or possibly to threaten to use nuclear weapons to stop the Soviets. Either course could escalate the conflict. The West might instead ignore the destruction, but this could encourage further Soviet aggression.

The Soviet Union is similarly vulnerable, with over 40 nuclear power plants of its own in operation or under construction. This figure does not include an unknown number of facilities contributing to its nuclear weapons program. Soviet vulnerability is enhanced by the absence of containment structures over most reactors. While current American strategic targeting doctrine already calls for destruction of Soviet electrical generators, rhetorical emphasis by the US Department of Defense on the implications of Soviet vulnerability—particularly for postattack recovery—could enhance the American deterrence posture. The targeting of facilities also could be used in coercive diplomatic and military strategy, if the Soviet plants are located near military bases.

Targeting of facilities might prove more attractive for countries that have limited nuclear aresnals and are less certain they can inflict unacceptable damage to deter Soviet coercion. Britain and France might enhance the credibility of their nuclear forces vis-a-vis the Soviets if they threatened to turn a portion of their atomic weapons arsenal against Soviet facilities, or if they acquired a cruise missile or highly reliable bomber capable of penetrating the Soviet Union's defenses with conventional warheads. More significantly, the nonnuclear deterrent capabilities of West Germany and the other NATO parties could be elevated with conventionally armed cruise and ballistic missiles capable of bombarding reactors. With such weapons, these countries for the first time would be able to inflict what may be unacceptable damage upon the Soviet Union, thereby reinforcing deterrence by making any Soviet notion of "winning" a European war seem less likely. In one sense this conventional capability might be less provocative than the present, politically sensitive alternative—the Pershing II missile, with its first-strike threat against Soviet command and control and other military facilities in the European USSR.

There may be several objections to this strategy, however. Some may argue that the simplest Soviet response would be to target nuclear installations within NATO as a counterdeterrent. Others may contend that the proposed policy is contrary to the 20-year trend of reducing collateral damage by developing nuclear weapons with lower yields and greater accuracy. Still others may assert that, given past Soviet objections to new weapons directed against the USSR, which resulted in East-West tension and fired local western populations into opposition, similar consequences are likely should the West, particularly nonnuclear states, adopt the policy and acquire the capacity to carry it out. Soviet concerns might be reinforced by the belief that though destruction of atomic plants has limited offensive utility, the threat could still change the psychological balance of power in Europe and limit Russian maneuverability in fulfilling its national objectives. Some might also be apprehensive about the proliferation of delivery systems to fulfill the targeting strategy, notably cruise missiles that could carry nuclear as well as conventional warheads.

Although worth considering, each of these objections is flawed. Collectively, they fail to appreciate that Soviet military doctrine denies nuclear war can be controlled. Although the Soviets could limit military operations in Europe to conventional weapons, both military exercises and developing capabilities suggest that the Soviets are also prepared to use large nuclear weapons "without any scruples concerning collateral damage and casualties," in the words of John Collins, a senior Congressional Research Service analyst (42). With three 150-kiloton warheads per launcher (the atomic bombs dropped over Japan were 10–20 kilotons), the ongoing Soviet SS-20 missile deployment now threatening Europe reinforces this contention. As for possible Soviet objections to the nuclear energy targeting strategy, it is incumbent on the West, not the Soviets, to determine its requirements for defense.

Still, renunciation of the strategy would be in the West's interest if the strategy appeared to undermine seriously relations with the Russians. An alternative would be to threaten adoption of the facility targeting policy to extract Soviet concessions, for example, to make Moscow more accommodating in European force reduction talks. The final option would be for NATO to embrace the strategy, notwithstanding likely Soviet objections. NATO is already hostage to Soviet nuclear weapons. For Western Europe, having limited nuclear weapons with which to respond and an American ally, that, Kissinger suggests, should not be too heavily relied on (43), adopting the nuclear reactor targeting strategy would introduce a new strategic element that could complicate Soviet risk calculations enough to enhance deterrence.

REMEDIES

Middle Eastern and European scenarios can apply to other regions as well, namely, Korea, South Asia, Taiwan, southern Africa, and the United States [see (16), ch. 3]. The implications of nuclear facility vulnerability in wartime should stimulate the search for alternatives that enhance stability. Equally compelling is the risk of early death, cancer, genetic disease, and associated traumatization. Assuming that nations continue to rely on nuclear power for their energy needs, the following discussion presents several nonexclusive avenues for controlling the dangers of nuclear facility destruction in war. These include controlling national behavior through international law, minimizing the consequences of facility destruction through civil defense, using different modes of facility siting, improving methods and procedures, and providing better international management of nuclear development.

INTERNATIONAL LAW

Legal restraint is clearly not the most reliable option for minimizing the military threat posed to nuclear energy facilities, but it does offer a relatively expeditious and inexpensive way to begin addressing the problem by establishing a clear standard of behavior where one does not now exist. This standard may offer some modicum of restraint, which in time can be supplemented by other more authoritative alternatives that require more time for implemetation.

The law of war prescribes norms of international conduct; it "attempts to reconcile minimum morality with the practical realities of war" (44, 45) through treaties, customary practice and perhaps resolutions by the United Nations General Assembly. At present, only the 1977 Protocol Addition to the Geneva Conventions of August 12, 1949 addresses the legitimacy of nuclear facility destruction in war. This document is a mixture of ambiguity and contradiction. Article 56 addresses the permissibility of attacks against "nuclear electrical generating stations" in the broader context of "installations containing dangerous forces," including dams and dikes. Paragraph 1 declares that such works and installations "shall not be made the object of attack, even where these objects are military objectives, if such attack may cause the release of dangerous forces and consequent severe losses among the civilian populations." The prohibition also extends to military objects located in the vicinity of the works "if such attack may cause the release of dangerous forces and consequent severe losses among the civilian population." Stipulations that the losses must be severe for the prohibition to be applicable raise the question of what constitutes severity. This point is further complicated in that irradiation might not result in death until years after exposure. Thus, rather than being a clear prohibition, the paragraph's ambiguity may conceivably be used to justify attack on a nuclear energy station.

Paragraph 2(b) further diminishes the strength of the prohibition: "The special protection against attack provided by paragraph 1 shall cease ... for a nuclear electrical generating station only if it provides electrical power in regular, significant and direct support of military operations and if such attack is the only feasible way to terminate such support."

In effect, this inclusion allows an adversary to decide whether a nuclear facility provides "regular, significant, and direct support of military operations." Such a rationalization can usually be found. Paragraph 3 appears to minimize the implications of this exceptional clause by stipulating, "If the protection ceases . . . all practical precautions shall be taken to avoid the release of the dangerous forces." This qualification is not an adequate safeguard.

The Geneva Protocol's treatment of nuclear facilities can be criticized further for a lack of comprehensiveness. It specifically addresses only one segment of the nuclear fuel cycle, namely, nuclear electrical generating stations. Large inventories of radioactivity, however, are located in other fuel cycle installations: nuclear spent fuel storage facilities, reprocessing plants, waste storage installations, and fuel fabrication facilities. If the prohibition is to be comprehensive, these plants must be included. Still another fault lies in the protocol's failure to address the permissibility of threats to destroy nuclear facilities. If their destruction is prohibited, consistency requires that the threat of destruction be prohibited in the codification.

Unfortunately, related conventions, namely the 1977 international environmental modification agreement and the treaty to ban the use of radiological weapons that is under negotiation in the Geneva-based United Nations Committee on Disarmament, do not contain such codifications. Since 1981, the Geneva Committee has attempted to remedy the situation. Sweden forwarded a proposal that attacks on all nuclear energy and weapons reactors be prohibited (46). The Israeli attack added further impetus to negotiate a convention in this regard (47, 48). Would this prohibition be effective? We cannot be overly sanguine. At the very least, such an agreement would provide a common standard in an area that would otherwise depend on prudent judgment. Such an accord would gain partisans inside and outside bureaucracies who would act as pressure groups to ensure observance (49).

MILITARY AND CIVIL DEFENSE

Although international law may provide some protection, there have been too many violations of other such measures for optimism. Therefore, diminishing the vulnerability of nuclear energy installations is critical. One way is to enhance military defense, including the ability to intercept and destroy hostile forces before they can inflict damage. Although nuclear facilities have not been constructed with wartime bombardment in mind, their massive containment structures do provide some protection against attack. Additionally, fences, alarms, cameras, and armed guards impede small groups of intruders bent on theft or sabotage. None of these prophylactic measures, however, is sufficient against lethal weapons.

Although not foolproof, alternative measures could improve point defenses. In a time of crisis, military units could be stationed around installations to prevent assaults. Antiaircraft and artillery could be situated to suppress bombardment. In the future, more sophisticated means of attack will require novel modes of defense. Homing missiles against cruise missiles, steel palings, and tons of steel pellets lofted by explosives to protect missile silos might be applicable in the defense of nuclear energy facilities (50).

Because military defense is imperfect—indeed, Iraq's reactor was defended—civil defense is critical. The programs of Sweden and Switzerland are exemplary (51–53). Most impressively, the well-organized Swiss program includes shelters, which during the 1980s will be able to accommodate the entire population, against many effects of nuclear weapons or facility releases. This program is supplemented by underground hospitals and stores of government and military supplies. Sweden's civil defense planning is notable for efforts to protect the economy by placing underground some industries, electrical power plants, and food stocks, as well as hospitals and command centers. The Swedes also have considered the removal of radioactive inventories from reactors and other installations to deep underground sites during crises. The population shelter program, although not as extensive as that of the Swiss, centers on a relocation plan to remove 90% of the urban population to areas up to 250 miles distant before or during the outbreak of war. Several successful relocation exercises have been carried out.

In comparison, programs of other countries are underdeveloped [(51), pp. 149–77; (52), pp. 75–86]. At the very least, governments should educate populations in the vicinities of nuclear plants about the dangers of radiation and about prophylactic measures. Evacuation routes should be planned, populations advised, and exercises practiced. Radiation shelters should be constructed for persons living close to nuclear installations. If shelters are not feasible, people should become acquainted with such expedient protective measures as remaining indoors and covering nose and mouth with a cloth during the passage of the radioactive cloud. In addition, every household should have a supply of potassium iodide tablets which will block the intake of radioactive iodine. They might also stock antinausea pills and breathing masks. Finally, civil defense planning should include steps to deal with long-term radiation contamination. If these measures were undertaken in conjunction with military defense, both the immediate and long-term casualties could be reduced, perhaps significantly.

FACILITY SITING

Most nations recognize facility siting as an important factor in protecting populations from accidental releases. Reactors are built 15 or 20 miles from urban areas, and support facilities are sited in even more remote regions. Residential populations are banned entirely for several miles around facilities. To minimize public exposure to contaminants, these distances could be extended. However, increased costs, especially in transmission of electricity, would result. Finding acceptable sites is another problem, and in small countries with high population densities, such as Israel or those in Western Europe, the problem is formidable. In others, including the Soviet Union, South Africa, Egypt, and Pakistan, all of which have remote regions, this option may be more attractive.

Remote siting need not be limited to land; facilities can be placed on large lakes, inland seas, or oceans. Power plants could be built on floating platforms surrounded by breakwaters, on floating vessels anchored to the marine floor, on artificial islands, or even undersea. However, there would be higher transmission costs for reactors, unique construction costs, and exposure to dangers peculiar to such siting environments, including ship collisions, accidental explosions, and naval bombardment.[11]

Underground siting is another option. Although it will not prevent a nuclear facility from releasing radioactivity via penetrations connecting the plant to the surface, it could reduce facility vulnerability. It affords several other advantages as well. Underground installations might be better able to withstand earthquakes and would also be immune to storms, explosions, and aircraft crashes. Rock formations able to contain releases would obviate the need for reinforced concrete containment vessels. Such facilities would not protrude from the landscape, thereby preserving aesthetic values, and land sites could be used for other purposes. Economic transmission costs can be realized if power plants are well situated and allow waste heat to be put to industrial uses. Finally, construction time can be shortened in the controlled environment of a cavern not exposed to the idiosyncracies of weather.[12]

Against such benefits, drawbacks must be weighed. Precise construction costs are unknown. One study ventured to say that burial of a power plant could add as much as 40% to its price (58). The siting may pose operational difficulties. Since subterranean installations are likely to be very compact, inspection and maintenance would be more difficult than in a surface facility. Furthermore, ground water seepage may pose unique corrosion problems.

[11] Studies that examine the marine option include (54, 55).

[12] There is some experience in siting research reactors underground in Norway, Sweden, France, and Switzerland (56, 57).

INCREASING PLANT SAFETY

Nuclear facilities and material can be made more resistant to radioactive releases. Some reactors are safer than others; heavy water reactors, high-temperature gas reactors, and molten salt breeder plants can withstand the loss of coolant better than the widely used light water reactors.[13] Improved containment would provide additional safety. For example, a filtration system connected via a duct to the containment building would significantly reduce radioactivity entering the atmosphere (59). Its performance in a wartime scenario would depend on the extensiveness of reactor destruction and the survivability of the filter system itself.

Inherent safety principles also can apply to other aspects of the nuclear fuel cycle, notably to high-level liquid wastes. Retention of wastes in liquid form makes them particularly susceptible to release when deprived of coolant. Further, there is a more acute problem in leaks resulting from containment rupture, a peril underscored by several such incidents due to corrosion. Fortunately, solidification into a glasslike substance makes migration unlikely except for water-induced leaching or vaporization of products as a consequence of a nuclear explosion. The commercial reprocessing plant built in Barnwell, South Carolina, called for glassification after one year's cooling. This period could be reduced to three months if the liquid were first converted to a granular state to allow better heat dissipation.

INTERNATIONAL MANAGEMENT OF NUCLEAR ENERGY

International management can provide some alternatives for addressing the nuclear facility wartime problem. Precedent exists in the International Atomic Energy Agency's responsibility to monitor nuclear material in most countries and in its advisory guidelines on methods to prevent sabotage and ensure safe operation. In addition, nuclear exporters have devised common criteria to ensure that their products are used for peaceful purposes.

Because destruction of nuclear installations in war affects global security, international institutions should exercise controls. The International Atomic Energy Agency should establish a working group to suggest guidelines that address the problem. A permanent standing committee might be created to advise nations about threats. Given the historical trend toward greater international institutional involvement to minimize nuclear dangers, creation of an authoritative international body may be warranted.

---

[13] See footnote 7, Taylor, pp. 11–13; private communication from Conrad Chester of the Oak Ridge National Laboratory, 9 April 1980.

Such a body would anticipate and regulate nuclear risks including national and subnational diversion of nuclear material for weapons purposes, the vulnerability of facilities to subnational sabotage, and the safe operation of plants, as well as wartime vulnerability (60). Whether more authoritative institutions are either desirable or attainable, in view of the reluctance of nations to surrender sovereignty to international bureaucracies which might not reflect their interests, is a matter that can be solved only in international negotiation and bargaining. Formulation of guidelines alone is probably the course of least resistance and is consistent with International Atomic Energy Agency practice.

CONCLUSION

Save for the elimination of nuclear energy—the practicality of which is hotly debated—none of the remedies to nuclear facility destruction in war is foolproof. All offer means of marginally reducing facility vulnerability in conventional weapons conflict and ameliorating the radiological consequences should destruction take place. They are likely to be ineffective in nuclear war. A conundrum therefore results, because the temptation to employ threats or actions against nuclear facilities adds a significant dimension to the problem of maintaining peace and minimizing the consequences of war. Given the potential for contaminating large areas, military destruction or threatened destruction may have significant implications for regional security. The Israeli bombardment of Iraq's nuclear facility (and Iraqi attacks in 1984 and 1985 on Iran's partially constructed power reactor along the Persian Gulf) suggests that the vulnerability of atomic plants to military action should be included in nuclear energy risk calculations and strategic planning.

ACKNOWLEDGMENT

The author wishes to thank Jan Beyea for technical advice in the preparation of this article.

Literature Cited

1. Permutter, A., Handel, M., Bar-Joseph, U. 1982. *Two Minutes Over Baghdad*, p. 142. London: Vallentin, Mitchell and Co
2. US Senate, Committee on Foreign Relations. 97th Congress, 1st Session. 1981. *The Israeli Air Strike*, p. 156. Washington, DC: USGPO
3. Ramberg, B. 1982–1983. Attacks on nuclear reactors: The implications of Israel's strike on Osirq, *Polit. Sci. Q.* 97: 653, footnote 1
4. Wilson, R. July 9, 1981. *Christian Science Monitor*, p. 23
5. Donnelly, W. H. 1981. Possible contamination of Baghdad from bombing of the

Iraqi reactor, *The Israeli Air Strike*, p. 156. Committee on Foreign Relations, US Senate, 97th Congress, 1st Session
6. Marshall, E. 1980. Iraqi nuclear program halted by bombing. *Science* 210: 508
7. *Physics Today*. 1981. 33: 53, 56
8. Am. Physical Society Study Group on Light-Water Reactor Safety. 1985. *Rev. Modern Physics*. Summer: Suppl. No. 1
9. Nuclear Energy Policy Study Group. 1977. *Nuclear Power Issues and Choices*. Cambridge, Mass: Ballinger
10. US Energy Res. & Dev. Admin. 1975. *U.S. Nuclear Power Export Activities*. ERDA 1542. Springfield, Va: Natl. Tech. Inf. Serv.
11. Chester, C. V., Chester, R. O. 1976. Civil defense implications of the U.S. nuclear power industry during a large nuclear war in the year 2000. *Nuclear Technol.* 31: 326–38
12. Fetter, S. A., Tsipis, K. 1981. Catastrophic releases of radioactivity. *Sci. Am.* 244: 41–47
13. Fetter, S. A., Tsipis, K. 1980. Catastrophic nuclear radiation releases. Report No. 5. Cambridge, Mass: Program in Science and Technology for Int. Security, Mass. Inst. Technol.
14. Cooper, C. L. 1978. Nuclear hostages. *Foreign Policy* 32: 125–35
15. Lewis, H. W. 1978. *Risk Assessment Review Group Report of the U.S. Nuclear Regulatory Commission*. NUREG/CR-4000. Washington, DC
16. Ramberg, B. 1980. *Destruction of Nuclear Energy Facilities in War: The Problem and the Implications*. Lexington, Mass: Lexington Books
17. Consultant Workshop, Sandia Labs. 1977. *Summary Report on Workshop on Sabotage Protection in Nuclear Power Plant Design*. SAND 76-0637. Washington, DC: US Nuclear Regul. Commission
18. Kaul, D. C., Sachs, E. S. 1977. *Adversary Actions in the Nuclear Fuel Cycle: I. Reference Events and Their Consequences*. SAI-121-612-7803. Schaumberg, Ill: Science Applications
19. US Nuclear Regul. Commission. 1975. *Reactor Safety Study*, Appendix VI, pp. 2-1–2-4
20. Hudson, C. I., Hass, P. H. 1977. New technologies: the prospects. In *Beyond Nuclear Deterrence: New Aims*, ed. J. J. Holst, U. Nerlich, p. 128. New York: Crane, Russak
21. Gervasi, T. 1977. *Arsenal of Democracy: American Weapons Available for Export*. New York: Grove
22. Cahn, A. H. 1977. *Controlling Future Arms Trade in the 1980s*. New York:

McGraw-Hill
23. Levenson, M. 1981. A perspective. *Nuclear Technol.* 53: 97–98
24. Levenson, M., Rahn, F. 1981. Realistic estimates of the consequences of nuclear accidents. See Ref. 23, pp. 99–108
25. Morewitz, H. H. 1981. Fission product and aerosol behavior following degraded core accidents. See Ref. 23, pp. 120–32
26. US Nuclear Regul. Commission. 1981. *Technical Bases for Estimating Fission Product Behavior During LWR Accidents*, NUREG-0772. Washington, DC
27. Beyea, J. 1980. *Some Long-Term Consequences of Hypothetical Major Releases of Radioactivity to the Atmosphere from Three Mile Island*. PU/CEES No. 109, pp. 12 and B-13. Princeton, NJ: Center for Energy and Environ. Studies, Princeton Univ.
28. Ramberg, B. 1984. *Nuclear Power Plants as Weapons for the Enemy: An Unrecognized Military Peril*. Berkeley: University of Calif. Press
29. Beyea, J. 1977. In the matter of Long Island Lighting Company. Jamesport Nuclear Power Station, Units 1 and 2. Direct Testimony of Dr. Jan Beyea. Tables 3c, 3d, and 6. New York: New York State Board on Electric Generation Siting and the Environment, Case No. 8003. Mimeo. Tables 3c, 3d, 6
30. Aldrich, D. C., Magrath, P. E. 1978. *Examination of Off-site Radiological Emergency Measures for Nuclear Reactor Accidents Involving a Core Melt*. SAND 78-0454. Albuquerque, NM: Sandia Labs.
31. US Environ. Protection Agency Task Force on Emergency Planning. 1978. *Planning Basis for the Development of State and Local Government Radiological Emergency Response Plans in Support of Light Water Nuclear Power Plants*. EPA 520/1-78-016. Springfield, Va: Natl. Tech. Inf. Serv.
32. Marshall, E. 1981. New A-bomb data shown to radiation experts. *Science* 212: 1364–65
33. Beyea, J. 1979. The effects of releases to the atmosphere of radioactivity from hypothetical large-scale accidents at the proposed Gorleben waste treatment facility, report to the government of Lower Saxony, Federal Republic of Germany, as part of the *Gorleben Int. Rev.*, p. 10
34. Rowen, H. S., Brody, R. 1980. The Middle East. In *Nonproliferation and U.S. Foreign Policy*, ed. J. A. Yager, pp. 203–37. Washington, DC: Brookings Inst.
35. El-Sayed Selim, M. 1982. Egypt, and

Mossavar-Rahamani, B. 1982. Iran. In *Nuclear Power in Developing Countries*, eds. J. E. Katz, O. S. Marwah, pp. 135–60 and 201–20. Lexington, Mass: Lexington Books

36. France: possibility of sale to Israel, and Iran may complete one reactor. 1982. *Nuclear News* 25: 62–63, 60

37. US Congress. 1981. Senate. Committee on Foreign Relations. *Hearings on the Israeli Air Strike*. 97th Congress, 1st Session

38. Betts, R. 1981. Nuclear proliferation after Osirak, *Arms Control Today* 11: 1–2, 8

39. Dunn, L. A. 1982. *Controlling the Bomb: Nuclear Proliferation in the 1980s*. New Haven, Conn: Yale Univ. Press

40. Feldman, S. 1982. The bombing of Osiraq—Revisited. *Int. Security* 7: 114–42

41. Collins, J. M., Cordesman, A. H. 1978. *Imbalance of Power: An Analysis of Shifting U.S.-Soviet Military Strengths*, p. 242. San Rafael, Calif: Presidio

42. Collins, J. M., Cordesman, A. H. 1978. See Ref. 41, p. 327

43. Kissinger, H. 1979. The future of NATO. *Washington Q.* 2: 3–17

44. Falk, R. A. 1976. Environmental warfare and ecocide: A legal perspective. In *The Vietnam War and International Law: The Concluding Phase*, ed. R. A. Falk, p. 290. Princeton, NJ: Princeton Univ. Press

45. Ramberg, B. 1978. Destruction of nuclear energy facilities in war: A proposal for legal restraint. *World Order Studies Occasional Paper 7*. Princeton, NJ: Center of Int. Studies, Princeton Univ.

46. Memorandum submitted by the delegation of Sweden on certain aspects of a convention prohibiting radiological warfare. 1981. CD/RW/WP.19. Geneva: Committee on Disarmament

47. United Nations. 1981. *Report of the Committee on Disarmament*, 36th Session, Supplement No. 27 (A/36/27), pp. 66–75, 106–7. New York: United Nations

48. Issraelyan, V. L., Flowerree, C. C. 1982. Radiological weapons control: A Soviet and US perspective, *Occasional Paper 29*. Muscatine, Ia: Stanley Foundation

49. Chayes, A. 1975. An inquiry into workings of arms control agreements, *Harvard Law Review* 135: 907–32

50. Garwin, R. L. 1976. Effective military technology for the 1980s, *Int. Security* 1: 53–56

51. Harvey, C. E. 1969. Civil defense abroad in review. In *Survival and the Bomb: Methods of Civil Defense*, ed. E. P. Wigner, pp. 159–62. Bloomington: Indiana Univ. Press

52. Murphey, W. H., Klinge, B. 1968. Civil Defense Abroad. In *Who Speaks for Civil Defense?* ed. E. P. Wigner, pp. 76–80. New York: Charles Scribner's

53. Swedish Government Committee on Radioactive Waste. 1976. *Spent Nuclear Fuel and Radioactive Waste*. SOU 32: pp. 64, 65. Stockholm

54. Yadigaroglu, G., Andersen, S. O. 1974. Novel siting solutions for nuclear power plants, *Nuclear Safety* 15: 654–57

55. Klepper, O. H., Anderson, T. D. 1974. Siting considerations for future offshore nuclear energy stations, *Nuclear Technol.* 22: 160–69

56. Watson, M. B. 1972. *Underground Power Plant Siting*. Pasadena, Calif: Environmental Quality Laboratory, California Institute of Technology

57. Crowley, J. H. 1974. Underground nuclear plant siting: A technical and safety assessment. *Nuclear Safety* 15: 531

58. *Nucleonics Week*, 28 July 1977

59. Okrent, D. 1977. *Post-Accident Filtration as a Means of Improving Containment Effectiveness*. Los Angeles: School of Engineering and Applied Sciences, Univ. of Calif.

60. Ramberg, B. 1979. Preventive medicine for global nuclear energy risks, paper delivered to the California Seminar on Arms Control and Foreign Policy

Ann. Rev. Energy. 1985. 10 : 515–56
Copyright © 1985 by Annual Reviews Inc. All rights reserved

MANAGING THE STRATEGIC PETROLEUM RESERVE: ENERGY POLICY IN A MARKET SETTING

R. Glenn Hubbard

Department of Economics, Northwestern University, Evanston, Illinois 60201, National Bureau of Economic Research, Cambridge, Massachusetts 02138, and John F. Kennedy School of Government, Harvard University, Cambridge, Massachusetts 02138

Robert J. Weiner

Department of Economics, Harvard University, Cambridge, Massachusetts 02138

INTRODUCTION

Ten years ago the US Congress authorized the creation of the Strategic Petroleum Reserve (SPR) with the intent of bolstering energy security. Today the Reserve stands at over 450 million barrels of oil—a large and potentially powerful policy instrument.

The questions surrounding the Reserve, however, are many, complex, and largely unresolved. This paper examines these questions, in the process reviewing the analytical approaches to resolving them.

The Problems of Oil Supply Disruptions

Supply shocks are a recurrent feature of the modern, petroleum-fueled economy. Six times since World War II the world has witnessed disruptions in the flow of crude oil from the Middle East. In 1953, 1957, and 1967, taking advantage of excess production capacity, governments and oil companies cooperated to patch the system together (1, 2).

515

0362–1626/85/1022–0515$02.00

By the 1970s, this excess capacity had all but vanished, save in the areas most subject to disruption. The 1973–1974 and 1978–1979 shocks caused oil prices to skyrocket, and were followed by recession in the industrialized countries. The damage in terms of lost economic growth in the OECD and oil-importing developing countries was staggering. The close historical relation between oil price increases and recession in the United States is documented in (3).

The development of buffer stocks can be seen as an attempt to restore the capability to deliver additional supplies rapidly to the oil market in periods of disruption. Such is the view adopted here. We shall have nothing to say about the military uses of strategic supplies; rather our paradigm is that of political economy. We approach the problem from the standpoint of the United States, but bear firmly in mind the international nature of the oil trade.

The implications of treating the SPR as an instrument of economic policy are several. First, a framework for analyzing how the oil market works is required in order to estimate the Reserve's potential impact on oil prices. Second, we need to understand the relationship between oil price shocks and macroeconomic performance. Finally, we must recognize the SPR's role as but one player in the market arena. The actions of others—oil-exporting countries, the domestic petroleum industry, other importers that hold buffer stocks—can make the difference between a potent SPR and an impotent one.

The Role of Policy

What can be done to alleviate the problems caused by oil supply disruptions? What should be done? These questions are meaningless in the abstract; no policy makes sense in the absence of a clearly articulated goal.

We assume that the goal of policymakers is to advance the economic welfare of the citizenry to the maximum extent possible. Here this implies protecting the economy from oil supply disruptions. Whether this goal stems from altruism or from the desire to remain in office is unimportant for our purposes.

According to the well-known "invisible hand" proposition first put forth by Adam Smith, national economic welfare is maximized when governments allow markets to operate freely. Only when some sort of market failure is present is there justification for public intervention. The case for any energy security policy, be it a Strategic Petroleum Reserve, energy conservation, or anything else, requires a demonstration of market failure.

The justifications for government intervention in the oil market on the grounds that the interests of private agents and the nation are not identical are four:

1. MACROECONOMIC LOSSES   These have been substantial in the past (4–6). A large, unanticipated increase in the price of oil generates economic losses for oil-importing nations through cyclical losses in aggregate demand, deteriorating terms of trade, and reduced potential output. The first effect is transitory, and is traceable primarily to downward inflexibility of nominal wages and nonoil prices, and to demand management problems. Related demand-side costs may stem from the redistribution of income among sectors, possibly affecting aggregate demand because of differences in propensities to spend.

The last two effects are more long-term. As oil is a major input to the economy's production process, an oil price increase will reduce potential output, i.e. the output attainable with all resources fully and efficiently employed. Conservation of energy and substitution for energy of other factors of production (e.g. capital and labor) in a free market will reflect adjustments to higher energy costs, but the reduction in energy consumption may lower capital and labor productivity (7).

2. MONOPSONY POWER   The United States is a "large player" in the world petroleum market; its actions affect market outcomes. An example is the price-control and entitlements program, which subsidized oil imports in the 1970s, thereby putting upward pressure on world oil prices. Imported oil thus has an additional cost associated with it (sometimes called the "monopsony premium"), inasmuch as increased US imports result in higher prices, and greater wealth transferred abroad.

3. VULNERABILITY   The level of national preparedness could affect the likelihood of a disruption (8, 9). This argument clearly applies only to the case of deliberate action taken by foreign powers against the United States.

4. NATIONAL SECURITY   The nation's security objectives may cause it to incur foreign-policy and military costs, regardless of whether a disruption is deliberately directed against the United States. An example of such a cost is the maintenance of the Rapid Deployment Force.

Two additional arguments for public intervention are based on imperfections in other markets. Information is the basis for the first; the government could have confidential access to intelligence regarding future supply conditions, and thus be in a position to make better decisions than private agents (10). The second is based on insurance; private agents may fear to act in their own best interests because they cannot insure against a possible public outcry for additional government regulation or taxation (e.g. windfall profits tax) during the disruption, and because of past behavior, the government cannot credibly commit itself to avoiding such actions (11).

The importance of each of these arguments is an empirical question, and will vary according to circumstance. Those who believe that the best energy security policy is to leave everything to the market, however, are obliged to reject them all.

Few analysts favor a public policy of pure laissez-faire. Support for a strategic reserve, possibly in conjunction with other government actions, is based on a mixture of theoretical and practical considerations. The alternatives to augmenting supply in order to lower price necessarily entail discouraging demand or price regulation. The unpalatability of these alternatives is an important factor in favor of maintaining a buffer stock.

Price controls are direct and highly visible, and thus politically attractive. The US experience in the 1970s resulted in unhappiness with price controls (12). Economists' arguments against price controls (apart from innate professional distaste) in a disruption divide into three. First, they discourage additional supply. This argument is weak unless supply is elastic in the short run, which seems unlikely. Second, they discourage conservation when it is most needed. Like the previous argument, this one depends on a short-run elasticity (in this case, of demand). There is an important additional consideration, however; the United States is not autarkic. Given imports as the marginal supply source, increased US demand exerts upward pressure on world prices, which in a disruption are very sensitive. No shortages need result, since only domestic prices are controlled, but demand pressure can raise world prices enough that even the controlled domestic price eventually exceeds the price that would prevail in a free market (13). Finally, price controls redistribute substantial income, creating beneficiary groups (in the past, US refiners and consumers) and making rescission difficult after the emergency has passed.

A short-term, or "disruption," tariff, tax, or quota is designed to exploit US monopsony power. By restricting domestic demand, downward pressure is exerted on world oil prices. In an economy free of macroeconomic rigidities (such as "sticky" wages), these are "first-best" policies for correcting market externalities, and thus improving social welfare. Although such policies are well suited to the objective of easing oil market conditions, they do so at the expense of raising domestic prices still further, thereby aggravating the macroeconomic harm associated with a disruption, and ought to be viewed with extreme skepticism. Nevertheless, they have been seriously proposed (14, 15). It should be noted that a buffer stock can also be a first-best policy response to the market-power externality.

Emergency mandatory conservation and fuel-switching combine the negative aspects of price regulation and demand restriction. That is, they are not only inefficient—since there is no reason that those who can most economically conserve will do so—but also have macroeconomic costs.

Finally, it is worth noting that the SPR is preferable to other forms of intervention on the grounds of political acceptability as well as economics. Past experience has demonstrated that measures that discourage consumption—long lines at gasoline stations, import tariffs, mandatory conservation—are tremendously unpopular as well as inefficient. Indeed, use of the Reserve is the only emergency policy endorsed by the current US administration (16).

Management Considerations

Acceptance of a government role in maintaining a buffer stock opens the door to a host of questions.

SIZE   The most obvious consideration is the size of the Reserve. The more oil stored, the more potent the SPR can be in a disruption. The chief criticism leveled at the Reserve, however, is its cost—each barrel must be purchased and stored. If the SPR is seen as a national insurance policy, the question arises: How much insurance is needed?

As important as size is the means by which the size is determined. What factors need to be taken into account? US oil consumption? Oil imports? Oil imports from potentially unstable areas? The likelihood of a supply shock? The behavior of oil prices, should one occur? Prospective damage to the economy? Reactions of other stock managers at home and abroad? Changing circumstances call for a reconsideration of the optimal size.

FINANCE   Like any large capital project, the SPR must be financed. To date, roughly $15 billion has been spent. Who should pay the cost? How should the financing be structured? Should those who pay have a say in its disposition?

INSTITUTIONS   Maintaining a reserve entails a grab-bag of institutional issues. Although these have little economic significance, they may be important nonetheless. Among such issues are the location and nature of the storage facilities, the type of oil to be stored, the means of procurement, the maintenance of physical and personnel infrastructure, and bureaucratic jurisdiction over planning and implementation.

USE   The drawdown decision is arguably the most important of all. A Reserve that is never expected to be used is no better than no Reserve at all.

How should the drawdown decision be made? By a policy rule or discretion? Should price serve as a trigger? Should the size of the shock? Should the Reserve be released immediately, or held back in case of catastrophe? Should releases be announced in advance? Should the size of the drawdown be a policy decision, or left to the market? Should it depend on the behavior of other importing countries?

How should sales of SPR oil be conducted? Who should be allowed to purchase? Should any groups be favored? If so, which, and how? By allocating a fraction of the Reserve? By giving them a subsidy?

HISTORICAL BACKGROUND ON STOCKPILING

Commodities have always been stored (17), both to facilitate smooth operation of distribution systems and in anticipation of future scarcity. Indeed it has become ritual to invoke the biblical tale of Joseph's storing grain against the seven lean years. The division of storage into "operational stocks" and "speculative stocks," although widely made in the literature (18), is not meaningful for analysis. Economists consider all inventory as held for the same purpose—profit maximization—with an associated "convenience yield" that is large for the initial units of inventory, and declines smoothly with size. This framework allows us to dispense with such nonsensical questions as "are stocks near minimum operating levels?"

Private Oil Stockpiling

Recalling that a public stockpile is justified only if market imperfections can be demonstrated, a review of past stockpiling behavior is in order. Table 1 presents data on the inventory-to-sales ratio (where both numerator and denominator have been seasonally adjusted) and the current-dollar and inflation-adjusted marginal cost of crude oil to US refiners[1] over the period 1960–1981. The data in the first column of Table 1 can be interpreted as "days of consumption." An examination of the table reveals a gradual decline in the inventory-to-sales ratio over the period 1960–1972, as stocks were not built as fast as demand increased. The almost flat nominal price indicates a declining real price over the period, and certainly a negative ex-post profit from holding speculative stocks when interest and physical carrying costs are taken into account.

Stocks were built up both absolutely and relative to sales during the period of the Arab oil embargo, and again during the second oil shock

---

[1] Various regulations have complicated the definition of P^{US} (defined briefly at the bottom of Table 1). A time series was constructed as follows. For the first quarter of 1960 (1960: 1) to the third quarter of 1973, when the average refiner's acquisition cost of domestic crude oil (PD) exceeded the average refiner's acquisition cost for imported oil (PM), P^{US} was equal to PD. Beginning in the fourth quarter of 1973, PM exceeded PD, which was kept down by price controls. From the fourth quarter of 1973, through the third quarter of 1974, PM was the marginal price of oil. From the fourth quarter of 1974 through the third quarter of 1976, the marginal cost faced by American refineries was measured as PM less the value of crude oil entitlements to refineries (N), which effectively acted to reduce the marginal cost. Beginning in the fourth quarter of 1976, the price of domestically produced stripper oil was uncontrolled. Therefore, from the fourth quarter of 1976 to the present, we used the price of stripper oil to refineries net of N as the marginal cost of oil. For more details, see (12).

Table 1 Quarterly crude oil inventory-to-sales ratio and prices

Year : quarter	I/S	1960–1981 P^{US}	P^{US}/P	Year : quarter	I/S	1960–1981 P^{US}	P^{US}/P
1960: 1	81	3.07	4.49	1971: 1	69	3.58	3.79
2	81	3.07	4.48	2	68	3.64	3.80
3	82	3.07	4.46	3	70	3.67	3.80
4	79	3.07	4.45	4	68	3.67	3.77
1961: 1	85	3.07	4.46	1972: 1	65	3.60	3.65
2	83	3.08	4.45	2	63	3.61	3.63
3	82	3.08	4.43	3	63	3.67	3.66
4	83	3.09	4.44	4	57	3.80	3.74
1962: 1	79	3.09	4.40	1973: 1	56	3.72	3.61
2	80	3.09	4.38	2	58	3.84	3.67
3	82	3.09	4.38	3	56	4.25	3.99
4	80	3.09	4.35	4	60	5.87	5.40
1963: 1	77	3.08	4.31	1974: 1	66	11.59	10.47
2	81	3.08	4.31	2	65	12.93	11.40
3	78	3.08	4.30	3	64	12.65	10.87
4	78	3.07	4.28	4	65	11.25	9.39
1964: 1	81	3.07	4.24	1975: 1	72	10.62	8.64
2	78	3.07	4.23	2	68	10.73	8.62
3	76	3.06	4.19	3	69	11.10	8.83
4	76	3.06	4.18	4	71	11.76	9.12
1965: 1	75	3.06	4.15	1976: 1	69	10.72	8.24
2	75	3.06	4.13	2	66	10.83	8.25
3	73	3.06	4.10	3	67	11.16	8.40
4	72	3.06	4.08	4	62	11.60	8.60
1966: 1	73	3.07	4.06	1977: 1	64	11.48	8.39
2	71	3.07	4.01	2	68	11.36	8.17
3	72	3.08	4.00	3	69	12.02	8.52
4	72	3.10	3.99	4	73	12.45	8.69
1967: 1	75	3.11	3.97	1978: 1	65	12.47	8.59
2	73	3.12	3.97	2	63	12.84	8.62
3	75	3.14	3.96	3	62	13.04	8.58
4	74	3.15	3.93	4	68	13.18	8.48
1968: 1	72	3.15	3.88	1979: 1	59	13.65	8.61
2	75	3.16	3.85	2	64	15.90	9.82
3	75	3.19	3.85	3	66	21.48	13.01
4	74	3.22	3.83	4	70	26.03	15.49
1969: 1	72	3.26	3.84	1980: 1	79	28.90	16.77
2	72	3.29	3.82	2	80	29.14	16.50
3	68	3.33	3.81	3	81	29.01	16.13
4	68	3.36	3.80	4	78	29.87	16.13
1970: 1	67	3.37	3.75	1981: 1	88	38.72	20.38
2	68	3.40	3.73	2	84	37.76	19.55
3	67	3.44	3.75	3	78	35.95	18.22
4	70	3.50	3.76	4	84	35.86	17.79

Notes: I/S = inventory-to-sales ratio (of seasonally adjusted quantities); P^{US} = marginal cost of crude oil to US refiners in dollars per barrel (construction described in footnote 1); P = GNP deflator (1972 = 1.000).

(Figures 1 and 2). The data for 1979–1981 point up difficulties in trying to isolate a speculative motive in oil inventory demand. It has frequently been stated that oil inventory levels were low prior to the interruption of oil supplies in Iran. Table 1 shows, however, that in the third quarter of 1978, the quarter before supplies were disrupted, inventory levels were not abnormal. The inventory-to-sales ratio did not fall to historically low levels until the end of the first quarter of 1979. This drawdown indicates not necessarily that oil companies were unprepared for a disruption, but rather that inventories may have been used to offset the supply reduction during the initial months. The recent downturn in oil prices, in combination with high interest rates, has encouraged oil companies to reduce their inventories.

This drawdown has occurred, but more slowly than many observers had expected—a likely consequence of a number of factors. First, oil demand has fallen short of what oil companies had predicted, leaving them with large quantities of unsold oil in inventory. Working through a standard oil demand function, companies may have underestimated the responsiveness of demand to higher prices (price elasticity) and the severity of the current recession. Second, if demand is weak, inventory reduction requires a reduction in oil purchases by oil companies. Such reductions may be limited in the short run by contracts or, in the case of the Aramco partners, by a special relationship with the producing country.

Public Oil Stockpiling in the United States

A brief review of the institutional history of the SPR will prove useful for subsequent analysis; a detailed discussion can be found in (19). The US government took its first step toward decreasing the growing dependence on foreign oil during the Eisenhower administration. After implementation of an unsuccessful voluntary program designed to decrease dependence, President Eisenhower acted under authority of the Trade Agreements Extension Act of 1955, establishing a quota system that remained in place until 1973.

In July 1973, the National Petroleum Council issued a report recommending that a stockpile of 540 million barrels of oil be secured in salt domes by 1978, enough to protect the United States from a supply disruption of the magnitude and length of 3.0 million barrels per day (bbl/d) for a six-month period (20). Also in 1973, Senator Henry Jackson introduced the Petroleum Reserves and Import Policy Act, calling for a strategic petroleum reserve of government-owned stocks held in salt domes.

During the debate over the Jackson proposal, events in the Middle East made apparent US vulnerability to foreign oil supply fluctuations. Soon after followed the Arab oil embargo. In 1974, President Nixon announced

Project Independence, which required the United States to be energy self-sufficient by 1980. The Project Independence report mentioned several fundamental concerns. First, the report listed the key considerations in designing a storage system: the type and location of the system and the size of the stockpile, in relation to the probability, magnitude, and duration of cutoffs. The report also included the two major costs, i.e. the economic costs of supply disruption and the costs of a stockpile necessary to buffer a disruption.

Finally, on December 22, 1975, President Ford signed into law the Energy Policy and Conservation Act (PL 94-163). The EPCA required that the Federal Energy Administration (FEA) submit a plan for implementation of the Early Storage Reserve within 90 days. Further, the FEA had to submit a comprehensive plan of the Strategic Petroleum Reserve by December 15, 1976. In April 1977, actual implementation of the SPR began. In addition, the Carter administration announced its plans to accelerate and expand the SPR program.

Difficulties in implementation came soon enough, however (20). Appeals to reduce SPR funding came from the Office of Management and Budget in 1978. In fact, not only was SPR oil not released following the Iranian revolution, but filling continued through August 1979. For a full year thereafter, there were no additions to the Reserve.

One of the reasons for this outcome was international political pressure. Though the SPR had little political opposition at home, it raised political problems at the international level. In the spring of 1979, US Government sources began leaking to the press that Saudi Arabia was threatening to cut its oil production by 1 million bbl/d if oil for the SPR were purchased. The Saudis claimed that US stockpiling purchases added to world oil demand and price pressure, undercutting their attempts to control OPEC's pricing policy. The US Department of Energy (DOE) was thus placed in the uncomfortable position of explaining away SPR purchases. It wanted neither to increase world demand or prices nor to aggravate the Saudis (whose 1 million bbl/d increase in production was helping to soften the shortfall caused by the collapse in Iranian production). Thus, the purchasing of reserve oil was halted or "delayed," as government sources claimed, until the world oil situation settled down.

The moratorium on filling the Reserve lasted from spring of 1979 through September 1980. Congress finally took steps to resume oil purchases through the Energy Security Act of 1980 (PL 96-294, 94 Stat. 932), urging the adoption of a fill rate of at least 100,000 bbl/d. By the end of 1980, the reserve contained only 107.8 million bbl of crude oil, well short of even revised schedules. Consequently, in 1981 Congress requested that the President "seek" to fill the reserve at an increased rate of 300,000 bbl/d. The

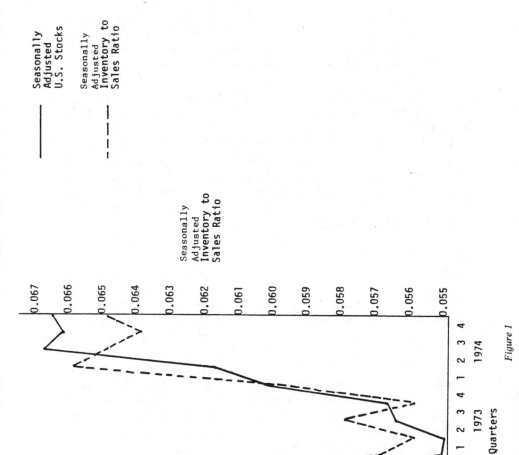

Figure 1

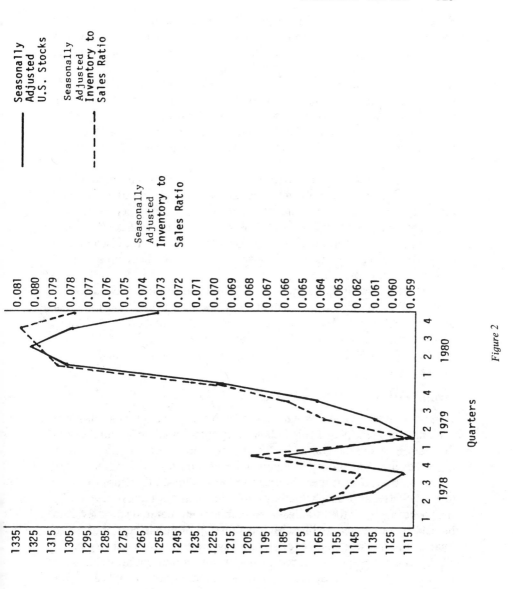

Figure 2

Reagan administration initially provided strong executive support for the SPR and oil was put into the reserve at an unprecedented rate. Oil in storage by the end of 1981 reached 230.3 million bbl. In addition, funding for the SPR was transferred to an "off-budget" Treasury account.

In 1982, to reduce federal expenditures, the administration proposed to slow down the fill rate from over 300,000 bbl/d to just over 200,000 bbl/d, moving the final completion date from 1989 to 1990. In addition, oil fill had rapidly caught up with available storage space. By the end of 1982, 300 million bbl were in the SPR. (By the end of 1983, 379 million bbl of oil had been accumulated.)

In 1985, with 450 million barrels in place, the SPR has become the principal tool with which the administration can deal directly with an oil disruption. In fact, it represents President Reagan's only activist energy policy, reflecting the administration's view that the energy market functions best with as little government intervention as possible [see the discussion in (21)]. Table 2 records the growth of the SPR from the beginning of 1978 through the end of 1983.

The SPR's current level and facilities permit a maximum drawdown rate of 2.1 million bbl/d for approximately 90 days, or 1.7 million bbl/d for around 150 days, after which the rate would progressively decline. Assuming a 6 million bbl/d import rate in the entire year of 1984, the drawdown would be equivalent to 35% of the imports for 90 days, or 28% for the 150-day period.

Public Oil Stockpiling Abroad

Primary OECD inventories, which stood at just over 3 billion barrels at the beginning of 1984, play a role in balancing short-run supply and demand fluctuations. As in the United States, strategic stocks are set aside (usually under government control) in Japan and Western Europe for potential use during a crisis. Government intervention has followed three approaches: (a) the US-style approach with the establishment and control of reserves by the government, (b) the establishment of minimum required stock levels to be held by private companies, and (c) creation of public corporations to finance and manage emergency stockpiling programs. A summary of the stockpiling programs employed in major consuming countries that are members of the European Economic Community (EEC) and the International Energy Agency (IEA) appears below in Table 3. A summary of the size of stocks in the OECD Big 7 (Canada, France, Germany, Italy, Japan, the United Kingdom, and the United States) is given in Table 4 below.

Only Canada and New Zealand have no official program; some others are implementing more than one of the three approaches. With the

Table 2 Size of the US strategic petroleum reserve

(millions of barrels)
1978–1984

Year : month	Stock level	Year : month	Stock level	Year : month	Stock level	Year : month	Stock level
1978 : 1	11.1	1980 : 1	91.2	1982 : 1	235.0	1984 : 1	384.0
2	14.3	2	91.2	2	241.0	2	387.0
3	17.1	3	91.2	3	249.0	3	392.0
4	21.8	4	91.2	4	256.0	4	397.0
5	25.6	5	91.2	5	261.0	5	404.0
6	30.1	6	91.2	6	264.0	6	414.0
7	35.2	7	91.2	7	267.0	7	424.0
8	41.0	8	91.2	8	274.0	8	429.0
9	47.1	9	92.8	9	278.0	9	431.0
10	53.1	10	96.6	10	285.0	10	438.0
11	59.3	11	102.3	11	290.0	11	443.0
12	66.9	12	107.8	12	294.0	12	451.0
1979 : 1	73.1	1981 : 1	112.5	1983 : 1	301.0		
2	78.2	2	116.1	2	306.0		
3	82.5	3	120.9	3	312.0		
4	83.9	4	134.2	4	318.0		
5	86.9	5	150.1	5	327.0		
6	88.6	6	163.1	6	332.0		
7	90.1	7	173.1	7	341.0		
8	91.2	8	184.7	8	352.0		
9	91.2	9	199.2	9	361.0		
10	91.2	10	214.8	10	367.0		
11	91.2	11	222.5	11	371.0		
12	91.2	12	230.3	12	379.0		

Source : US Dept. Energy, *Monthly Energy Rev.* (various issues).

Table 3 Government emergency reserve programs of EEC and IEA member nations

	Industry compulsory	Government owned	Public corporation	No program
Australia	×	—	—	—
Austria	×	—	—	—
Belgium	×	—	—	—
Canada	—	—	—	×
Denmark	×	—	×	—
France	×	—	—	—
Germany	×	×	×	—
Greece	×	—	—	—
Ireland	×	—	—	—
Italy	×	—	—	—
Japan	×	×	—	—
Luxembourg	×	—	—	—
Netherlands	×	—	×	—
New Zealand	—	—	—	×
Norway	×	—	—	—
Portugal	×	—	—	—
Spain	×	—	—	—
Sweden	×	×	—	—
Switzerland	×	—	×	—
Turkey	×	—	—	—
United Kingdom	×	—	—	—
United States	—	×	—	—

Source: (22).

exception of the United States, all have used the compulsory program for industry, mandating the amount of stocks that companies must maintain for use during emergencies. The public stockpiling policies of some of the major oil-importing countries are discussed in more detail in Appendix A.

Institutions for International Stockpile Coordination

A broad consensus holds that international cooperation in meeting oil shocks is at the same time essential and damnably difficult. Among the OECD countries, cooperation is under the aegis of the International Energy Agency (IEA).

It is not our task here to provide a detailed critique of past IEA actions; suffice it to say that consumer cooperation has not always been a resounding success. Indeed it has sometimes proved difficult to detect. The relevant regulations are codified in the International Energy Program,

Table 4 Stock-consumption ratios for the OECD (Big 7)

Year: quarter	Canada	France	Germany	Italy	Japan	United Kingdom	United States
			(days of consumption) 1973–1983				
1973	93	91	NA	NA	61	80	58
1974	101	115	89	111	76	104	64
1975	104	124	82	97	82	100	69
1976	95	111	85	94	82	103	64
1977	101	121	91	109	82	89	71
1978	85	97	92	99	81	93	68
1979	85	107	102	101	89	100	72
1980: 1	85	100	118	89	79	98	73
2	103	119	126	108	104	122	85
3	108	176	131	123	123	144	90
4	96	120	144	109	104	117	82
1981: 1	96	110	139	83	90	117	81
2	110	143	163	109	126	131	91
3	111	172	145	119	118	127	94
4	103	116	140	103	104	104	93
1982: 1	98	115	129	90	96	90	87
2	104	122	151	101	128	111	88
3	93	146	144	115	121	110	94
4	93	120	140	109	109	95	95
1983: 1	97	104	132	95	97	89	95
2	89	102	132	106	115	95	94
3	85	131	140	111	110	99	96
4	80	93	127	85	103	95	95

Source: Calculated from *Monthly Energy Rev.*, various issues.

signed by the United States in 1974. The details are too involved to present here [see (23)], but the salient points are three. First, countries are required to hold buffer stocks in proportion to their imports. Second, the agreement is dormant until a determination of emergency is made. (The emergency is signaled by a shock large enough to reduce supply by 7% compared to its preshock value. In practice the time unit is the quarter, and the preshock value is a moving average of the previous four quarters.) Third, the agreement calls for countries to "restrain demand" by 7% (through taxes, tariffs, regulation, exhortation, etc.) and substitute buffer stock releases in making up any remaining loss in supply (e.g. a 10% reduction in quantity supplied calls for 3% to be made up by stockpile releases in addition to the 7% demand restraint). The scheme's monopsonistic intent is clear.

Various technical problems with such a program have been pointed out in the literature; here, we take note of two broader difficulties. First, the 7% threshold corresponds to a severe disruption. Assuming oil consumption of roughly 50 million bbl/d in the noncommunist world, and taking the IEA share of consumption as constant,[2] a loss of 3.5 million bbl/d (net of increased exports by other producers) is necessary to trigger the emergency mechanism. The Iranian crisis, during which oil prices more than doubled, was of considerably lesser magnitude. Second, demand restraint proved easier said than done; the March 1979 agreement to cut consumption by 5% was honored more in the breach than in the observance.

Among the lessons to come out of the 1979 and 1980 supply shocks was that while high stockpile levels are a sine qua non for the functioning of international sharing agreements, it is the drawdown (or buildup) behavior that is likely to spell the difference between containment and disaster. Another is that actions taken in a so-called sub-trigger disruption (one falling beneath the threshold) may serve to avert a 1979-style catastrophic price run-up. Demand restraint's having failed, the economic damage attending a sub-trigger disruption has called forth proposals for coordinated drawdown programs.

In evaluating SPR drawdown strategies, international coordination considerations must be kept in mind. How effective would SPR draw be in relieving pressure in the world oil market if other IEA members do not do likewise? Or, even worse, what if some countries fill while others draw?

ANALYSIS OF STOCKPILE POLICY

The Strategic Petroleum Reserve represents a source of supply (or demand, depending on whether it is being released or filled) in the oil market. To evaluate its role, it is essential to have a framework for describing the behavior of the other participants—and hence of prices—in the market. This is not the place for a treatise on oil market modeling, but a quick review will be helpful in understanding the discussion of stockpile policy that follows.

Modeling the World Oil Market

Models of the world oil market fall into two categories. Long-run models are based on dynamic optimization; producers maximize the present value of income subject to demand conditions. While well-suited to examining depletion and long-term price evolution, these models perform systematically poorly in accounting for short-term price perturbations (24).

---

[2] This will be strictly true only if the elasticities of demand in the IEA and the noncommunist world as a whole are equal.

The question of "what causes oil price shocks?"—that is, how does the market behave in the very short term?—has not been resolved satisfactorily (25). Modelers are confronted with a world wherein inventory fluctuations can play a critical role, spot prices and official government selling (i.e. contract) prices may diverge widely, and information disseminates slowly compared to the speed of market events.

The "facts" to be explained are the enormous price increases of 1973–1974 and 1978–1979, and the absence of one in 1980, despite the fact that the three shocks [OAPEC (Organization of Arab Petroleum Exporting Countries) embargo, Iranian revolution, Iran-Iraq war] were of the same order of magnitude. It should be clear that the causes of an oil price shock must be addressed in order to evaluate the SPR's efficiency in alleviating it.

The explanations offered for the price shocks are several. First, the oil market was cartelized by the Organization of Petroleum Exporting Countries (OPEC) in the early 1970s (26–29). Second, the decline in new discoveries, pessimistic perceptions regarding alternatives, and ever-increasing demand forced prices higher; OPEC merely validated the new scarcity (30, 31; 32 agrees for 1973–1974, but cites monopoly power for 1978–1979). Third, the market is competitive, but the supply curve is backward-bending; the higher the price received, the less oil will be produced (33–35). Fourth, the market is competitive, but the replacement of the multinational companies by nationals with lower discount rates resulted in a decline in production (because oil in the ground would now be worth more), and hence, higher prices (36, 37). Although each of these theories contains elements of plausibility, and each has its vocal adherents and detractors, none provides a completely convincing characterization of the market, and none can explain price behavior over the last 10 years.

Some assumption about market structure is needed if the SPR is to be evaluated. The exercise is usually undertaken through the use of a "stockpile premium"—the value to society of a barrel added to the Reserve, above and beyond the purchase price of this additional barrel.

While the idea of such a premium is legitimate (although not necessarily important for policy), the methodology employed to calculate it is not. The premium—along with its similarly questionable companion, the "import premium" (the cost to society of the marginal barrel of imported oil above and beyond its purchase price)—is typically arrived at by assuming that the oil market is competitive (38, 39). As noted above, the competitiveness of the oil market is far from manifest. In such cases, it is necessary to establish the robustness of the calculations under various market structures. That this is not done presumably owes to the fact that the problem must be set up differently under the different market regimes.

The second oil shock could not be predicted or explained by these theories, spawning further attempts to characterize oil market behavior in the short run. A prototype model had been built by Gately, Kyle & Fischer (40), in which unused production capacity resulted in exporters' lowering prices; prices were raised as producers approached their capacity constraints. Although the pricing rules offered were ad hoc, this approach succeeded in formalizing the notion of a "tight" market. It did not, however, address the critical role of inventories.

Attempts were also made to capture the dynamics of spot and long-term contract pricing, and their interaction. The "ratchet theory" (41, 42) described the process this way: In the absence of supply shocks, spot and contract prices are flat. A supply shock raises spot prices, whereupon oil-exporting countries respond by increasing their official contract prices. When the shock eases, the exporters cut production to maintain the higher prices. Thus prices evolve by remaining constant most of the time, but "ratcheting upward" periodically. The implications for SPR policy are clear: quashing the initial ratchet lowers oil prices permanently.

This model has two serious problems. The first is suggested by this "free lunch" aspect of SPR release: The ratchet theory is lacking in economic rationale, since the behavior ascribed to exporters is internally inconsistent. The second objection is even more basic: Under the ratchet, oil prices can never fall. That the theory appears foolish today could easily have been anticipated, since all commodity prices have risen and fallen periodically since time immemorial. Nevertheless, the ratchet appears to have been taken seriously for a while.

A related, somewhat more sophisticated approach to spot and contract pricing was first suggested in the copper market (43), and later adapted to the oil market (14). Under this analysis supply shocks register first in the spot market, and are reflected in contract prices only with a lag. No motivation is offered for this behavior, however; thus the recommendation for SPR policy—release as much as possible as soon as possible in order to damp the initial spot price run-up—stands on shaky ground. The model is at least internally consistent; OPEC does not exert market power, but rather acts competitively. Although the model is unsatisfactory as an explanation of pricing behavior in the oil market (25), it can be made to fit the data well, and has been used in a number of empirical studies (44–46).

The faults in our models of the oil market run deeper, however. Once we admit the possible divergence of spot and contract prices, no longer can we assert the traditional economist's view that "the world oil market, like the world ocean, is one big pool" (47). Would that it were, for then arbitrage would immediately eliminate spot-contract price differentials, making the modeler's job easier.

The other extreme assumption—that embargoes against particular countries, and "access" to oil supplies, make sense—is simply unbelievable in a market as large and liquid as world oil. The 1973–1974 shock was spread roughly evenly across importing countries (48). Warnings of dire consequences due to increasing "rigidities" since then (49) have a curious Chicken Little aspect to them; since an analytical framework is lacking, it is unclear even what would constitute supporting evidence. Unfortunately, although the effects of buffer stock policy are sensitive to assumptions about the ease of reshuffling trade patterns, we are just beginning to model the case intermediate between the "no transaction cost" and "infinite transaction cost" poles (50).

Inventory behavior presents additional difficulties for oil-market modeling. The importance of oil inventory demand, which fell during the last shock but climbed sharply during the previous two, is widely acknowledged. A satisfactory explanation for inventory behavior remains elusive, yet is critical in assessing the private sector's response to public stockpile policy. For example, the SPR will be rendered impotent if its releases are hoarded by private stockholders. Without a model of their behavior, however, it is impossible to predict how these agents would react to an SPR release.

Inventories always smooth shocks in a well-functioning market (51), by reallocating supplies to periods when they are valued most. Claims that stock accumulation ("hoarding") by private agents exacerbated the 1979 shock (14, 41, 52) depend on these agents' acting irrationally—i.e. "buying dear and selling cheap" (not a very promising explanation)—or on some unspecified market imperfection that allows inventory accumulation in the anticipation of higher prices, yet prevents arbitrage from eliminating these speculative profits immediately.

Size

How large should the Reserve be? Answering this question entails solving three related problems: First, how will the Reserve be used? Drawdown policy is discussed below; size studies typically assume the SPR will be used optimally. Although the assumption itself is questionable, it is useful methodologically, and serves to calculate an upper bound on the optimal size of the stockpile. Inferior drawdown strategies will reap fewer benefits, which implies a smaller Reserve, assuming decreasing marginal returns to size.

Second, how will releasing the SPR benefit the economy? Answering this question requires a model of the oil market, as discussed above, as well as a model of energy-economy interaction. Simply treating gross national product as a function of oil prices (53) is inadequate; a production-function (microeconomic) or income-expenditure (macroeconomic) approach, or both, is required.

The final, and hereunto most difficult, problem with which to grapple is the characterization of uncertainty. The correct approach is stochastic dynamic programming, which entails calculation of the best fill-cum-draw strategy at each point in time, given expectations of the entire future path of oil prices. Such a calculation is prohibitively difficult, and has been solved only by reducing the oil market to a very small number of states—typically two (54) or three (55, 56) that correspond to a disruption's being "on" or "off" (or "small" or "large" in the latter case)—and introducing a Markov transition matrix describing the probabilities of moving between states.

Unfortunately, while Markov matrixes are appropriate for discrete systems, the oil market is not characterized by being "disrupted" or "normal"; supply and demand are continuous, and thus so are prices. Envisioning it this way is apt to be misleading, and misses entirely the dynamic issues discussed above. Nevertheless, in the absence of more realistic models, this approach is useful in generating estimates of optimal stockpile size.

The benefits stemming from a Reserve must be compared with its costs in order to determine the optimal size. These costs comprise the purchases of oil to be stored, the increase in world oil prices resulting from these purchases, the interest paid on capital borrowed to make the purchases, the purchase or development costs of storage capacity, and the operation and maintenance costs of the storage program.

Given the diverse assumptions and methodologies used to attack the three problems listed above, it is encouraging, as well as surprising, that the recommendations regarding SPR size fall in a relatively narrow band. A survey by the National Petroleum Council (57) found that of 20 studies conducted in the 1970s, all but two suggested figures between 500 million and 1000 million barrels, with the more recent recommendations in the upper half of this range. Analyses conducted in the early 1980s, when the climate appeared most menacing, recommended sizes of 750–2000 million barrels (55, 56, 58, 59).

These analyses assume a perfectly competitive oil market. That the dynamic programming models returned verdicts similar to those of the most "naive" models, which employ a fixed supply (e.g. 58), suggests that future research should adopt a simplified characterization of uncertainty, and instead focus on alternative models of oil market structure. Indeed, even stockpile models with no uncertainty (8, 9, 60) afford considerable insight into the roles played by market power and strategic behavior. One approach is to have the shocks follow a low-order autoregressive process, thereby capturing much of the complicated transition matrix apparatus in one or two parameters (50).

Finance

Demonstration of the public good aspects of public strategic stockpiling of oil points up the need to consider ownership of SPR oil in the context of the economy's portfolio of assets. Modern theories of the valuation of capital assets (61, 62) emphasize the importance of not only the mean return on an asset, but also the covariance of that return with that on the market portfolio as a whole. Given that oil price increases have been associated with losses in real income and declines in aggregate wealth, the return on SPR oil is likely to be negatively correlated with the return on financial assets. That is, claims to SPR oil have a high payoff in states of nature in which other sources of income are low. Ownership of SPR oil may thus be advantageous for purposes of portfolio diversification.

Apart from the issue of the optimal size of a public oil stockpile, the question of how the stockpile is to be financed arises. That is, given that the economy as a whole gains from the existence of an SPR, but that individual risk preferences may be different, are taxpayers better off with compulsory or voluntary participation in an SPR investment program? Certainly to the extent that public oil stockpiling shifts ownership of stocks from private interests demanding a positive risk premium to taxpayers who because of macroeconomic costs of oil supply disruptions demand a negative risk premium, the overall allocation of risk is improved. [See (63) for further discussion.]

Complete financing of SPR acquisitions by the government may be unnecessary. In addition to taxpayer financing of public benefits, private benefits can also be financed through oil-denominated equity or bonds. By allowing a market allocation, these types of private ownership would act to equalize the risk premium across agents in the economy.

Under the former, or public capitalization, approach, the government would sell to the public certificates entitling the bearer to a set quantity of oil (in the SPR). These certificates would be bought, sold, and traded in secondary markets, much as common stocks are traded today. Investment returns would be determined exclusively by changes in the market price of oil.[3] Under the latter (debt financing) option, the federal government would

---

[3] Institutionally, the public capitalization proposal allows for maximum government control, since the SPR administration is left with sole charge of drawdown management. Possible pressure to draw down the SPR when certificate holders wish to "cash in" at (what they consider) peak value should be compensated for in an active secondary market. Certificate holders would be free to sell at any time, and as long as investors remain confident in the SPR's ultimate survival, prospective buyers should be plentiful.

issue a new series of bonds whose sole purpose would be to raise funds to finance the SPR. Various suggestions have been made as to the potential yield of such bonds, including return tied to the rate of oil price appreciation, and return connected with the prevailing market interest rate.[4]

In addition to the options outlined above, two dirigiste alternatives are possible—development of an Industrial Petroleum Reserve (IPR)[5] or mandatory private contributions to the SPR.[6] Regardless of the alternative, the proper way to view the financing issue is through an economic analysis of the benefits (public and private) of ownership of SPR oil as an asset. Discussions of whether the SPR should be financed "on budget" or "off budget" merely reflect the fact that "current account" and "capital account" components are not distinguished in the federal government budget.[7]

Drawdown: The SPR and the Oil Market

The issues surrounding drawdown are perhaps the most difficult analytical facet of a study of the Strategic Petroleum Reserve. Before considering some specific options for stock release, we examine three related areas of inquiry that will determine the effectiveness of the SPR as an economic policy instrument—(a) market structure, (b) expectations (about the likelihood of future

[4] New bond issues may or may not provide a steady flow of income for the SPR. If the bond return is linked to some fixed long-term interest rate, the inflow of capital should be stable. However, a return based on oil price appreciation would make budget planning more difficult (as oil price forecasts would be necessary). Should the sale of bonds collect less than the expected revenue, direct federal outlays would be required to meet the shortage. Furthermore, upon drawdown of the SPR oil, oil prices would have to have increased by more than the rate of interest paid, or the sale of reserves would not cover the cost of establishing and administering the SPR. The difference would be absorbed in the government budget deficit.

[5] Numerous proposals have been put forth to outline how the government could elicit industry participation in an IPR.

(a) The President could require all importers and refiners of foreign oil to store up to 3% of their annual consumption in an "emergency inventory" ("decree" option).

(b) The government could provide financial incentives (in such forms as tax credits and direct subsidies) to firms if they would increase their inventories of stock crude oil ("incentives" option).

(c) The government could require that each firm be responsible for storing a set amount of oil. Firms could store the oil themselves or see to it that somebody else stores it for them. So long as a firm could account for its required share, the government would not distinguish between the two (the "sufficient evidence" option).

[6] This agenda calls for firms that import, refine, or domestically produce oil to be directed by the government to make specified amounts of oil available for use by the SPR. The government could either allow the firms to bear this cost (to the degree that they would be unable to pass the costs on to the oil-consuming public), or it could subsidize them.

[7] This is not to deny the political importance of the issue.

shocks or the duration of present ones), and (c) the reaction of domestically held private stocks and foreign public stockpiles to announced and unannounced SPR policy. We consider below these three areas together in a simple theoretical model emphasizing the joint determination of oil prices and private and public inventory decisions. While the framework is simple (and described in more detail in Appendix B), it yields predictions for the response of private stocks to public stock changes and for the chances of success of international stockpiling agreements. We then take an example of the model and combine it with an econometric model of the US economy to test some of the predictions in simulation exercises.

I. PRIVATE STOCKS AND OIL PRICES   To study the interaction of public and private stocks, we begin with a stylized model of the short run. Detailed description of the model is presented in the Appendix; only a summary is given here. Throughout, we suppose that oil production and consumption are responsive to the current period price, and are subject to transitory supply and demand shocks.

Of course, total demand is the sum of consumption and inventory demand. Speculators trade in inventories on the spot market in anticipation of changes in price and are assumed to maximize expected profit from speculation.[8] Holding stocks is assumed to be costly—in fact, increasingly costly—in the size of the stock due to payments to factors fixed in the short run, such as storage facilities, tankers, and pipelines. Thus changes in price expectations cannot be fully acted upon instantaneously. The optimization problem described here yields speculative holdings as a function of the expected increase in price, taking into account the cost of adjusting stock levels.

Since quantities demanded and supplied in the market must be equal in equilibrium, the assumptions about the behavior of consumption, speculative stockpiling, and production can be used to determine the market (spot) price. Under simplifying assumptions of linear supply and demand functions, the rational expectations solution for the price P can be written as

$$P_t = \psi P_{t-1} + \beta(\varepsilon_{Dt} - \varepsilon_{St}), \qquad 1.$$

where ε_D and ε_S represent the transitory demand and supply shocks, respectively.[9] ψ and β are functions of the parameters of the system—the

[8] That is, they are assumed to be risk-neutral. The assumption of risk neutrality is not necessary for the results that follow; it merely simplifies the exposition.

[9] That is, ε_D and ε_S are independently and identically distributed with mean zero and variance σ_D^2 and σ_S^2, respectively.

price responsiveness of production and consumption and the cost of adjusting stock levels.

Explaining the persistence effect of oil shocks is important for understanding the impact of energy policy interventions. Here even transitory shocks exhibit persistence effects on the price because of the behavior of inventories. It can be shown (see Appendix B) that the more price-responsive is consumption, the smaller is the initial increase in price from a supply shock and the lower is the persistence. Hence, policies such as oil import tariffs or certain types of buffer stock stabilization policies (which effectively raise this price responsiveness) can mitigate both the impact and the long-run effects of transitory shocks on prices.

Equation 1 can be used to motivate empirical work and simulation exercises, as we demonstrate below. As a qualification, while this discussion illustrates the importance of private speculative stockpiling for analyzing the behavior of oil prices during supply disruptions, consideration of the SPR and of realistic disruption scenarios requires an examination of serially correlated quantity shocks. Suppose that demand and supply shocks follow first-order autoregressive processes, so that

$$\varepsilon_{Dt} = \rho_D \varepsilon_{Dt-1} + v_{Dt}, \quad \text{and} \qquad\qquad 2.$$

$$\varepsilon_{St} = \rho_S \varepsilon_{St-1} + v_{St}, \qquad\qquad 3.$$

where v_{Dt} and v_{St} are white noise and $\sigma_{v_D v_S} = 0$.

Given this structure of shocks, we can rewrite the solution for price as

$$P_t = \psi P_{t-1} + \gamma(1 - \gamma \rho_D) v_{Dt} + \gamma(1 - \gamma \rho_S) v_{St}. \qquad\qquad 4.$$

Note that in this situation persistence effects come also from the serial correlation parameters ρ_D and ρ_S.

As long as supply shocks are purely transitory, inventories will be drawn down in response to a negative supply shock. Speculative accumulation requires either serially correlated shocks or the expectation that the disruption will get worse. Given a negative supply shock ($\varepsilon_{St} < 0$), the likelihood of stock accumulation in response to a supply disruption is greater the higher the serial correlation of the shocks (ρ_S) and/or the higher is the intertemporal correlation of price changes (ψ).

Quantifying the impact of SPR releases on oil prices requires a set of assumptions about the behavior of private stocks. The discussion above indicates that the consideration of market structure, the characteristics of the shocks (e.g. "transitory" versus "permanent"), and expectations of future shocks must be central elements in any attempt to model that linkage.

II. PUBLIC STOCKS   In addition to understanding the role of domestic private stocks, consideration of the precise way in which (domestic and foreign) stocks are to be used is important for modeling efforts. For example, the simplest sort of intervention is an exogenous injection of supplies at the onset of a shock, a move tantamount to reducing the magnitude of a supply shock ε_{St}. Such a move will dampen private stock drawdown if the shock is perceived to be transitory, or blunt speculative accumulation if the shock is perceived as likely to get worse in the future.

A second type of public stock response is to substitute a rule for discretion. For example, one countercyclical rule would be to determine stockpile releases S as a function of deviations of prices from trend, i.e.

$$S_t = \omega P_t, \qquad \omega > 0. \qquad\qquad 5.$$

A rule of the form of Equation 5 is analytically equivalent to an *ad valorem* tariff. The price responsiveness of demand is effectively heightened, reducing the price, as well as the persistence effects of the shocks.[10]

These examples of stockpile intervention, however, beg the question of optimal public stockpile behavior. Focusing on the optimizing process of the public authority facilitates consideration of the government's objectives, specific sources of market failure, and the potential benefits from international stockpile coordination (e.g. the IEA agreements).

To illustrate the formulation of such an optimizing process, suppose that the public stockpile is used in accordance with an assumed economic policy of maximizing real income (output less payments for imported intermediate goods). More specifically, suppose that the stockpile authority is risk-neutral, and that its objective is to maximize real income (by minimizing oil price increases) less the cost of carrying out the stockpile program and of adjusting stockpile levels, subject to the constraint that stockpile releases not exceed the amount of oil held in the reserve.

There is a clear distinction between the optimization problem for the public stockpile authority and the problem for the private firm discussed earlier. The public authority pays attention to aggregate output. Private firms do not consider the macroeconomic effects of their stockpiling behavior; that is, they do not consider the impact of their transactions on the world oil price.[11]

---

[10] See (64) for a discussion of stockpiling rules.

[11] Hence the division of stocks between the private and public sectors is important. In addition, in the argument in the text, we have implicitly assumed that there is a single stockpiling authority in consuming countries. In reality, there are strategic stocks in each country. Because the stockpiling decisions of other countries affect the price of oil, they can affect the optimal release strategy of the domestic authority.

Ceteris paribus, the larger the oil consumption, the greater will be the stockpile release because of the benefits of lowering the price paid on inframarginal imports. As with the case of private stocks, the expected persistence of shocks is an important factor in determining the optimal public stockpile policy. The greater the persistence of the shocks, *ceteris paribus*, the smaller the release at the onset of the shock. We pursue this issue in more detail in the next section, in the context of international cooperation in drawing down strategic stocks.

III. INTERNATIONAL COORDINATION OF PUBLIC STOCKPILES   Since the oil market is internationally integrated, the use of a buffer stock by one country has spillover effects on others. The possibility of international policy coordination thus becomes important in attempting to reduce the impacts of oil shocks. While our example below finds merits (in terms of lower prices) of international stockpile coordination, issues of whether such an outcome would occur in the absence of an agreement and of what types of institutional mechanisms might facilitate cooperation have been largely ignored.

That cooperation can reap benefits begs the question of how it might be achieved. Regulation at the international level is difficult to enforce; since there is no regulator with the power to require compliance, the incentive question naturally arises. While import restriction is clearly in the interest of the group as a whole, the effectiveness of the regulatory rule in attaining the cooperative outcome is not evident.

As with the case of private-public interaction, the characteristics of shocks and of expectations play an important role in modeling international policy coordination. Those characteristics in turn may depend on elements of market structure—e.g. the use of long-term contracts in the oil market in addition to spot transactions. More specifically, in a related paper (50), we show that, following a negative supply shock, the anticipation of higher oil prices in the future (i.e. serial correlation of the effects of the shock) leads to a higher rate of public inventory accumulation (lower optimal stockpile release) in the current period.

During a crisis in which the (now higher) oil price is expected to decline, countries are willing to draw down their stockpiles at the onset of a shock, even in the absence of a coordinating agreement. If the oil price is expected to increase further, however, a drawdown in the current period mandated by a stockpile coordination agreement is not in the interests of the individual members.

Given a specification of the optimizing behavior of the public stockpiles of consuming countries, we can estimate the benefits of stockpile coordination using "game-theory" methods in economics. In our analysis of public stockpile behavior as an international game (50), we found that whether member countries to an agreement drew down more stocks at the onset of a shock than if there had been no agreement depended on (a) the expected persistence of the shocks and (b) the heightened monopsony power made possible by coordinated buyer behavior.

Our discussion of persistence in the previous sections assumes new importance here because given monopsony power, the greater the persistence effect of shocks on oil prices, the greater the cooperative optimal drawdown relative to noncooperative optimal drawdown. That is, the benefits (in terms of lower world oil prices and recouped GNP loss) of participating in an international agreement are greatest when the impact of oil shocks on oil prices exhibits substantial persistence.

The simple example of the previous section used inventory-smoothing behavior as a motivation for persistence. Perhaps more important in the oil market are long-term contracts. Given that only a fraction of oil trade occurs on spot markets, oil prices adjust only gradually to even transitory shocks. This institutional complication influences the benefits from agreements. By examining the structure of contracts in the oil market, we found (50) that persistence depended on, among other things, the fraction of trades carried out through contracts and on the price responsiveness of demand in consuming countries. Because these factors changed the persistence effects of oil shocks, the relationship between the best stockpile responses of the SPR alone and those of an international agreement also changed.

Those results have clear implications for the operation of the IEA agreement. The use of current quantity loss as a regulatory signal is misdirected, since it ignores the critical influence on national optimizing behavior of market dynamics. Loosely speaking, whether the shock is anticipated to "improve" or "worsen" determines the relationship between the cooperative and noncooperative solutions.

Of course, any policy not in effect at all times requires a "trigger" to activate it. A natural candidate, used in buffer stock schemes for some commodities, is price. In a market characterized by short-run contract rigidities, however, a supply shock leads to at least two prices' prevailing at any given time. However, treating the spot price as the marginal cost of acquiring oil is in general unwarranted; the usefulness of this price as a signal depends on the fraction of trades carried out in the spot market. It is the marginal acquisition cost that is relevant.

Finally, scattered evidence suggests that the market is becoming more flexible (i.e. the share of contract trade is declining), implying that noncooperative behavior will be less costly in the future than in the "high persistence" regime of the past. If such is the case, we should concentrate on developing guidelines for using the Strategic Petroleum Reserve and not be preoccupied with other nations' incentives to cooperate.

IV. SPR RELEASES AND OIL PRICES: AN EXAMPLE   We use a short-run "price reaction" function, based on capacity, following (13, 38, 44, 45), in order to formalize the notion of market "tightness." We employ two prices as proxies for the many prevailing in the market at any given time. Crude oil is sold under term contracts at the contract price. The spot price (labeled P^s below) is paid for oil purchased on a single-cargo basis. The contract price is set by OPEC in accord with its production decisions and demand estimates. Given the difficulty in forecasting demand and the numerous minor shocks inherent in any market, the contract price will not in general equate supply and demand. The spot market serves to satisfy the excess, and thus acts as a signal of market disequilibrium to OPEC, which adjusts the contract price. The process is then repeated.

The spot price increases when the market tightens. Two forms of tightening are possible—demand can increase due to changes in consumption or stock buildup, and supply can decrease in response to disruption in a producing country or deliberate production cuts.

When a disruption occurs, capacity is removed from the market, and the output-to-capacity ratio of the nondisrupted producers rises. At higher prices, these producers are willing to accelerate output, thereby bumping up against their own capacity constraints. When excess capacity no longer exists (output/capacity = 1), even large increases in the spot price can elicit little further supply response; hence the nonlinearity of the curve. In equation form,

$$P_t^s = \psi P_{t-1}^s + f(X_t/X_t^*), \qquad\qquad 6.$$

where t indexes the time period, f is a function (f′ > 0), ψ indicates the persistence effects of shocks on the spot price, and X and X^* are OPEC output and capacity output, respectively.

In this simple example, capacity decisions are assumed to be determined by longer-term considerations outside the scope of the model and are taken as exogenous. X is obtained from the conditions that supply and demand be equal:

$$Q^{US} + Q^F + S^{SPR} + S^{US} + S^F = X + X^D + X^{NO}, \qquad\qquad 7.$$

where US stands for United States, F for foreign, Q for consumption, S for stock change, S^{SPR} for Strategic Petroleum Reserve fill or draw, X for the production of nondisrupted OPEC producers, X^D for the (reduced) output of disrupted producers, and X^{NO} for non-OPEC production.

US consumption is assumed to depend on the domestic refiners' acquisition cost (P^{US}), income (Y^{US}), and a vector of structural variables, including past prices (Z^{US}). Foreign consumption is defined similarly:

$$Q^{US} = j^{US}(P^{US}, Y^{US}(P^{US}), Z^{US}) \qquad 8.$$

$$Q^{F} = j^{F}(P^{F}, Y^{F}(P^{F}), Z^{F}). \qquad 9.$$

The US refiners' acquisition cost is taken to be an average of spot and contract prices (plus transport costs), and is assumed to adjust to the spot price. The domestic price abroad is defined similarly:

$$P_t^{US} = a_0 + a_1 P_{t-1}^{US} + a_2(P_t^s - P_{t-1}^{US}) \qquad 10.$$

$$P_t^{F} = b_0 + b_1 P_{t-1}^{F} + b_2(P_t^s - P_{t-1}^{F}). \qquad 11.$$

In general, the a's and b's will differ for institutional as well as tax reasons.

To obtain X in terms of consumption, stock change, and production by other countries, rearrange Equation 7 to obtain

$$X = (Q^{US} + Q^{F}) + (S^{SPR} + S^{US} + S^{F}) - (X^{D} + X^{NO}). \qquad 12.$$

In this framework, the objective of stocks policy is clear—to lower the second of the three terms in order to reduce demand for OPEC output, thus moderating increases in the output-to-capacity ratio and reducing pressure on the spot price. Stock policy yields more "bang for the buck" as a disruption worsens, because of the nonlinearity of the price-reaction function.

In this framework, SPR releases have three effects on the spot price. The direct effect is to ease pressure as the SPR release reduces demand for OPEC output. A feedback effect occurs because holding down the spot price serves to hold down domestic prices at home and abroad as well, thus reducing the cutbacks in US and foreign consumption. The feedback effect clearly works against the direct effect. The interaction effect depends on the reaction of domestically held private stocks and foreign stocks to SPR releases. In equation form,

$$\frac{dP^s}{dS^{SPR}} = \underbrace{\frac{f'}{X^*}}_{\substack{\text{direct} \\ \text{effect}}} \underbrace{\left[1 - \frac{f'}{X^*}\left(a_2 \frac{dQ^{US}}{dP^{US}} + b_2 \frac{dQ^{F}}{dP^{F}}\right)\right]^{-1}}_{\text{feedback effect}} \underbrace{\left(1 + \frac{dS^{US}}{dS^{SPR}} + \frac{dS^{F}}{dS^{SPR}}\right)}_{\text{interaction effect}}. \qquad 13.$$

The sign of the direct effect is positive. The term in brackets is larger than one; the feedback effect partly offsets the direct effect. The effect of international interaction depends on the sign of the interaction effect. Cooperation (on the part of private stocks or foreign stockpile authorities) serves to magnify the benefits of the SPR release, while competition serves to mitigate them. The magnification (or mitigation) effect is more than proportional, due to the nonlinearity of the price-reaction function. Thus for a given SPR drawdown, a higher value of $dS^{US}/dS^{SPR} + dS^F/dS^{SPR}$ not only hits the spot price proportionally through the interaction effect, but also works through the direct effect (by lowering the argument of f') to exert additional downward pressure. Hence cooperation provides more than proportional benefits.

Finally, we should note that the benefits of stock draw are likely to be underestimated by the above analysis, which covers only the "within-drawdown-period" effects. Insofar as future spot prices are determined by an autoregressive process such as Equation 6, an SPR drawdown will be felt in future periods as well.

In related papers (45, 65), we used a simulation model of the world oil market similar to that outlined above in conjunction with a quarterly econometric model of the US economy to examine the ability of stockpile policy to mitigate the economic costs of oil supply disruptions. In one case, we simulated a reduction in OPEC capacity of 7 million bbl/d for one year, starting in the first quarter of 1983.

Our analysis revealed that given a US laissez-faire policy (no SPR draw or fill) during the disruption, it made a noticeable difference whether the rest of the OECD followed a "cooperative" (draw) or "noncooperative" (build) path. Even under the moderate assumptions employed, the difference in the spot prices was substantial. The loss in US GNP was of the order of 30–40% greater in the "noncooperative" case in the first two quarters, and 15–20% greater in the next two quarters.

The effects of using the SPR were two. First, the decreased demand for OPEC output exerted downward pressure on spot prices. Second, SPR draw substituted for imported oil almost entirely; this import reduction improved the trade balance and hence the US GNP. In the case of unilateral drawdown by the United States, use of the SPR recouped about one-third of the loss in GNP traceable to the shock, yielding a value of the SPR oil of about $45/bbl. This figure measured economic benefits over and above the revenues accrued from SPR sales. In an intertemporal optimization calculation, these revenues must be compared with the costs of buying and storing oil for the reserve.

In addition, we found the claim that government stock drawdown is impotent due to countervailing actions taken in the private sector to be without foundation. Government releases damped spot price increases, serving to reduce private inventory accumulation, not increase it. The response of domestically held private stocks to the SPR release (as well as the response of foreign stockpile authorities) was an important factor in determining the effectiveness of the stock policy in reducing prices.

Releasing SPR oil on the spot market is not the only way in which stockpile intervention might occur. Devarajan & Hubbard (66) consider the case of selling SPR oil through futures contracts. Their simulation results, based on the model discussed above, indicate that futures sales (*a*) achieved much of the price-reducing benefits in the early stages of a disruption and (*b*) led to a lower price trajectory overall when compared with spot market sales. They also discuss some institutional benefits of using futures markets.

CONCLUSION

The economic losses accompanying the oil supply shocks of the 1970s provided the impetus for the search for beneficial energy security policies. In one development, the US Congress created the Strategic Petroleum Reserve. Currently storing over 450 million barrels of oil, the reserve is a potentially powerful policy instrument. Our discussion addresses economic issues in managing the SPR.

Taking as the goal of national economic policy the maximization of economic welfare, policy intervention is beneficial only in the presence of some sort of market failure. The case for any energy security policy requires a demonstration of market failure. Potential sources of market failure in the oil market include macroeconomic externalities, monopsony power, vulnerability to disruptions, and national security. The SPR is a politically acceptable and economically efficient—in some cases, first-best—option to address those imperfections.

Considered as an economic policy instrument, the SPR represents a source of supply (or demand, depending on whether it is being released or filled) in the oil market. Hence, evaluating its role requires a framework for describing the behavior of other players—and thus of prices—in the market. We review several modeling strategies, stressing the importance of market structure and the behavior of private stocks. By way of illustration, we discuss a simple simulation model of the SPR and the world oil market. In that exercise, use of the SPR during a simulated oil shock obtained substantial benefits in terms of reduced oil prices and increased GNP. The results obtained are particular to our input assumptions and are by no means definitive.

It is customary to conclude with recommendations for policy. Since this is a review article, our recommendations address SPR research policy, rather than SPR policy itself. It is clear that we have raised more questions than we have been able to answer.

Specifically, it should prove fruitful to assess oil supply disruptions in a general equilibrium framework, and dispense with the partial equilibrium premia that are characteristic of the present literature. There must be two-way interaction between the oil market and the rest of the economy.

The sensitivity of the wisdom of SPR policy to the structure of the oil market has been virtually ignored. This consideration is particularly important for analyzing drawdown decisions, where the responses of other players are crucial. For example, without a model of expectations formation, the desirability of announced versus unannounced releases, and of futures versus spot sales, cannot be addressed.

Much has been learned in a short time about the subject of managing the Strategic Petroleum Reserve, yet much work remains to be done. We close with this challenge to economists and energy policy analysts.

ACKNOWLEDGMENTS

We thank James Hamilton, Chris Hope, and Henry Li for comments, and Brett Parsons, Scott Streckenbach, and Lynne Wiekert for assistance. We acknowledge financial support from Northwestern University, the US Department of Energy, and the National Science Foundation. The opinions expressed are ours alone.

APPENDIX A

Stockpiling Programs Outside the United States

GERMANY   Until recent years, the German government refrained from regulating its domestic oil market closely. In 1965, however, the government imposed a storage requirement on refiners and dependent importers. The independents were excluded. In 1975, the requirement was raised to the present level of 90 days for refiners—i.e. the refiner must keep the equivalent of 90 days of his average production of finished products from imported crude during the previous year. Dependent importers were required to keep a 70-day supply of their average product imports, and independent importers a 25-day supply (but increasing to a 40-day supply by 1980), of their average imports during the previous year.

After 1975, that program was superseded by a joint proposal from the majors and independents to create a quasi-public corporation to manage oil stockpiles. All the companies refining and importing oil in Germany would be obliged to become members of this corporation. The corporation would hold the obligatory stocks for the companies, and by being able to take advantage of economies of scale otherwise not available to small firms, the corporation would reduce the costs of stockpiling to the industry as a whole and therefore to consumers. The corporation's debt would be financed by the capital market, and the running costs would be paid by the members—i.e. the oil companies. This would remove the grounds for the fears of the independents that the financial burden of holding stocks would be too high.

After a year of negotiations, the various segments of the oil industry and the government reached an agreement on the structure of this corporation and in the fall of 1978, the Erdolbevorratungsverband (EBV) was established. The most important feature of the EBV program is that it removes obligatory stocks from the balance sheets of the oil companies. It also causes emergency reserve to be segregated administratively, and to some extent physically, from commercial inventories.

As a parallel effort, the German government decided in 1970 to create a Federal Reserve of 8 million tons (about 70 million barrels, equivalent to approximately a 25-day supply of current net imports). This government stockpile was scheduled to be completed by 1980, but the 1979 Iranian crisis has caused a delay. The Federal Reserve is stored in underground cavities near Etzel in northern Germany.

The development of the Federal Reserve and the EBV indicates that the German government is willing to intervene in the oil industry for energy security purposes. Its policies for the use of these stocks, however, are very conservative. The government is reluctant to use its emergency stock as a first line of defense to disruptions (for example, as buffers to stabilize prices). The Germans support the view that subcrisis management can be handled by the market.

The Germans feel that the price of oil could climb above $40/bbl before they would attempt to draw down stockpiles to restrain price increases. Moreover, German law states that the German government has widespread control over the oil industry only in the case of a "real" (IEA designated) emergency; thus, it is not clear that the West German government could easily manage the drawdown of stocks in a less than severe crisis. On the other hand, the Germans feel that when a real crisis arrives, their government is better prepared to use their crisis instruments than the United States or any other country (67).

JAPAN   The government agency responsible for energy regulations in Japan is the Ministry of International Trade and Industry (MITI). MITI's authority over oil affairs derives from the Basic Petroleum Law of 1962, which gave MITI the right to grant licenses for the construction of new refineries, recommend modifications in refinery production plans, and establish a "standard price" for oil products during the crises. MITI does not rely strictly on specific laws for its authority, however, but depends more generally on the traditional understanding that exists between government and industry.

In terms of oil stockpiling policy, MITI first "guided" Japanese oil companies toward holding a 60-day supply of oil in reserve in 1972. At that time the average stock level held by the companies for commercial operating requirements was about a 45-day supply. MITI hoped to meet the 60-day target by 1975, but the oil crisis of 1973–1974 upset this schedule. Japan's decision to join the IEA in 1974 established a new storage target; a 90-day supply of net oil imports by 1980. This much higher target would have a far greater impact on the oil industry, persuading MITI to seek formal, legislative approval and authority to require the oil companies to increase their stock levels in a schedule designed to meet the 90-day supply target. The Petroleum Stockpiling Law of 1975 was the result.

This law empowers MITI to "set an objective in terms of the stockpiling of oil for the following four fiscal years." MITI must detail its objectives for each year in ordinances that define the stockpiling obligations of oil refiners, marketers, and importers. Later in 1975, a MITI ordinance stated that the national storage target for 1976 would be a 70-day supply. This national obligation was parceled out of the various kinds of companies as follows: refining companies were obliged to store a 55-day supply based on their average product output in the preceding year; marketing companies, a 15-day supply calculated from their sales volume in the previous year; and product importing companies, a 45-day supply based on their import volume in the previous year. The plan was that MITI would adjust these company obligations yearly as it gradually increases the national stockpile target to a 90-day supply based on the previous year's oil sales by 1980, and perhaps a 120-day supply by 1985.

The storage target, being expressed in days, will contain a different amount of oil each year, in contrast to the American reserve, whose quantity will be fixed. The volume of oil in the stockpile will depend upon the level of Japan's oil consumption.

Most of the Japanese National Oil Company's (JNOC) stockpiling efforts interface with private Japanese companies. JNOC provides funds and low-interest loans to local governments to secure storage facilities and to private companies to subsidize the maintenance of higher inventory levels. JNOC has also created government-industry joint stockpiling companies. These companies have been designated to construct and operate additional storage facilities under leasing arrangements.

By law, MITI has established oil inventory levels to be maintained by each private oil company. During 1979, mandated targets for oil stocks were to reach 90 days of historical consumption levels. Certain companies were granted exemptions if their crude supplies were tight. By December 1979, Japan's total oil stocks were at record levels, and MITI instructed Japanese companies to draw down their inventories in stock levels to serve general economic objectives (68).

The analysis in (67) suggests that Japan will seriously consider a new oil emergency agreement to supplement its existing oil emergency policies and its participation in the IEA. Although the Japanese government is concentrating on improvements in economy and efficiency in energy use, it does have large oil stocks (as do the United States and West Germany) and realizes that, under the present IEA framework for stock management, no formal policies exist for their use.

FRANCE   The French government has been an important factor in the operation of the petroleum industry at least since 1928, when the Loi Poincare gave the government powers to regulate the oil market. Although in 1978 the government announced it would liberalize the domestic petroleum market, it still remains under firm government controls.

Stockpiling policies have been imposed since 1928 also. In 1951, refineries were required to maintain the equivalent of 10 days of their average crude oil imports at all times. Another requirement instituted in 1958 and modified in January 1975 required all importers to maintain stockpiles equivalent to 25% of their inland sales during the preceding twelve months. This obligation is widely referred to as the "90-day" requirement.

Unlike the German storage program, the French program does not impose variable obligations on oil refiners and importers. The same obligation falls on any company licensed to import crude oil or finished products.

The French storage obligations today are less controversial than the storage regulations in Germany for a number of reasons. First there are not as many product importers. Second, the requirements have been in force since 1958; the companies built up their stocks when oil was cheap. Third, the French system of regulations gives the government control and also enables the companies to earn a "fair" return on their investment.

The French government has not agreed to remove the obligatory stock debt from the oil industry's balance sheets as the German government has done, but it does attempt to equalize the impact of the storage expenses on company cash flows.

Only insofar as the demand for French oil increases are there formal plans to increase the size of the national stockpile. There are no government requirements that stocks be located in centralized storage facilities; thus, stocks tend to be widely dispersed and refineries tend to shift their excess product stocks to downstream bulk terminals. This dispersal is explicitly encouraged by the technical specifications of the regulation (69).

ITALY   The Italian government's intervention into oil affairs dates back to the 1920s, when the government created a number of state holding companies that evolved into Ente Nazionale Idrocarburi (ENI)—the fully integrated and international energy corporation in operation today.

Decrees in 1961 and 1976 established Italy's national storage target at stocks equivalent to 90 days of the previous calendar year's domestic sales. The burden of providing these stocks is divided between the owners of storage facilities (nonrefiners) and the refining industry. The owners of storage facilities are required to maintain as a minimum level of fill 20% of the capacities of their tanks. This 20% fill provision is a unique feature of the Italian program. It applies not only to oil businesses such as wholesalers, but also to any company or individual owning the requisite storage capacity. Thus, trucking firms, electrical utilities, petrochemical companies, and other large-scale oil users are subject to the storage regulations.

The practice of including part of the nonoil industry stocks in the national storage target, and hence of treating these stocks as a kind of emergency reserve, is the most extreme form of amalgamation of emergency reserves with commercial inventories. However, (69) argues that these reserves would exist even without regulation.

APPENDIX B

Private Stocks, Public Stocks, and Market Equilibrium

As in the text, to study the interaction of public and private stocks, we begin with the following stylized model of the short run. Let production Q^S and consumption demand Q^D be responsive to the current period price P. Production and consumption are subject to random additive disturbances ε_{St} and ε_{Dt}, respectively, which are assumed to be independently and identically distributed with mean zero and variances σ_S^2 and σ_D^2, respectively. Total demand is the sum of consumption and speculative inventory demand.

Speculators trade in inventories on the spot market in anticipation of changes in price, and are assumed to be risk-neutral, so that they maximize expected profit. Let the objective of the speculators over the period $(t, t+1)$ be

$$\max_{I_t} E_t \left\{ \left((1+\delta)^{-1} P_{t+1} - P_t \right) I_t - \frac{h}{2} I_t^2 \right\}, \qquad \text{B1.}$$

where I represents the end-of-period stock level and δ is the discount rate. E_t denotes the expectation operator conditional on information available at time t. The first term represents speculative gains on the stock held, the second, holding costs. Holding stocks is assumed to be costly—in fact, increasingly costly—in the size of the stock due to payments to factors fixed in the short run, such as storage facilities, tankers, and pipelines. Thus changes in price expectations cannot be fully acted upon instantaneously. We follow the literature in modeling such costs as quadratic, the simplest specification of "diminishing returns"; these costs are indexed by the parameter h.

Maximizing the quantity B1 with respect to I_t yields the following demand function for stocks:

$$I_t = h^{-1}((1+\delta)^{-1}E_t P_{t+1} - P_t).$$ B2.

As with most other specifications since the original development by Muth (1961), the holdings of risk-neutral speculators are a function of the expected increase in price, taking into account the cost of adjusting stock levels.[12] Inventory demand (stock change) is just

$$I_t - I_{t-1} = h^{-1}[(1+\delta)^{-1}(E_t P_{t+1} - E_{t-1} P_t) - (P_t - P_{t-1})].$$ B3.

The spot price solves the following equation for market equilibrium:

$$Q^D(P_t) + h^{-1}(1+\delta)^{-1}(E_t P_{t+1} - E_{t-1} P_t)$$
$$-h^{-1}(P_t - P_{t-1}) + \varepsilon_{Dt} = Q^S(P_t) + \varepsilon_{St}.$$ B4.

Under simplifying assumptions of linear responses of supply and demand to price, we have

$$Q^D(P_t) = A - aP_t, \quad \text{and}$$ B5.

$$Q^S(P_t) = B + bP_t.$$ B6.

Hence Equation B4 can be rewritten as

$$A - aP_t + \varepsilon_{Dt} + h^{-1}[(1+\delta)^{-1}E_t P_{t+1} - P_t$$
$$-(1+\delta)^{-1}E_{t-1}P_t + P_{t-1}] = B + bP_t + \varepsilon_{St},$$ B7.

or

$$[b + a + h^{-1}]P_t = A - B + \varepsilon_{Dt} - \varepsilon_{St}$$
$$+ h^{-1}[(1+\delta)^{-1}(E_t P_{t+1} - E_{t-1} P_t) + P_{t-1}].$$ B8.

For simplicity, consider the case in which $\delta = 0$. If we define the long-run average price obtained when expectations are realized ($E_t P_{t+1} = E_{t-1} P_t = p_t$) by $\hat{P}$, then it follows that

$$\hat{P} = \frac{A - B}{a + b}.$$ B9.

Let lower-case variables be defined in deviation form (i.e. $P_t = P_t - \hat{P}$). Under the assumptions of rational expectations, we solve the second-order inhomogeneous difference Equation B8 by standard methods to yield

$$p_t = \alpha p_{t-1} + \frac{\varepsilon_{Dt} - \varepsilon_{St}}{a + b + 2h^{-1} - h^{-1}\alpha},$$ B10.

---

[12] For a more general intertemporal optimizing model of oil inventory behavior under uncertainty, see (50).

where α is the root within the unit circle of the quadratic equation $h^{-1}\alpha^2 - (a+b+2h^{-1})\alpha + h^{-1} = 0$. Equation B10 corresponds to Equation 1 in the text, where $\beta = (a+b+2h^{-1}-h^{-1}\alpha)^{-1}$.

Hence even transitory shocks exhibit persistence effects on the spot price because of the behavior of inventories. Moreover, both the one-period and asymptotic variances of the spot price increase with α, since

$$\sigma_p^2(1) = \frac{\sigma_D^2 + \sigma_S^2}{(a+b+2h^{-1}-h^{-1}\alpha)^2}, \quad \text{and} \qquad \text{B11.}$$

$$\sigma_p^2(\infty) = \frac{\sigma_p^2(1)}{1-\alpha^2}. \qquad \text{B12.}$$

Note that since $d\alpha/da < 0$, the steeper is the demand curve for oil, the smaller is the initial increase in price and the lower is the persistence. Hence, policies such as oil import tariffs or certain types of buffer stock stabilization policies (which effectively raise a) can mitigate both the impact and the long-run effects of transitory shocks on prices.

We can easily extend the above analysis to the case of serially correlated quantity shocks. Suppose that demand and supply shocks follow first-order autoregressive processes (AR(1)):

$$\varepsilon_{Dt} = \rho_D \varepsilon_{Dt-1} + v_{Dt}, \quad \text{and} \qquad \text{B13.}$$

$$\varepsilon_{St} = \rho_S \varepsilon_{St-1} + v_{St}, \qquad \text{B14.}$$

where v_{Dt} and v_{St} are white noise and

$$\sigma_{v_D v_S} = 0.$$

Using Equations B13 and B14 for the demand and supply shocks, we can rewrite Equation B8 as

$$E_t\{P_{t+1}[h^{-1}-(a+b+2h^{-1})L+h^{-1}L^2]\} = -(\varepsilon_{Dt}-\varepsilon_{St}), \qquad \text{B15.}$$

where L denotes the lag operator. The solution to B15 given rational expectations is just

$$p_t = \alpha p_{t-1} + \gamma(1-\gamma\rho_D)v_{Dt} - \gamma(1-\gamma\rho_S)v_{St}, \qquad \text{B16.}$$

where α is the root within the unit circle of the quadratic equation contained within the brackets in B15 and γ is the root outside the unit circle.

Now persistence effects come also from the serial correlation parameters ρ_D and ρ_S. Equation B16 points up the need to consider the structural parameters determining α (and γ). As α and γ tend toward unity, the existence of serial correlation amplifies the price effects of shocks. The variance of the spot price is also higher when the shocks are serially correlated. Finally, if the shocks are serially correlated, then changes in α induced by policy changes have all the more impact.

It is clear from Equation B13 that as long as supply shocks are purely transitory, inventories will be drawn down in response to a negative supply shock, since

$$\frac{\mathrm{d}I_t}{\mathrm{d}(-\varepsilon_{St})} = \Omega(\alpha-1) < 0, \quad \text{where} \quad \Omega = h^{-1}(\mathrm{d}p_t/\mathrm{d}(-\varepsilon_{St})) \qquad \text{B17.}$$

Speculative accumulation requires either serially correlated shocks or the expectation that the disruption will "get worse."[13]

Following this stylized model when shocks follow AR(1) processes, we note that the response of private stocks to a supply shock in the current period is

$$\frac{\mathrm{d}I_t}{\mathrm{d}(-v_{St})} = \gamma(\rho_S-(1-\alpha)). \qquad \text{B18.}$$

For negative supply shocks ($\varepsilon_{St} < 0$), Equation B18 implies that private shocks will increase if $\rho_S > 1-\alpha$. That is, the likelihood of stock accumulation in response to a supply disruption is greater the higher the serial correlation of the shocks (ρ_S) and/or the higher the intertemporal correlation of price changes (α).

In the text, we considered the impact of optimal public stockpile behavior on private stockpiling and the world price. As in the text, suppose that the public stockpile is used in accordance with an assumed economic policy of maximizing real income (output less payments for imported intermediate goods). Output Y of a single final good is produced from oil Q^D and other factors $\bar{X}$ according to the production function

$$Y_t = \mathrm{F}(Q_t^D,\bar{X}); \quad \mathrm{F}_1,\mathrm{F}_2 > 0, \quad \text{and} \quad \mathrm{F}_{11},\mathrm{F}_{22} < 0. \qquad \text{B19.}$$

All oil is imported, and oil is the only imported intermediate input. Nonoil factor supplies are fixed.

The stockpile authority is assumed to be risk-neutral, and its objective is to maximize real income (by minimizing oil price increases) less the cost of carrying out the stockpile program and adjusting stockpile levels, subject to the constraint that stockpile releases not exceed the amount of oil held in the reserve. The problem is to choose the stockpile level I^p in period t so that

$$\max_{I_t^p}\mathrm{E}_t\left\{Y_t-P_tQ_t^D+((1+\delta)^{-1}P_{t+1}-P_t)I_t^p-\frac{h}{2}\,I_t^{p^2}\right\}, \qquad \text{B20.}$$

---

[13] One example might be an AR(2) process in which the first lagged coefficient is greater than unity and the second is negative.

subject to the constraint that

$$I_t^p = I_{t-1}^p + S_t, \qquad I_t^p \geqslant 0, \qquad\qquad \text{B21.}$$

where δ is the discount rate. Again, the quadratic term is a proxy for the cost of adjusting stock levels. Assuming that the nonnegativity constraint does not bind, the solution to B20 can be written as

$$\begin{aligned} I_t^p &= h^{-1}((1+\delta)^{-1}E_t P_{t+1} - P_t) - h^{-1}\omega Q_{t-1}^D \\ &= (h + (\alpha - 1)\omega)^{-1}((1+\delta)^{-1}E_t P_{t+1} - P_t - \omega Q_t^D), \qquad \text{B22.} \end{aligned}$$

where $\omega = dP/dI^p$.

There is a clear distinction between the optimization problem for the public stockpile authority and the problem for the private firm stated earlier. The public authority pays attention to aggregate output. Private firms do not consider the macroeconomic effects of their stockpiling behavior; that is, they do not consider the impact of their transactions on the world oil price. The last term in Equation B22 captures this market power effect. *Ceteris paribus*, the larger the oil consumption, the greater will be the stockpile release because of the benefits of lowering the price paid on the inframarginal barrels. Equation B22 makes clear the role of persistence in determining the optimal public stockpile policy. That issue is pursued in more detail in the text in the context of international cooperation in stockpile drawdowns.

Literature Cited

1. Frankel, P. H. 1958. Oil supplies during the Suez crisis. *J. Industrial Econ.* 6:85–100
2. Kapstein, E. B. 1984. Alliance energy security: 1945–1983. *Fletcher Forum*
3. Hamilton, J. D. 1983. Oil and the macroeconomy since World War II. *J. Pol. Econ.* 91:228–48
4. Fried, C. R., Schultze, C. L., eds. 1975. *Higher Oil Prices and the World Economy.* Washington, DC: Brookings
5. Hickman, B. G., Huntington, H. G. 1984. *Macroeconomic Impacts of Energy Shocks: An Overview.* Mimeo, Energy Modeling Forum. Stanford
6. Mork, K. A., Hall, R. E. 1980. Energy prices and the U.S. economy in 1979–1981. *Energy J.* 1:41–54
7. Jorgenson, D. W. 1984. The role of energy in productivity growth. *Am. Econ. Rev. Pap. and Proc.* 74:26–30
8. Balas, E. 1980. Choosing the overall size of the strategic petroleum reserve. In *Energy Policy Modeling: U.S. and Canadian Experiences*, ed. W. Ziemba, S. Schwartz, E. Koenigsberg, pp. 144–58. The Hague: Martinus Nijhoff
9. Kinberg, Y., Shakun, M. F., Sudit, E. F. 1978. Energy buffer stock decisions in game situations. *TIMS Studies in the Management Sciences* 10:109–27
10. Marchand, M., Pestieau, P. 1979. Tarification et anticipations divergentes. *Annals Pub. Coop. Economy* 50:111–36 (In French)
11. Wright, B. D., Williams, J. 1982. The roles of public and private storage in managing oil import disruptions. *Bell J. Econ.* 13:341–53
12. Kalt, J. P. 1981. *The Economics and Politics of Oil Price Regulation.* Cambridge, MA: MIT Press
13. Hubbard, R. G., Fry, R. C. 1982. The macroeconomic impacts of oil supply

disruptions. *Energy and Env. Pol. Cent. Disc. Paper* E-81-07, Kennedy School of Govt., Cambridge, MA

14. Verleger, P. K. 1982. *Oil Markets in Turmoil.* Cambridge, MA: Ballinger
15. Hogan, W. W. 1982. Policies for oil importers. In *OPEC Behavior and World Oil Prices,* ed. J. M. Griffin, D. J. Teece, pp. 186–206. London: Allen & Unwin
16. US Dept. Energy. 1981. *Domestic and International Energy Preparedness.* DOE/EP-0027. Washington, DC
17. Devarajan, S. 1982. Commodity stockpiles: a summary of the literature and the experience. *Oil Stockpiling: An International Perspective.* Putnam, Hayes & Bartlett, report to the US Dept. of Energy
18. Colglazier, E. W., Deese, D. A. 1983. Energy and security in the 1980s. *Ann. Rev. Energy* 8:415–49
19. Weimer, D. L. 1982. *The Strategic Petroleum Reserve.* Westport, CT: Greenwood
20. Paris, D. L. 1982. *Energy Politics.* New York: St. Martin's
21. Hogan, W. W. 1984. Energy Policy in the Reagan Era. Cambridge, Mass: Energy and Environ. Policy Cent., Kennedy School of Government
22. Exxon Corp. 1981. *World Oil Inventories.* Background Series. New York
23. US Senate, Comm. Interior and Insular Affairs. 1974. *Int. Energy Program Hearing.* Washington, DC
24. Stanford Univ. Energy Modeling Forum. 1982. *World Oil.* Summary Report. Stanford, Calif.
25. Bohi, D. R. 1983. What causes oil price shocks? *Res. for the Future Disc. Paper* D-82S. Washington, DC
26. Adelman, M. A. 1982. OPEC as a cartel. In *OPEC Behavior and World Oil Prices,* ed. J. M. Griffin, D. J. Teece, pp. 37–63. London: Allen & Unwin
27. Singer, S. F. 1983. The price of world oil. *Ann. Rev. Energy* 8:451–508
28. Hope, C. W. 1984. *The Oil Price Without OPEC.* Presented at Ann. Int. Meet. Int. Assoc. En. Econ., 6th, Cambridge, UK
29. Adams, F. G., Marquez, J. 1984. Petroleum price elasticity, income effects, and OPEC's pricing policy. *Energy J.* 5:115–28
30. MacAvoy, P. W. 1982. *Crude Oil Prices.* Cambridge, Mass: Ballinger
31. Roumasset, J., Isaak, D., Fesharaki, F. 1983. Oil prices without OPEC: a walk on the supply side. *Energy Econ.* 5:164–70
32. Chasseriaux, J.-M. 1982. Une interpretation des fluctuations du prix du pet-

role. *Revue d'Econ. Industrielle* 22:24–38 (In French)
33. Cremer, J., Salehi-Isfahani, D. 1980. Competitive pricing in the oil market: how important is OPEC? *University of Pennsylvania Working Paper.* Philadelphia
34. Teece, D. J. 1982. OPEC behavior: an alternate view. In *OPEC Behavior and World Oil Prices,* ed. J. M. Griffin, D. J. Teece, pp. 64–93. London: Allen & Unwin
35. Reza, A. M. 1981. An analysis of the supply of oil. *Energy J.* 2:77–94
36. Mead, W. J. 1979. An economic analysis of crude oil price behavior in the 1970s. *J. En. Dev.* 4:212–28
37. Johany, A. D. 1980. *The Myth of the OPEC Cartel.* Chichester, UK: Wiley
38. Plummer, J. L., ed. 1982. *Energy Vulnerability.* Cambridge, Mass: Ballinger
39. Bohi, D. R., Montgomery, W. D. 1982. *Oil Prices, Energy Security, and Import Policy.* Baltimore: Johns Hopkins Univ. Press
40. Gately, D., Kyle, J., Fischer, D. 1977. Strategies for OPEC pricing decisions. *Eur. Econ. Rev.* 10:209–30
41. Danielsen, A. L., Selby, E. B. 1980. World oil price increases: sources and solutions. *Energy J.* 1:59–74
42. Jacoby, H. D., Paddock, J. L. 1980. Supply instability and oil market behavior. *En. Systems and Policy* 3:401–23
43. Fisher, F. M., Cootner, P. H., Baily, M. N. 1972. An econometric model of the world copper industry. *Bell J. Econ. Manage. Sci.* 3:585–609
44. Nordhaus, W. D. 1980. Oil and economic performance in industrial countries. *Brookings Pap. Econ. Activity,* pp. 341–88
45. Hubbard, R. G., Weiner, R. J. 1983. The "sub-trigger" crisis: an economic analysis of flexible stock policies. *Energy Econ.* 5:178–89
46. US Gen. Account. Office. 1983. *Oil Supply Disruptions: Their Price and Economic Effects.* RCED-83-135, Washington, DC
47. Adelman, M. A. 1984. Sense and nonsense about world oil. *Energy J.* 5:169–71
48. Stobaugh, R. B. 1976. The oil companies in the crisis. In *The Oil Crisis,* ed. R. Vernon, pp. 179–202. New York: Norton
49. Neff, T. L. 1981. The changing world oil market. In *Energy and Security,* ed. D. A. Deese, J. S. Nye, pp. 23–46. Cambridge, Mass: Ballinger
50. Hubbard, R. G., Weiner, R. J. 1985. Oil

supply shocks and international policy coordination. *Eur. Econ. Rev.*

51. Muth, J. F. 1961. Rational expectations and the theory of price movements. *Econometrica* 29:315–35

52. Badger, D. B. 1984. The anatomy of a "minor disruption": missed opportunities. In *Oil Shock: Policy Response and Implementation,* ed. A. L. Alm, R. J. Weiner, pp. 33–53. Cambridge, Mass: Ballinger

53. ICF Incorporated. 1982. *Modeling the Effect of Strategic Petroleum Reserve Drawdown on the Oil Market and the Macroeconomy.* Report to the SPR Office, US Dept. of Energy. Washington, DC

54. Teisberg, T. J. 1981. A dynamic programming model of the U.S. strategic petroleum reserve. *Bell J. Econ.* 12:526–46

55. Chao, H.-P., Manne, A. S. 1982. An integrated analysis of U.S. oil stockpiling policies. In *Energy Vulnerability,* ed. J. L. Plummer, pp. 59–82. Cambridge, Mass: Ballinger

56. Hogan, W. W. 1983. Oil stockpiling: help thy neighbor. *Energy J.* 4:49–71

57. Natl. Pet. Council. 1981. *Emergency Preparedness for Interruption of Petroleum Imports into the United States.* Washington, DC

58. Rowen, H. S., Weyant, J.-P. 1982. Reducing the economic impacts of oil supply interruptions: an international perspective. *Energy J.* 3:1–34

59. US Dept. Energy. 1982. *Report to the President and the Congress on the Size of the Strategic Petroleum Reserve.* DOE/EP-0036. Washington, DC

60. Nichols, A. L., Zeckhauser, R. J. 1977. Stockpiling strategies and cartel prices. *Bell J. Econ.* 8:66–96

61. Sharpe, W. F. 1964. Capital asset prices: a theory of market equilibrium under conditions of risk. *J. Finance* 19:425–42

62. Lintner, J. V. 1965. The valuation of risky assets and the selection of risky investments in stock portfolios and capital budgets. *Rev. Econ. Stat.* 47:13–37

63. Bouhdili, C., Montgomery, W. D. 1983. Risk analysis of the strategic petroleum reserve: theory and preliminary evidence. *Res. for the Future Disc. Paper* D-82U. Washington, DC

64. Newbery, D. M. G., Stiglitz, J. E. 1981. *The Theory of Commodity Price Stabilization.* Oxford: Oxford Univ. Press

65. Hubbard, R. G., Weiner, R. J. 1984. Inventory optimization in the U.S. petroleum industry: empirical analysis and implications for energy emergency policy. Mimeo, Harvard University

66. Devarajan, S., Hubbard, R. G. 1983. Drawing down the strategic petroleum reserve: the case for selling futures contracts. In *Oil Shock: Report of the Harvard Energy Security Program,* ed. A. L. Alm, R. J. Weiner, pp. 187–96. Cambridge, Mass: Ballinger

67. Gibson, C. S. 1984. Cooperative oil stockpiling: evaluating a three-country agreement and designing a negotiation strategy. Mimeo, Kennedy School of Government, Harvard Univ.

68. US Dept. Energy. 1981. *Energy Industries Abroad.* DOE/IA-0012. Washington, DC

69. Krapels, E. N. 1980. *Oil Crisis Management.* Baltimore: Johns Hopkins Univ. Press

Ann. Rev. Energy 1985. 10 : 557–87
Copyright © 1985 by Annual Reviews Inc. All rights reserved

A DECADE OF UNITED STATES ENERGY POLICY[1]

Mark D. Levine

Energy Analysis Program, Applied Science Division,
Lawrence Berkeley Laboratory, Berkeley, California 94720

Paul P. Craig

Department of Applied Science, University of California, Davis,
California 95616

In 1965, a high level government committee reported that "no grounds exist for serious concern that the Nation is using up any of its stocks of fossil fuel too rapidly; rather, there is the suspicion that we are using them up too slowly... rather than fearing a future day when fossil-fuel resources will be largely exhausted..., we are concerned for the day when the value of untapped fossil-fuel resources might have tumbled... and the nation will regret that it did not make greater use of these stocks when they were still precious" (1).

At about the same time, warnings were appearing that US oil and gas production would soon peak (2). Few then took Hubbert seriously. Yet his predictions have since turned out to be remarkably prescient. US natural gas production peaked in 1973, and has been dropping since. Oil production in the contiguous United States peaked in 1970, and it too is declining. (More recently oil from the North Slope of Alaska has come on stream, thus holding total US production roughly constant in recent years.) These developments provided the best support in the early 1970s for the idea that the United States and possibly the world were entering a period of decline in the growth of oil supply over the long term.

In December 1973, in the middle of the Organization of Arab Petroleum Exporting Countries (OAPEC) embargo and only months after the Yom

[1] The views expressed in this paper are those of the authors and do not represent views of their institutions.

0362–1626/85/1022–0557$02.00

Kippur war of September 1973, the Nixon administration issued the "Dixy Lee Ray report" on energy R&D. The report, commissioned in the summer of 1973, included a figure (Figure 1) showing oil imports to the United States—then about 6 million barrels per day (mmbpd)—dropping to zero by 1980 (3). President Nixon stated that "by the end of the decade we will have developed the potential to meet our own energy needs without relying on any foreign energy sources" (4).

Events proved otherwise. Imports increased to more than 8 mmbpd, then dropped due to conservation and a slumping economy (though hardly at all due to the new supply technologies that were the dominant focus of the Ray report) to about 3 mmbpd today, when total self-sufficiency is neither in sight nor a national goal.

The year 1973 is remembered primarily for the oil embargo and the Yom Kippur war between Israel and Egypt. It was also the year, however, in which the Texas Railroad Commission, responsible for allocating ("pro-rationing") oil production among Texas producers, found that even at 100% production it was no longer possible to meet national demand. Thus, in 1973, the United States no longer had the technical capacity to achieve self-sufficiency in oil. That this possibility existed in 1973 was known to some. That the shift of power to OPEC could occur so rapidly was a surprise to almost everyone—probably including the OPEC nations themselves.

As a result of the embargo in 1973, the nation became aware of its growing energy vulnerability. Crash efforts were mounted to develop new energy technologies. Legislation was passed in abundance. An army of people (ourselves included) became instant energy experts. The federal energy establishment was constantly being reorganized. Despite the continuing confusion, most participants felt themselves part of an important process with vital implications for the nation's future.

Today much of this feeling has dissipated. Oil supply is more than ample to meet demand, oil prices have eased, and energy issues have receded from national prominence. In the course of a decade, the energy problem was transformed in the public consciousness from a minor issue concerning a few experts to a major national crisis threatening the nation and back again to a peripheral issue. (It is of interest to note that a compendium of editorial cartoons for 1983 (5) ignored energy entirely.)

Yet, attitudes have changed irreversibly, and no survey of the energy decade can fail to take note of this fact. Consumers and industry alike are aware that energy costs have gone up, and are likely to go up in the future. Purchase decisions for new houses, cars, appliances, and industrial plants often take energy into account. Thus, energy conservation is becoming a part of our normal social planning process. Citizens remain aware of the nation's vulnerability to oil interruptions. Though this awareness is receding as years go by, the consciousness of energy as a critical element of vulnerability continues to emerge in public opinion polls. Energy is far from a dead issue, and it is easy to envision ways it could soon again occupy center stage on the national agenda.

The energy decade has been responsible also for deep changes in the ways

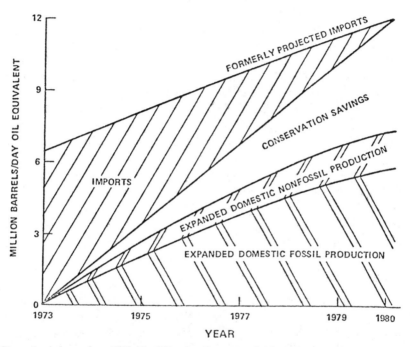

SELF-SUFFICIENCY BY 1980 THROUGH
CONSERVATION AND EXPANDED PRODUCTION

Figure 1 A dream from 1982. The "Dixy Lee Ray report," submitted to President Nixon in the month the first OAPEC embargo occurred, included a graph showing how the United States could achieve energy self-sufficiency by 1980 (3).

that the technical community thinks about energy and the economy. At least three critical factors in the relationship between energy and the economy were given relatively little weight a decade ago: the effect of energy prices on demand, the potential for increasing the efficiency of energy use, and the ability to make substitutions both among energy forms and between energy and other goods and services. All of these factors are related in some way to a better understanding of energy demand, an area of knowledge given little attention during times of abundant supplies. This lack of attention to energy demand and factors affecting it led to a widespread belief that energy demand moved more or less in lock step with gross national product. The energy decade has shown that such an iron linkage between economic growth and energy growth does not exist. Demand patterns are subject to change. Energy efficiency can increase. Energy prices matter. Substitutions take place. The energy system is recognized to be both more complex and more flexible than most writers and analysts of a decade ago perceived.

Among the multitude of events that formed the energy decade, we single out in this review those we believe to have had the greatest impact on the performance of the US energy system and on people's ways of thinking about it. We give attention to the US response to the crisis caused by the oil embargo, pricing policy, research and development, energy conservation, and factors affecting energy supply and production. For reasons of space we barely touch many important issues, such as land management and leasing, environmental policy, and energy data collection. We also exclude international energy issues although it is clear that other nations—particularly developing nations—have been adversely affected far more than the United States, and the implications for the developing nations of the energy decade are far greater than for us.

PREEMBARGO HISTORY

Three areas of government regulatory policy dominated the US energy picture from the 1930s until World War II: oil, gas, and electrification policy. A fourth area, the development of a new industry—atomic power— consumed the largest fraction of federal funds for energy R&D from World War II until the embargo.

Oil

In the 1930s, oil producers, mostly in Texas, recognized the problem of multiple owners' rights to single underground oil pools. In the interest of maintaining optimal rates of oil extraction, they restructured the Texas Railroad Commission to give it authority to determine oil production rates.

This procedure, proportional rationing or prorationing, served a conservation function. At the same time it provided authority for limiting total output, and thereby keeping prices high.

Later, as cheap and abundant international oil threatened to drive prices down, oil import quotas were instituted to restrict the amount of foreign oil entering the country. It is a curious accident of history that legislation designed to keep internal prices high was, in the energy decade, turned on its head as foreign prices rose above those of price-controlled US oil. [For a brief overview, with a focus on the influence of the oil companies, see (6).]

Gas

Gas policy in the 1930s was closely related to oil production. The reasons are largely technical. Associated gas is gas found during drilling for oil. For years there was no market for it, and it was flared—burned and wasted. A capital intensive pipeline distribution system was required to transport the gas to markets where it could be productively used. Federal laws were passed to regulate interstate gas prices, with the idea of holding them high enough to justify the large capital investment. The justification was one used many times since—the developmental cost is so large, and the risk so great, that the private sector cannot do the job alone. The mechanisms put in place to hold gas prices high from the 1930s through the 1960s have, because of the changing circumstances in the 1970s and thereafter, achieved the opposite purpose: they have held natural gas prices artificially below those of competing fuels.

As early as the 1950s there was general recognition that natural gas price regulation had already served its purposes and was no longer needed. During the Eisenhower administration a bill deregulating natural gas prices passed both houses of Congress. Eisenhower, who supported the bill, chose to veto it because an attempt to bribe members of Congress to vote for the bill was revealed. Thus, the artificially low prices of natural gas that persisted throughout most of the decade are the result of a system originally designed to keep prices high and a fluke of history that maintained the system too long.

Electricity

The third strand of US energy policy in the early period relates to the development of electricity. The government undertook a massive program to (a) subsidize electrification of the nation and (b) regulate what was clearly a natural monopoly. Rural electrification was highly successful. In some parts of the nation the federal government did the job itself. The Bonneville Power Administration and the Tennessee Valley Administration were the successful results. States set up public utility commissions and the federal

government established the Federal Power Commission to regulate interstate transfers of electricity.

Creation of a Nuclear Power Industry

While regulatory policies that preceded the embargo were to become
increasingly important (because their existence at the time of the embargo
partly defined the government policy reaction), the main focus of federal
involvement in energy before the 1970s was the establishment of a viable
industry to produce electricity from the atom. The pressures associated
with the development of military use of atomic energy led directly to the
decision that the government should take responsibility to develop "the
peaceful atom."

The promise of the peaceful atom was clear to almost everyone who
looked into it after World War II. Early enthusiasm led to the famous
remark that nuclear electricity would be "too cheap to meter." This phrase,
a rallying call at the time, is today still quoted as a reminder of how wrong
experts can be. It is, however, a reflection of the high optimism of the time.

The US Atomic Energy Commission (AEC), under the not-too-critical
Joint Committee on Atomic Energy of the US Congress, viewed its
mandate as the promotion of nuclear energy. The atomic bureaucracy had a
single goal—to make the nuclear electricity business happen. Little thought
was given to checks and balances, a problem that would later emerge with a
vengeance. The AEC established a multitude of laboratories, new divisions
in many large corporations, and a vast bureaucracy. Though the AEC was
arrogant and narrow-minded, it did get things done. [An especially good
example of the attitude of the AEC was the US Supreme Court decision that
the AEC's "crabbed interpretation" of its responsibilities for examining the
environmental impacts of nuclear power plants made "a mockery" of the
environmental impact process required under the National Environmental
Policy Act of 1969 (7).]

Today, nuclear energy supplies about 15% of all the electricity generated
in the United States. Despite many recent difficulties—including Three
Mile Island and the total collapse of new orders—nuclear plants now under
construction will result in a growing percentage of US electricity being
supplied by nuclear power plants for at least a decade. Gaining a sizable
fraction of the electricity generation market for a fundamentally new
technology is no small achievement, whether one views nuclear power
favorably or not. Widespread commercialization of new energy technologies has occurred very slowly and infrequently in industralized nations.

The goal of the AEC when it began in 1946 was to foster a new industry
where none had hitherto existed. The costs and risks were clearly large, and
it was believed that private industry would not absorb the risk. The AEC

subsidized American industry to learn the reactor business. AEC management knew from the outset that a large amount of highly technical knowledge would be required if nuclear power was to prove a success. The areas were vast: heat transfer, nuclear cross sections, material properties under radiation insult, effects of radiation on people, on animals, on biota in general. A competent basic and applied research establishment was developed, and much of that core remains to this day, although with the lack of clear DOE objectives the program has developed a schism between basic and applied research, and many of the national laboratories are functioning without clear priorities.

THE TRANSITION

Most accounts of the oil embargo of 1973–1974 stress the lack of preparation for, and awareness of, the changes in the supply-demand balance for oil that could permit both an embargo and a successful international cartel. The accounts are generally correct; nonetheless, key individuals in government, the energy industries, and other institutions realized that important changes were taking place. The Ford Foundation, recognizing that energy would become a major national policy issue, commissioned a multimillion-dollar Energy Policy Project in early 1973 (8). The federal government, aware that energy would loom larger, initiated an interagency study with the major task of reevaluating the government role in supporting energy R&D. The Dixy Lee Ray report was completed in December, 1973, just at the time of the Yom Kippur war and the OAPEC oil embargo. The Ray report called for a dramatic shift in federal energy research priorities toward a balanced program that would include significant efforts in virtually every energy technology. Of particular interest was the appearance of program elements devoted to the use of energy and to renewable energy sources. These were significant and controversial shifts, and they remain controversial to the present.

Supply and Demand

At the time of the OAPEC embargo, demand for energy had for many years been growing rapidly and consistently, at an average rate of about 3% per year. Demand for electricity had been increasing even faster, at 8% per year. The stability of energy and electricity growth rates had been so great that forecasters typically used the simple technique of drawing straight lines on semilog graph paper.

Today, demand has leveled; US energy use was 74.2 quads in 1973, peaked at 78.8 quads in 1979, and was 70.0 quads in 1983 (9). This new behavior may be emphasized by placing it in contrast with a typical forecast made in 1971 (10), which showed historic growth rates continuing through the end of the century, with about 110 quads projected for 1983, 55% above actual use.

Even by the middle of the energy decade forecasts were wrong in virtually every particular. In an article emphasizing the need for nuclear power, Hans Bethe (11) quoted from President Ford's 1975 State of the Union message in which oil imports in 1985 were projected at nearly 35 quads, with imports accounting for well over half of US oil use. The United States actually imported 6.7 quads, or about 3 million barrels per day (mmbpd) in 1983.

Some individuals and groups were convinced in the early 1970s that energy demand growth would eventually decline. Foremost among individuals with this view was Amory Lovins, who foresaw the achievement of a steady state zero energy demand growth and increasing small-scale renewable technologies to meet this demand. However, even Lovins appeared to expect demand growth through 1985. In his book *Soft Energy Paths: Toward a Durable Peace* published in 1977, Lovins presented a figure entitled "An alternative illustrative future for U.S. Gross Primary Energy Use" that showed energy demand reaching 90 quads by 1985 and 100 quads by 2000 (12). Lovins made clear in the text that he believed lower energy growth to be possible; it seems, however, that it was not until the late 1970s that Lovins expected and put together a case for declining energy growth. Similarly, the Ford Foundation Energy Policy Project, in its zero energy growth scenario (8) published in 1974, indicated energy demand equaling 93 quads in 1985. At the time of publication, the zero energy growth scenario was widely viewed as implausible.

In fairness to these early studies, it should be noted that they all anticipated or assumed a much higher growth rate in the economy than actually occurred. Nonetheless, the fact that responsible parties who foresaw or advocated low energy growth in the early 1970s were publishing growth rates much higher than actually occurred underscores the magnitude of the failure of the mainstream forecasting community to perceive and take account of factors leading to reduced energy demand growth.

Electricity production was 1.6×10^{12} kWh in 1973, and increased to 2.3×10^{12} kWh in 1983. This is a growth rate of 3.7% per year, far below the rates anticipated in virtually all electricity forecasts early in the decade (as exemplified by the expectations expressed in the Starr (10) and Bethe (11) articles). Although the total output from nuclear power increased from 83×10^9 kWh in 1973 to 294×10^9 kWh in 1983, production is far below forecasts. At the same time orders fell dramatically. The category "total design capacity" (plants in operation and startup, plus those with construction permits, those on order, and those announced) fell from 212×10^9 net kWh per year in 1973 to 125×10^9 net kWh per year in May 1984. There have been no new orders for nuclear power plants in the United States in the last half decade.

Since electricity growth rates were higher than total energy growth, the energy input to electricity increased, from 27% of total demand in 1973 to 36% in 1983.

Coal use has increased more slowly than expected, from 8.6 quads per year in 1973 to 13.2 quads per year in 1983. The often sought goal of production in excess of 1000 million tons (24 quads) per year has yet to occur.

In the early parts of the energy decade large contributions from synthetic fuels were anticipated. These expectations were not realized. Despite the establishment of a federally funded Synthetic Fuels Corporation and extensive research, development, and demonstration, the synthetic fuels business today is moribund, the victim of rising costs of large-scale engineering, technical problems, the cost of money, and the slump in world petroleum prices.

One bright spot on the new technologies scene is a group of energy systems often referred to as "alternative technologies," dominated by cogeneration and wind. For a variety of reasons (discussed below), these technologies are advancing rapidly. California appears to be the leader in implementing alternative technologies, and reports submitted by the California utilities in mid-1984 showed 1190 MW of capacity installed, and a total of almost 10,000 MW under construction or at an advanced stage of negotiation (13). Current and planned production of electricity in California from cogeneration, wind, geothermal, and small hydropower facilities, in conjunction with the reduced electricity growth rate, virtually ensures that California will require no new central station power plants of any type in the rest of this century.

Immediate Reaction to the Embargo

The government could not avoid responding to the OAPEC embargo in 1973. Even if the policymakers were convinced at the time (which they were not) that no policy action was the best response to the embargo, the intense fears of the population and accompanying political pressure made action unavoidable. The nation was not prepared for a rapid loss in oil supply. No policies were in place or even analyzed to deal with the potentially devastating economic consequences of a prolonged oil embargo. The options at the time were limited. Mandatory allocation or rationing were probably the only two possibilities for the government to (*a*) respond in a highly visible manner to the embargo (required by the political process) and (*b*) establish measures that would avoid the most serious adverse effects of a prolonged embargo.

MacAvoy (14) analyzed the impacts of the federal allocation programs in 1973 and 1979 and stated "the gasoline crises of 1974 and 1979 were the products of decisions of regulators to hold inventories off the market. The results were shortages, extended beyond what could be tolerated on grounds that they were necessary for price stability." He estimates that during the 1973–1974 embargo, virtually the entire shortage resulted from government action, and virtually none of it was caused directly by the embargo. The 2 mmbpd shortfall was caused, in MacAvoy's analysis, by controls on oil prices (0.25–0.33 mmbpd), interrefinery crude allocations that reduced imports (1 mmbpd), and government requirements to increase stock of heating oil (0.8 mmbpd).

Considering the seriousness of the economic impacts of the shortages, it is worthwhile to understand some of the reasons for the federal government's difficulty in dealing with the embargo. An important purpose of the government response was to protect the economy from potential disaster—represented, for example, by an embargo that both was effective and persisted for six months to one year or longer. During the embargo, no one could state with certainty how long or effective the embargo would be. However, the possibility existed of devastating impacts—either economic or as a cause of war.

Thus, some of the government actions—especially the measures to increase inventories of fuel oil at the expense of motor gasoline—were basically paying a price in the short term for longer-term insurance. The price was high (perhaps higher than it needed to be) but the insurance was necessary, considering the risks of much more serious damage.

The actions of the federal government to hold oil prices down can be understood in the context of the times. The Cost of Living Council had established wage and price guidelines for the economy. If the government had not applied these guidelines to oil, the results would have been politically suicidal in the face of the popular belief that the oil companies were causing the higher prices that accompanied the embargo.

The government had an enormous stake in the outcome of the embargo. One of the costs of a government that is so responsive to political pressures is an overreaction to events that capture the public imagination (especially if the events are threatening). The benefits of responsive—sometimes over-responsive—government need to be weighed against the costs sometimes incurred by hasty decisions to avoid unacceptable political consequences.

As an example, the allocation of crude from refineries that were rich in domestic crude (which was less expensive than imported oil as a result of the steep price rise during the embargo) to refineries more dependent on imported oil had disastrous consequences. We have already noted that these provisions, adopted in February 1974, exacerbated the shortfall considerably. The reason for this was that the allocation rules caused domestic oil to be favored over imports; refiners that chose to import additional oil were effectively required to subsidize refiners that reduced imports.

If an interrefinery allocation system were judged to be either desirable or politically necessary, the government could have implemented one that did not discourage imports during the time when they were most needed. The crisis atmosphere and the lack of prior government experience in or knowledge of energy contributed to this failure of policy.

New Institutions of Government

FEDERAL GOVERNMENT REORGANIZATION   The embargo and subsequent energy price escalation combined with growing fears of shortages to produce what became a long series of energy "czars" and government reorganizations. The Federal Energy Office (FEO) was created to manage the shortage and, after the embargo, to administer price controls. The AEC became ERDA (the Energy Research and Development Administration) under the Energy Reorganization Act of 1974; AEC regulatory functions were assigned to the Nuclear Regulatory Commission (NRC). In 1974 the FEO became the Federal Energy Administration (FEA), the change signifying increasing importance and permanence within the federal government. The FEA absorbed the Federal Power Commission (FPC), the agency that had administered natural gas price regulations. New energy programs in solar energy and conservation were started in the National Science Foundation, while older programs in coal housed in the Department of Interior grew in size. All of these were moved to ERDA.

As the decade progressed and the national sensitivity to the implications of oil shortage grew, the idea of ERDA, devoted simply to research and development, grew obsolete. A cabinet-level Department of Energy (DOE) was proposed and approved (Department of Energy Organization Act, 1977). DOE was the product of a national sense of crisis. This new entity brought together staff from many institutional backgrounds, with many persuasions. From the outset the DOE was engulfed in continuing controversy, relating to both its objectives and its managerial competence. Contributing to its problems was an inability to draw on or to build a national constituency. Among the many problems faced by DOE were the self-induced problem of defining success in virtually every program in terms of "barrels per day of oil delivered." Goal-oriented criteria for success are helpful in virtually any project. However, a goal specified in terms of product delivered is inappropriate to a government agency that cannot itself produce the product. This type of goal is even more inappropriate when one is dealing with research and development projects, where the intrinsic nature of the problems renders the goal of product delivery virtually without meaning.

Philosophically opposed to DOE's objectives, the Reagan presidential campaign identified the agency as a candidate for abolition. The battle between Congress (generally supportive of DOE) and the Reagan administration resulted in disarray and demoralization in many programs (particularly conservation, renewable energy, and fossil energy research; data collection and analysis within the Energy Information Administration; and energy regulation throughout the agency and in the Economic Regulatory Agency of DOE) during the first two years of administration. An agreement between Congress and the executive branch was achieved during 1983 and 1984, and the administration permitted stronger leadership within DOE in those years. However, the future of DOE remains uncertain, and the promise of the organization for national leadership has (for the time being at least) vanished.

STATE LEVEL REORGANIZATION   At the state level the story is mixed. Some states accomplished a great deal. California was probably the leader, with its California Energy Commission (CEC). The CEC plays two major roles. First, the CEC established an independent forecasting organization capable of highly sophisticated analysis. The utilities were forced to respond in kind, building staffs with high levels of expertise. The net result is that forecasting techniques and institutional skills today are vastly improved over those of preembargo times. The second area was in development and implementation of conservation regulations. California led the way in analyzing the possibilities of conservation measures, and in developing regulations for energy use in buildings. These measures, now on the books, will have impact for years to come.

While California has been a leader in developing expertise in energy, many other states have also developed expertise in analysis and implementation of energy programs. It is interesting to note that the state-level activities tend to devote considerable attention to energy use, largely because the state often has more ability to influence decisions about demand than about supply.

FEDERAL LEGISLATION   Energy legislation enacted in the energy decade ultimately included incentives for virtually all energy supply and conservation technologies. There was something for everyone, yet the incentives were not well balanced. The prevailing view was that competition would lead to the best of the technologies flourishing, while the losers would fall by the wayside.

The problem is that incentives, once introduced, are hard to remove. The intangible drilling expenses long enjoyed by the oil and gas business provided a subsidy to that sector not enjoyed by solar energy. Since political factors made it impossible to remove the existing subsidy, the compromise was new subsidies. For example, lobbying by solar groups led to incentives for active solar systems. As chance would have it, Congress at that time was not interested in conservation incentives. Since no one could come up with a good way to distinguish passive solar devices from conservation, only active solar systems were included—a singularly poor decision from a cost-effectiveness perspective.

The lengthy list of energy legislation enacted during the energy decade has been summarized by the US Congress. We present in the Appendix an abbreviated annotated list of the key laws. Virtually all aspects of the US energy system were subjects for legislation. The modern history of federal energy regulation is summarized in McKie (15).

The chaos in energy regulation was captured by former FPC Commissioner William O. Doub (16): "...energy policymaking literally resembles a Rube Goldberg production with an abundance of action and few real results...the cost of complying with federal regulations represents about $14 billion annually." A similar gloomy conclusion characterized the government's energy production incentives (17): "We feel secure in concluding that both past and present energy policy has not had the objective of promoting a more efficient use of our scarce energy resources." And, in a generally optimistic report on the eventual prospects for a synthetic fuel industry in the United States, Perry & Landsberg (18) express reservations about the ability of the government to induce this new industry: "This program will either become part of the everyday economy, or it will fail."

These quotations hint at some of the dilemmas posed for the nation in attempting to deal with the "crisis" following the OAPEC embargo. In the largest sense the debate was over the allocation of increasingly scarce resources—the root stuff of economics. Rapid price rises induced by an emergent monopoly, and shortages induced in large part by poor allocation programs imposed by our own government, led to new incentives and to new winners and losers. Many of the government regulations were designed to minimize adverse impacts on special groups (e.g. low-income weatherization programs and lifeline utility rates intended to protect low-income persons from fast-changing energy prices). Other policies encouraged new technologies on grounds that government investment would speed their entries into the marketplace. Some, like the windfall oil profits tax, were designed to capture some of the unanticipated gains to those who happened by chance or good planning to own oil.

Energy Price Regulations

The laws with the most profound effects on energy during the decade were those that continued or extended energy price regulations. Oil and natural gas prices were regulated by the federal government throughout most of the energy decade. Electricity prices are regulated by state utility commissions. Coal and uranium prices are not regulated.

OIL PRICES   The origin of the oil price regulations was twofold: (a) In the early 1970s, the Cost of Living Council regulated prices throughout the economy in an effort to control inflation; oil, like many other products, was price-controlled. (b) The embargo and subsequent control of world oil prices by a functioning cartel motivated the government to continue price controls on oil long after controls on other products were removed.

The rapid increases in oil prices in 1974, combined with the government commitment to reduce inflation and the widespread belief that oil prices contributed greatly to inflation, resulted in oil price controls that became increasingly complex as the decade progressed. Strong political currents also supported price controls: the oil companies were widely seen as gaining enormous profits from the operations of the OPEC cartel. There were powerful popular pressures to keep prices of oil products as low as possible. For example, John Sawhill, Administrator of the Federal Energy Administration under President Ford, was forced to resign his position because of his public support for a five-cent-per-gallon gasoline tax. (In 1983, a five-cent-per-gallon tax on gasoline was enacted into law with little controversy or notice.)

Toward the end of the Carter administration, legislation was passed that gradually phased out oil price controls. A windfall profits tax was included in the political compromise that effected oil price decontrol. The Reagan administration has taken steps to hasten the process of full decontrol of oil prices set in motion in the previous administration.

NATURAL GAS PRICES   Natural gas price regulation persisted throughout the decade (with the exception of the intrastate market, where prices were unregulated). While there was interest in deregulating prices, the influence of interest groups that benefited from the price controls (especially consumers and pipeline companies that were not vertically integrated) was sufficient to maintain the controls.

The consequences of the price controls were particularly visible for natural gas. A significant shortage in natural gas persisted from 1975 through 1978 throughout large parts of the country. The shortage is not surprising, as the price of natural gas in the interstate market was far lower than that of competing fuels. Partially in response to the shortage, the Federal Power Commission in the late 1970s tripled the price of new gas contracts in the regulated interstate market. There was relatively little outcry at this very large price increase because prices on new contracts could be folded into those on existing contracts to yield a much smaller increase in average gas prices. Adding transport costs and profits to the selling price of natural gas still left it much lower in price than competing fuels.

The Natural Gas Policy Act (NGPA), passed in 1978, continues gas price regulation until 1985 (or later, for some types of natural gas). The objective of the Act is to allow higher natural gas prices to achieve greater parity with competing fuels and to spur development of new gas reserves. The NGPA coincided with and contributed to the end of the natural gas shortages. Although the Reagan administration favors natural gas price deregulation, it has not actively pursued this policy.

ELECTRICITY PRICES   Electricity prices continue to be regulated at the state level. There is little support for deregulation (although reform of the regulatory process is much desired by many groups), primarily because electric utilities are recognized to be natural monopolies for which price regulation is appropriate.

Important changes have occurred in the areas of wheeling and of marginal cost pricing. The Public Utilities Regulatory Policies Act (PURPA, 1978) requires utilities to wheel power at reasonable prices. PURPA also requires utilities to calculate the marginal cost of electricity as a function of time of day, season, and reliability, and to pay marginal costs for electricity to any supplier who wishes to sell it. This part of the legislation has had a significant effect on the sources of electricity supply in California, creating a substantial role for small power producers. (Probably the most important element in the present success of small power producers in California has been the development of standard, long-term contracts between the California utilities and the independent power producers.) The PURPA legislation has not yet had the effect of bringing in small power producers throughout most regions of the United States, probably largely because of the lack of contracts for power with the independent producers that are adequate to raise the investment capital for the projects. Nonetheless, the possibility exists of a shift over time in the roles of at least some utilities toward distribution rather than generation. The chief planning executive of one major utility, Pacific Gas and Electric Company, has gone on record as saying that his company is, for planning purposes, assuming that all new electricity will be purchased from external sources during the next 20 years.

OBSERVATIONS   Price regulation that keeps prices down reduces supplies and encourages demand, creating substantial potential for shortages and fostering inefficient and uneconomic allocation and use of energy resources.

The past decade has provided strong evidence of the disadvantages of energy price regulations. While the government has a legitimate role in protecting the population (and groups within the public) from serious shocks outside their control, energy price regulations are not the means of achieving these ends.

Nonetheless, price deregulation remains a complex issue. A major policy matter is the treatment of economic and monopoly rents under oil and gas price deregulation. The problem, stated simply, is that producers of a scarce commodity such as oil and gas can receive considerably more than the cost of production and fair returns on risk. The problem is compounded by the strong influence of world oil prices, established by a cartel of oil producers, on domestic prices under price deregulation. Effectively, deregulated US prices for crude oil operate under the protection of a partial world monopoly on oil supply. Furthermore, because natural gas and oil compete in many markets, deregulated gas prices will be higher because of OPEC.

At times, substantial portions of the public have accused the oil companies of causing shortage and/or higher prices. Although the evidence does not support this view, the notion that the oil companies have benefited from OPEC is defensible; any major economic actor will attempt to gain from a new and changing economic environment. The more serious analytic issue is the degree to which the public should capture a portion of the economic and monopoly rents derived from oil and gas production. Congress decided that such public capture of these rents was appropriate in passing the windfall profits tax in 1980.

Although the windfall profits tax was strongly opposed by the oil industry prior to its passage, it appears to have worked relatively smoothly in the past few years. Many members of Congress and the public believe that the government should capture a portion of the rents because the relative scarcity of the natural resources can cause a considerable discrepancy between costs of production (including risk) and consumer prices. Many others believe that the rents rightfully belong to the owners of the rights to the resources, in the same way that owners of other resources are rewarded through market processes.

In short, energy price deregulation is likely to cause a far more efficient allocation and use of energy resources than regulated prices as well as to encourage the development of new supplies. Protection of consumers from high prices, if the market is working effectively, is in our judgment better achieved through means other than price control. There are, however, reasons to believe that the energy market is less than fully competitive or that the magnitude of the income transfers within the present system requires careful attention to distributional issues. The existence of a world oil cartel and the ability of owners of energy resources to capture scarcity rent from their holdings are two examples of justifications that have been used for public policy involvement in oil issues. The effects of rising energy prices on low-income groups is an important public policy concern. Energy is an essential commodity to all groups and makes up an increasing portion of discretionary income among the poor. Many of these issues are, we believe, justifications for a significant government role in energy markets, but are not sufficient to support government control of energy prices. Actions such as residential weatherization programs (to support the poor by reducing energy bills) and windfall profits taxes are specific social policies to achieve defined goals. Directed policies to remove undesirable or unacceptable consequences of rising energy prices and to respond to failures of the market are preferable to across-the-board price regulations.

Industry Structure

Changes in the law, and in political and economic realities, have led to changes in the structures of the energy industries. Before the embargo the oil industry controlled a major portion of the international oil it sold. The embargo showed the OPEC nations that they could control their own resources. The authority of the international oil companies was circumscribed, so that their roles as contracting and distributing agents increased while their direct control over sources of oil supply diminished. This change, together with the growing realization that in the long run oil was a declining source of income, has led the major oil companies to diversify. Today, the large oil companies are extensively involved in virtually every aspect of the energy sector (and many have even moved outside of energy, at least to a degree). The oil companies were for years a classic example of vertical integration. They have since become integrated horizontally as they have purchased coal, oil shale, and uranium industries.

While the structure of the electric utilities did not change in any basic way during the energy decade, almost everything else about the industry did change. At the beginning of the energy decade, electric utilities still enjoyed the benefits of electricity price regulation during times of declining real costs of generating and distributing the product. Ever since the 1920s, when Samuel Insull, a one-time assistant of Thomas A. Edison, demonstrated that electric utilities constitute a natural monopoly, the utilities were granted territorial rights in exchange for regulation of their rate of return on investment.

An ensured return on investment had many effects on the industry. Because return was based on investment, there was a tendency to build too well or too much (the Averch-Johnson effect, which encourages "gold plated" systems). The regulated returns on investment made the utility industry a standard of stability and ensured good return on investment with risk almost as low as that of government securities. Utility stocks were favorites of "widows and orphans."

All of this changed during the energy decade. Starting in 1973, the real costs of producing electricity began to rise. The electric utilities not only faced rising energy costs, but also a whole series of other problems. High interest rates—in the most capital-intensive industry in the United States—plagued utilities. This made the cost of construction both higher and more uncertain. Unanticipated and dramatically increasing costs of nuclear power led to added costs for many utilities. Environmental concerns became important in the siting of new generation facilities and distribution lines, and also resulted in government requirements for control of emissions and effluents from power plants. An industry accustomed to steady growth (at more than twice the growth rate of energy demand) was not able to anticipate the reduction in demand growth during the energy decade. This led to the construction of large amounts of capacity—at a time of high-cost construction—for which there was no market. (Capacity well in excess of demand has caused severe economic problems for many utilities. Utilities producing more than one third of the electricity in the nation—with a large number concentrated in the Midwest—have capacity well in excess of demand for power.) In the face of rising prices, citizen protests to regulatory agencies (as well as perceptions by many regulators that utility managements could have avoided some of the problems leading to rising costs) often led to decisions in rate cases unfavorable to the utilities. The regulated environment was no longer an environment that guaranteed financial success to the utility industry.

All of these factors and others have led to a severe deterioration in earnings of utilities. They also lead to considerable uncertainty in the planning environment. Uncertain future demand, combined with long and uncertain lead times for large power plants, has made decisions about new power difficult. If a large power plant is delayed—and many have been in the recent past—interest expenses during construction and engineering cost overruns can be substantial. There is no longer an assurance that the regulatory agencies will permit charging all of these expenses to the consumer. In addition, if demand does not materialize, then underutilized capacity adds to the overall burden to the utility and the consumer.

This discussion makes it evident that electric utilities are in a state of flux. Solutions to the problems are not obvious. It is possible that electric utilities will become more involved as distributors of electricity than as producers. Small-scale power generation may play a large role in supplying electricity—as is occurring in California—to avoid some of the financial risks of the larger plants. These changes, if they occur, are likely to be evolutionary. The utility industry, with its enormous infrastructure and expertise, is moving slowly to adapt to rapidly changing times.

Energy Conservation

During the energy decade conservation grew from nonexistent to pervasive. Federal programs covered virtually every aspect of the economy. A partial list includes the following:

1. Mandatory fuel economy standards for automobiles and light trucks, which established progressively tighter fleet average fuel economy requirements for new vehicles from 1978 to 1985
2. Building energy performance standards for new buildings, designed to reduce energy use for heating and cooling in all new buildings
3. Residential conservation service, requiring utilities to provide low-cost (subsidized) energy audits for houses
4. Solar and energy conservation tax credits
5. Mandatory minimum energy-efficiency standards for consumer products (household appliances and heating and cooling systems), and energy labels for these products
6. Funds to support a wide range of state and local activities to reduce wasteful energy use, including support for state energy agencies, energy extension services at universities, and related activities
7. Grants for low-income weatherization, to provide subsidies for residential conservation investments among low-income groups
8. Schools and hospitals programs, to provide support for energy conserving capital investments in these institutions
9. A federal energy management program that established criteria for energy conservation measures in all federal buildings
10. Some (albeit limited) consumer information programs about energy conservation
11. Targets for reductions in energy use in industry.

Consistent with the national goal of achieving energy independence, the actions to reduce energy demand were easily justified as the easiest and most cost-effective measures in the short run to reduce energy supply pressures.

Energy conservation has proven remarkably successful. One measure is the ratio of economic output to energy input. The energy to GNP ratio has fallen each year since 1975. The dramatic reduction of energy demand growth during the decade has reduced the nation's vulnerability.

A major change in perceptions about energy conservation and the economy took place during the decade. Early in the decade, there was widespread belief that energy and economic growth were inextricably linked, so that a 3% growth in the economy meant a 3% growth in energy demand. It is now widely recognized that increased efficiency in energy use means that the economy can grow considerably more rapidly than energy demand. Rather than being seen as a threat to the economy, energy conservation is now widely recognized as a means of improving economic performance as well as providing protection from future energy difficulties.

Because most of the federal conservation programs were regulatory in nature, we turn to a discussion of two of these programs—the automobile fuel economy standards and the building energy performance standards—to illustrate some of the issues raised by these kinds of programs.

AUTOMOBILE FUEL ECONOMY STANDARDS   The mandatory fuel economy standards required an increase in fuel efficiency from a fleet average of 18 miles per gallon (mpg) in 1978 to 27.5 mpg in 1985, an increase of slightly more than 50% in less than seven years.

At the time the standard was being considered by Congress, the automobile industry had endorsed a voluntary guideline that called for 20 mpg by 1980. The industry strongly opposed a mandatory standard. In testimony in 1976, all three major US manufacturers stated that the 1985 standard was not attainable.

The government analysis, based in part on confidential data supplied by the automobile manufacturers, contradicted the testimony of the manufacturers. The Department of Transportation (DOT) presented analysis showing not only that the standards were attainable but also that they were cost-effective (i.e. the higher automobile price would be recovered in reduced fuel expenditures). The government analysis also indicated that no change in the percentage of cars in each size class would be required to meet the standards, and suggested that technological improvements alone could permit the manufacturers to achieve the standards.

In early 1979, Ford and General Motors (GM) reopened the issue of federal fuel economy standards, maintaining that DOT had overestimated the benefits and underestimated the cost of fuel economy improvements. The Chase Automotive Division of the Chase Manhattan Bank released a report "finding...conclusively that the current fuel economy standard for passenger cars should be lowered from...27.5 in 1975 to 26 mpg or less..." [quoted in Difiglio (19)].

DOT's refusal to reconsider its rulemaking evoked continued opposition from the automobile industry. Ford claimed that the standards would throw the nation into an economic recession.

The Iranian revolution and the subsequent gasoline shortages dramatically altered the conflict between the automobile industry and the federal government over fuel economy standards. By August of 1979, GM announced that it expected to exceed the standards. During 1980, all three major manufacturers released projections of fuel economy in excess of the standards. (The GM projections were for each year through 1985; Ford and Chrysler each presented a projection for one year, 1985 and 1981 respectively.)

Judged in terms of its own goals, the fuel economy standards have been successful. A rapid increase in fuel economy has been achieved. Neither the industry nor the consumer appear to have suffered as a result of the standards. It is apparent that the government's analysis of the automobile market and the ability of the industry to produce more efficient automobiles was more accurate than that of the industry—or at least more accurate than published industry reports.

A full assessment of the costs and benefits of the fuel economy program depends critically on what would have happened without the program. Considering available public knowledge of industry plans, it is likely that the standards speeded up the production of more efficient automobiles by several years (and guaranteed that efficiencies would continue to improve through 1985). Each mile-per-gallon improvement in the fleet average during the eight years of the standards reduces total gasoline consumption by about 1.5 billion gallons in the last year of the program (1985), resulting in a gross savings to the consumer of about $1.8 billion annually. (The net saving is lower, as the added cost of the more efficient automobile needs to be deducted from this amount (20).) Thus, if the standard resulted in the fleet average's being 2 mpg more efficient than otherwise, the nation will spend about $3.6 billion per year less on gasoline in 1985.

The fuel economy standards probably helped the US auto industry. Because of the perception of gasoline shortages in 1979, the demand for more efficient automobiles increased faster than the industry had expected. Because it was required to design and produce more efficient automobiles, the industry was better able to respond to the changed market.

BUILDING ENERGY PERFORMANCE STANDARDS   The story of the building energy performance standards (BEPS) is much shorter than that of the automobile fuel economy standards, because BEPS was never implemented. The standards were legislated under EPCA in 1976, with a key provision that the final rulemaking required Congressional approval (the now outlawed legislative veto). The legislation recognized that buildings consumed one third of the energy in the United States and had substantial potential for cost-effective reductions in energy use. The framers of the legislation had little confidence that the market would achieve this conservation without mandatory standards.

Responsibility for developing the standards was transferred from the Department of Housing and Urban Development to the Department of Energy when DOE was established. In early 1981, after the rulemakings were complete, DOE presented the results to the Senate. The analysis of the standards was based on establishing building practices that minimized life-cycle costs of energy use, using only widely available conservation measures. The standards were estimated to save about one quad per year by the year 2000 (a reduction in fuel bills at today's price of about $8 billion per year). Furthermore, DOE estimates indicated a substantial net present benefit to consumers (totaling $6 billion over 20 years in 1978 dollars) if the standards were implemented.

Testimony before the Senate in 1981 revealed very strong industry opposition to the proposed standards. Some of the opposition was questionable: for example, some testimony maintained that builders would meet the standards without federal action while arguing at the same time that the standards would raise new housing costs so much that builders would be put out of work.

However, there were also legitimate concerns about the viability of the standards that DOE developed. In spite of a substantial research effort, DOE was not able to present a clear and simple description of what builders needed to do to comply with BEPS. (A large part of this problem derived from the legislative requirement that mandated a performance standard, in which the builder was given a choice of many different ways of achieving compliance with an energy budget for a building. Considering the relatively limited knowledge of building energy use at the beginning of the analysis, the development of a comprehensive energy performance standard was a very ambitious task.) A second major issue was fair treatment of different fuels. The electric utility industry was strongly opposed to the standards; it believed that BEPS favored gas heating. (For commercial buildings, their concern was justified.) Finally, at the time Congress reconsidered BEPS, the housing industry was facing the worst time it had seen since World War II. Some 30% of builders were eventually to go out of business in the early 1980s. Under these economic conditions, an additional regulation perceived as encumbering builders was not popular.

BEPS has been changed into a voluntary program (mandatory for federal buildings). This voluntary program is being developed in concert with industry groups. One objective of the present approach is to encourage the American Society of Heating, Air Conditioning, and Refrigeration Engineers (ASHRAE), the industry body responsible for building energy standards, to use research results from the DOE effort to update their codes.

A comparison between the development of energy-efficiency standards for buildings and those for automobiles is instructive. If the automobile industry had had serious economic problems in 1979, the automobile standards might well not have been promulgated. If the building industry had been in good economic condition in 1981, BEPS might have been adopted.

There are other major differences. The automobile industry is highly concentrated, so that the implementation of a regulation is more straightforward. A compliance procedure as complex as BEPS would present no difficulties to the automobile manufacturers. At the same time, because of the concentration of the automobile industry and its importance to the economy, errors in regulations can have disastrous economic consequences. The building industry is diverse and scattered. The industry is not as easily reached and has far fewer resources with which to deal with new and potentially complex requirements than the automobile industry.

It is clear that research in support of the regulation advanced the state of knowledge about energy use in buildings. It also developed high quality information in a form that can be made much more widely available to builders. It is too early to know whether builders will respond to this information under the auspices of a voluntary program or whether ASHRAE will adopt more stringent standards as a result of the BEPS effort. The need for improvement in the energy efficiency of buildings remains substantial, in light of the findings of the BEPS analysis that cost-effective measures can reduce energy use in new buildings by 25–40%.

ENERGY CONSERVATION POLICIES TODAY   At present, most of the energy conservation policies initiated during the energy decade are caught between an administration that opposes them and a Congress that provides some support. Policies whose implementation awaited the new administration have been stopped. Existing regulatory programs have been pursued with little vigor. Few new initiatives on energy conservation policy have been taken either by the administration or by members of Congress.

Justification for federal energy conservation policies continues to be strong. The basic reason for energy conservation policy during the energy decade was to reduce vulnerability to energy shortfalls. This problem is less serious today—because of lower oil imports and the strategic petroleum reserve—but it remains an important issue for the future. A more basic issue is the economic benefits of improving energy efficiency in all sectors of the economy. Even though energy use per unit of GNP has declined throughout the decade, more of this decline is due to short-term behavioral changes than to investments in higher energy efficiency. Appropriate policy

approaches—to encourage investment in increased energy efficiency—
could provide considerable economic benefits to the nation and reduce
energy vulnerability in the long term.

Research and Development (R&D) During the Energy Decade

RESEARCH ACTIVITIES   The rapid rise in energy R&D expenditures during
the energy decade (up until 1981) diversified R&D activities away from the
earlier overwhelming emphasis on nuclear energy. The budgets reflect the
objective of ERDA (as stated by its two major planning documents, ERDA
48 and ERDA 76) to support a diverse array of projects with the intent that
sufficient numbers would succeed and thereby provide protection particularly against vulnerable oil supplies.

The success of the federal R&D has been mixed, at best. Judged from the
perspective of the Dixy Lee Ray report and later major planning documents
such as ERDA-48 and ERDA-76, achievements are very disappointing. The
investments in new energy technologies have not succeeded in establishing
new major sources of energy for the country. The expectations that the
federal research, development, and demonstration (RD&D) activities
would produce in the middle or late 1980s commercial technologies to
support new, large, and viable industries were not realized. Technologies to
convert coal into liquid and gaseous fuel forms, to extract and process shale
oil and possibly tar sands, to develop a viable breeder reactor (to ensure a
virtually inexhaustible source of uranium for electricity), to convert wastes
into liquid or gaseous fuels, and to provide solar energy for space heating,
industrial process heat, and electricity were seen in the early 1970s as
candidates for widespread commercial success in the 1980s.

It is now clear that none of these technologies had a chance of widespread
commercial application within a short time. Many of the programs were
oversold. R&D contributed very little to increasing energy supplies during
the decade. Some key R&D projects did hasten the ability to reduce
demand growth by speeding up commercialization of cost-effective energy
conserving end-use technologies (e.g. energy-efficient lighting systems and
improvements in heat pumps and other heating systems).

Much of the RD&D budget was devoted to very large-scale demonstration programs. The large increases in the budgets necessitated some
ways of spending money in large chunks. The demonstration programs met
this need admirably. They also made fine "pork barrel" projects. Large
facilities can be spread around the country, an approach taken for coal
conversion facilities.

The research on environmental, safety, and health aspects of energy
technologies did achieve some successes during the decade: a considerably

improved understanding of measures to improve the safety of nuclear power plants (or, from a different perspective, a recognition of the inadequacies of current design in ensuring safety), increased knowledge about indoor air quality and some control measures, advances in understanding (and in some cases reduction) of environmental and possible health problems of fossil fuel conversion.

FEDERAL ENERGY R&D EXPENDITURES   The federal budgets for energy R&D demonstrate the rapid change in research priorities that has taken place since 1980. Conservation and solar energy research have been slashed to 15–20% of fiscal year 1980 levels; fossil energy research is funded at less than 40% of 1980 levels, energy research (high-energy physics and basic research support) has increased by more than 20% in nominal dollars, and nuclear energy and defense-related research have grown very rapidly. During the past three years, the federal involvement in energy is rapidly shifting back to the priorities of the vintage AEC days, as displayed vividly in Figure 2.

The current budget represents a compromise between the proposals of the Reagan administration, which had originally committed itself to dismantling DOE and eliminating the conservation and solar—and possibly the fossil fuel—research programs altogether, and Congress, which has acceded only partially to the administration budget requests and has preserved at least a skeleton of these programs. The very rapid budget cuts combined with strong administration policy in opposition to these research programs caused widespread demoralization of DOE staff in these areas during 1981 and disrupted many longer-term research activities throughout the nation. Subsequent agreements between Congress and the administration and changes in the high level staff at DOE have eased the problem in more recent years.

Thus, we have seen two dramatic shifts in R&D priorities during a single decade. The approach of the 1970s, in which the federal government became involved in virtually all aspects of R&D, has been replaced by a program of greatly enhanced weapons and nuclear energy research, somewhat expanded basic research, and tremendous cuts in research on conservation, solar energy, environmental questions, and fossil fuels.

OBSERVATIONS   A major misconception that guided the federal research program during most of the energy decade was the assumption that energy R&D could solve the nation's energy problems in a short time. Not only was the program oversold—to Congress and the American people—but the overselling led to poorly designed research activities. The government acted as if it believed that it could achieve commercialization of new energy technologies by spending large sums of money. Perhaps the experience of

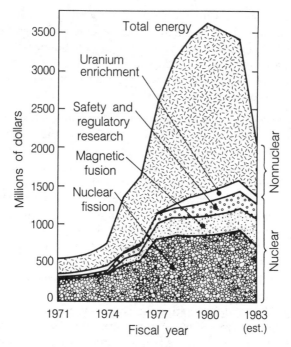

Figure 2 Federal energy R&D expenditures through the energy decade. In the middle of the energy decade nonnuclear activity boomed. Today the budget mix is remarkably like that prior to the 1973 embargo. (Source: *Science*, July 29, 1983, page 443.)

the AEC contributed to this misconception: the federal government had been a major factor in the creation of the nuclear power industry, and its efforts were crucial in achieving commercialization of nuclear power plants.

Without very large subsidies, new energy technologies cannot achieve commercial acceptance until they are cost-effective. This is obvious, but was often ignored throughout the past decade. The demonstration plants to convert coal to liquid or gaseous fuels, the Barstow thermal solar electricity facility, oil shale projects, and the Clinch River Breeder Reactor were premature. The technologies are not ready to compete in the market; they do not yield sufficient nonmarket benefits to justify large expenditures on full-scale demonstrations; and they are so unlikely to be picked up by the private sector that the federal expenditures for demonstrations cannot in our judgment be justified.

While the government cannot commercialize a technology before it is ready, federal research can be essential to carrying out longer-term research to identify new and promising technologies and overcome important technical problems; to treat environmental, safety, and health issues that will not be dealt with by the private sector; and to support small-scale demonstration projects that will enhance the viability of energy technologies that are close to economic, particularly in risk-sharing ventures with the private sector.

These views are similar to the rhetoric, but not the actions (i.e. budget proposals), of the present administration. The administration is, in our view, pursuing an unbalanced set of priorities. Long-term research in areas of energy that will remain critical to the nation, relatively inexpensive near-term support for small demonstration projects to encourage new technologies, and dedication to important environmental research in the energy area are all lacking. We believe the nation will pay the price in reduced diversity of supply and end-use technologies, in less efficient development and use of energy, in higher costs for energy in the long run, and in a reduced ability to deal with possible future supply interruptions.

CONCLUSIONS

In reviewing the energy decade, we generally found more mistakes than successes. Those who believed that the federal government could solve the major energy problems generally found themselves chastened, while those who believed in R&D were discouraged by the failure of our governmental processes to provide consistent support. For the time being at least, the seemingly insuperable energy problems that confronted the nation in the 1970s appear to have been reduced to a manageable size. The same uncertainty in events and factors outside our control will—with certainty—cause unexpected changes in the US energy system, affecting the population in important ways. We do not presume to know what these unexpected changes will be. The energy decade provides a useful basis for assessing a governmental role that can best cope with an uncertain future.

From this complex web of history we would draw some conclusions about government involvement in the management of a critical mineral, during times of real or potential disruption:

1. The response to the first embargo, of 1973, cost the nation heavily. The nation was clearly unprepared for the embargo, and political considerations made effective response very difficult. Some developments during the energy decade—particularly the success in building a strategic

petroleum reserve—will ease the pain of dealing with another embargo, if one should occur.

2. A critical issue for the future is the rapidity of the decline of US oil production capability. Drilling rates in recent years have been high, and this has kept total production from falling. Yet there is evidence that this situation cannot continue, and drops in production of 20% or more by the end of the century are likely (21).

3. Research and development with a long time to payoff is important for the national good, yet hard for industry to support. There is a continuing need for government to become involved, and to remain involved. The precedent for this approach is well developed in basic sciences, especially as these are managed in the National Science Foundation and in the Basic Energy Sciences programs in the DOE. In many areas the federal government never learned either the importance of basic science, nor the need for long-term sustained support for applied and basic research. This lack of continuity in research support is particularly apparent in conservation, solar energy, and fossil fuel research.

4. Government policy to foster energy conservation can provide near- and longer-term economic benefits. The approach during the energy decade relied solely on mandatory federal regulations. While the regulatory approach can work in many cases, there is a major need to explore a variety of measures to improve the working of the market. New initiatives are needed to reestablish an effective national energy conservation policy (even with deregulated energy prices, which is a first essential step to foster conservation).

5. Government involvement in the energy pricing system and in allocations has caused a long series of problems. While there are certainly occasions where such involvement is important—as when there is a prospect for a long-term shortage—the major lesson learned is that the problems involved in government intervention in the price system are far larger than many people believed, and that the costs of intervention can be substantial.

6. Notwithstanding the problems of government involvement in energy pricing, there remain vital roles for the government in energy. Our account of the past decade has convinced us that energy was and is an important national problem with major societal consequences. The impacts of disruptions in the energy system, the need over the long term to develop new energy sources, inefficient and costly use of energy, environmental impacts of energy production and use, and the enormous effects of the energy system on all aspects of the economy—affecting both rich and poor—all require that the government remain involved in protecting the public interest in the face of continuing energy issues.

ACKNOWLEDGMENTS

The authors gratefully acknowledge Carmen Difiglio of the US Department of Energy for his contributions to the discussion of the automobile energy economy standards. The discussion of this subject relies almost entirely on Mr. Difiglio's work on this subject (19).

APPENDIX

Excerpts from a compendium developed by the Congress (1981) indicate the scope of congressional energy legislation during the decade.

1. The Emergency Petroleum Allocation Act (1973) authorized the government to allocate fuel in time of shortage. There was a widespread feeling that federal allocation procedures led to more disruption than would have occurred had industry carried out the allocations.
2. The Trans Alaska Pipeline Act (1973) overruled environmentalists' NEPA-based objections, and permitted this major oil supply project to move ahead.
3. The Energy Supply and Coordination Act (ESECA, 1974) was a first step in a continuing effort to encourage fuel switching away from oil and gas. It included controversial provisions for suspension of air pollution requirements.
4. The Federal Nonnuclear Energy Research and Development Act (1974), based on the Dixy Lee Ray report (see above), was the first major legislation focusing on nonnuclear R&D.
5. The International Energy Agency (IEA, 1974) was established in the hope of providing an organization of oil users that might work together to counter OPEC.
6. The Energy Policy and Conservation Act (EPCA, 1975) focused on energy conservation. It mandated fuel-efficiency standards for automobiles, labels and efficiency standards for a number of consumer products, and reporting requirements for industrial conservation. It also provided funds for conservation in public buildings and required the government to collect and report data relating to energy use.
7. The Energy Conservation and Production Act (ECPA, 1976) introduced electric utility rate design initiatives and conservation standards for new buildings.
8. The National Energy Conservation Policy Act (NECPA, 1977) required utilities to undertake and finance residential conservation programs, provided weatherization grants for low-income families, provided secondary financing for conservation, and enhanced federal

action regarding conservation and solar energy in federally owned buildings.

9. The Public Utilities Regulatory Policies Act of 1978 (PURPA) encouraged cogeneration and alternative energy forms by providing incentives to utilities to purchase power and to provide wheeling services.

10. The National Gas Policy Act of 1978 built on much previous gas regulation, and began a process of price deregulation, designed to allow gas prices to move upward toward parity with oil.

11. The Energy Tax Act (1978) established tax exemptions for energy conservation, as well as gas guzzler taxes, and expanded intangible drilling cost tax credits to geothermal energy and geopressurized brine.

12. The Power Plant and Industrial Fuel Use Act of 1978 (PIFUA) was designed to encourage fuel switching from oil and gas to other fuels, especially coal.

13. The Crude Oil Windfall Profit Tax Act (1980) recovered a portion of the oil companies' windfall profits.

14. The Pacific Northwest Electric Power Planning and Conservation Act (1980) restructured the extensive federal energy holdings in the Pacific Northwest (especially the Bonneville Power Administration) to encourage conservation and alternative resources.

Literature Cited

1. Energy Study Group. 1965. Energy R&D and national policy. Quoted in The Clouded Crystal Ball section of *The New Yorker*, May 1977
2. Hubbert, M. K. 1969. Energy resources. In *Resources and Man*, a report of the Committee on Resources and Man, Natl. Res. Council. San Francisco: Freeman
3. US Atomic Energy Commission. 1973. The nation's energy future: a report to Richard M. Nixon, president of the United States. Submitted by Dixy Lee Ray, Chairman, US Atomic Energy Commission. USGPO Stock Number 5210-00363
4. Stobaugh, R., Yergin, D. 1979. *Energy Future*. Balantine
5. Brooks, C. 1984. *Best Editorial Cartoons of the Year*. Pelican
6. McFarland, A. S. 1984. Energy buddies. *Ann. Rev. Energy* 9:501–27
7. Ford, D. 1982. *The Cult of the Atom: the Secret Papers of the Atomic Energy Com-*

mission. Simon and Shuster
8. Ford Foundation Energy Policy Project. 1974. *A Time to Choose: America's Energy Future*. Cambridge, Mass: Ballinger
9. US Dept. Energy. 1984. *Monthly Energy Rev*. May
10. Starr, C. 1979. Energy and power. Reprinted from *Sci. Am.* in *Energy*, ed. S. F. Singer. San Francisco: Freeman
11. Bethe, H. 1979. The necessity of nuclear power. Reprinted from *Sci. Am.* in *Energy*, ed. S. F. Singer. San Francisco: Freeman
12. Lovins, A. 1977. *Soft Energy Paths: Toward a Durable Peace*. Cambridge, Mass: Ballinger
13. Hamrin, J. 1984. Independent Energy Producers Association, Sacramento, Calif. Private communication
14. MacAvoy, P. W. 1983. *Energy Policy: an Economic Analysis*. Norton
15. McKie, J. W. 1984. Federal energy regulations. *Ann. Rev. Energy* 9:321–49

16. Doub, W. O. 1976. Energy regulation: a quagmire for energy policy. *Ann. Rev. Energy* 1:715–25
17. Kalt, J. P., Stillman, R. S. 1980. The role of government incentives in energy production: an historical overview. *Ann. Rev. Energy* 5:1–32
18. Perry, H., Landsberg, H. H. 1981. Factors in the development of a major US synthetic fuels industry. *Ann. Rev. Energy* 6:233–66
19. Difiglio, C. The automobile fuel economy standards. Unpublished draft
20. Craig, P. P. 1984. Fuel efficiency for cars: oil import vulnerability reduction. *Energy J.* 5(1):141–48
21. Kerr, R. A. 1984. How fast is oil running out? *Science* 226 (Discussion of Congressional Research Service Report No. 84-129 SPR)

Ann. Rev. Energy 1985. 10:589–611
Copyright © 1985 by Annual Reviews Inc. All rights reserved

INTERNATIONAL ENERGY RESEARCH AND DEVELOPMENT COOPERATION

John E. Gray, Edward F. Wonder[1]*, and Myron B. Kratzer*

International Energy Associates Limited, Washington, DC 20037

International research and development cooperation has been a hardy theme in energy policy. The energy "crisis" has intensified the attention given to cooperation, and promoting energy R&D cooperation has been one of the principal functions of the International Energy Agency (IEA). Cooperation has figured prominently in strategies for developing and deploying especially costly technologies, most notably fusion and the fast breeder reactor.

At the same time, international R&D cooperation is not a natural development, which probably helps explain why it is often difficult to negotiate cooperative agreements, and also why the record of success in R&D cooperation is decidedly mixed. Technological nationalism remains potent at both the governmental and corporate levels. Concern that the international competitiveness of advanced industrial economies depends on their capacities to innovate, worries about the availability of foreign technology, and strong government and corporate convictions that a strong national technological base improves one's bargaining position all contribute to the tendency to seek national solutions to R&D challenges first.

In addition, governments tend to associate technological strength with diplomatic strength. The need for technological strength is not confined to

---

[1] This article was written and submitted in December 1984, when the author was employed as a senior consultant at International Energy Associates Limited. The views are those of the author and not IEAL.

589

0362-1626/85/1022-0589$02.00

the military sector. A belief that national technological strength in the so-called commanding heights of advanced industrial economies (e.g. energy, communications, and information technology) is essential to being a leader in world politics has shaped government R&D policies. Where governments have pursued R&D cooperation, the reason has frequently been to acquire capabilities otherwise unavailable, as a last rather than a first resort.

To the above factors must be added bureaucratic resistance. To a program manager, international R&D cooperation can mean loss of control over a project. Even in projects organized on a "lead country" basis, cooperation may mean a sharing of program control, as well as additional layers of joint decision-making structures to which a program manager is subject. Even within the same bureaucracy, greater intra- and interagency coordination may be required before decisions can be made, lessening the control of an individual office. Often, when program managers favor cooperation, the project in question has insufficient support at home, so that cooperation offers an opportunity to keep it alive.

Consequently, international R&D cooperation has frequently required initiative and resolve at the highest levels (in both government and industry) to establish a commitment to a joint effort. This pattern of high-level commitment making will be apparent in the analysis of energy R&D cooperation that follows. Because there often is strong resistance to cooperation, it is necessary to look at not only how, and at what level, a commitment was made, but whether and how the commitment was institutionalized so as to outlast, if necessary, the individuals who first made it.

Where R&D cooperation has occurred, one or more of a number of reasons have proven persuasive. Among these are beliefs that cooperation can:

1. Allow costs to be shared and eliminate unnecessary duplication of effort.
2. Speed the development of the conceptual base for a particular technological area by pooling national technical resources.
3. Accelerate the development and demonstration of a particular technology.
4. Make possible a wider range of approaches.
5. Enhance the scientific and engineering competence of the participating parties.
6. Serve broader political objectives.

The Organization for Economic Cooperation and Development (OECD), in a 1975 report on energy R&D, reiterated these points. Collaboration in energy R&D was essential, it maintained, in that the economic and political security of the OECD states depended on energy security. The report recommended "systematic, large-scale research and development covering the totality of the energy field," of such a scale as to exceed the means of any one state, to make up for the time lost during the period of "cheap oil" (1).

International R&D cooperation also clearly has disadvantages, including the following:

1. Opportunity costs may be high if resources allocated to a joint project could have been better spent elsewhere, especially if joint projects tend to be "second best" ones that have failed to justify dedicated national R&D efforts.
2. Development costs may not be adequately compensated because the results of the project are not fully appropriated or translated into competitive gains. At the very least, patent rights, licensing, and marketing arrangements are often difficult to resolve.
3. More unwieldy cooperative projects may be less sensitive to changing user requirements, jeopardizing the marketability of the technology.
4. Control over R&D resources can be lost to an international bureaucracy, or at least shared with others.
5. Management may be cumbersome because of joint decision-making, the application of different national legal systems, possible additional staffing requirements, and other features of the agreements.
6. Governments frequently seek to ensure business opportunities for their own national companies, giving rise to problems of fair return in which procurement contracts are guided by government-set quotas rather than the market.
7. Projects may be vulnerable to subsequent withdrawals by one or more parties, or to cutbacks in funding.
8. Not only may national control over projects decline, but projects can take on lives of their own because of the political costs of terminating them.

While carefully crafted contractual or intergovernmental agreements can effectively address some of the above problems, some of these disadvantages and costs are unavoidable. These potential disadvantages mean that the process of cooperation—the organizational and procedural arrangements, and the balance struck between the efficiency of the effort and equity (division of labor) considerations—takes on added importance.

International cooperative R&D programs tend to take one of three broad forms. First, information exchanges entail the sharing of experience and results without a formal pooling of effort. Nonetheless, such exchanges can lead to an informal division of labor if the parties avoid duplicating programs to which they have access via an exchange agreement. Second, formal joint planning involves allocating responsibilities in advance of the actual R&D, explicitly to avoid duplication. Such planning may or may not involve control over final national program decisions. Third, joint facilities or enterprises can be established to centralize R&D effort in one or more mutually agreed upon locations; they clearly entail the two preceding forms of cooperation as well. The degree of political and economic commitment (and dependence) involved increases not only with the complexity of the arrangement, but also with the dependence of each party on the outcomes of joint measures. The matter of how much dependence is acceptable or necessary is critical; as one might expect, information exchanges are far more common than more comprehensive forms of cooperation.

The discussion of international energy R&D cooperation presented here is not meant to be a comprehensive history. Rather, we have examined selected components of that history to analyze why cooperation occurs, how it has been carried out, its reasons for success or failure, and what can be learned from the experiences. In doing so, we will very briefly look at the European experience prior to 1973 and then analyze the record of the IEA in this area, US attitudes and policies regarding R&D cooperation, and the experience with R&D cooperation on fast breeder reactors as a brief case study of cooperation in a field with a potentially high payoff.

Our inquiry is limited to actual research and development, although we must point out that international technological cooperation goes far beyond R&D to include technology transfer, training, expert missions, and the like. Our discussion also focuses on cooperation as it affects the relations between advanced industrial states. Expectations of mutual gain are critical to R&D cooperation, and as the following discussion shows, R&D cooperation, unlike technology transfer between stronger and weaker parties, implies technological parity, or, at least, a comparable level of development, in the affected area.

THE EUROPEAN EXPERIENCE

Despite the fillip the Arab oil embargo of 1973 gave to the idea of international R&D cooperation, there was substantial cooperation in energy R&D before the onset of the "energy crisis." Unlike the case in the post-1973 period, energy security concerns, except during the immediate post-Suez period in 1956–1957, were only one factor, and not necessarily the most important, in encouraging this cooperation. Technological cooperation was most pronounced in Western Europe, where it was valued not only as a way to acquire capabilities, but as an instrument of European political integration and as a means of fending off what was perceived to be an "American challenge" based on American domination of high technology fields. Consequently, there was a strong political and diplomatic dimension to cooperation, much as there would be in the post-1973 period, but with different underlying concerns.

The European response to a perceived "technology gap" with the United States took the form of both "European" project undertakings at the intergovernmental level and companies with ownership and top-level managerial control spread among several countries, as well as budgets administered by Euratom and other organs of the European Community to fund R&D in Community-run laboratories. Joint entities in the nuclear, aerospace, and computer sectors were common, but enjoyed very mixed results. Resistance, among the European states themselves, to greater technological interdependence, integration of markets, or even agreed upon European R&D objectives resulted in the failure of many of these efforts.

The European experience is well chronicled elsewhere, and that discussion is not duplicated here (2–4). Rather, we present a summary of general trends and lessons learned from that experience.

Regarding trends, for the most part functions related primarily to information and regulations developed more smoothly than those related to hardware. Once they moved into the hardware stage, projects tended to be caught up in competitive national and private interests. Fair return proved to be a serious problem.

Disenchantment with government-inspired cooperative projects grew, because of repeated cost overruns and the primarily political inspirations of many projects. Government-conceived projects often paid too little attention to technical merit and objectives. What should be done internationally and what was best carried out at the national level were often ill thought out. Financial stability often suffered where project budgets depended on annual parliamentary action. Industry-level initiatives appeared more promising, particularly in nuclear power projects, where the utilities took an increasingly greater hand in guiding R&D cooperation.

One lesson learned was that the process of creating commitments is often critical. While much can be made of the difficulty of overcoming legal, tax, and financial obstacles, a more complex and fundamental resistance may stem from a preference for national solutions until their impracticality is beyond dispute.

It was also demonstrated that R&D cooperation is likely to be stronger and more sustained when the initiative comes from the users and producers of technology, rather than government. The commitments in such cases arise from commercial and industrial, rather than political, interests and negotiations. If government is the single initiator or political interests are paramount, the probability of good projects is lower.

The European experience with institutionalizing cooperation is also instructive. European-level bodies often found themselves at too low a level to make key political decisions (i.e. assert executive authority over the programs of member states), or prevented from doing so by decision-making structures, while at the same time being at too high a level to address managerial questions. The European approach to R&D cooper-ation consequently evolved in the direction of an a la carte strategy, in which participation was voluntary and project-specific institutional ar-rangements were established. Institutions with "programs" declined in importance compared with "project" groupings.

THE INTERNATIONAL ENERGY AGENCY AND R&D COOPERATION

The International Energy Agency (IEA) emerged from a fusion of several strands of US energy and foreign policies. The establishment of the IEA and its program for emergency oil sharing was a major US diplomatic achievement, despite the French decision not to participate. The R&D role of the IEA, however, has been less successful. The reasons for this lie in both the inspiration of the IEA and the difficulty of institutionalizing inter-national R&D cooperation, as earlier European experience has shown.

Recognition that energy supply would require more conscientious government attention antedated the Arab oil embargo. The first US presidential energy message (in June 1971) preceded the embargo by more than two years and drew attention to the role of research, development, and demonstration, particularly with respect to the breeder reactor and coal gasification. While non-R&D issues (such as oil and gas price decontrol) dominated President Nixon's second energy message in April 1973, R&D again received important attention. In June 1973, President Nixon announced a much expanded energy R&D program ($10 billion over a five-year period, largely for nuclear programs and coal liquefaction and gasification, and to a lesser extent for conservation and renewable energy sources) and proposed to establish an Energy Research and Development Administration to coordinate and direct this program.

In parallel to this expansion of the domestic energy R&D program, President Nixon advanced the idea of international R&D cooperation in his April 1973 message, and directed the State Department, in coordination with the Atomic Energy Commission and other agencies, to "move rapidly" to develop a program of international energy R&D cooperation and to develop international mechanisms to cope with energy shortages. An interagency task force was created to identify opportunities for inter-national R&D. The NATO Committee on the Challenges of Modern Society began cooperative initatives in solar, geothermal, and conservation technologies, and the OECD began a study of its members' energy policies, including proposed plans to coordinate national R&D. Negotiation of a broad energy R&D agreement with Japan began in 1973.

To the enhancement of the domestic energy R&D program and the greater attention paid to international R&D cooperation must be added a third element: serious US concern, prior to the energy crisis, that the western alliance might unravel under the pressure of divergent economic interests, a false sense of security in a period of US-Soviet detente, and a perceived decoupling of the United States' security fate from that of Europe. The growing intra-alliance tensions were recognized in Secretary of State Henry Kissinger's April 1973 "Year of Europe" speech, which was poorly received in Europe. The open differences among the allies that emerged during the Yom Kippur War in October 1973 further reinforced the apparent urgency of the need to restore alliance cohesion. The oil embargo and the resulting high oil prices threatened to rend the alliance even further by encouraging separate consumer-state dealings and special relationships with oil suppliers and beggar-my-neighbor economic and trade policies intended to shift to other nations the economic consequences of higher oil prices.

The concept of energy consumer cooperation, incorporating both measures to deal with oil supply shortages and R&D to develop new energy options, pulled together all three of the above strands of US policy thinking. The United States offered to share its energy technology, via R&D cooperation and direct transfer, as a lure to hold together a fragile coalition threatened by spiraling oil prices and vulnerability to physical shortages of supply. Restoring alliance cohesion was an objective of US policy before the oil embargo; what the oil embargo provided was a crisis atmosphere that focused attention on collective responses with an urgency that previously was lacking.

The US initative to establish a multilateral framework for dealing with the energy crisis was launched in December 1973, when Secretary Kissinger proposed that the United States, Europe, and Japan form a high-level Energy Action Group to develop a program of collaboration in all areas of the energy problem, including, as a high priority, international energy R&D. Kissinger's call for a comprehensive program went far beyond previously stated US positions on the subject. Undersecretary of State for Economic Affairs William Casey had only in the June before the embargo observed that R&D cooperation would be ad hoc, and for the most part within the OECD context.

The Washington Energy Conference opened on February 11, 1974. Opening the conference, Kissinger repeated the US offer to share technology.

> The United States is...willing to share American advances in energy technology to develop jointly new sources of supply and establish a system of emergency sharing; and we are prepared to make a major contribution of our most advanced energy research and development to a broad program of international cooperation in energy.

Kissinger went on to delineate a seven-point approach to energy cooperation, three points of which related directly to energy R&D. Among the technical areas cited were coal-based technologies, the development of oil shale resources to produce synthetic fuels, and centrifuge nuclear fuel enrichment technology (where the offer to share US technology marked a major departure from the 1971 US position, which had been limited to the older gaseous diffusion technology when it had become apparent that European countries might build their own enrichment plants using their own technology). The offer to share enrichment technology was not intended merely to demonstrate US sincerity; the United States was very concerned at the loss of its control over free world enrichment supply, and the implications of that loss for US influence in the nonproliferation area. Use of US technology could secure a US role in foreign enrichment plants, especially with regard to the retransfer of that technology.

When viewed in the context of US policy objectives, this offer to share technology appears as largely instrumental in character, serving primarily short-term diplomatic objectives, and only secondarily directed toward long-term technical objectives. Clearly, achievement of the latter, as they promised a shift from dependence on oil, would be desirable. However, the pressing problems at the time were not seen as technical in nature. Indeed, progress came much faster in establishing the International Energy Agency's oil sharing scheme, which spoke to the immediate problem, than in defining and implementing a joint R&D program, which, due to its limited scope and resources, would have less impact on the structure of energy supply than an intensive US domestic R&D program.

Despite the multilateral thrust of the initative, the United States moved forward to sign bilateral R&D agreements (with West Germany, Japan, and the United Kingdom) even as it discussed multilateral cooperation. These were the three leading countries, with comparable technical capabilities, with which the United States wanted to cooperate on a bilateral basis. Multilateral cooperation where the distribution of technical capability among the participants was very uneven offered less immediate technical return to the United States.

The International Energy Agency itself emerged from the work of the Energy Coordinating Group (ECG), which was the follow-on mechanism to the Washington Energy Conference. The ECG contained an ad hoc group on R&D (as well as another group on the oil sharing scheme), which formulated the plan for R&D cooperation. The R&D working group defined areas of potential interest, with a view toward identifying specific programs or projects. Although a fairly broad spectrum of areas of interest was identified, radioactive waste management and nuclear safety, coal, fusion, and conservation drew the greatest substantive attention, while solar energy, waste heat utilization, bioconversion, and production of hydrogen from water drew less interest, despite the activities of particular countries in these areas (5).

In addition to defining areas for cooperation, the group addressed the issue of the institutional framework. The framework eventually chosen represented a compromise between a separate high-level steering group to monitor and coordinate R&D, which the United States supported, and an OECD restructured to deal more effectively with energy, which the Germans and others advocated. The IEA itself was established within the framework of OECD, but with its own governing board, management committee, and standing groups, one of which was for R&D.

The problems that would continue to handicap the IEA in this area were apparent from the start. The research program itself was determined on the basis of suggestions made by lead countries as to what technologies they believed cooperation should promote. The end result resembled a smorgasbord of projects rather than a coherent strategy as to what, from a collective perspective, should be promoted. Moreover, there was a definite effort to minimize the proportion of the program devoted to nuclear issues, when in fact nuclear R&D loomed proportionally much larger in the national R&D programs of the major R&D performers in the group. This created a mismatch between the IEA's research agenda and what national R&D budgets revealed to be the real priorities of the key member states. Of fourteen working parties in existence in mid-1977, only one—high-temperature reactors—dealt with a major area of national nuclear R&D, and, even then that reactor technology occupied a somewhat anomalous position between the fast breeder and the light water reactor. Energy conservation, coal technology, and other alternative sources, including fusion, accounted for the vast majority of cooperative agreements.

Beyond these considerations, the initial IEA experience indicates a high level of working party activity, but less actual project activity. Indeed, many of the participants in these groups actually had limited meaningful R&D programs, and the working parties apparently provided a means for them to monitor the R&D of others. This "free rider" activity would be expected in this type of setting.

The kinds of cooperation that occurred also suggested the possibly limited effectiveness of cooperation in the IEA context. While the number of multilateral agreements increased, in large part because of the IEA, most were limited to information exchanges. Moreover, the key role of operating agent in actual specific projects tended to be dominated by the United States, the United Kingdom, and West Germany. In effect, this gave these countries a de facto veto over what the IEA did, as their nonparticipation in projects going beyond information exchanges effectively denied such projects the requisite financial and technical resources.

By the end of 1977, there was a growing recognition in the United States and elsewhere that the IEA R&D program lacked a clear sense of purpose. Moreover, the potential conflict between energy independence and multilateral interdependence had never been resolved. In particular, the United States lacked a clear perspective on how national and cooperative R&D related to each other, and on how to set priorities between the two.

Other impediments to IEA cooperation going beyond information exchanges had become apparent. Possible conflicts with other organizations, which some members preferred to the IEA in particular areas, existed. This was true with respect to the OECD's Nuclear Energy Agency in the areas of nuclear safety and radioactive waste management, and NATO's Committee on the Challenges to Modern Society, which had its own programs in waste heat, waste utilization, conservation, and solar heating and cooling, and about whose possible association with the IEA the Japanese were concerned on political grounds.

In addition, there was the problem of how to address the natural desire of technically weaker members to secure roles for their industries in projects. For example, the Belgians sought the addition of a high-temperature gas-cooled reactor project to the list of areas of cooperation, despite Belgium's weaker capabilities in this area. As a general rule, the United States and other technically advanced countries were reluctant to let weaker parties have full access to projects and emphasized reciprocity of contributions, which would tend to limit cooperation to parties of equal or complementary capabilities.

Japanese participation in actual IEA undertakings was also limited. Although Japan was clearly interested in solar energy and other alternative energy technologies, and had launched its own Project Sunshine in 1974 to develop them, Japan was reluctant to enter into multilateral undertakings because of proprietary considerations, and preferred bilateral dealings in which Japan might have more control.

Finally, basic differences between the US and European approaches to government-industry relationships limited the IEA's ability to mount projects involving joint facilities or enterprises. The European pattern of closer industry-government collaboration, including the use of restrictive marketing covenants on occasion, and the US practice of licensing government-developed technology on a royalty-free basis posed problems where joint facilities were involved.

As a result of these impediments, the IEA program took a not unexpected shape. Cooperation largely took the form of information exchanges. Most of the technologies involved were remote from commercialization, and many were of second order priority in relation to national R&D programs. Lastly, "software" (e.g. regulations and standards) rather than "hardware" became of increasing prominence. This was true of environmental protection, nuclear safety, and nuclear waste management. Not only were there more definable collective interests and needs in these areas, but internationally developed standards and practices often were perceived to have greater credibility (especially political credibility) than purely national ones.

This is not to say that the IEA made no effort to focus R&D cooperation more sharply. Two major attempts have been made to develop a common framework for energy strategy. Rather than attempt to devise a specific IEA-sponsored and IEA-administered R&D program, each effort has aimed at providing a common framework for national R&D programs in member states, which might in turn help identify specific opportunities for new cooperation.

The first strategy document, *A Group Strategy for Energy Research, Development, and Demonstration* (1980), identified which new energy technologies the IEA countries would need, and considered how IEA-country R&D programs could be structured to provide these technologies. By approaching the task this way, the study challenged the prevailing "IEA mentality" of emphasizing joint projects by showing that more national activity in certain common technological areas would yield joint benefits by decreasing the dependence of individual member states on imported oil.

The study assigned the highest priority to conservation; advanced converter and breeder reactors; coal conversion, tar sands, and oil shale; and enhanced oil and gas recovery. Lower priority was accorded most renewable energy sources. However, there was no collective will to act on the results of the study. Few accepted the implication of the study that it could make sense for one country to support financially and technically R&D on, say, oil shale or tar sands in the United States or Canada even though those technologies would not be deployable widely outside those countries, because their deployment would reduce US or Canadian oil imports, and thus lessen pressure on world oil markets.

A strategy document prepared in 1984, the *Energy Technology Policy Study* (ETPS), argues that in R&D, electricity production is over-emphasized compared to that of synthetic fuels and to enhanced oil recovery technologies, in view of the continued vulnerability associated with some liquid fuels for transportation. The ETPS, like previous studies, identifies a range of technologies in which more R&D is necessary, without attempting to assign priorities, as the 1980 study did.

Developing a common strategy will clearly require a major effort. Not only are countries reluctant to lend financial or technical support to R&D projects whose results might be of little immediate use to themselves, but the IEA countries also pursue very different R&D priorities. Three countries—the United States, Japan, and West Germany—accounted for 73% of the total IEA expenditure on energy R&D in 1982. These three countries, collectively, spent 67.3% of their R&D funds on conventional and advanced nuclear technologies. For the six countries spending the least on energy R&D—Denmark, New Zealand, Ireland, Portugal, Turkey, and Greece—56% of total R&D goes to renewable energy sources and conservation. These disparities reflect the constraints low funding levels placed on R&D and indicate that the burden of developing large-scale supply technologies will fall on a handful of countries, and that international trade in these technologies will be a key element in their wider deployment in the IEA countries. The differences in priorities also provides additional basis for expecting that the scope for joint IEA projects will remain relatively limited.

THE US APPROACH TO R&D COOPERATION

The US approach to international cooperation in energy R&D has gone through a number of phases since 1973. The Nixon and Ford administrations dangled access to US technology as bait to encourage institutionalizing cooperation of oil consumers through the IEA. As described earlier, the US objective here was more political than technical in nature. If the United States actually saw something technical to be gained via cooperation, the emphasis was given to signing bilateral agreements with the states technically most advanced. Moreover, most US cooperative activity was related to nuclear power reflecting the low priority given at the time to nonnuclear R&D. This priority changed, and the Ford administration began to place greater weight on synthetic fuels and other substitutes for oil.

The US approach to R&D cooperation entered another phase during the Carter administration. The objective of R&D policy in the Carter period was to reduce western dependence on imported oil by aggressively pursuing nonnuclear technology. The framework, or paradigm, governing this policy was one emphasizing energy security. Under this paradigm, government should not only finance basic research, but actively participate in R&D up to and through the demonstration stage. The principal justification for this expansion of the role of government was that energy security was an externality to which industry could not be expected to respond positively in allocating its R&D resources. The responsibility to pursue R&D vigorously, at a pace faster than industry alone might proceed, therefore fell to government.

The Carter administration's guarded approach to nuclear power left little or no ground for international R&D cooperation in the one area where a strong case could be made for it—fast breeder reactors. The Carter policy, for essentially political reasons, in this case nonproliferation, sought to shift the technical direction of US nuclear energy R&D away from plutonium producing and toward programs to develop so-called proliferation-resistant nuclear technologies, incorporating technical barriers to the diversion of sensitive nuclear material. The Carter administration urged other advanced nuclear countries to follow suit, but those countries, lacking domestic energy resources comparable to those of the United States, refused to deviate from a technical path they considered essential to their energy security. This attempt to persuade Europe and Japan to follow a course they considered to be imprudent, together with the major revision of US nonproliferation policy and law at the time, severely impeded collaboration in nuclear areas.

The energy security paradigm affected US domestic and international R&D policies in several ways. The most visible manifestation of an energy R&D strategy based on energy security considerations was the synthetic fuels program. The legislation creating the Synthetic Fuels Corporation in 1980 contained near-term production goals and a mandate to demonstrate a range of synthetic fuels technologies. The program allowed for foreign investment, subject to the requirement that control over projects remain securely in the hands of US companies.

The international counterpart to the domestic synthetic fuels program was the SRC-II project, which actually originated in a DOE-administered synthetic fuels program in 1977. The Department of Energy, Gulf Oil, and government-industry consortia from West Germany and Japan were to participate as 25% partners in the SRC-II coal liquefaction project in West Virginia. The original estimated cost of the plant, which was to produce 20,000 barrels per day of synthetic oil, was $600 million; the rise of this estimate to $2 billion was in large measure the reason for the plant's cancellation.

The US concept of an International Energy Finance Corporation took the Carter administration's emphasis on the imperative of energy security and on the early commitment to demonstration plants, particularly in synthetic fuels, one step further. The inspiration for the concept was the second oil price shock, which followed the Iranian revolution. The objective was to accelerate the commercialization of new energy technologies, largely synthetic fuels, by creating an international entity, capitalized at $10 billion, to finance demonstration plants. The concept originated in the Treasury

Department, which saw it as a way to spread the costs of commercializing US synfuels technologies and, via the inflow of foreign money, ease the US balance of payments position.

The concept did not find favor abroad, however; the British, French, and West Germans were unwilling to commit funds to specific, costly US proposals, while technically less strong countries like Italy and the Netherlands feared being excluded from participation. Because the United States proposed that participation be limited to a small number of nations, including France, the impression that the United States was trying to establish another arena of cooperation, separate from the IEA, was inescapable.

The record of the Carter period was not entirely one of setbacks for major initiatives, however. One of the most important aspects of the Carter period was the growing importance of bilateral cooperation with Japan. Although an agreement for an information exchange with Japan in the breeder reactor area had been in effect since 1965, a more conscious US effort to court Japan began in the Carter period. This effort culminated in a US-Japanese agreement on fusion cooperation, and an "umbrella" science and technology agreement intended by the United States largely to encourage the Japanese to spend more money on basic research. This growing emphasis on ties with Japan reflected the strength of Japanese technical capabilities, a high-level US desire to promote better relations with Japan, and a US desire to tap Japanese financial resources.

This relationship with Japan continues to the present day. Bilateral agreements exist in high energy physics (information exchange), fusion (information exchange, testing, cross-investment), geothermal energy (information exchange), reactor criticality experiments (information exchange and joint planning), and the fast breeder (information exchange and joint experiments). In addition, the United States and Japan are partners in a number of IEA-sponsored projects.

Under the Reagan administration, another phase in US policy began. The administration refocused the government role in energy R&D to emphasize high-risk high-cost areas remote from commercialization. Budget priorities emphasized basic research and nuclear energy, at the expense of coal and other alternative energy sources, while the administration believed the private sector should assume financial responsibility for R&D. In addition, government would no longer finance or share the investment risk in demonstration plants. Cuts in the DOE R&D budget disrupted planning of IEA projects, such as the hot dry rock geothermal project, the small solar power project, and pressurized fluidized bed combustion tests in the United Kingdom. The threat to dismantle DOE further impeded US international cooperation.

The reasons for this redirection included a number of diverse considerations. Dismantlement of DOE had been a campaign promise. Budget pressures made R&D an easy mark. A free market economic philosophy justified restricting government involvement in R&D. Decreases in oil prices and apparent erosion of OPEC market power diminished the apparent urgency of energy R&D, and certainly of demonstration plants. Moreover, the whole notion of an energy security strategy was redefined to place much greater weight on diplomacy and military postures (e.g. the ability to keep open the Persian Gulf) and much less on a "forced march" approach to deployment of high-cost energy technologies. This revised approach to energy security reinforced the withdrawal of government from many of the R&D areas the Carter administration had supported, such as synthetic fuels.

Following the appointment of Donald Hodel as Secretary of Energy, the situation at DOE has stabilized. Indeed, there is greater recognition than before of the importance of energy R&D and the value of international cooperation in R&D. Part of the explanation for this may be idiosyncratic; the Secretary was significantly more internationally minded than his predecessor. However, an important part of the reason for this subtle shift in policy was recognition that, in a context of budget austerity the costs of some undertakings can be formidable, making cooperation more attractive, and that industry's approach to R&D in some key energy sectors remains largely short-term and risk averse.

DOE is attempting to formulate a comprehensive international R&D strategy. The Energy Research Advisory Board has been asked to develop a rationale and strategy. DOE also regards the Energy Technology Policy Study in the IEA as an important opportunity to shape international strategy. International cooperation is beginning to occupy a prominent position in US strategy for developing advanced energy technologies, particularly in the nuclear area, as the financial costs involved place greater pressure on rationalizing expenditures via closer interaction with comparable foreign programs. This is especially true of fast breeder reactors.

FAST BREEDER REACTORS—HISTORY AND ISSUES

Six nations—France, West Germany, Japan, the United Kingdom, the United States, and the Soviet Union—have had major breeder development programs for more than two decades. Each of these programs has focused mainly on the plutonium-fueled, liquid-metal-cooled fast breeder reactor (LMFBR). Italy has also developed a major program of cooperation with other European nations. Several additional nations have some

604   GRAY, WONDER & KRATZER

involvement with breeder reactors—either through activities in breeder-related R&D, or through cooperation with one of the major breeder development nations. These include Belgium (cooperating with West Germany), the Netherlands (also a partner with West Germany), Switzerland (cooperating with West Germany and the United States on gas-cooled breeder technology), and India (with its own program initiated through cooperation with France).

Comparison of the major programs indicates that the United States had the early lead in developing experimental fast breeder reactors (FBRs). It has put no new breeder reactors in place for some time and at present has no firm plans for doing so. Measured in terms of successful FBRs of progressively increasing scale, France and the Soviet Union have clear leads now. Moreover, both nations have plans for continued demonstrations of scaled-up reactors. Measured on the same scale, the British were, for a time, not far behind the French and the Soviets. However, the British program faltered; it does not currently have a reactor in construction and so may not keep pace. West Germany and Japan started programs more recently than France, the United States, the United Kingdom, and the Soviet Union. West Germany progressed rapidly at first but experienced some difficulty with its SNR-300 schedule. Japan built its first breeder reactor, JOYO, in a relatively short time; its second effort, MONJU, is in construction and is expected to progress rapidly.

Of central importance to the future prospects for international cooperation on breeders are the following: the close existing association between France and West Germany, together with Italy, Belgium, and the Netherlands; the intent of the United Kingdom to join in a collaborative European program; the US loss of interest in a domestic demonstration program; and the continuing interest of Japan in both a domestic demonstration program and international cooperation.

The close European association came about principally as a result of a new determination, which emerged in Europe in the early 1970s, to proceed cooperatively. The motivation was in part the continuing search by European countries, especially France and West Germany, for concrete ways to be bound together politically through common projects and programs. A more fundamental motivation, however, was the conventional but powerful one of sharing the costs and risks of what was increasingly recognized as a long, difficult, and expensive development effort.

In contrast to the intensive cooperation on the European continent, the US and Japanese programs remain relatively unattached. The United States, beginning in 1975, took steps to review and expand its bilateral exchange arrangements. New arrangements were concluded with the United Kingdom in 1976 and with France and West Germany in 1977. The scope of Japanese exchange arrangements was expanded in 1976 to include new areas. These arrangements, however, are limited, basically providing for an exchange of certain R&D information without rights to the full portfolio of each country's technology, patents, and know-how. They do not provide for the close links and genuine sharing of technology and results that characterize European cooperation.

Five major issues have historically been central to breeder policy in all nations considered: resource adequacy; assurance of energy supply; health, safety, and environmental considerations; economics; and non-proliferation. France, West Germany, the United Kingdom, and Japan have generally comparable overall positions on these issues. Although each nation clearly represents a spectrum of views, with respect to these five major issues the dominant views at policymaking levels show significant similarities, and there is a strong conviction that energy security will eventually require breeder deployment, even though at the present uranium resources appear sufficient to postpone breeder deployment.

At the same time, the economics of both the capital plant and the fuel cycle for a commercial deployment scenario over the next 20–30 years are still speculative. A recurrent theme of national breeder policy is that, to decrease the economic uncertainties, one must build at least a few commercial-scale demonstration breeders. This may be an expensive proposition, especially if the economics do not prove out and the breeder is therefore (or for some other reason) not deployed. However, the potential economic payoff over the years ahead, if the breeder becomes a commercial success, is so large as to dwarf the cost of development.

Nonproliferation is also a concern, especially from the technical standpoint of ensuring the safeguard ability of the designs of breeder reactors and related fuel cycle facilities. However, deferral of the breeder is not viewed by France, West Germany, the United Kingdom, Japan, or the Reagan administration as simplifying proliferation problems. On the contrary, serious efforts by one nation to inhibit breeder development may be viewed in another country as threatening to both national security and energy supply.

Based on compatibility of technology, programs, views, and economic and industrial ties, broadening cooperation beyond the existing European arrangements is judged the most promising course of action. In particular, an arrangement between Japan and the United States could be useful and workable because the two countries currently have close and constructive relations, neither is yet committed to other international breeder programs, and their breeder technologies may be compatible.

No major unresolved technical issues throw into question the basic viability of the breeder. (There is, of course, the choice between "loop" and "pool" designs, with the pool design, which holds certain advantages, expected by many to be the final choice for further demonstration and eventual commercialization.)

The most serious issues have to do with financing—both for current national programs and possibly for prospective international cooperation programs. The US Clinch River Breeder Reactor (CRBR) program was abandoned in mid-course because of lack of the financing to complete it; and how the planned French program (e.g. Super Phenix II) will be financed is currently in question. All national or cooperative programs anticipate eventual large in-country demonstration plants, requiring multibillion dollar financing—with major doubt about the near-term economic viability of the FBR. Utilities outside of Japan appear to have lost interest in near-term FBR commercialization. Thus, financing development and demonstration-scale plants is regarded as unusually difficult at this time. This difficulty is, of course, a major reason for interest in international cooperation.

There appear to be no fundamental obstacles in the domestic licensing and regulatory area. However, there is an important need to ensure that the safety design criteria being developed in joint cooperative programs are consistent with, and will meet, the requirements for licensing imposed by the national governments contributing to such ventures and wishing to make use of the results.

The economics of producing electricity with an FBR are uncertain and will remain so at least until a true prototype of the first generation of commercial plants has operated for a satisfactory period of time. No FBR in the advanced design stage is considered competitive with the current generation of light water reactors (LWRs). Correspondingly, there is too little experience with a prospective FBR fuel cycle for sound economic estimates. Major efforts are under way in France, as well as the United States, to design FBRs with significantly lower capital costs than those of such current designs as Super Phenix II, to improve their prospective economic competitiveness. Any domestic or cooperative program should continue this emphasis, as well as improve total fuel cycle costs.

These economic uncertainties are to be expected in the development of any large energy supply system. Precise understanding of economics follows the development and demonstration phases; it cannot precede them. Accordingly, too much belief in the precision of economic performance forecasts may confuse decision-makers about the wisdom of proceeding or not proceeding. In the case of FBRs, the economic and energy security benefits are prospectively so enormous as to warrant very large expenditures to find out what the economic costs really are and how the benefits may be realized.

Institutional issues, bearing on the roles and responsibilities of governmental, industrial, and electric utility organizations in each nation, also loom large. This is a critical problem area in organizing both national and international cooperative efforts, and in some instances nations have not sorted out the roles that would be played by different sectors. Apart from their participation in the French-led SERENA consortium, France and the United Kingdom have simpler institutional infrastructures than the United States and other nations, since their utilities are nationalized, as in effect are their nuclear plant suppliers, and both work closely and cooperatively with government agencies. The nuclear plant industries in West Germany and Japan include strong and influential electric utilities and suppliers and enjoy reasonably good relations with government bodies in the planning and implementation of all major development programs related to electric power supply. West Germany has developed one national supplier for LWRs. The same supplier, Kraftwerk Union (KWU), is also designated as the national supplier for FBRs. Also, Rheinisch-Westfalisches Elektrizitaetswerk AG (RWE) has been the de facto accepted leader among the utilities in all West German FBR planning.

In the United States, the electric utilities are divided in interest, and in recent years FBR development and demonstration have not been among their highest priorities. DOE continues to support an ambitious FBR base technology program. However, neither utility nor DOE support for CRBR sufficed to protect the project from congressional budget cutters and antibreeder activists. Moreover, DOE has not been aggressive in financing near-term energy supply development and demonstration programs of any kind. Its attitude has been influenced by an unenthusiastic Congress and a budget-cutting administration that wishes the private sector to assume greater financial responsibility for demonstration plants.

The Electric Power Research Institute (EPRI) has been a US leader in exploring international cooperation, especially with the United Kingdom and Japan. These initiatives have been sponsored principally by DOE, rather than by the US electric utilities, which have not thus far shown willingness to contribute financial support to current FBR development, design, and construction programs. However, EPRI, through the Consolidated Management Office (COMO) for the LMFBR, provides a focal point for developing an understanding of technical programs and options that could become the subject of either US national programs or international cooperation programs in which the United States might participate. COMO also provides a vehicle for preliminary consideration on the part of the United States and other nations of how to collaborate internationally on large-plant development, design, financing, construction, and operations for FBRs and on related fuel cycle programs.

Also, there are now signs that the US DOE will take a more assertive role than it has in the recent past in defining the US strategy and policy toward international cooperation, with the US emphasis being initially focused on competitive designs to push down costs, and with the base program itself focused on the same objective.

These institutional issues are regarded as dominant, since they control the decision-making and the decision-makers. In turn, the institutional issues are markedly influenced by political factors. The FBR itself is not especially favored by the domestic politics of the day. Energy, especially electric energy, is not a key political issue in the United States, France, or the United Kingdom at present, except for the perception that the excess capacity of generation is very high and that some utilities have major overcommitments to nuclear power. There is thus insufficient support for fast breeders in both the electric utilities and governments, and these countries appear increasingly unable or unwilling to continue to support their own national FBR programs on a "go-it-alone" basis. International cooperation may be newly popular because it may be the only, or the most productive, way left in the near term to generate support for national FBR programs.

Thus, perhaps what is needed is a new politics for FBRs (or, more to the point, a new political rationale for financing their development), based on international issues such as "comprehensive security," alliance energy security, and nonproliferation (i.e. breeder developers should coordinate their work to ensure that appropriate nonproliferation measures are built into the design and development of the technology). The new politics might be able to sustain interest in research, development, and demonstration investments in the FBR until utility investment can be justified on economic grounds. It could also serve to broaden the political constituency inside and outside of legislatures.

Some effort may be required to produce a consensus in the United States on a strategy for international cooperation, since a number of different actors with different private agendas are involved. The Office of Science and Technology Policy (OSTP) in the White House is promoting the Williamsburg summit follow-on breeder study group, as much to provide political support for breeder cooperation as to produce something new and dramatic. DOE is trying to develop an overall coherent policy approach for the United States to follow and is seeking to improve the general environment and promote active cooperation with Japan through concrete proposals aimed at near-term cooperation that will appeal to both the United States and Japan. Concurrently, DOE is looking seriously at the longer-term picture, including options for association with Europe as well as Japan.

The US Congress does not yet appear to have a clear or coherent attitude on breeder cooperation, and those likely to be most vocal on the issue on Capitol Hill will be the hard core of antibreeder activists who led the fight against the CRBR. Greater coherence in the US scene may depend very much on either the quality of leadership in the executive branch or the emergence of attractive proposals from foreign countries.

The Europeans evidently view the United States with respect because of the strength of its base technology program, but also with concern because of past signs of instability in US nuclear policies and commitments. They also view the Japanese with high regard and have a continuing interest in Japanese (as well as US) participation in the next European FBR project, but their prime current focus appears to be on deciding whether the French Super Phenix II or the West German SNR-2 will be the next object of collaboration as a large (1200 MW) demonstration plant; the roles of the French, West German, and Italian utilities and suppliers; and the accommodation of the British interest in one or both of the projects. Sorting these issues out is expected to consume much of the balance of the decade, leading to an improved FBR design and plans for a supporting fuel cycle, both of which will be constructed in the late 1990s. It is anticipated that the United States and Japan will play supporting roles in the European program, as well as carrying out other cooperative programs on a bilateral basis.

CONCLUSIONS

Although the obstacles to effective international collaboration in energy R&D are numerous, a compelling case can be made for well-crafted undertakings. The interdependence of the energy situations of nation states remains. If, as some experts expect, the balance of oil market power shifts back toward OPEC at the end of this decade, vulnerability and energy security will remain valid concerns. Not only will nations continue to learn from one another's efforts, but the costs and risks of some projects will leave them little choice but collaboration.

The scope of R&D needed in the energy field is large and complex, and the uncertainties are numerous, so that courses of action cannot be based on a single or even a few credible scenarios; rather, R&D program policy needs to be robust over a wide range of future possibilities. Moreover, the national benefits of developing broad technological capability can be large; they cannot, however, be appropriated by individual private companies. Consequently, much of the burden of these programs may continue to fall on government.

However, a robust energy R&D program will be costly and technically risky. Few nations can afford such a program, and even the United States has not yet demonstrated the ability or willingness to implement one. Consequently the arguments that encourage cooperation—exchanging information to reduce learning costs, avoiding needless duplication, and sharing large technological and cost risks—remain valid and may grow in force.

Cooperation need not mean primarily joint projects, however, and both the Euratom and IEA experiences indicate the limited effectiveness and greater institutional problems sometimes associated with cooperation consisting principally of joint projects. If a technology, to be of interest to all the prospective participants in a joint project, must be usable in each of their home markets, invariably the range of possible areas for cooperation will be narrow to begin with, and the potential for encountering "second best" technologies higher. Moreover, a joint project approach will most likely reflect the preferences of a handful of countries, each of which gains a de facto veto over cooperation where its participation is needed.

Joint planning, accompanied by an agreed upon division of labor, on balance might make more sense in many cases. Joint planning offers a means of rationalizing expenditures in a period of budget constraints while providing a mechanism for some specialization in what each country does best. It is not without its pitfalls. Each party depends on the others to keep their commitments (which raises the issue of how to ensure the stability of funding), and each must accept the fact that with the advantages of specialization comes the need to accept that, once the technology is commercialized, industry in the other nations might hold a competitive edge over one's own in certain areas of the technology. The alternative, however, is continued technological nationalism, which is beyond the reach of all but a few (as the French learned after the de Gaulle period) if it remains within the grasp of any at all.

This is not to say that joint projects are no longer viable. There remain high-cost, high-risk technologies for which there is no alternative to joint projects. However, the problems with joint projects are large, and much can be gained from collaboration that falls short of joint projects, which should be reserved for areas where the advantages clearly outweigh the very real problems.

While the purpose of this paper is not to propose specific technical areas for collaboration, a brief review of a few demonstrates where collaboration could be advantageous. The development of fast breeder reactors is such a case. The high cost of demonstration plants and the uncertainties regarding commercial deployment make the case for international collaboration particularly compelling.

The thrust of much current energy R&D will have only marginal impact on the vulnerability stemming from the use of oil in the transportation sector. National R&D programs in the leading IEA states concentrate largely on the electricity sector. While this might back out most oil from the electricity sector, the transportation sector would remain oil-dependent and there is thus still need to develop technologies offering substitutes for oil. Until these technologies (largely synthetic fuels, oil shale, and electric cars) mature, this vulnerability will remain. Despite the current softness of the oil market and the adverse economics of commercial-scale synthetic fuels plants, development of direct substitutes for oil in transportation can reduce technological uncertainty and promote a learning process that could enable deployment to proceed if appropriate market signals are sent. The cost of such development (and demonstration) and the relevance of the availability of such substitutes to many nations would appear to provide a basis for collaboration.

Finally, the approach taken to defining an agenda for collaboration is of major importance. Thinking in terms of energy problems and how R&D can help solve them, rather than simply listing every nation's pet technology, might produce a more rational agenda. The temptation simply to produce lists of technologies is strong; it is politically the course of least resistance, since it avoids the issue of picking winners and losers; yet it provides not a strategy for collective action, but only a blueprint for diffusion of effort.

Literature Cited

1. Organ. Econ. Coop. Dev. 1975. *Energy R&D*. Paris
2. Nau, H. R. 1974. *National Politics and International Technology: Nuclear Reactor Development in Western Europe.* Baltimore: Johns Hopkins Univ. Press
3. Williams, R. 1975. *European Technology: The Politics of Collaboration.* New York:

Halstead, Wiley
4. Pavitt, K. 1971–1972. Technology in Europe's future. *Res. Policy* 1:210–73
5. Lester, J. P. 1976. Energy R&D: U.S. technology transfer to advanced western countries. In *Technology Transfer and U.S. Foreign Policy*, ed. H. R. Nau. New York: Praeger

Ann. Rev. Energy. 1985. 10 : 613–88
Copyright © 1985 by Annual Reviews Inc. All rights reserved

AN END-USE ORIENTED GLOBAL ENERGY STRATEGY

Jose Goldemberg

President, Companhia Energetica de São Paulo (CESP), São Paulo, Brazil

Thomas B. Johansson

Professor of Energy Systems Analysis, Department of Physics and Environmental Studies Program, University of Lund, Lund, Sweden

Amulya K. N. Reddy

Professor of Inorganic and Physical Chemistry, Indian Institute of Science, Bangalore, India

Robert H. Williams

Senior Research Scientist, Center for Energy and Environmental Studies, Princeton University, Princeton, New Jersey, USA

It is generally believed that, at the present rate of consumption, the world's remaining oil resources that can be ultimately recovered amount to less than a 100-year supply. Two thirds of these resources are located in the Middle East and North Africa and in countries with centrally planned economies; the amount left in the rest of the world is about a 40-year supply. These numbers indicate the ephemeral nature of the present oil glut and highlight the need to begin a global transition away from oil.

The half of the world's people who are dispersed in rural areas of developing countries use little oil, but they are also caught up in an energy crisis. These people depend heavily on biomass—mainly fuelwood used for cooking. A fuelwood crisis has arisen because in many areas increased fuelwood demand associated with population growth is exceeding the rate

613

0362–1626/85/1022–0613$02.00

of fuelwood regeneration. Some 100 million human beings now suffer acute scarcity of fuelwood, and about 1 billion a fuelwood deficit (1).

These oil and fuelwood supply considerations show clearly that energy is a major global problem. But energy is only one of several important global problems, the solutions of which are necessary if mankind is to achieve a sustainable world society for the long term. The most pressing such problems include the global economic crisis, North–South conflicts, widespread poverty in developing countries, population growth, food scarcity, the risk of global nuclear war, nuclear weapons proliferation, man's role in changing the global climate, environmental degradation, and deforestation and desertification. This review recognizes that all of these problems have strong links to energy and that pursuing solutions to the energy problem without considering these links might aggravate the other problems.

THE ROLE OF ENERGY IN SOLVING OTHER GLOBAL PROBLEMS

The Economic Crisis

The last decade has been a period of rampant inflation, major global recessions, widespread unemployment, soaring real interest rates, and an associated international debt crisis which could lead to collapse of the global financial system if the debts of developing world debtors are not discharged.

Costly energy has been a major contributing factor to these problems. On average, low- and middle-income developing countries in 1981 spent, respectively, 61% and 37% of their export earnings on oil imports (2). Also, energy from new sources is generally far more costly than energy now being used. For example, in the United States, between 1972 and 1982, capital expenditures on energy supply rose from 26% to 39% of all new plant and equipment expenditures (3), with essentially no net increase in domestic energy production in this period. In developing countries, investments committed to energy supply expansion increased during the 1970s on average from 1–2% to 2–3% of Gross Domestic Product (GDP) (4), requiring in 1982 some $25 billion of foreign exchange—over one third of the foreign exchange required for all kinds of investments (4). Committing so much capital to energy supply makes capital scarcer for other economic activities.

If expanding of energy supplies continues to be emphasized as the primary means of providing energy services, the economic burden of providing energy in the future could be even greater than indicated by this recent experience.

Because of the concentration of the remaining oil resources in the Persian Gulf region, even modest increases in world oil demand could lead to very large increases in the world oil price. The US Department of Energy projected in 1983 that the world oil price would increase from $34 per barrel in 1982 to some $55–111 per barrel in 2010, associated with a 13–24% increase in world oil demand (5) that would once again bring the production of the Organization of Petroleum Exporting Countries (OPEC) up to nearly the level of 1979, when tight market conditions made possible the second world oil shock with the outbreak of the Iranian revolution.

For developing countries, energy supply expansion strategies to meet development goals would be especially costly. The World Bank in 1983 estimated that, to increase per capita use of commercial energy from 0.54 to 0.78 kW between 1980 and 1995, investments in energy supplies for all developing countries would have to average some $130 billion a year (in 1982 dollars) between 1982 and 1992 (4% of aggregate GDP). Half of this investment would have to come from foreign exchange earnings, requiring an average annual increase of 15% in real foreign exchange allocations to energy supply expansion in this period. And despite this targeted effort at energy supply expansion, the World Bank projected that oil imports by oil-importing developing countries would still increase by nearly one third, to nearly 8 million barrels per day by 1995 (4). The staggering costs of providing these energy services would lead many to believe (but rarely to state) that it is not feasible to improve living standards substantially in developing countries.

But the costs of providing energy services need not be as high as these projections suggest. The costs of energy can be brought under control, if efforts to provide energy services (cooking, lighting, space conditioning, mechanical work, mobility, etc) at the least total social cost are stressed in energy planning. The costs of providing such services can often be reduced by investments in energy efficiency improvement, in industrialized (6) and developing (7) countries alike. This can also mean a lower world oil price, if world oil demand is low enough that tight world market conditions can be avoided (Figure 1).[1] Prices would be lower for other energy forms as well, to the extent that costly new energy sources are not needed. In such ways

[1] Here a distinction should be made between the world oil price and consumer oil prices. It may be desirable for governments to put a tariff on imported oil or tax oil products so that consumer prices are constant or slowly rising in real terms, as an alternative to the bumpy path of the price of world oil in the last decade.

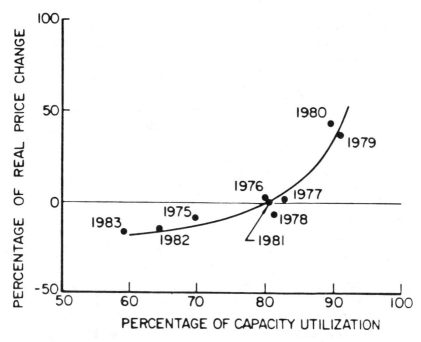

Figure 1   OPEC pricing behavior: percentage change from previous year in the real world oil price vs percent of OPEC production capacity used. The percentage of capacity used is equal to crude oil production divided by maximum sustainable production for that year. Source: (101).

energy planning could contribute to improved economic efficiency and well-being.

North–South Conflicts

Poor countries of the "South" account for three fourths of the world's population but have per capita incomes on average only one tenth as large as in the rich countries of the "North." This income disparity is a crucial factor responsible for the grave and worsening crisis characterizing the world economy. Problems such as deteriorating commodity prices, Northern protectionist barriers against the emerging manufacturing industries of the South, and the vulnerability of southern debtors to rising interest rates because of their dependence on variable rate loans have put developing countries at a disadvantage in the global marketplace.

The solution to North–South conflicts is to work toward eradication of the disparities that give rise to the conflicts, by means of policies that foster the development of developing countries—including policies that would help make affordable the energy needed to meet development goals.

Poverty in Developing Countries

Not only are disparities between North and South large, but within developing countries there are enormous disparities between the elites, who typically account for 10% of the population and one third to one half of all income, and the rest of the population, who live in abject poverty. The traditional approach to tackling this problem of widespread poverty by maximizing economic growth and expecting the benefits of growth to trickle down to the poor has failed.

A policy targeting the satisfaction of basic human needs for food, shelter, sanitary services, health care, education, and meaningful employment is more promising (8–10). Because satisfying these needs requires energy services, providing such services must be a goal of development efforts.

Population Growth

The population explosion is closely linked to the problem of poverty, since large families tend to be economically favorable for the poor (2). Thus efforts to solve the population problem will be assisted by targeting the satisfaction of the basic human needs of the poor.

Food Scarcity

Whatever success is achieved in slowing population growth, feeding an expanding population will remain a major global challenge for decades. Modernization of agriculture and the associated increased energy inputs are vital to meeting increased food production goals. The Food and Agriculture Organization has estimated that food production in developing countries must double by 2000, for which additional energy equivalent to 2.8 million barrels of oil per day is required (11). As this is less than the amount of oil saved by the United States alone between 1978 and 1982, it is clear that the challenge has less to do with the quantity of energy required than with ensuring that supplies are available to meet agricultural needs.

The Risk of Nuclear War

That conflict in the Middle East can draw in the superpowers and threaten nuclear war is indicated by the experience of October 1973, when the Soviet Union threatened to intervene in the Arab–Israeli War, and the United States, in response, raised the alert status of its nuclear forces (12). The creation of the US Rapid Deployment Force to assure the industrialized market economies of continued access to Persian Gulf oil, and the presence of mobile Soviet forces in the region indicate the continuing potential for US–Soviet conflict arising from Middle East turmoil. The potential for superpower conflict can be reduced if the industrialized market economies become less dependent on Persian Gulf oil.

Nuclear Weapons Proliferation

In 1964 the United States, the Soviet Union, France, and Great Britain were the only nuclear weapons states in the world. Since then, China and India have acquired nuclear weapons, and several other countries either have already achieved or soon will have the capability to produce nuclear weapons. In the coming decades many more countries could join the "nuclear club"—particularly if nuclear power comes to be a major energy resource.

An indissoluble link between nuclear weapons and nuclear power arises from the fact that plutonium, a material usable in nuclear weapons, is produced in substantial quantities in nuclear power reactors. There is no technical fix for eliminating this link, short of avoiding dependence on the troublesome technologies involved.

The proliferation risk increases enormously if plutonium is recovered from spent reactor fuel and is recycled in fresh fuel. If a state without nuclear weapons acquires plutonium recycle technology, it obtains nearly all the technology and materials needed to make nuclear weapons quickly, without ever having to make an explicit decision to acquire such weapons. This route to nuclear weapons, which has come to be called "latent proliferation" (13), is a particularly dangerous route to proliferation, involving low risk to the would-be proliferator.

The risk of proliferation can be greatly reduced by avoiding nuclear fuel cycles that involve reprocessing spent reactor fuel and recycling recovered plutonium. To deter proliferation effectively, such fuel cycles would have to be avoided in all countries, including the nuclear weapons states, as any two-class system that discriminates against countries judged to be proliferation-prone would prove ultimately unstable (14).

The incentive to reprocess spent fuel and recycle plutonium will remain low if uranium prices do not rise too much, which in turn will be the case if worldwide nuclear power development is slow enough—specifically, if nuclear power is regarded as an energy technology of last resort.

Technological constraints on the scopes and characters of nuclear power programs by themselves could not halt this "horizontal proliferation" of nuclear weapons capability to many countries. As long as there is "vertical proliferation" by the superpowers, brought about by the feeling that their security is enhanced by having nuclear weapons, other nations will want nuclear weapons as well. The only way to avoid a world in which nuclear weapons are much more widely proliferated, both horizontally and vertically, is to couple the avoidance of dangerous nuclear power technologies with superpower efforts to move away from dependence on weapons of mass destruction (14).

Global Climatic Change

In a matter of decades man could bring about major changes in the global climate because of activities leading to the buildup in the atmosphere of carbon dioxide and the resulting "greenhouse effect." The problem is closely related to energy because the major source of this buildup is the burning of fossil fuels. Already in 1979 the atmospheric CO_2 level was 1.15 times the preindustrial level, or 334 ppm. Climatologists believe that with a doubling of the CO_2 level there would be an increase in the global temperature of $3 \pm 1.5°C$ and perhaps a two- to threefold greater warming at the poles. The resulting slowdown of the "atmospheric heat engine" associated with the differential equatorial/polar heating rates would, it is estimated, lead to significant changes in global weather patterns (15). The contributions to atmospheric heating of other radiatively active trace substances (RATS) (15) amplifies the concerns about atmospheric carbon dioxide buildup.

There appear to be no feasible technical fixes for the CO_2 problem (16). The magnitude of the prospective climatic change can best be reduced by reducing dependence on fossil fuels.

Deforestation and Desertification

Between 1952 and 1972 the world's forests were lost at an average rate of some 30 million hectares per year (17), while losses of cropland and rangeland to desertification averaged 6 million hectares per year (18). Deforestation arises from the permanent clearing of forest land for agriculture, from fuelwood gathering, and from other overuse of the forest resource; desertification, in large part from the overuse of marginal lands for agricultural purposes, especially the grazing of livestock. Efforts are needed to reverse these trends—for both environmental and economic reasons. Fuelwood resources could be used renewably if demand levels were maintained below the regeneration rate via more efficient use of biomass and via increased production through better forest management. The trends toward both deforestation and desertification could be eased if agricultural production, including the grazing of livestock, were shifted from marginal to better lands, via use of modern energy-intensive, yield-increasing agricultural techniques—a hopeful prospect since only one third of all cropland is under heavily mechanized production (personal communication to R. Williams from D. Pimentel, Cornell University, February 19, 1985) and in developing countries fertilizer inputs per hectare average only 40% of the levels in industrialized countries (19).

Trade-Offs

Most people would agree that it would be desirable to pursue global energy strategies consistent with or supportive of the solutions of the other global problems we have described above, or, more fundamentally, that energy strategies should contribute to the achievement of broad societal goals of economic efficiency, equity, environmental soundness, human welfare in the long term, and peace—in essence the goals that must be pursued to achieve a sustainable society.

Our ongoing end-use-oriented global energy project, some of the major results of which we report here, leads us to be optimistic that it is feasible to evolve long-run global energy strategies consistent with the achievement of a sustainable world society.

This optimism is not widely shared. Many are skeptical that it is possible to satisfy all such objectives simultaneously, because of unavoidable trade-offs. Concerns about equity, for example, suggest to some that large global increases in energy use are needed to bring the majority of the world's population up to a decent standard of living—the achievement of which must take precedence over environmental and security concerns. Others have argued that the environmental and security risks of modernizing global society outweigh the benefits.

The notion that we must learn to live in a much more troublesome world as the price of human progress stands out as a theme implicit in many recent studies of the long-term energy problem.

SOME GLOBAL ENERGY STUDIES

There are several reasons for studying the long-term energy future:

1. As an input for all economic activity, energy is needed to realize the development goals of developing countries and to provide for the continued economic well-being of the industrialized countries.
2. Insights about long-term energy requirements are needed as part of the information base for making investment decisions that are characterized by long lead times and long investment lifetimes—for energy exploration, energy supply development, energy transport and conversion technology, and energy-using equipment.
3. Information about the long term is also needed to help formulate long-term energy policy and research and development programs.

4. Since remaining oil resources are quite limited, a transition from oil must begin in the decades ahead. This fact prompts interest in understanding how the world's future energy system will differ from that at present.
5. Recently there has been considerable interest in understanding better one of the troublesome side effects of energy production and use—the atmospheric carbon dioxide problem—and the options for coping with this problem.

In this section we shall briefly review some of the more recent global energy studies, each of which was motivated by one or more of these concerns. We focus on those studies published since the second oil price hike in 1979. Some key features of the studies we review and of our own analysis are presented in Table 1.

The IIASA Study, 1981

Between 1973 and 1979 the Energy Systems Program Group at the International Institute for Applied Systems Analysis (IIASA) carried out a major study of the global energy problem under the direction of Wolf Häfele. More than 140 scientists from 20 countries participated in this effort, which had a budget exceeding $6.5 million. The results of this study have been reported in a large number of reports and articles, and a 1981 book (20). Of all the global energy analyses that have been carried out, this one has attracted perhaps the most attention, not just from energy analysts but also from the broader community of citizens. In the heated energy debates of the late 1970s, this energy project came to symbolize the "hard" energy future, as distinct from the "soft" future advocated largely by Lovins et al (21).

The IIASA study makes projections to the years 2000 and 2030 by aggregating projections for seven regions of the world, focusing mainly on commercial energy. The point of departure for the study is an analysis of present and future energy demands, highly disaggregated by energy end uses. The analysis was based on the MEDEE-2 model, developed at the University of Grenoble (22) and adapted to the IIASA global energy assessment by Lapillone (23). Future energy demands for various end uses were estimated by correlating activity levels for particular energy-using activities (e.g. passenger-km for travel), with various economic and demographic parameters and by specifying the energy performances of the technological systems involved. Given a set of energy demand levels determined via the MEDEE-2 model, a model called MESSAGE was used to compute the least expensive energy supply mix that meets the specified consumption levels. Then another called IMPACT was used to calculate

Table 1 Comparisons of some global energy studies

	1980	IIASA (20) High	IIASA (20) Low	WEC (29) High	WEC (29) Low	IEA/ORAU (32)	IEA/ORAU (31)	MITEL (33) A	MITEL (33) J	Nordhaus & Yohe (35)	Colombo & Bernardini (36)[j]	Lovins et al (20)[k]	This study
Projection year	—	2030		2020		2025	2025	2025		2025	2030	2030	2020
Population (billion)	4.43	7.98		7.72		7.36	7.36[a]	7.36		7.82	7.98	7.98	6.95
Per capita GDP growth rate to year of projection (% per year)		2.1	1.1	2.0	1.1	1.6	1.6[b]	1.6		1.9	1.2	1.1	—[c]
Primary energy[d,e]													
TW, World	10.3	35.2	22.0	24.7	19.2	18.8	26.6	18.1	12.1	24.5	14.4	5.2	11.2
Industrialized	7.0	20.1	13.5	14.8	12.5	12.7	15.8				7.2	3.6	3.9
Developing	3.3	15.1	8.4	9.9	6.7	6.1	10.8				7.2	1.7	7.3
kW/cap, World	2.33	4.41	2.76	3.20	2.49	2.55	3.61	2.46	1.64	3.14	1.81	0.65	1.61
Industrialized	6.3	12.9	8.64	9.67	8.15	8.52	10.6				4.64	2.28	3.15
Developing	1.0	2.35	1.32	1.59	1.08	1.04	1.84				1.12	0.26	1.28
Energy supply (TW)													
Oil	4.18	6.83	5.02	5.81	4.26	4.01	5.81	2.44	1.57		2.71[f]		3.23
Gas	1.74	5.97	3.47	4.59	3.42	3.60	3.59	1.64	1.69		—		3.23
Coal	2.44	12.0	6.45	7.74	6.06	9.47	11.1[g]	8.60	3.23		4.95		1.95
Hydropower	0.19	0.52	0.52	0.70	0.52	1.04[h]	1.09[h]	0.93[h]	1.22[h]		0.76[h]		0.46
Nuclear power	0.22	8.09	5.17	3.21	2.31	0.69	5.02	1.81	2.99		1.74		0.75
Other	1.49	1.81	1.33	2.65	2.65	—	—	2.67[i]	1.36[i]		4.28		1.58

[a] Inferred from (32).

[b] As no value is reported in (31), this is assumed to be the same as in (32), since growth rates for the period 1975–2050 are the same for the two IEA/ORAU studies.

[c] No global GDP growth rate was explicitly assumed. However, the energy demand projection is consistent with per capita GDP increasing 1.5–2-fold in industrialized countries and up to 10-fold in developing countries. These assumptions are consistent with a global per capita GDP growth rate of up to about 3% per year.

[d] Here TW is an abbreviation for TW-years/year and kW an abbreviation for kW-years/year. Nuclear power is counted as the fossil fuel equivalent for producing the same amount of electricity. Hydroelectricity and other solar electric sources are counted as the electricity produced. Includes noncommercial energy, feedstocks, and bunkers. Only the IIASA and WEC studies and our study include noncommercial energy, although the MITEL study attempts to deal with noncommercial energy in the model used by assigning very low cost to part of the biomass fuels consumed.

[e] Commercial energy and fuelwood consumption data for Eastern Europe and the Soviet Union are from (89). Bioenergy data for developing countries are from (90). In 1980 bioenergy consumption by the US pulp and paper industry was 1.1 EJ (91), while wood consumption for household fuel use was 0.87 EJ (92). Noncommercial energy use in other industrialized market economies was obtained from (93).

[f] Oil plus natural gas.

[g] Includes biomass.

[h] Includes solar electric.

[i] Synthetic fuels.

[j] May not include feedstocks and bunkers.

[k] May not include feedstocks and bunkers. All electricity is provided by renewable energy sources or cogeneration.

requirements for capital investment, labor, land, water, materials, equipment, and additional energy. Finally, IMPACT modified the original economic assumptions to ensure consistency.

The high- and low-energy scenarios generated for the year 2030 in the IIASA analysis involve primary energy use levels of 35 and 22 terawatts (TW)—$3\frac{1}{2}$ and 2 times the 1980 energy consumption level—associated with global GDP growth rates of 3.4% per year and 2.3% per year, respectively (Table 1).

Such levels can be met only with massive expansions of both fossil fueled and nuclear energy systems. Indeed, achieving the targeted energy supply levels would require:

1. New fossil fuel production capacity equivalent to bringing on line a new Alaska pipeline (2 million barrels of oil equivalent per day) every one to two months
2. One new 1-GW(e) nuclear power plant every 2–3.5 days.

Anyone familiar with the problems of building new energy supply capacity will realize how difficult it would be to meet these production targets.

Furthermore, developing energy supply on a scale envisioned by these projections would be in strong conflict with many of the other global concerns discussed in the preceding section.

Both scenarios involve such high levels of global oil demand in the year 2030 that the Middle East–North Africa region must produce at a level estimated by the IIASA analysts to be the maximum potential production level for the region, some 34 million barrels per day—more than twice the average 1982–1984 production. This implies a return to the tight oil supply conditions that provided the basis for the world oil price hikes in the 1970s, and a condition of global insecurity once more as a consequence of overdependence on Persian Gulf oil.

The projections for future fossil fuel use imply that the atmospheric carbon dioxide level would double in the latter half of the twenty-first century, with consequent major climatic change (15).

The nuclear power projections in these studies imply a serious nuclear weapons proliferation risk; for example, the nuclear projections imply that by 2030 some 2.6–4.0 million kg of plutonium would be recovered from spent reactor fuel and circulated each year in global commerce; some 5–10 kg are required to make a nuclear weapon.

Despite these disturbing implications of the IIASA energy projections, the IIASA Energy Systems Program Leader Wolf Häfele and Hans-Holger Rogner have argued that concerns about externalities must be subordinated to the more important goal of bringing about a decent standard of living to the world's impoverished majorities (24):

> Finally, let us address the heart of the issue in question: the old controversy "soft versus hard energy paths." Yes, we are not soft enough to suggest to the Have-Nots of the world who currently live with an average per capita energy consumption of 0.2 kW-yr/yr that they can expect no more than, say, a per capita consumption of 0.6 kW-yr/yr, while in North America the average per capita consumption is some 10 kW-yr/yr and in Europe some 5 kW-yr/yr. This is no basis for healthy global politics. We refuse to prescribe to the Have-Nots how to live, especially under these circumstances. Nor do we wish to live with a per capita consumption of 0.6 kW-yr/yr. In fact, we believe that all people should be given the energy they need and be allowed to choose for themselves their way of living. We do not want to transform or change societies. We do want free development constrained only to the degree that is unavoidable. Indeed, with eight billion people in the year 2030, a per capita energy consumption of 2 kW-yr/yr would result in 16 TW-yr/yr, 3 kW-yr/yr in 24 TW-yr/yr, and 4 kW-yr/yr in 32 TW-yr/yr. Yes, we consider an average global per capita consumption between 3 kW-yr/yr and 4 kW-yr/yr a reasonable figure. This is in line with the IIASA energy scenarios.

Indeed considerations of global equity must play an important role in the formulation of long-run energy strategies, if the world is to evolve toward a more stable state of North–South politics. However, by not distinguishing between energy services and energy use per se, the IIASA analysts have not provided a compelling case that it is necessary to increase energy supplies in the manner and to the extent they envisage, in order to bring about a world of widespread prosperity. Moreover, the IIASA projections do not indicate a more equitable pattern of global energy use in 2030 than in 1980: more than half the increment in annual global energy use during 1980–2030 would be accounted for by the industrialized countries (Table 1), a recipe for even greater North–South tensions than at present.

While the IIASA analysts insist that their scenarios are robust, the range of feasible energy futures is certainly much broader than the range spanned by their scenarios. A recent analysis of the IIASA methodologies by Keepin (25) shows that the optimal supply mixes arrived at in the IIASA analysis should be regarded cautiously. Keepin reports that "the important dynamic contents of the scenarios are effectively prescribed (before the computer is turned on) in the form of assumed projections that are supplied as inputs to the mathematical energy models. Meanwhile, the computerized models themselves perform a simple heuristic analysis that reproduces various input assumptions with few alterations." Furthermore, ". . . the IIASA scenarios are not robust with respect to minor changes in various input data . . . it is found that the energy supply scenarios are strongly dependent on arbitrary (and in some cases, unlikely) assumptions about the future costs and availability of energy sources and supply technologies. Small changes in these assumptions (such as increased costs that have already been observed in reality) can yield extremely different scenarios from the models."

As our own analysis, presented in later sections, indicates, much lower energy demand levels than those projected in the IIASA scenarios can be consistent with the achievement of ambitious economic goals. Such low-energy futures are feasible in part because of the ongoing shift to less energy-intensive economic activities in industrialized countries—a shift that is discussed by the IIASA study but is not adequately taken into account in the study's numerical analysis. The other important factor that makes such futures feasible is the wide range of opportunities for making more efficient use of energy in industrialized and developing countries alike.

While the IIASA analysts claim that strong energy conservation trends were built into the energy demand projections from the beginning, scrutiny of the energy intensities assumed for various important energy-using activities in the IIASA analysis[2] has convinced us that in fact very little attention was given to opportunities for improving energy efficiency in this study.

World Energy Conference, 1983

The Conservation Commission of the World Energy Conference (WEC) in 1978 issued an initial analysis of energy supply and demand to the years 2000 and 2020 (28), and in 1983 it issued a new study (29) for presentation to

---

[2] Consider residential space heating and automotive examples in IIASA's North American (NA) region. Since Canada's population, residential energy use, and number of cars are only one tenth as large as in the United States, comparisons of IIASA calculations for the NA region with the US situation give an indication of the emphasis on energy efficiency in the IIASA analysis.

In the IIASA scenarios the average space heating load (the useful thermal energy output of the heating system) is 52 GJ per household in the NA region in 2030. Also an efficiency of 80% for fossil fuel furnaces and a coefficient of performance (COP) of 2 for electric heat pumps (assumed to account for half of electric heating) are assumed for space heating equipment (26). For comparison the average space heating loads in the United States in 1978–1979 were 65 GJ for gas-heated houses and 49 GJ for electrically heated single family dwellings (27). Savings of 30% have been demonstrated as being cost-effective in retrofits of existing houses (see below). The average heat load for new single family dwellings in the United States is in the range 35–43 GJ (27), but with cost-effective, super-insulated designs the heat load can be reduced below 10 GJ (27). Also, new condensing furnaces have efficiencies of 90–95%, and the best air-to-air heat pumps available on the US market in 1982 had seasonal average COPs of 2.6 (27).

For the automobile in 2030 the IIASA analysts project an average fuel economy in the range 31–33 mpg (7.1–7.5 liters per 100 km) (26), which is slightly higher than the 27.5 mpg mandated for new cars in the United States by 1985. With present technology cars with fuel economies of 60–100 mpg are feasible.

the 12th Congress of the World Energy Conference held in September 1983 in New Delhi.

The 1983 analysis was the result of a cooperative effort of one central team and 10 regional working teams involving 50 experts (30 from developing countries and 20 from industrialized countries) with varying backgrounds (11 represented energy authorities, 9 oil companies, 9 electric companies, 14 international and regional organizations, and 7 academia and research organizations). The work was carried out in 17 international meetings and supplemental correspondence.

The analysis, carried out for 10 global regions, focused on the supply of and interregional trade in primary energy. Both commercial energy (coal, oil, natural gas, nuclear power, hydropower, and new sources) and noncommercial energy (fuelwood and animal and vegetable wastes) were taken into account.

High and low projections were made for energy demand, based largely on GNP projections that were correlated with energy demand via the use of income elasticities. The underlying global GNP projections averaged 3.3 and 2.4% per year during 1978–2020, and the corresponding global energy demand levels for 2020 were $2\frac{1}{2}$ and 2 times the 1980 level for the high and low growth cases, respectively (Table 1).

Despite the fact that the 1983 WEC global energy demand projections were 6 TW less than the 1978 WEC projections (28), the heavy dependence on oil, fossil fuels generally, and nuclear power in the 1983 projections would pose problems similar to those discussed for the IIASA study. Likewise, the continued concentration of demand for global energy resources envisaged for the industrialized countries [which would account for more than half the net increase in global energy demand during 1980–2020 (Table 1)] would not be conducive to better North–South relations.

The IEA/ORAU Model, 1983–1984

A detailed, long-term global energy-economy model has been developed for the US Department of Energy by the Institute for Energy Analysis/Oak Ridge Associated Universities (IEA/ORAU) to study the determinants of future carbon dioxide emissions from the burning of fossil fuels (30). A baseline global projection was made in 1983 for the years 2000, 2025, and 2050 (31), and in 1984 three alternative projections were made for the years 2000, 2025, 2050, and 2075 (32). The 1983 and 1984 baseline projections for the year 2025 are indicated in Table 1.

The world is divided into nine regions in the IEA/ORAU model, but far more detail is provided for countries belonging to the Organization for Economic Cooperation and Development (OECD) than for the rest of the world. The model is highly disaggregated on the supply side, taking into account nine primary energy sources, although noncommercial energy is not considered. On the demand side, there is little disaggregation. For OECD countries, demand is disaggregated only to the gross sector level: residential/commercial, industrial, and transport. There is no demand disaggregation for the rest of the world.

In the model, the major determinants of the level and composition of energy demand in a given region are population, GNP, the relative prices of different energy carriers, and energy price and income elasticities. In addition, a modest allowance is made for non–price-induced energy conservation. GNP is specified initially and then allowed to adjust to energy price changes. Using various supply and demand elasticities, energy prices for each of the globally traded fuels (oil, gas, and coal) are adjusted from their initially specified values until global energy supply and demand are brought into balance.

The baseline primary commercial energy demand projection presented in the 1983 study (31) for the year 2025 is 26.6 TW, implying that commercial energy use would grow only slightly more slowly than the assumed GNP growth rate of 2.9% per year. Energy supply would be dominated by coal (42%), oil (22%), and nuclear energy (19%), at levels that would lead to all the same problems described earlier in reference to the IIASA scenarios.

In the 1984 report (32), coauthored by the authors of the 1983 study (31), the baseline primary demand projection B for 2025 is 18.8 TW.[3] Additional

---

[3] The baseline projection for 2025 in the 1984 report is lower than that in the 1983 study by an amount nearly as great as total global commercial energy demand in 1980, for essentially the same assumptions about population and global GNP growth in the two scenarios. No explanation is given for this striking difference. However, comparison of the stated assumptions and outcomes for the two cases indicates that the difference is due at least in part to changes in assumptions about energy prices, income elasticities, and non–price-induced energy conservation. One difference is that in the later report the assumed cost of delivered nuclear electricity is double that in the earlier report. Also, a much higher price is assumed in the later report for the cost of oil from shale.

About two thirds of the global demand reduction is attributable to developing countries, where per capita primary energy use was cut almost in half—from 1.8 kW in the 1983 study to 1.0 kW in the 1984 report. Nearly two thirds of the reduction for developing countries appears to be associated with a change in the assumptions for income elasticity. In the 1983 study the income elasticity for developing countries was assumed to be 1.4; in the 1984 report it was assumed to be reduced continually from 1.4 in 1975 to 1.0 (the value assumed throughout for OECD countries) by 2050.

In addition, non–price-induced conservation is assumed in the 1983 study to take place at arithmetic rates of 1% per year in the industrial sector of OECD countries and 0.4% per year for the entire economy in non-OECD regions. In the 1984 report geometrical rates are assumed, and the rate for non-OECD regions is increased to 0.5% per year:

scenarios A and C, for which global energy use rates for 2025 are 58% higher and 20% lower, respectively, are intended to bracket likely outcomes for energy and carbon emissions.

The estimated dates for a doubling of the atmospheric carbon dioxide level predicted in the 1984 report are between 2025 and 2050 for scenario A, near 2050 for scenario B, and near 2075 for scenario C. The parameters varied to bring about the alternative scenarios are population growth, GNP growth, the degree of non–price-induced conservation, and prices for different energy carriers.

As in the IIASA and WEC analyses, these projections do not adequately reflect the possibilities for decoupling energy and economic growth. This shortcoming can be described in terms of the characteristic parameters used in the model.

While the 1984 report allows for the possibility of reducing the income elasticity in developing countries over time, the income elasticity for OECD countries is maintained at 1.0 in these applications, meaning that each percentage increase in GNP, at constant energy price, would be associated with a 1% increase in energy demand. Yet, as shown below, there is strong evidence that industrialized countries are entering a post-industrial era characterized by new economic activities that are far less materials-intensive, and thus less energy-intensive, than economic activity of the past.

Moreover, these scenarios make only modest allowance for non–price-induced improvements in energy efficiency. In the base case scenarios for OECD countries allowance is made for a 1% annual increase in energy efficiency for the industrial sector and no improvement whatsoever for other sectors. Even in the "extreme" case C, the rate of improvement is only 1.5% per year for the industrial sector and 0.5% per year for the other sectors.

Non–price-induced efficiency improvements of two types are important. The first of these is reflected, at least in part, in these analyses—namely the long-term trend toward improved energy efficiency in the industrial sector, associated with process innovation, even during periods of declining energy prices. In addition there are major opportunities for public policy–induced improvements in energy efficiency, which should be considered seriously both for the economic benefits they would bring and for the flexibility such efforts would provide in dealing with other important global problems such as the carbon dioxide problem. The failure to consider such policy-induced improvements is a major shortcoming of all the energy–economy modeling studies that take into account energy-efficiency improvements largely via elasticities, to account for the impacts of energy price changes on energy demand.

The MITEL Report, 1984

Analysts at the MIT Energy Laboratory (MITEL) have used the IEA/ORAU energy–economy model to explore further the opportunities for coping with the carbon dioxide problem by pursuing alternative long-run energy strategies (33). They designed 11 scenarios, all variations of the base case presented in the 1983 IEA/ORAU study (31). All of the scenarios involve higher end-use efficiencies, and most of them higher synthetic fuel costs than were assumed in the 1983 IEA/ORAU study. Various changes in energy supply conditions were assumed.

The MITEL analysts concluded that improving energy efficiency offers the single most important opportunity to ameliorate the CO_2 buildup, and that it also appears attractive in its own right, both economically and environmentally.

Scenario J, shown in Table 1, is one of several low-energy-demand scenarios produced in the MITEL analysis, which the authors identified as "CO_2 benign." In this scenario overall energy use in 2025 is 12.0 TW, only 17% more than in 1980. Fossil fuel use in 2025 is 0.78 times as large as in 1980, and the atmospheric carbon dioxide "doubling time" would be about 250 years. Oil use would be reduced to 0.38 times the 1980 level, as a consequence of an assumed cutoff of Middle East oil supplies in 2000 and high prices of unconventional substitutes for oil. Non–price-induced energy efficiency improvements are assumed to increase geometrically at an average rate of 1% per year, for all end-use sectors and for all countries.

The authors describe this and their other low-energy scenarios as follows:

> The relatively CO_2-benign scenarios ... are not low-energy in a Draconian sense.... They would require global awareness and collaboration, starting very soon. While perhaps at the lower limit of possible realities, these scenarios do not appear to us impossible; recall that energy projections for the early 2000's now being made are much below what people believed possible only a decade ago.

The global energy demand levels implied by the MITEL low-energy projections are comparable to what we believe are achievable by emphasizing improvements in energy efficiency, in a manner consistent with the achievement of economic goals, as we shall show in later sections.

While the deemphasis on fossil fuels implied by the MITEL scenarios would be an effective way to deal with the carbon dioxide problem, the MITEL scenario would involve a 14-fold increase in the use of nuclear energy during 1980–2025 [corresponding to an average annual growth rate of 6% in this period, and an installed capacity of 1300 GW(e) in 2025]. With this level of nuclear power development, the nuclear weapons proliferation problem could be exacerbated.

The rapid growth envisaged for nuclear power in the MITEL low-energy-demand scenarios is driven in large part by the very low cost assumed for delivered nuclear electricity—some $6.83 per GJ (1975 $) or $0.035 per kWh (1980 $)—which is far less than even optimistic estimates of future nuclear power plant costs by US DOE officials.[4]

Nordhaus & Yohe, 1983

In connection with the 1983 carbon dioxide–climatic change analysis of the US National Academy of Sciences (15), Nordhaus & Yohe developed a global model aimed at bracketing the uncertainty in estimates of future levels of carbon dioxide in the atmosphere (35).

Nordhaus & Yohe constructed a simple and transparent but highly aggregated model of the global energy economy. The starting point of their analysis involves the use of a production function for the global economy, relating global GNP to population, labor productivity, and inputs of fossil and nonfossil commercial energy. The model will generate some $3^{10} = 59,049$ alternative projections of future global economic activity, future global energy demand (both fossil and nonfossil fuel), and future carbon dioxide concentrations in the atmosphere, by assigning high, medium, and low values of each of ten important variables needed in the model: (a) ease of substitution between fossil and nonfossil fuels (measured by cross-price elasticities); (b) general productivity growth; (c) ease of substitution between energy and labor (measured by price elasticities of demand); (d) extraction costs of fossil fuels; (e) trends in the real costs of producing energy; (f) airborne fraction for carbon dioxide emission; (g) fuel mix among fossil fuels; (h) population growth; (i) trends in relative costs of fossil and nonfossil fuels; and (j) total resources of fossil fuels.

The authors used the means and variances (expanded, where the authors

---

[4] For comparison, the US Department of Energy in 1982 (34) estimated that the busbar cost of new nuclear power plants that would come on line in the United States in 1995, at a capital cost of $1550 per kW, to be $0.043 per kWh (1980 $). Taking into account 8% transmission and distribution (T&D) losses and adding costs of $0.013 per kWh would bring the cost of delivered electricity to $0.06 per kWh. The cost would be even more if recent US experience were to persist. For the 35 nuclear projects in the United States, which as of late 1983 were scheduled for completion after 1981, the average capital cost, exclusive of interest during construction, was about $1700 per kW (1980 $) (private communication from Charles Komanoff, March 1983). With 5% real interest and a 7.8-year construction period (34), the installed cost would be $2000 per kW, with interest during construction included. For this capital cost, the delivered cost of electricity from a nuclear power plant would be $0.070 per kWh (1980 $).

thought it was necessary, to correct for systematic underestimation) for the published estimates of each of the random variables identified in the model as a basis for constructing normal, judgmental probability distributions for each variable.

The results of a sample of 1000 runs gives (a) a probability-weighted mean commercial energy consumption in the year 2025 of 24.5 TW (with a standard deviation of 8.5 TW) and (b) 2070 as the most likely date by which the atmospheric carbon dioxide level would double. The authors found that the parameters to which the atmospheric carbon dioxide level are most sensitive are: (a) the ease of substitution between fossil and nonfossil fuels (i.e. the cross-price elasticities); (b) general productivity growth; and (c) the ease of substitution between energy and labor (related to price elasticities of energy demand).

While the Nordhaus-Yohe model provides these insights on the most important parameters influencing the carbon dioxide buildup, inherent shortcomings prevent the model from accomplishing what its authors probably hoped it would: bracketing the range of uncertainty of future carbon dioxide emissions. First, the model does not allow income elasticity of energy demand to vary with level of economic development, and thus cannot take into account shifts to inherently less materials-intensive and hence less energy-intensive economic activities at high levels of economic development. Second, the model does not allow for improvements in energy end-use efficiency that are not driven by price via the assumed elasticities.

Also, because of its highly aggregated nature, the Nordhaus-Yohe model does not offer guidance as to how public policy might facilitate major changes in energy efficiency or in the mix of energy supplies.

Colombo & Bernardini, 1979

In 1979 Umberto Colombo, director of the Italian Atomic Energy Commission, and Oliviero Bernardini prepared a report to the Panel on Low Energy Growth, of the Commission of the European Communities, exploring the dimensions of a global energy future, to the year 2030, in which the global primary per capita energy consumption rate remained fixed at the 1975 level (36). The study was motivated by the authors' concerns about the great difficulties they felt would be involved in bringing about the energy supply–intensive IIASA scenarios.

For industrialized countries, Colombo & Bernardini stress the importance of ongoing structural shifts toward less materials- and energy-intensive activities, and opportunities for energy-efficiency improvements. For developing countries they call for less centralized settlement patterns with a better balance between centralized and decentralized energy supply systems. The authors conclude that under these conditions an energy future with no increase in the global average level of per capita commercial energy use would be compatible with satisfactory economic growth (Table 1).

The Colombo–Bernardini scenario is based largely on qualitative arguments, supported by some important observations about the structure of economic growth in the longer term, and the opportunities for more efficient use of energy.

Lovins et al, 1981

A 1981 study carried out for the German Federal Environmental Agency by Amory Lovins and his collaborators was perhaps the first to stress the importance of alternative energy strategies in coping with the CO_2 problem (21). This analysis emphasizes renewable, decentralized, and nonelectric energy sources, but sees improving energy efficiency as the single most promising way to alleviate the CO_2 problem.

The authors project a global primary energy use level for the year 2030 of only 5 TW, about half the 1980 global energy use level (Table 1), assuming that (a) GDPs will grow as in the IIASA low scenario; (b) there is a continuing shift to less energy-intensive economic activity in industrialized countries and, after the year 2000, in developing countries; and (c) the fourfold improvement in energy efficiency the authors identified as feasible for West Germany can be extrapolated to the whole world by 2030.

For industrialized countries, the authors' analysis indicates a primary energy use of 2.3 kW per capita for 2030. The economic criterion specified for the selection of the alternative technologies underlying this scenario is that these technologies must be able to provide the same energy service at less cost than that for a nuclear power plant or a synthetic fuel system ordered now. This is an appropriate marginal cost criterion when aggregate demand is rising, and nuclear power plants and synthetic fuel facilities are the supply alternatives to investments in energy efficiency. But if demand were to fall as much as the authors indicate, it is likely that there would be cheaper marginal energy supplies.

For developing countries, the authors' analysis projects a per capita primary energy use of 0.26 kW for 2030—about one fourth the present level. Potential energy savings were estimated by applying to developing countries an overall commercial energy demand reduction factor of four, derived from an analysis of the opportunities for improving energy efficiency in West Germany. Such an analogy is of questionable validity, since a large fraction of the savings achievable in West Germany are in energy-using activities of little relevance to developing countries: space heating (not needed in most developing countries) and use of the automobile (of which there are presently very few in developing countries).

The authors have not shown that development goals could be met with such a low level of per capita energy use.

A Retrospective on Global Energy Projections

Most global energy projections involve expanded use of particular energy sources that would lead to greatly aggravated problems that tend to be accepted as the "price of progress," or they give scant attention to the energy problems of developing countries, or they are concerned largely with projecting the range of plausible future outcomes implicit in historical trends, neglecting the possibilities for influencing events by policy initiatives.

ACCOMMODATION TO GLOBAL RISKS   Most of the studies we have reviewed (the MITEL and Lovins analyses are notable exceptions), and other global energy studies as well, either ignore the major global risks of overdependence on Persian Gulf oil, on fossil fuels generally, and on nuclear power, or they argue for risk accommodation, i.e. social adaptation to these risks.

The IIASA analysts, for example, were well aware of these risks but chose not to deal with them as part of their energy analysis. On Persian Gulf dependency and related "hard energy sources" they argue (20):

> Some of the present energy systems are already "hard" and global in nature. The Persian Gulf is nearly a point source of energy, yielding 1.7 TW-yr/yr, which are supplied across global distances. Discarding hard options and limiting our choices to local resilient forms of energy, as suggested by Meadows and especially by Lovins, would deprive mankind of many of its cheapest energy sources.

Here the IIASA analysts fail to come to grips with the problem of overdependence on Persian Gulf oil by suggesting that the choice is all or nothing (i.e. the high level of Persian Gulf dependence implied by the IIASA scenarios or a shift to "local resilient forms of energy").

They recognize that because of the atmospheric carbon dioxide buildup, major climatic change could occur in a matter of decades, that "the implications of these climatic changes are potentially large," and that (20):

> We are faced with the dichotomy of having a highly disaggregated policial power on the globe and a truly global problem of carbon dioxide buildup. Are we doomed to encounter this dilemma? Probably, yes. Nevertheless, maintaining flexibility in designing energy supply policies to delay the buildup of carbon dioxide beyond certain levels is prudent and advisable.

While recognizing the need to be flexible in energy planning, the IIASA analysis, which looks to a doubling or a tripling of fossil fuel use by 2030, does not reflect this need. The IIASA analysts are not atypical in this regard, and in fact carbon dioxide policy planning today tends to be oriented to strategies of accommodation. For example, one of the major recommendations of the 1983 US Environmental Protection Agency report on the carbon dioxide problem is to expand research on improving our ability to adapt to a warmer climate (37).

Similarly, while recognizing that concerns about proliferation could limit the buildup of nuclear power in the coming decades, the IIASA analysts again chose not to incorporate these concerns in their energy analysis, arguing (20):

> The IIASA Energy Systems Program did not deal with these [nuclear waste and proliferation] issues explicitly for two reasons. The first reason is a pragmatic one. A large number of capable and sizeable groups are already studying these problems, and we did not judge it practical for the relatively small groups of IIASA scientists from many nations to compete with these efforts.... The second and more commanding reason for abstaining is that we regard the problems of nuclear waste handling and proliferation as primarily political ones.... We hope that indirectly we have contributed to the debate on this subject by clarifying the factual lines of the energy problem as a whole. Is civilian nuclear power needed or not? This question was upmost in our minds throughout our investigations as we sought to provide decision- and policymakers with the information they need to make strategic choices.

The tendency to ignore such energy-related risks or stress accommodation, is commonplace among energy analysts who view the energy problem primarily as the relatively narrow engineering challenge of bringing forth new energy supplies.

INADEQUATE ATTENTION TO ENERGY PROBLEMS OF DEVELOPING COUNTRIES   The second striking feature of most global energy studies (the Colombo & Bernardini study being the exception among the studies reviewed here) is the scant attention given to the energy problems of developing countries.

Despite the fact that nearly half the energy consumed by the three fourths of mankind living in developing countries is noncommercial energy, such energy has not been an important consideration in most global energy studies.

To the extent that they do deal with developing countries, global analyses have tended to see as the desirable future for developing countries a retracing of the path taken by the already industrialized countries. Specifically, the tendency has been to focus on centralized energy systems. While the centralized energy technologies emphasized in conventional energy planning may be applicable to certain urban situations, solving the fuelwood problem and meeting the needs of rural industry require quite different strategies, with greater emphasis on decentralized solutions.

And finally, the IIASA, WEC, and IEA/ORAU studies all envisage that the increment in energy use in industrialized countries will be 10–100% more than in developing countries (Table 1) in the period to 2020 or 2030.

The large increases in demand projected for the already industrialized countries would put strong upward pressure on the prices of energy, making it increasingly difficult for developing countries to meet their development goals. We believe a world with such disparities would suffer severe North–South tensions.

NEGLECT OF OPPORTUNITIES FOR POLICY INTERVENTION   Most global energy analysts (the notable exceptions among the studies we have analyzed are the Colombo & Bernardini and Lovins analyses) tend to see their roles as understanding the range of plausible outcomes implicit in historical trends. Here the tacit assumption is that the future is determined to within a relatively narrow range of outcomes, with little or no room for changes induced by human intervention.

This perspective is in fact inherent in any modeling effort that projects future energy demand by using elasticities to characterize the responses of energy consumers to changing prices.

Our Approach

In the rest of this paper we discuss some of the findings of our own study of the global energy problem, the results of which are discussed at length elsewhere (38).

Unlike other analyses, ours has not assumed that there are unavoidable trade-offs between solutions of the energy problem and those of other important global problems. We have taken a normative approach to the global energy problem, seeking to identify and describe global energy strategies that contribute to or are at least consistent with solutions of other important global problems.

We have also given as much attention to the energy problems of developing countries as to those of industrialized countries—focusing on the energy needs of the poor, on the energy aspects of the unemployment problem in developing countries, on noncommercial energy, and on decentralized as well as centralized energy systems.

Finally, we have tried to orient our analysis toward providing a more informed basis for public policy decisions that could change the course of the evolving energy system in ways that would avoid or mitigate the problems that would arise if "business as usual" persisted.

The use of energy is not an end in itself. Energy is useful only insofar as it provides energy services like cooking, lighting, heating, refrigeration, mechanical work, or transport. The focus of our analysis has been on understanding better the role of energy in society by examining in detail the patterns of energy end uses, how and by whom different forms of energy are used today, and how the energy end-use system might look in the future.

This end-use approach was also adopted in the IIASA study, via the MEDEE model, but our analysis makes fuller use of the end-use approach. In the IIASA analysis the end-use approach was adopted largely to provide a data base for historical trends in energy consumption patterns dis-aggregated by end use. We have adopted the end-use approach to explore the feasibility of modifying the evolution of the energy system in ways compatible with the achievement of a sustainable society.

We have been able to identify feasible energy futures far outside the range normally considered in long-term energy projections, because the end-use approach to the energy problem facilitates the discovery of problems (e.g. whether progress is being made in eradicating poverty), trends (e.g. structural shifts in the economy), and opportunities (e.g. more energy-efficient end-use technologies) that are obscured in analyses of the energy problem based on highly aggregated descriptors of the energy system.

This flexibility in the energy system largely reflects the fact that, for many energy-using activities, there seem to be wide ranges in the amounts of energy use required to provide given amounts of energy services, with little corresponding variations in the life-cycle cost, i.e. the discounted present value of all capital and operating costs associated with providing that service.[5]

The use of more energy-efficient technologies for providing energy services involves lower operating costs, but higher initial investments. In many cases, the individual will pay less overall for the more efficient option. Even where the individual does not gain directly, the collective societal benefits of many individual decisions in favor of the more efficient options may be significant. For example, a widespread shift to cars with high fuel economy (Figure 2) would lead to reduced oil imports, a lower world oil price, and enhanced security.

Unfortunately, however, experience shows that energy consumers, for a variety of reasons, tend to avoid purchases that require extra first costs unless the expected "payback" time from operating cost savings is exceedingly short. Put another way, the discount rates implicit in invest-ment choices relating to improving energy efficiency tend to be far in excess of market interest rates (40–42).

The large societal benefits that would result if energy consumer

---

[5] In the case of new natural gas–heated houses in the United States, for example, it has been shown that the energy use for space heating can vary by a factor of three, depending on the degree of investment in energy efficiency, although the life-cycle costs for the various technological options involved vary by less than 10% from the mean (27). Similarly, in the case of the automobile, Figure 2 shows that the total cost per km of owning and operating a car changes very little over the entire range of fuel economies, 8–3 liters per 100 km (30–80 mpg), that can be achieved with present automotive technology (39).

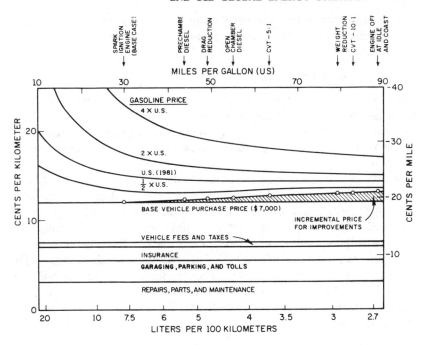

COST OF DRIVING (1981 U.S. CENTS)

Figure 2 The total cost of driving as a function of the average fuel efficiency, for four fuel prices. The shaded area represents the added initial cost for fuel economy improvements. Based on a computer simulation, with the base car taken to be a 1981 Volkswagen Rabbit with a gasoline engine, for which the fuel economy is 7.9 liters per 100 km (30 mpg). Source: (39).

investment decisions were instead characterized by discount rates close to market interest rates provides a powerful motivation for policy initiatives aimed at reducing these discount rates to more reasonable levels. Later we shall discuss some of the policy choices relating to such interventions.

Such considerations lead us to believe that, in looking to the long-term future, it is less important to try to predict the energy future than to understand the full range of technically and economically feasible energy futures, including those that require market interventions.

THE DEMAND FOR ENERGY SERVICES

Two of the most useful applications of the end-use approach have been in understanding the extent to which energy planning is and can be effective in supporting development goals in developing countries, and in understanding ongoing shifts away from energy-intensive activities in industrialized countries.

Energy and Development

ENERGY FOR MEETING BASIC HUMAN NEEDS   In the 1950s, when development strategies were first being articulated, it was generally felt that maximizing economic growth was the best way to eradicate poverty, but experience shows that the benefits of rapid economic growth have not trickled down to the poor.

Rapid growth is necessary for successful development, but not sufficient. A more effective way of dealing with poverty is by allocating resources directly to the satisfaction of specified basic human needs, with emphasis on the needs of the poorest, ensuring that minimum standards for nutrition, shelter, clothing, health, and education are met (8). There is no empirical evidence that targeting the satisfaction of basic human needs would lead to slower economic growth (9), and there are theoretical grounds for believing that a basic human needs policy would lead to higher growth because of the resulting increase in worker productivity (10). The allocation of sufficient energy to basic needs programs is of crucial importance in energy planning.

ENERGY AND EMPLOYMENT GENERATION   Employment generation is a development challenge closely related to the eradication of poverty. Because technologies used for industrialization in developing countries today are far more labor-saving than the technologies used at the similar stage of development in the now industrialized economies, the challenge is daunting. While there is no going back to the primitive industrial technologies of yesteryear, it is desirable to pursue those development strategies most capable of providing employment, which has acquired the status of a basic human need. Energy is a key factor in addressing this problem, because energy and labor tend to be substitutable inputs for industrial activity (43).

The importance of employment generation has major implications for the industrial mix and the choice of technologies for a given mix, both of which are often shaped by public policies. In countries where labor is cheap, overall production costs would often be lower if labor-intensive technologies and industries were emphasized, but planners are often tempted to use subsidies to attract large-scale, energy-intensive industries that provide little direct or indirect employment.

NONCOMMERCIAL ENERGY   While there is poverty in urban slums, most of the poor live in rural areas, and a significant fraction live outside the market economy. The importance of rural poverty reflects a population distribution between rural and urban areas very different from that in industrialized countries. In the latter only 30% of the population live in rural areas, whereas 70% live in rural areas in developing countries (2).

In rural areas people depend for energy largely on biomass, mainly fuelwood, used mostly for cooking. But in parts of the developing world fuelwood consumption exceeds the rate of regeneration. Fuelwood-gathering involves many hours of drudgery each day, particularly for women and children. The ecological effects of deforestation created by excessive fuelwood use amplify this human toil.

The vast scale of rural poverty, the weakness of market forces in being able to deal with it, and the central importance of cooking and its relation to the fuelwood crisis are all factors that must figure prominently in energy planning efforts.

DECENTRALIZED AS WELL AS CENTRALIZED ENERGY   Inadequate attention to the problems of rural areas is causing the rural poor to flee to urban slums, which offer access to services that are unavailable in rural areas. The urban population in developing countries is growing more than twice as rapidly as the population as a whole (2). The urbanization trend is becoming increasingly unmanageable, however, as the crowded and polluted cities are unable to offer enough jobs to support the number of job seekers. Urban migration could be slowed and the cities made more livable thereby, if living conditions were improved in rural areas. In particular, rural-based, labor-intensive industries are needed.

Providing the energy for such industries requires an emphasis in energy planning in which centralized energy production, which is essential for meeting urban energy needs, is complemented by decentralized energy production in rural areas, where it is often uneconomic or otherwise impractical to provide energy services from centralized sources. To this end biomass used renewably is a promising feedstock for providing solid, gaseous, or liquid energy carriers or for making electricity in small-scale operations for many rural areas.

It would be highly fortuitous if significant contributions to meeting development goals resulted from conventional energy planning, which mainly involves the engineering challenge of expanding conventional, centralized energy supplies to the extent suggested by energy–GDP correlations. Just as the trickle-down approach to economic development has failed to improve the lot of the poor, we fear that an energy trickle-down approach to energy development is likely mainly to expand the energy services available to the affluent, while leaving the poor little or no better off.

To be supportive of development efforts, energy planning in developing countries must emphasize the provision of energy for the satisfaction of basic human needs, employment generation, cooking, and the general problems of rural areas, seeking an appropriate balance between centralized and decentralized energy sources.

Structural Changes in the Industrialized Countries

Changing consumer demand patterns in industrialized countries are reflected in the growing importance of services production at the expense of goods production and, within the goods-producing sectors, the growing importance of fabrication and finishing. Both of these shifts are in the direction of less energy demand per dollar of value added.

THE GROWING IMPORTANCE OF SERVICES   The shift to services (e.g. finance, insurance, education, and communications, as well as marketing, information, medical, and recreational services) in industrialized countries has been proceeding for decades, as is evident from the long-term trends in employment in Sweden and the United States (Figure 3). In the early years of industrialization the shares of employment accounted for by both manufacturing and services grew at the expense of employment in

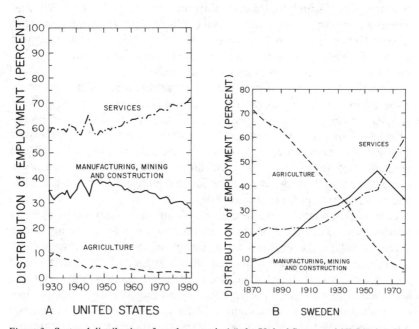

Figure 3 Sectoral distribution of employment in (*A*) the United States and (*B*) Sweden. For the United States the employment measure is the number of full-time equivalent employees. For Sweden it is number of employees working more than half time.

agriculture. More recently, services have grown at the expense of manufacturing, mining, and construction as well (Figure 3).

The shift to services is also reflected in the slower growth of goods production. The output of the goods-producing sector [measured by gross product originating (GPO), or value-added] grew just 0.83 and 0.60 times as fast as GNP in the period 1970–1980 for the United States and Sweden, respectively.

THE GROWING IMPORTANCE OF FABRICATION AND FINISHING   Within the goods-producing sector there is a continuing shift away from the energy-intensive processing of basic materials toward fabrication and finishing activities—processes which involve much fewer inputs of energy per dollar value-added than the processing of basic materials (Figure 4).

Consider the situation in the United States, which has a largely closed economy, so that the consumption of goods and services is approximately equal to production in most sectors. The industrial sector here can be disaggregated into mining, agriculture, and construction (MAC); the basic materials processing (BMP) subsector of manufacturing; and "other" manufacturing. In 1978, these three sectors accounted respectively for about 25%, 25%, and 50% of industrial output and for 15%, 73%, and 11% of final energy use in industry; they required 3, 14, and 1 units of energy per dollar of output. Similarly, in Sweden, these sectors accounted for about 35%, 37%, and 28% of industrial output; 10%, 82%, and 8% of final energy use in industry; and 1, 7.5, and 1 units of energy per dollar of output, respectively. Thus while "other" manufacturing, which involves the fabrication and finishing of basic materials, is economically very important in both countries, the BMP subsector of manufacturing dominates energy use.

Shifts in output among these sectors toward less materials-intensive activities have been pronounced (Figure 5). In the 1970s the rate of growth of industrial output (GPO) for fabrication and finishing activities in the United States was 4.3% per year on average, compared to 3.0% per year for the BMP subsector and 1.2% per year for MAC. Similarly, in Sweden there is strong evidence of the declining importance of basic materials as contributors to economic growth (44): fabrication and finishing activities grew in this same period at an annual average rate of 2.0%, compared to 1.1% for industry as a whole, 1.2% for the primary metals sector, and a 1.4% rate of decline for the cement industry.

There is strong evidence that the shift towards fabrication and finishing is associated with saturation in the use of materials (e.g. a cessation of growth in per capita consumption), as indicated in a recent analysis of the long-term history and future outlook for a representative sampling of basic materials in the United States and some other industrialized market

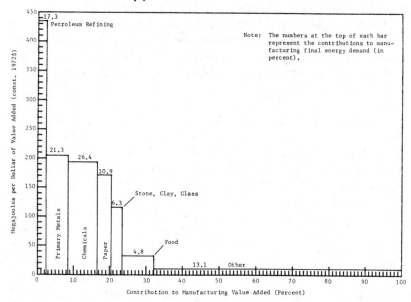

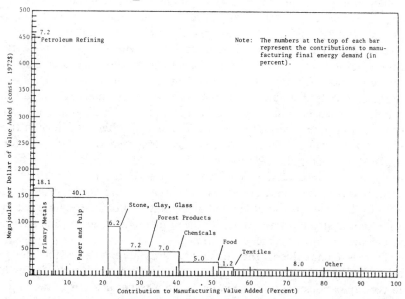

Figure 4 Final energy intensity vs manufacturing value added in 1978 for (*A*) the United States and (*B*) Sweden.

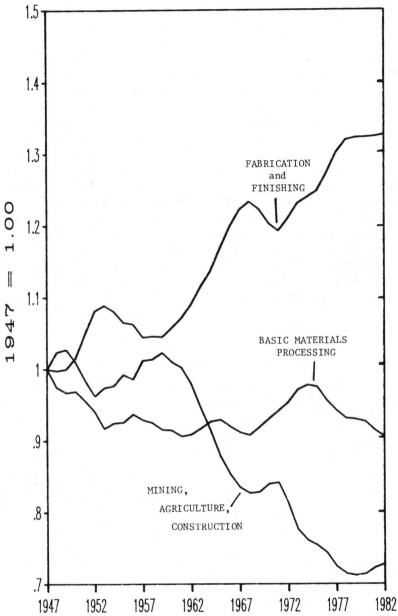

Figure 5 Trends for the United States in the mix of industrial output (gross product originating) for mining, agriculture, and construction; the basic materials processing subsector of manufacturing; and the fabrication and finishing subsector of manufacturing, all relative to total industrial output.

economies (Figure 6) (45). For both traditional materials (steel, cement, and paper) and modern materials (aluminum, ethylene, ammonia, and chlorine) per capita consumption stopped growing in the United States in the 1970s and in most cases actually began to decline. Similar trends have been found for the same materials in France, West Germany, and the United Kingdom (46). The trends appear to be due to a combination of factors, including more efficient use of materials in providing essentially the same materials services, materials substitution, and market saturation. In all cases, the outlook for volume growth in consumption was found to be poor, largely because of market saturation. Only markets for high-value-added, specialty products appear promising. It is unclear whether growth in these markets will be adequate to offset the declines in markets for high-volume bulk products, but the evidence for at least saturation is strong.

Shifts to fabrication and finishing, which typically require an order of magnitude less energy per unit of output than the processing of basic materials, can have profound effects on industrial energy use. For the United States, such shifts accounted for an annual rate of decline in industrial energy use per dollar of GNP of 1.6% during 1973–1984, out of the total annual rate of decline of 3.6% in this period (46).

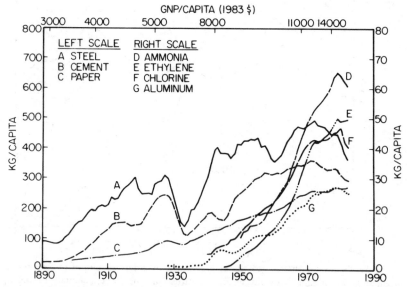

Figure 6 Basic materials use in the United States. From (45). The data are five-year running averages for apparent consumption (production plus net imports, corrected for stock changes).

The trends toward reduced materials intensity (and hence usually energy intensity) of the economies of highly industrialized countries began well before the onset of the energy crises of the 1970s. These trends can be expected not only to continue but perhaps also to accelerate, as a response to the sharp increases in energy prices that have taken place. This response may be complemented by the increased use of technologies that make much more efficient use of energy supplies in providing energy services.

OPPORTUNITIES FOR ENERGY PRODUCTIVITY IMPROVEMENT

Opportunities exist for large improvements in energy productivity for all major energy-intensive activities—space conditioning, lighting, cooking, and major household appliances; automobile, air, and truck transport; and the production of basic materials. The energy price increases of the 1970s have led to the commercialization of much more efficient energy-using technology than that now in wide use and to major R&D efforts that will lead to even more efficient technology in years to come. We illustrate the possibilities with examples from each energy-using sector, for both industrialized and developing countries.

Industrialized Countries

RESIDENTIAL   Space heating, accounting for 60–80% of final energy use in residential buildings in industrialized countries, warrants focused attention, and indeed has been the focus of ongoing residential energy conservation programs. There are two major routes to increase energy productivity for space heating: improvements in building shells and in heating equipment.

Shell improvements   Large improvements in the energy performance of the building shell are possible for both new construction and existing housing.

Table 2 lists the energy performance of various groups of new houses that incorporate major energy saving features. In Sweden there has been considerable experience in building thermally tight houses. For example, the energy performance (corrected for climate and floor area) of houses routinely built to conform to the 1975 Swedish building standard would be nearly as good as the performance of some of the better houses that have been built in the United States (e.g. the Minnesota and Oregon houses shown here). Houses that perform considerably better than the 1975 standards are being built routinely in Sweden, as indicated by the Skane examples in Table 2.

While one would expect the very energy-efficient houses to be very expensive, there is a growing body of evidence suggesting that the net extra cost of a very energy-efficient house may not be very large in comparison to the cost of a conventional house, because the added costs of shell improvements may be offset to a considerable degree by savings in heat generation and distribution systems (27). Particularly good data on the cost

Table 2 Space heat requirements in single family dwellings (kJ per m^2 per degree-day)[a]

United States	
Average, housing stock[b]	160
New (1980) construction[c]	100
Mean measured value for 97 houses in Minnesota's Energy Efficient Housing Demonstration Program[b]	51
Mean measured value for 9 houses built in Eugene, Oregon[d]	48
Calculated value for a Northern Energy Home in New York City area[e]	15
Sweden	
Average, housing stock[f]	135
Homes built to conform to the 1975 Swedish Building Code[g]	65
Mean measured value for 39 houses built in Skane, Sweden[h]	36
Measured value, house of Mats Wolgast, in Sweden[i]	18
Calculated value for alternative versions of the prefabricated house sold by Faluhus[j]	
Version #1	83
Version #2	17

[a] The required output of the space heating system (i.e. heat losses less internal heat gains less solar gains) per unit floor area per heating degree-C-day (base 18°C).

[b] See (27).

[c] As reported by the National Association of Home Builders (94).

[d] See (94).

[e] The Northern Energy Home (NEH) is a super-insulated home design sold in New England and based on modular construction techniques with factory-built wall and ceiling sections mounted on a post and beam frame. The energy performance was estimated using the Computerized Instrumented Residential Audit computer program (CIRA) (personal communication from D. Macmillan of the American Council for an Energy Efficient Economy, Washington, DC). The house has 120 m^2 of floor area; triple glazed windows with night shutters; 20 cm (23 cm) of polystyrene insulation in the walls (ceiling); 0.15 ACH natural ventilation plus 0.35 ACH forced ventilation via 70% efficient air-to-air heat exchanger; and an internal heat load of 0.65 kW. The indoor temperature is assumed to be 21°C in the daytime, set back to 18°C at night.

[f] In 1980 the average values for fuel consumption, floor area, and heating degree-days were 98.5 GJ, 120 m^2, and 4474 degree-days respectively, for oil-heated single family dwellings (95). To convert fuel use to net heating requirements, a furnace efficiency of 66% is assumed.

[g] According to (95), and assuming a single-story house with 130 m^2 floor area, no basement, electric resistance heat, an indoor temperature of 21°C, and 4010 degree-C-days.

[h] The average for 39 identical, 4 bedroom, semidetached houses (112 m^2 of floor area; 3300 degree-days).

[i] The Wolgast house has 130 m^2 of heated floor space, 27 cm (45 cm) of mineral wool insulation in the walls (ceiling), quadruple glazing, low natural ventilation plus forced ventilation via air preheated in ground channels. Heat from the exhaust air is recovered via a heat exchanger. See (44).

[j] The Faluhus has a floor area of 112 m^2. The more energy-efficient Version #2 (with extra insulation and heat recuperation) costs 3970 Swedish kronor (Skr) ($516) per m^2 compared to 3750 Skr ($488) per m^2 for Version #1. The electricity savings for the more efficient house would be 8960 kWh per year. The cost of saved energy (assuming a 6% discount rate and a 30-year life for the extra investment) would be 0.20 Skr per kWh (US $0.026 per kWh). For comparison, electricity rates for residential consumers in Sweden consist of a large fixed cost independent of consumption level [about 1200 Skr ($156) per year] plus a variable cost of 0.25 Skr per kWh ($0.032 per kWh).

of very energy-efficient houses are provided by the two different versions of Swedish prefabricated houses offered by Faluhus. These two versions are identical except for their energy performance characteristics, the more energy-efficient version being one of the most energy-efficient houses available. The associated cost of saved energy (the annualized cost of the added investment divided by the annual energy savings) is less than the present Swedish price of electricity, even though the present hydropower-based Swedish electrical rates are low and far below marginal costs of electricity from new sources [see note (j) of Table 2].

Because the building stock turns over so slowly, energy use for space heating will be dominated by existing buildings for decades. Although the opportunities in retrofitting buildings are not as large as for new construction, a great deal can still be done.

For example, the Swedish government's ten-year plan for retrofitting buildings aims at a reduction of energy use in the 1978 building stock by about one third. The plan was initiated in 1978, and is optimized for an energy price 30% below 1981 prices. In the United States, the gas utility-based Modular Retrofit Experiment (MRE) demonstrated state-of-the-art possibilities for exploiting low-cost unconventional opportunities that are identified via sophisticated diagnostic equipment (47). In the MRE, the measured savings associated with the one-day two-person ("house doctor") visit averaged 19% of the gas use associated with space heating. Subsequently, more conventional shell modification retrofits raised the average fuel savings to 30%, for an average total investment of about $1300; the associated real internal rate of return in fuel savings was nearly 20%, for an assumed life-cycle gas price of $8 per GJ (the heating oil price in 1982) (27).

The achievements demonstrated in the MRE do not represent the limit of what can be achieved with shell improvements of existing dwellings. One important experiment exploiting additional unconventional retrofit opportunities resulted in a two thirds energy savings in a US house that, prior to modification, was regarded as thermally tight by US standards (48). Over a period of several decades the energy savings potential of retrofits is generally much greater than what can be achieved immediately. Over time various structural modifications of houses will be needed, and some important energy reducing shell improvements are much more cost-effective if carried out in conjunction with such structural changes (e.g. putting in energy-efficient windows when new windows are needed anyway) than if carried out for the energy savings alone. Moreover, new technical opportunities for energy demand reductions will continually be developed.

Energy conversion equipment   The last several years have shown that the efficiencies of space heating equipment can be much improved. For gas furnaces conversion efficiencies have increased in the United States from an average of about 69% in new units sold in 1980 to more than 90% in new "condensing furnaces," so-called because the improved performance involves extracting heat from flue gases past the point where the water vapor condenses out (27). Heat pumps with coefficients of performance (COPs) up to 2.6 for air-to-air units and up to 3 for water-to-air or water-to-water units have also come on the market. For comparison the average COP of heat pumps in the existing US stock is less than 2 and that of resistive electric heating units is 1 or less.

Total residential final energy use An indication of the overall potential for energy savings is given in Table 3, which shows per capita final residential

Table 3 Final energy use in the residential sector (watts per capita)

End use	Average households at present		All electric, 4-person, households with the most efficient technology available in 1982–1983[d]	
	United States (1980)[a]	Sweden (1978–1982)[b,c]	United States	Sweden[b]
Space heat	890	900	60[e]	65[f]
Air conditioning	46	—	65[g]	—
Hot water	280	180	43[h]	110[i]
Refrigerator	79	17	25	8
Freezer	23	26	21	17
Stove	62	26	21	16
Lighting	41	30	18[j]	9[j]
Other	80	63	75	41
Total	1501	1242	328	266

[a] The total consists of 24% electricity and 76% fuel.

[b] For details, see (49).

[c] This total consists of 28% electricity and 72% fuel. 50% of the electricity is for appliances and 50% is for heating purposes.

[d] With 100% saturation for the indicated appliances, plus dishwasher, clothes washer, and clothes dryer.

[e] For the average-sized, detached, single-family house (150 m² of floor space); and average US climate (2600 degree-C-days); a net heating requirement of 50 KJ/m²/degree-day (Table 2), and a heat pump with a seasonal average COP of 2.6 (that of the most efficient air-to-air unit available in 1982).

[f] For a Faluhus (Table 2) in a Stockholm climate (3810 degree-C-days). This house uses a heat exchanger to transfer heat from the exhaust air stream to the incoming fresh air.

[g] For the average cooling load in air-conditioned houses in the United States (27 GJ per year) and a COP of 3.3 (the COP on the cooling cycle for the most efficient heat pump available in 1982).

[h] For 59 liters of hot water (at 49°C) per capita per day (corresponding to 910 kWh/year/capita) and the most efficient (COP = 2.2) heat pump water heater available in 1982.

[i] Per capita hot water energy use is assumed to be 1000 kWh per year, provided by a resistance heater. Ambient air-to-water heat pumps are not competitive at the low Swedish electricity prices.

[j] Savings are achieved by replacing incandescent bulbs by compact fluorescent bulbs having an efficacy four times as large.

energy use, first for average households in the United States and Sweden at present, and second for hypothetical all-electric households having a full set of major energy-using amenities and the most efficient technologies commercially available in 1982. While these hypothetical households have a higher level of amenities than average households today, they would use only about 300 W per capita, much less than present levels of household energy use. With the more efficient technologies under development, energy use could be reduced even further.

COMMERCIAL BUILDINGS   The energy budgets of commercial buildings, like those of residential buildings, are dominated by the requirements for space conditioning; but for commercial buildings shell improvements other than for daylighting and sun control are much less important. Most of the opportunities for improved energy performance involve the use of more energy-efficient equipment and a better matching of energy supplies to energy service requirements via better control technology. The importance of controls is suggested by the fact that in US commercial buildings energy is often wasted on heating air in summer and cooling it in winter.

While new commercial buildings are less energy-intensive than the existing stock in both the United States and Sweden, the energy performance of some new buildings is far better than that of typical new construction (Table 4). Perhaps the most energy-efficient commercial building constructed in the late 1970s is the Folksam Building in Farsta, near Stockholm, Sweden. With an ordinary design the building would become overheated (from lighting and other internal heat loads) and would require cooling in the daytime and heating at night. But with the Folksam design excess heat produced in the daytime is stored for use at night (or for heating up the building in the morning). Storage is accomplished via the "Thermodeck" concept, which involves passing the office ventilation air through long tubular cores in the massive concrete floor slabs on its way to the offices. With this storage scheme, the air temperature rise in the offices during the day is only about 2°C, so that cooling is unnecessary. In summer the system stores heat in the slabs during the day, as in winter, but the slabs are cooled with outside air at night.

The most energy-efficient commercial building in Sweden today is the Harnosand Building in northern Sweden, constructed in 1981. By using the Thermodeck principle, preheating ventilation air with solar panels, and using microprocessor controls to better match energy supply and demand, the building uses only about half as much energy per square meter as the Folksam Building (Table 4).

Improved energy management, involving little or no capital investment (e.g. night setbacks of thermostats, adjustments in ventilation to better match needs) typically leads to savings of 20–30% in existing commercial

Table 4 Site energy intensity factors for commercial buildings (GJ per m^2 per year)

	Fuel use	Electricity use	Total
United States			
Average 1979 building stock[a]	0.82	0.49	1.31
Current US practice[b]	0.16	0.57	0.73
Am. Inst. Architects (AIA) Res. Corp. redesigns[b]	0.07	0.40	0.47
AIA life-cycle cost minimum designs[b]	0.04	0.28	0.32
Enerplex South, Princeton, NJ[c]	—	0.31	0.31
Sweden			
Average 1982 building stock[d]	0.66	0.38	1.04
Swedish norm for new construction[b]	0.57	0.19	0.76
Folksam Building, Farsta[e]	0.07	0.39	0.46
Harnosand Building, Harnosand[f]	0.12	0.13	0.25

[a] For an average of 2700 heating degree-C-days. See (51).

[b] Table 1.12 (p. 39) and Figure 1.61 (p. 156) in (52).

[c] For 2700 heating degree-C-days. Calculated, not measured values. See (96).

[d] Consumption corrected to normal weather (4010 heating degree-C-days). (Personal communication from L.-G. Carlsson, Natl. Energy Admin., Sweden, 1984.)

[e] Measured values for the representative period Dec. 1978 to Dec. 1979 (3810 heating degree-C-days). "Fuel consumption" is the energy actually delivered by the district heating system (97).

[f] Measured values for 4600 heating degree-C-days. (Personal communication from K-Konsult, a Swedish architect-engineering consulting firm, 1984.)

buildings in Sweden (49). In the United States, the average measured savings in 184 buildings was 23%, and the corresponding cost of saved energy for 56 buildings where cost data were available was $2.8 per GJ (1982 $), assuming a 10-year retrofit life and a 10% real discount rate (50). These savings fall short of the economic potential, however, since there is probably much more that can be done for a cost of saved energy less than the average price of energy, some $8 per GJ for US commercial buildings in 1979 (51). In a survey of experienced architects and engineers conducted by the Solar Energy Research Institute the consensus judgment was that a 50% reduction in energy use per square meter on average was an achievable target for US commercial buildings by the year 2000 (52).

TRANSPORTATION   In 1982 transportation accounted for 53% of all oil consumption in the OECD nations (53). Automobiles and light trucks, which accounted for over 60% of all oil use in transport, warrant special attention.

There are opportunities to improve the fuel economy of automobiles and light trucks from the present typical values of 12–8 liters per 100 km (20–30 mpg) to the range 4–2.3 liters per 100 km (60–100 mpg) in the decades immediately ahead, both by increasing engine and drive train efficiency and by reducing vehicle weight and aerodynamic and rolling resistances.

Engine efficiencies are typically low. For example, the model year 1981 gasoline-powered Volkswagen Rabbit with manual transmission has an average engine–drive train efficiency of only 13.5% in converting fuel to mechanical energy at the wheels (39). One possibility for improving the efficiency is to shift to a diesel engine. The diesel version of the Rabbit has an energy performance on the US EPA combined driving cycle of 5.3 liters per 100 km (45 mpg), compared to the 7.9 liters per 100 km (30 mpg) for the gasoline version. The cost of saved energy for this $525 engine switch, assuming a 10% real discount rate and annual savings of 420 liters (110 gallons) of fuel for the average driving distance of about 16,000 kilometers (about 10,000 miles) per year, would be just $0.22 per liter ($0.78 per gallon) of gasoline equivalent.

Additional improvements in the Volkswagen Rabbit, based on proven technology, such as reducing aerodynamic drag and rolling resistance, shifting from a prechamber to a direct-injection diesel, using a continuously variable transmission, reducing weight, and adding an engine-off feature during coast and idle, would improve its fuel economy to 2.6 liters per 100 km (89 mpg) (39), without increasing the total cost of owning and operating the car (Figure 2). Many of these features have been incorporated in prototypes, of which the Volkswagen Experimental Car 2000 and the Volvo Light Component Project 2000 (LCP 2000) car with fuel economies of 3.8 and 3.6 liters per 100 km (62 and 65 mpg), respectively, are notable examples (Table 5).

Among further improvements possible with advanced technology, the efficient adiabatic diesel engine is especially promising. The Volvo LCP 2000, with a three-cylinder, heat-insulated, direct-injection, turbocharged engine is an advance in this direction (54).

Researchers at Ford Motor Company describe what an "average" vehicle in the late 1990s could be like (55) as:

> ... a four- or five-passenger vehicle in the 2000-pound (900 kg) inertia weight class with an aerodynamic drag coefficient of 0.20 or less. ... Electronics would control a turbocharged, ceramic, adiabatic diesel engine and continuously variable transmission to provide smooth effortless performance and fuel economy in excess of 100 miles per gallon on the highway....

One concern raised about super-mpg cars is their safety. That a lightweight car need not be unsafe, however, is indicated by the safety features built into Volvo's LCP 2000 (54). Moreover, researchers at the Cummins Engine Company and the NASA Lewis Research Center have described the design of a "heavy" super-mpg automobile: a 1360-kg (3000-lb) passenger car, using an adiabatic diesel engine with a turbo-

Table 5 Fuel economy for 4-passenger automobiles[a]

Car	Status	Fuel economy [liters/100 km (mpg)]	Engine power (kW)	Curb weight (kg)	Drag coefficient
1981 Volkswagen Rabbit (gasoline)	commercial	7.9 (30)	55	945	0.42
1981 Volkswagen Rabbit (diesel)	commercial	5.3 (45)	39	945	0.42
Honda City Car (gasoline)	commercial	5.0 (47)	46	655	0.40
Volkswagen Experimental Car 2000[b]	prototype	3.8 (62)	33	786	0.25
Volvo LCP 2000[c]	prototype	3.6 (65)	66	707	0.27
Volvo LCP (potential)[d]	design	2.75 (85)	—	—	—
Cummins/NASA Lewis Car[e]	design	3.0 (79)	51	1360	—
Pertran Car (diesel version)[f]	design	2.2–2.4 (100–105)	—	545	0.25

[a] The 1978 world average automobile fuel economy was 13 liters per 100 km (18 mpg).

[b] 3-cylinder, direct injection, turbocharged diesel engine; more interior space than the Rabbit; engine off during idle and coast (98).

[c] 2-passenger + cargo or 4-passenger; 3-cylinder, heat-insulated, direct-injection, turbocharged engine with multifuel capability (54).

[d] The Volvo LCP 2000 plus CVT and engine off during idle and coast features. (Personal communication to Frank von Hippel from Rolf Mellde, Volvo Car Corporation, Goeteborg, Sweden, 1985.)

[e] 4–5 passenger, 4-cylinder, direct-injection, spark-assisted, multifuel capable, adiabatic diesel with turbo-compounding; 1984 model CVT; Ford Tempo body (56).

[f] Prechamber diesel engine with supercharger; CVT; flywheel for energy storage in braking (99).

compound bottoming cycle, with a fuel economy of 3.0 liters per 100 km (79 mpg) (56).

Another concern is the air pollution from diesel engines. One solution is that spark-assisted versions of diesel engines (e.g. the Cummins/NASA Lewis car) would be able to use a wide range of fuels including gasoline and methanol (57), without loss of efficiency (58).

For trucks it appears feasible to reduce energy use per tonne-km by half, relative to the present US average for long haul trucks, via a combination of measures such as the development of adiabatic diesel engines and bottoming cycles, reductions in aerodynamic drag, and tire improvements (59). Additional savings might be achieved through increased load factors.

For passenger aircraft, the high cost of fuel, accounting for as much as 30% of the operating costs of US commercial airlines, provides a strong incentive to implement fuel economy improvements. It appears feasible to reduce fuel intensity by 50% relative to 1977 levels in the United States by a combination of measures such as completing the shift to wide-bodied jets with high-bypass turbofan engines, improving wing design, and reducing weight through use of composites (59).

INDUSTRY   The energy price shocks of the 1970s resulted in relative price increases for industry much larger than those for other energy-consuming sectors. Within the industrial sector, many of the energy-intensive basic materials processing industries, which accounted for 70% of industrial energy use in OECD countries in 1979, experienced much larger relative energy price increases than the average for all industry. A measure of the relative economic impacts of high energy prices on different manufacturing activities is the ratio of energy costs to value added. In the United States this ratio in 1980 ranged from 15 to 76% for various basic materials processing sectors and subsectors, but averaged only 3% for other manufacturing activities (60).

Economic conditions thus provide a powerful motivation for seeking improvements in energy productivity. Fortunately, as in other sectors, there is a wide range of technical opportunities for making such improvements. It is useful to classify these opportunities as good housekeeping measures, fundamental process changes, product changes, and new energy conversion technologies.

Good housekeeping   Improved management measures such as plugging leaks in and insulating steam lines or turning off energy supply systems when not in use have accounted for much of the energy savings in industry since the energy price increases. The potential for energy productivity improvements here are typically of the order of 10–20% at little or no capital cost.

Process innovation   The objective of process innovation is not to reduce energy demand or minimize the cost of providing energy services but to minimize the total cost of production. The history of modern industry tells us that new processes for providing familiar products are most likely to overcome resistance to technical change and displace existing processes if they offer opportunities for simultaneous improvements in several factors of production—reduced labor, capital, materials, and energy requirements (61, 62). This has been such a powerful phenomenon that energy requirements have often been reduced in the process of technological innovation even during periods of declining energy prices.

New processes are continuously being developed. Important R&D areas from which industrial process innovations are likely to continue to emerge include powder metallurgy, plasma metallurgy, computer-assisted design and manufacturing, laser processing of chemicals, biotechnology, membrane separation technology, and the use of microwaves for localized rather than volumetric heating. Improvements in all such areas will make it possible to do more with less—to produce more value added with lower inputs of production factors, including energy.

Consider steel. About five sixths of all steel production takes place in industrialized countries, where it accounts for a major fraction of all manufacturing energy use; e.g. one sixth in Sweden and one seventh in the United States. The theoretical minimum amount of energy required to produce a tonne of steel from iron ore is 7 GJ (63); and 0.7 GJ is required if steel is produced from scrap. At present, steel-making in Sweden and the United States is based on a 50–50 mix of iron ore and scrap, so that the thermodynamic minimum is about 3.9 GJ per tonne of raw steel. For comparison, the actual energy used to produce raw steel was 27 GJ per tonne in the United States in 1979 and 22 GJ per tonne in Sweden in 1976, where the evaluations are done on a comparable basis (6).

The potential for practical energy productivity increases in steel production is illustrated with iron-making processes now under development in Sweden—Plasmasmelt and Elred. In both cases, the objective is to reduce overall costs and reduce environmental problems: by using powdered ores (concentrates) directly, without agglomeration of the ore into sinter or pellets; by using ordinary steam coal instead of the much more costly coke; and by integrating what are now individual operations.

Energy requirements are 8.7 GJ/tonne (of which 4.2 GJ is electricity) for Plasmasmelt and 11.9 GJ/tonne (1.3 GJ electricity) for Elred (44). The Plasmasmelt process would be especially appealing in coal-poor, hydropower-rich countries, while in countries where electricity prices are high (e.g. the United States) it may be preferable to focus on less electricity-intensive processes like Elred or other iron-making processes that produce not hot molten metal but rather solid direct-reduced iron. Direct reduction processes convert iron ore in various forms into sponge iron at temperatures much below the melting point, using a variety of reductants other than metallurgical coke. Other promising advanced processes that attempt to integrate now separate operations to save on capital, labor, and energy costs include continuous casting, direct steel-making, and dry steel-making (64).

The dry steel-making process, which yields a final product in powder form (and avoids melting), holds the promise of very low capital costs and suitability for small-scale operations, as well as major energy savings compared to conventional processes (64).

Product change Product design can lead to reduced energy use if it facilitates materials recycling; this is especially important for metals. Recycled steel requires only 35% as much energy to become finished steel relative to iron ore, and recycled aluminum less than 10% as much. Product design can also lead to reduced energy use if it extends product life—facilitating repair, remanufacture, and reuse.

Reducing product weight often leads to energy savings. But a shift to lightweight materials can increase manufacturing energy use, for example, when aluminum is substituted for steel in cars. But such increases are usually offset by much greater reductions in operational energy use, as would be the case with the Volvo lightweight car, the LCP 2000 (54).

Some of the most exciting possibilities for energy-saving substitutions involve developing entirely new primary materials that may be more appropriate for the new era of high cost energy. One candidate is a "super-cement" now under development.

Ordinary cement is a primary building material that can be made from commonplace resources—limestone, clays, and sands—and has a relatively low energy intensity; it takes 6 times as much energy to produce a cubic meter of polystyrene and 29 times as much to produce a cubic meter of stainless steel. It would seem desirable therefore to be able to substitute cement for such energy-intensive primary materials. The substitution possibilities are quite limited today, largely because cements tend to have low tensile strength and low fracture toughness. But recent research and development has led to the discovery of ways to improve cements dramatically in these respects (65).

The new super-cement is a macro-defect-free (MDF) cement, which differs from ordinary cement in that the pores in the cement are reduced from millimeter to micrometer size. This dramatically increases tensile strength and fracture toughness; super-cement can be made highly resistant to impact by reinforcement with fibers. These fibers can be inexpensive organic materials because cement is manufactured at low temperatures. Strips of fiber-reinforced MDF cement can be made pliable enough to bend like strips of metal (65).

New energy conversion technology   While process and product change innovations often generate severalfold energy savings, the technologies involved tend to be of limited applicability. There are also energy savings opportunities involving energy conversion devices, which typically yield smaller energy savings of 20–50%, but which are important in aggregate because of their wide applicability throughout industry. The possibilities here include more insulation on furnaces, radiation reflectors, heat recovery devices, induction heating of metals, microwave heating, cogeneration (66), and better mechanical drive systems. To illustrate the possibilities here, we briefly discuss mechanical drive technology.

Mechanical drive accounts for a major share of industrial electricity use in industrialized nations. In both the United States and Sweden, for example, industrial motor drives accounted for about three fourths of total industrial electricity use. A study of mechanical drive systems in British light industry has shown that typically less than half of the input power to a plant is delivered to the tool tip, while about one third is lost in the gearboxes and in throttling (67).

Constant-speed, oversized motors are typically used to move gases, and the gas flow is regulated by baffles; similarly, throttling valves are used for controlling liquid flows. The matching of power demand to supply via throttling involves considerable energy waste. Alternating current variable speed drive (VSD) technology is an energy-efficient alternative of wide applicability to variable load situations involving pumps, compressors, fans, etc. One estimate is that it would be economical to associate half of alternating-current motor use in the United States with VSD controls by 1990, with an average savings of 30% for the motors affected (68). Paybacks of 1–3 years are possible in a wide variety of applications. Due to improvements in solid state technology, reliability of VSD devices has improved in recent years, and costs have tumbled—trends that can be expected to continue.

Developing Countries

THE MODERN SECTOR   Most of the opportunities for more efficient energy use relevant to industrialized countries are relevant to the modern sectors of developing countries as well—in buildings (except for space heating, which is not needed in most areas), transportation, and industry. Such opportunities may often be even more attractive to developing countries, for the following reasons.

First, capital generally tends to be scarcer in developing countries. While energy-efficiency improvements usually require increased investment at the point of end use, the required investments are often less than the investments in an equivalent amount of new energy supply that would be used with less efficient end-use equipment, so that total capital requirements for the energy system as a whole would typically be reduced by investments in energy efficiency. For example, in São Paulo State in Brazil, the total life-cycle (50-year) cost of delivering 1 kW of baseload electricity from a new hydroelectric facility to an industrial customer is $3250, while the corresponding life-cycle cost of saving a kW by installing variable speed drives for industrial motors (25% electricity savings) is $900–1800 per kW, assuming a 10% discount rate in both cases (7).

Second, the severe strain on export earnings in developing countries caused by oil import bills provides a powerful incentive to reduce oil import requirements and thereby become more self-reliant. Energy-efficiency improvement (e.g. for cars and trucks) can be an especially cost-effective way of doing this.

Third, the potential for major growth in energy-intensive activities (e.g. the processing of basic materials) is a condition conducive to major process and product innovations. In the new era of high-cost energy, technologies introduced to reduce total cost will often be far more energy efficient than the corresponding technologies now in place in industrialized countries, most of which were introduced in the era of low-cost energy. Because of saturated markets and thus a less favorable climate for innovation in the stagnating basic materials processing industries of industrialized countries, it could turn out that in some areas major industrial innovations will take place first in developing countries. The successful Brazilian programs (a) to shift cars from gasoline to ethanol derived from sugar cane (68a) and (b) to produce high-quality steels using charcoal derived from eucalyptus instead of coal, prove that such technological leaps forwards are possible.

THE TRADITIONAL SECTOR   There are also major opportunities for energy-efficiency improvements in the traditional sector, where fuelwood and other forms of biomass dominate energy use, mainly for cooking. The inefficiency of present woodstoves is highlighted by comparing the energy use in fuelwood stoves (used mainly for cooking) in developing countries today— some 0.25–0.6 kW per capita [0.4–1 tonne of wood per capita per year (69)]—with the corresponding rate of using liquefied petroleum gas (LPG) or natural gas for cooking in developing countries, and in industrialized market economies as well, typically some 0.05 kW (Figure 7).

While as recently as 1983 only marginal gains were being made in stove efficiency improvement programs (70), recent successes in applying scien-

PER CAPITA ENERGY USE RATES FOR COOKING

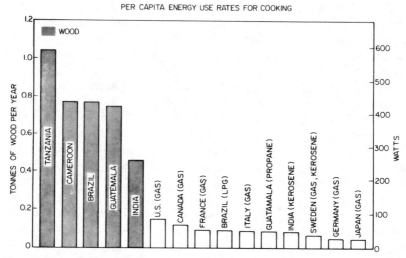

Figure 7 Per capita energy use rates for cooking. For both wood stoves and stoves involving high-quality energy carriers the per capita energy use rate for cooking is expressed in watts. The wood consumption rate is also given in tonnes of dry wood per year. Assuming 1 tonne = 18 GJ, 1 tonne per year = 570 watts. From (69).

tific principles of heat transfer and combustion to stove design, combined with standardized testing methods and production techniques, have made it possible to introduce a variety of high-efficiency, low-cost wood cooking stoves that appeal to users in a wide range of cultural contexts. With such stoves fuel requirements for cooking can be reduced by one third to one half, and the fuel savings are complemented by other important benefits such as reduced cooking time and reduced labor requirements for fuelwood gathering (71, 72).

Looking ahead, further improvements would be feasible if gaseous energy carriers (LPG, natural gas, biogas, or producer gas) were made available for cooking, as simple gas stoves can be 50% efficient, while even the most efficient wood stoves that have been made have efficiencies of only 30–40%. Still further improvements would be possible via the use of advanced 70% efficient gas stoves that have been developed recently (73).

All such opportunities could free biomass resources for other purposes (74), such as energy for transport, agriculture, and rural industry.

A WORLD ENERGY SCENARIO

We now construct a long-term global energy demand and supply scenario, based on the end-use analysis summarized in the previous sections and

	DEMAND		SUPPLY
INDUSTRIALIZED COUNTRIES	Energy Use Per Capita	Future Activity Levels Based on Projections of Trends	Normative Supply Choices, Within Constraint of Tech., Econ. Feasibility (Avoid Over-Dependence on Oil, Fossil Fuels, Nuclear Power)
		Future Energy Intensities Based on Normative Choice of "Best/Advanced" Technologies Which Are Tech., Econ. Feasible	
	Population	UN Projection	
DEVELOPING COUNTRIES	Energy Use Per Capita	Future Activity Levels Based on Normative Assumption of Large Increase in Amenities, Ranging Up to W. European Level of 1970s	Normative Supply Choices, Within Constraint of Tech., Econ. Feasibility (Modernize Bioenergy, Promote Self-Reliance)
		Future Energy Intensities Based on Normative Choice of Improved Efficiencies, Ranging Up to Efficiencies of "Best/Advanced" Technologies	
	Population	UN Projection	

Figure 8   Schematic summary of the methodology used to develop the world energy scenario.

presented in detail elsewhere (38). The purpose is to show that it is both technically and economically feasible to evolve an energy future compatible with the achievement of a sustainable world. Our scenario is not a forecast but a normatively constructed energy future that we believe could evolve with appropriate public policies.[6] The argument that we now present as the basis for the construction of the scenario is summarized schematically in Figure 8, highlighting the normative aspects of the analysis.

The first step is to understand present and future needs for energy services, such as cooking, lighting, domestic hot water, passenger and freight transport, and basic industrial materials. Fortunately, most energy use is concentrated in just a few activities in each energy-using sector

---

[6] We are skeptical that there can be long-term energy projections or forecasts that are not normative. Energy markets are not free, but are shaped by existing systems of taxes, subsidies, and regulations, the acceptance of which as the basis for a modeling exercise represents a normative judgment about the way the world ought to work. Even if markets were free, the acceptance of free market conditions would represent the normative judgment that various externalities such as those we have described in this paper must be ignored in energy planning. Thus our analysis is by no means unique in being "normative," although it is not customary to call one's analysis normative explicitly.

(residential, commercial, transportation, and industry), so that the list of important end-use activities that must be scrutinized is readily manageable in most instances. Estimated levels of energy services associated with alternative economic development paths can be based on extrapolations of historical trends, taking into account ongoing shifts and saturation effects [as we have done in developing energy scenarios for Sweden (44) and the United States (3)], or departures from historical trends can be specified to conform to feasible societal goals (e.g. shifts from historical trends in production and consumption may be necessary in particular developing countries to ensure that basic human needs are satisfied).

With energy service levels specified, the next task is to obtain estimates of the energy intensities for these service activities, i.e. the energy required per unit of services provided (e.g. kJ of kerosene per passenger-km of air travel), associated with the cost-effective ways of providing services. Here consideration is given both to potential improvements in energy efficiency and to the use of alternative energy carriers.

With these assumptions, future aggregate demand estimates are obtained by summing (over all activities) the products of the activity levels for energy services and the corresponding energy intensities. Then the demand levels so obtained can be matched to estimated available energy supplies.

Because of resource, climatic, and cultural variations from one country and region to another, a comprehensive perspective on future energy demand and supply should be constructed "from the bottom up," with an aggregation of country studies into regional studies, which are then aggregated into a global picture. While detailed country studies and end-use strategies have so far been developed for only a few countries,[7] it is nevertheless feasible to formulate a preliminary global perspective.

In constructing a global energy scenario we focus on the year 2020. By that time (a) it should be feasible both to satisfy basic human needs and to bring about considerable additional improvements in living standards in developing countries; and (b) there should be time for the widespread adoption of improved energy-using technologies. Yet 2020 is close enough that it has an important bearing on long-range energy planning today.

A Global Energy Demand Scenario

FUTURE PER CAPITA ENERGY DEMAND IN INDUSTRIALIZED COUNTRIES   Our analyses for two countries, Sweden (44) and the United States (3), indicate that a large reduction in the energy intensity of economic activity is feasible. Specifically, we have found that in the period of interest, it would be technically and economically feasible to reduce per capita final energy use

[7] In addition to the studies on which this paper is based, see also (52, 75–79).

in Sweden from 5.4 kW in 1975 to 2.8 (3.3) kW, and in the United States, from 9.0 kW in 1980 to 4.2 (4.6) kW, along with a 50% (100%) increase in per capita GDP (Table 6). These scenarios are based on matching future activity levels obtained from extrapolations of historical trends, taking into account ongoing structural changes, with efficient energy end-use technologies, such as those described in the previous section, which are judged to be cost-effective on a life-cycle cost basis and which are assumed to be introduced at the normal rate for new capital stock. The energy efficiencies of the new technologies assumed for these scenarios are comparable to the efficiencies of either the best available units on the market today or "advanced" technologies that could lead to commercial products over a period of about a decade. In all cases technical efficiencies are far from thermodynamic limits, and, in most instances, possibilities for further improvement are apparent.

Much of what we have learned from our analyses of the Swedish and US situations is probably applicable to most other industrialized countries as well—especially to many of the other countries in the OECD. In light of the paucity of data available to us concerning patterns of energy use in industrialized countries with centrally planned economies we are less certain about the extent to which our findings are relevant to the Council

Table 6 Alternative scenarios for per capita final energy use in Sweden and the United States (kW)

	Sweden[a]			United States[b]		
		Consumption of goods and services			Consumption of goods and services	
	1975	up 50%	up 100%	1980	up 50%	up 100%
Residential	1.4	0.34	0.41	1.5	0.57	0.57
Commercial	0.53	0.15	0.16	0.88	0.34	0.34
Transportation				2.9	1.33	1.53
Domestic	0.89	0.55	0.65	—	—	—
International bunkers	0.18	0.10	0.11	—	—	—
Industry	2.4	1.6	2.0	3.7	2.01	2.16
Totals	5.3	2.8	3.3	9.0	4.2	4.6

[a] Based on "advanced technology" (end-use technology in an advanced stage of research and development and judged to be in a cost bracket of interest). If introduced at the rate of capital turnover, this technology could become average technology by 2015–2020. See (44).
[b] For the year 2020. It is assumed that as the capital stock turns over and grows, investments are made in the most energy-efficient technologies available that are judged cost-effective. Most of the technologies considered are commercially available today; a few are advanced technologies that could be available within about a decade. See (3).

for Mutual Economic Assistance (CMEA) countries. The fact that energy use in CMEA countries continued to follow long-term historical trends in the 1970s, in contrast to the sharp break with the trend in OECD countries (Figure 9), suggests either that the pursuit of energy-efficient futures is more difficult in CMEA countries or that some kind of lag is at work. But since average per capita primary energy use levels are comparable [5.8 kW and 5.7 kW for OECD and CMEA countries, respectively in 1982 (80)], while the levels of amenities made possible by energy are probably higher, on average, in the West than in the East, it seems reasonable to assume that what can be achieved in a few countries like the United States and Sweden provides an "existence proof" of what can be achieved in most industrialized countries.

We thus assume for our global scenario that our findings for Sweden and the United States can be extrapolated to all industrialized countries, so that, in the period 1980–2020, per capita final energy use in industrialized countries is cut in half, from 4.9 kW to 2.5 kW, associated with a 50–100% increase in per capita GDP.

For this scenario the final energy use/GDP ratio would have to decline at an average rate of 2.7% per year if there were to be a 50% increase in the per capita consumption of goods and services during 1980–2020, and at 3.5% per year for a 100% increase. This is a much more rapid trend than that of the pre–energy crisis era, when energy demand grew only slightly more slowly than GDP, but it is not so different from recent experience in OECD countries: this ratio actually declined there at an average rate of 2.5% per year for the period 1973–1982, when GDP increased about 20% [Figure 9, (53)].

Despite the fact that the indicated trend in the energy use per dollar of GDP would not be markedly different from the recent experience in OECD countries, this recent period was an unusual one, characterized by very large price increases. Although the present stock of capital equipment is still far from optimized for the new energy prices, there are numerous institutional obstacles to economically optimal choices that involve major changes from the present energy system, so that new public policies are probably needed to bring about such an energy-efficient future. We shall return later to a discussion of these issues.

FUTURE PER CAPITA ENERGY DEMAND IN DEVELOPING COUNTRIES   The challenge for energy planners in developing countries is to ensure that energy services needed for satisfying basic human needs, for building infrastructure, and generally for substantially raising the standard of living are available in affordable, environmentally sound, and sustainable ways.

To indicate how emphasis on improving energy efficiency would

PRIMARY ENERGY CONSUMPTION, NET OIL IMPORTS,
AND GDP, 1973-1982   (1973 = 1.00)

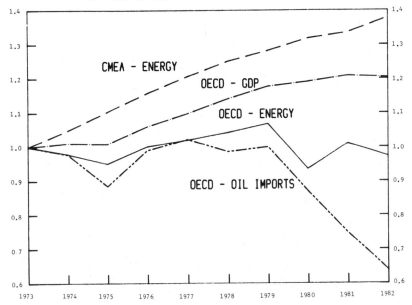

Figure 9 Primary energy consumption, net oil imports, and gross domestic product for OECD and CMEA countries, 1973–1982 (1973 = 1.00).

facilitate the achievement of these goals, we present in Table 7 an energy budget for a hypothetical future developing country with an activities mix similar to that for Western Europe[8] in the 1970s (excluding space heating, which is not needed in most developing countries) but matched to much more efficient end-use technologies than those now in common use in Europe.

The activity levels for this scenario are far in excess of present values in developing countries. For an "average" developing country to retrace a historical development path to such a state by 2020 might require that during 1975–2020, per capita GDP increase tenfold (the ratio of the average per capita GDP in Western Europe to that in developing countries in 1975), or at an average rate of somewhat more than about 5% per year. To achieve by 2020 the indicated levels of consumption of those basic materials explicitly highlighted in Table 7 (steel, aluminum, cement, paper, and nitrogen fertilizer) would require sustained per capita growth rates of 4–6%

[8] Strictly speaking, most of the data discussed here and shown in Table 7 as being characteristic of Western Europe are average values for the WE/JANZ region (Western Europe, Japan, Australia, and New Zealand).

Table 7 A hypothetical final energy use scenario (in watts per capita) for an affluent, energy-efficient developing country[a]

	Activity level	Technology, performance	Electricity	Fuel	Total
Residential[b]	4 persons/household (HH)				
Cooking	Brazilian cooking level with LPG stoves	70% efficient gas stove		34	
Hot water	50 liters of hot water/capita/day	heat pump water heater, COP = 2.5	29.0		
Refrigeration	1 315-liter refrigerator-freezer/HH	Electrolux, 475 kWh/year	13.5		
Lights	New Jersey (US) level of lighting	Compact fluorescent bulbs	3.8		
Television	1 color TV/HH, 4 hours/day	75 Watt unit	3.1		
Clothes washer	1/HH, 1 cycle/day	0.2 kWh/cycle	2.1		
Subtotal			51	34	85
Commercial					
Commercial	5.4 m²/capita floor space (WE/JANZ ave, 1975)	Harnosand Building, Sweden (all uses but space heating)	22	—	22
Transportation					
Automobiles	0.19 autos/capita, 15,000 km/car/yr (WE/JANZ ave, 1975)	Cummins/NASA Lewis Car at 3.0 liters/100 km		107	
Intercity bus	1850 p-km/capita	0.45 MJ/p-km, 3/4 1975 ave		26	
Passenger train	3175 p-km/capita	0.45 MJ/p-km diesel, 3/4 1975 ave[c]	4.5	32	
Urban mass transit	520 p-km/capita (WE/JANZ ave, 1975)	0.85 MJ/p-km diesel, 3/4 1975 ave[d]	2.0	8	
Air travel	345 p-km/capita	1.9 MJ/p-km, 1/2 US ave in 1980		21	
Truck freight	1495 t-km/capita	0.67 MJ/t-km, 2/3 of Swedish ave		32	
Rail freight	814 t-km/capita	Electric, 0.18 MJ/t-km, Sweden ave	5		
Water freight (including bunkers)	1/2 OECD European ave, 1978 (reduced reliance on oil)	60% of ave OECD energy intensity		50	
Subtotal			12	276	288

Manufacturing					
Raw steel	320 kg/capita (OECD Eur ave, 1978)	ave of Plasmasmelt, Elred processes	28	77	
Cement	479 kg/capita (OECD Eur ave, 1980)	Swedish ave in 1983	6	54	
Primary aluminum	9.7 kg/capita (OECD Eur ave, 1980)	Alcoa process	11	26	
Paper and paperboard	106 kg/capita (OECD Eur ave, 1979)	Ave of 1977 Swedish designs	11	24	
Nitrogenous fertilizer	26 kg N/capita (OECD Eur ave, 1979–1980)	Ammonia via steam reforming of methane	—	36	
Other[e]			65	212	
Subtotal	Swedish industrial mix with 1975 W. European level of GDP/capita (55% of Swedish level)	Energy intensity for Swedish industry with 1975 level of goods and services and advanced technology (44)	121	429	550
Agriculture	WE/JANZ ave, 1975	3/4 of WE/JANZ ave energy intensity	4	41	45
Mining, construction	WE/JANZ ave, 1975	3/4 of WE/JANZ ave energy intensity	—	59	59
Totals			210	839	1049

[a] For a country in a warm climate, with a level of amenities (except for space heating) comparable to that in the WE/JANZ region (Western Europe, Japan, Australia, New Zealand, and South Africa) in the 1970s, but with currently best available or advanced energy utilization technologies.
[b] Activity levels for the residential sector are estimates, owing to lack of data for the WE/JANZ region.
[c] Here 30% of all passenger-km are via electric trains, for which the final energy intensity is one third that of diesel trains.
[d] Here 40% of passenger-km are via electric systems, for which the final energy intensity is about one third that of diesel buses.
[e] Here "other" is the difference between the manufacturing total and the sum of the items calculated explicitly. Energy usage associated with "other" for the nonmanufacturing sectors is negligible and thus is not shown explicitly in this table.

per year for these materials. These are ambitious but not inconceivable growth schedules. A number of developing countries achieved average per capita GDP growth rates of 5% per year or more in the period 1960–1982, including China, Thailand, South Korea, Brazil, Yugoslavia, and Singapore (2). In the United States, per capita consumption grew at average rates of 5–10% per year for each of the above-mentioned basic materials during the 20–40-year initial rapid growth period, while per capita GDP was growing at an average rate of only about 2% (45).

To illustrate what can be achieved with efficiency improvements, we have multiplied these activity levels by energy intensities corresponding in energy efficiency to the best available technologies on the market today or to advanced technologies that could be commercialized over a period of about a decade. The result (Table 7) is that total final energy use per capita would be only 1.0 kW, or only slightly more than the actual 0.9 kW average final energy use rate in 1980!

It is possible to achieve such large improvements in living standards without increasing energy use in part because enormous increases in energy efficiency arise simply by shifting from traditional, inefficiently used, noncommercial fuels (which at present account for nearly half of all energy use in developing countries) to modern energy carriers, as the above comparison of cooking with fuelwood and with gaseous fuels shows clearly. The importance of modern energy carriers is also evident from the fact that for Western Europe in 1975 per capita final energy use for purposes other than space heating was only 2.3 kW, about $2\frac{1}{2}$ times that in developing countries, even though per capita GDP was 10 times as large.

In addition to the savings associated with the shift to modern energy carriers, considerable further savings can be gained by adopting more energy-efficient technologies that have recently become available, such as those indicated in Table 7 and discussed in detail elsewhere (38). Some of the technologies assumed here for the domestic sector illustrate how large increases in amenities can be achieved without approaching present energy consumption levels of Western Europe. A new 315-liter refrigerator–freezer with the energy performance of the most energy-efficient unit available in 1982 requires just 475 kWh per year—or less than one third of the electricity required by the average refrigerator–freezer in United States. Similarly, new compact fluorescent light bulbs use just one fourth as much electricity as incandescent bulbs.

While most of the technologies shown in Table 7 are commercially available today, a few are still in an advanced state of development—for example the 3.0 liter per 100 km (79 mpg) automobile and the Swedish Plasmasmelt and Elred steel-making processes (see the previous section).

The high level of average performance in end-use technology characteristic of this scenario could in principle be achieved more quickly in a developing country than in an already industrialized country. In the former there is such a large demand for new energy-using capital stock that the rate of introducing new efficient technology is not limited by the rate of turnover of the existing stock, as would be the case in industrialized countries.

The set of activities indicated by this scenario should not be construed as targets to be achieved by 2020 or any other date. The appropriate mix and levels of activities for, say 2020, may well have to be different to be consistent with overall development goals. But this analysis suggests that it is possible to provide a standard of living in developing countries anywhere along a continuum from the present one up to a level of amenities typical of western Europe today, without departing significantly from the average per capita energy use for developing countries today, depending on the level of energy efficiency that is emphasized.

That development goals can be achieved with little change in the overall per capita level of energy use should not obscure the challenge of bringing this about. As in the case of development, large amounts of capital would generally be required to bring about a shift to modern energy carriers and to efficient end-use technology. But our analysis suggests strongly that for a wide range of plausible sets of activity levels and for a wide range of end-use technologies, it would be less costly to provide energy services using the more efficient end-use technologies than to provide the same services with conventional, less efficient end-use technologies and increased energy supplies. [For examples in the Brazilian context, see (7, 80a)].

On the basis of such considerations we assume for our global energy scenario an average per capita final energy use of 1 kW for developing countries in 2020—a level that, with emphasis on energy-efficiency improvement and modern energy carriers, would be adequate to ensure that basic human needs are satisfied and to allow for considerable further improvements in living standards.

TOTAL GLOBAL ENERGY DEMAND   Combining the above projections of per capita final energy use for the year 2020 for industrialized and developing countries, with populations 1.24 billion and 5.71 billion for industrialized and developing countries, respectively,[9] leads to a global final energy use in

---

[9] Whereas population is usually treated as an exogenous variable in conventional energy projections, it is reasonable to expect that population growth would be slower with than without a basic human needs policy, since large families tend to be economically attractive for the poor (2). Since the quantification of the impact of a basic human needs policy on population has not been carried out, we reflect this effect by the adoption of the United Nations' 1980 Low Variant Projection of world population—7.0 million people by 2020 (vs 7.8 million for the Medium Variant) (81).

Table 8 Global energy demand scenario

	Industrialized countries		Developing countries		World	
	1980	2020	1980	2020	1980	2020
Population (billion)[a]	1.11	1.24	3.32	5.71	4.43	6.95
Final energy use, in TW[b]						
Fuel	4.77		2.77		7.54	7.23[c]
Electricity	0.70		0.13		0.83	1.58[d]
Totals	5.47	3.10	2.90	5.71	8.37	8.81
Per capita final energy use (kW)	4.92	2.5	0.87	1.0	1.89	1.27

[a] The 1980 UN low variant population projection (81) is assumed.

[b] In these energy balances final energy use is defined as the total fuel (including bunkers) and electricity consumed by "final consumers." Excluded are losses in the generation, transmission, and distribution of electricity, and the consumption of petroleum fuels by refineries.

[c] Assuming 10% oil refinery losses and 30% average losses in converting biomass into final energy carriers, the total consists of some 6.33 TW of fossil fuels and 0.90 TW of fuels derived from biomass.

[d] In 1980 10% of final global energy use was accounted for by electricity, which was 0.8 times as large as the percentage for the United States. For this scenario the percentage assumed for 2020 is 18%, which is 0.8 times as large as what we project for the United States in 2020 (3).

2020 of 8.8 TW, differing only slightly from the 1980 level of 8.4 TW (see Figure 10 for a schematic summary of the global energy demand scenario construction and Table 8). But the demand distribution would be markedly different; the per capita levels of energy use of the North and the South would converge, and developing countries would account for about two thirds of total world energy use, up from one third in 1980. Figure 11 shows our scenario in terms of primary energy use, and places it alongside the corresponding projections made by the International Institute for Advanced Systems Analysis (IIASA) (20) and the World Energy Conference (WEC) (29).

A SIMPLE ENERGY DEMAND MODEL   To express our energy demand scenario in terms more familiar to most analysts in the energy modeling community, we have constructed a simple model relating commercial final energy demand per capita (FE/P) to gross domestic product per capita (GDP/P), the average price of final energy (P_e), and a rate of energy-efficiency improvement (c) which is not price-induced:

$$FE/P(t) = A \times [GDP/P(t)]^a \times [P_e(t)]^{-b}/(1+c)^t,$$

where A is a constant, a is an income elasticity, and $-b$ is a long-run final

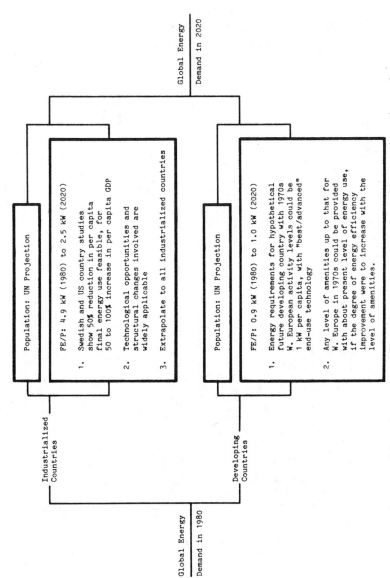

Figure 10 Schematic summary of the assumptions underlying the world energy demand scenario.

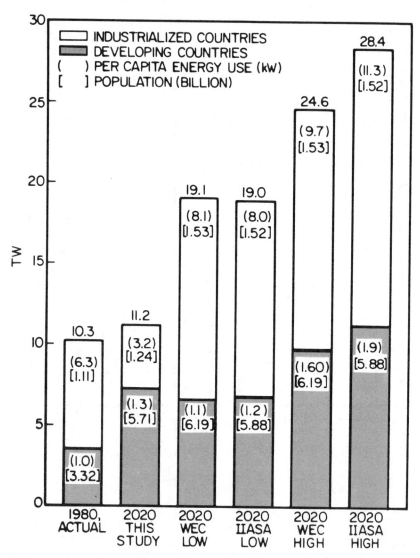

Figure 11   Primary energy use, in TW-years per year by region. Historical data for 1980 and projections to 2020 according to the IIASA study (20), the WEC study (29), and the present study.

energy price elasticity. This is esssentially the energy demand equation underlying the IEA/ORAU (32) and MITEL (33) analyses.

We have applied this equation separately to industrialized and developing countries, relating PE/F, GDP/P, and P_e values in 2020 to those in 1972, as indicated in Figures 12 and 13, for illustrative values of the parameters (a, b, and c).

We have chosen 1972 as the base year for this modeling exercise, because

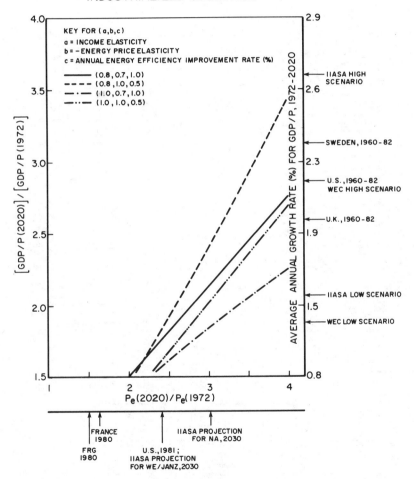

Figure 12 GDP and energy price parameters for 2020 consistent with the energy demand scenario for industrialized countries, under alternative assumptions about income elasticities, price elasticities, and the non–price-induced energy-efficiency improvement rate.

this is the last year before the first oil price shock, so that it is reasonable to assume that the economic system was then in equilibrium with the existing energy prices (unlike the situation in 1980, say). For this base year the values of PE/P were 4.7 kW and 0.38 kW, compared to our 2020 scenario values of 2.5 kW and 1.0 kW, for industrialized and developing countries, respectively.

For income elasticities we have chosen alternative values of 0.8 and 1.0

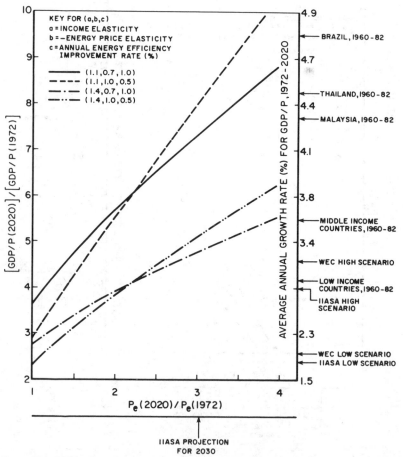

DEVELOPING COUNTRIES

Figure 13 GDP and energy price parameters for 2020 consistent with the energy demand scenario for developing countries, under alternative assumptions about income elasticities, price elasticities, and the non–price-induced energy-efficiency improvement rate.

for industrialized countries and 1.1 and 1.4 for developing countries, respectively. We believe that the value of 0.8 would tend to capture the ongoing shift to less energy-intensive economic activity in industrialized countries (and may well understate the extent of the shift), but we have also included the value of unity assumed in many modeling efforts for comparison. A value of 1.4 was used for developing countries in the 1983 IEA/ORAU study (31) and may be roughly characteristic of the historical situation in developing countries. However, as developing countries modernize in the decades ahead, the income elasticity can be expected to decline. The two assumed values may span the range of uncertainty for the income elasticity in developing countries for the period of interest here.

For the long-run price elasticity, Nordhaus has reviewed various studies and has concluded that the range of plausible values is from -0.66 to -1.15 (82). The illustrative values chosen here (-0.7 and -1.0) span most of this range.[10]

The assumed non–price-induced energy-efficiency improvement rates are 1.0 and 0.5% per year. The higher value is the one assumed for the "CO_2-benign" scenarios developed in the MITEL analysis (33), while the lower value approximately reflects the contribution from such energy-efficiency improvements in the 1984 IEA/ORAU analysis (32). We have coupled the higher energy-efficiency improvement rate with the lower price elasticity and the lower rate with the higher price elasticity in this modeling exercise, to reflect the tendency of non–price-induced energy-efficiency improvement policies to diminish the efficacy of prices in curbing energy demand (83).

While we have not made explicit assumptions about energy prices in constructing our scenario (since most of the end-use technologies underlying our analysis would be economic at or near present prices on a life-cycle cost basis, with future costs discounted at market interest rates), we expect that there would be continuing final energy price increases to reflect rising marginal production costs, an expected continuing shift to electricity (see below), and the levy of some energy taxes to take into account externalities (see below). Prices have already risen substantially above 1972 values; in West Germany and France average final energy prices in 1980 were 1.5 and 1.6 times the 1972 values in real terms, respectively (84), and in the United States the average price in 1981 was 2.3 times the 1972 price (85). Looking to the future, the IIASA study projects that by 2030 final energy prices will be 3 times the 1972 value in all regions but the WE/JANZ region (Western Europe, Japan, Australia, and New Zealand), for which a 2.4-fold increase is

[10] These elasticities appear high but are not. Long-run price elasticities are much larger than short-run elasticities. Likewise, final demand elasticities are greater than secondary demand elasticities, which in turn are greater than primary demand elasticities (83).

projected instead (20). The 1983 projection of the US Department of Energy is for much larger (3.6–5.7-fold) average final energy price increases for the United States during 1972–2010, associated with an 11–17% reduction in final energy use per capita in this period (5). It is reasonable to associate an average increase in the final energy price by 2020 in the range 2–3 times the 1972 value with our scenario.

For industrialized countries (Figure 12) and the cases $(a, b, c) = (0.8, 0.7, 1.0)$, $(0.8, 1.0, 0.5)$, and $(1.0, 1.0, 0.5)$, our energy demand projection is consistent with a 50–100% increase in per capita GDP (comparable to the values assumed in the IIASA and WEC low scenarios) and 2020 energy prices of 2–3 times the 1972 values. For the case in which there is no structural shift to less energy-intensive activities and a low price elasticity $(1.0, 0.7, 1.0)$ the energy price in 2020 would have to be about $3\frac{1}{2}$ times the 1972 value to be consistent with a doubling of per capita GDP.

For developing countries (Figure 13) our scenario with 2020 prices 2–3 times 1972 prices would be compatible with the per capita GDP growth rates assumed in the IIASA and WEC high scenarios, for the high income elasticity cases $(a = 1.4)$. For the lower income elasticity cases $(a = 1.1)$, our results would be consistent with much more rapid GDP growth.

This modeling exercise shows that while our global energy demand projection for 2020 is far outside the range of most other projections, it appears consistent with plausible values of income and price elasticities, and with plausible expectations about energy price and GDP growth, if the non–price-induced energy-efficiency improvement rate can be in the range 0.5–1.0% per year. While this energy-efficiency improvement rate is a measure of the public policy effort that would be required to bring about this energy future, not all of this efficiency improvement would have to be public policy–induced. As we have pointed out, energy-efficiency improvements associated with technological innovation have often been made even in periods of declining energy prices, a phenomenon that led IEA/ORAU analysts to include the non–price-induced technological improvement factor in their model for the industrial sector in the first place (30).

At the same time, however, the energy-efficiency improvement factor may not represent the full extent of needed public policy effort, if low energy demand levels were to lead to stable or even falling energy prices. If that were the case then energy taxes could also be needed to keep gradual upward pressure on final (consumer) prices.

A Global Energy Supply Scenario

The higher the energy demand, the less flexibility there is in planning an energy supply mix. Conversely, when energy demand is low, one can often plan the energy supply mix with considerable flexibility, avoiding over-dependence on troublesome energy sources.

Of the many possible energy supply futures that can be matched to the demand patterns described above, one future, involving only very minor shifts from the present situation, is shown in Figure 14, alongside the supply projections made in the IIASA and WEC studies. The primary energy use level in 2020 for our scenario would be 11.2 TW, only slightly higher than

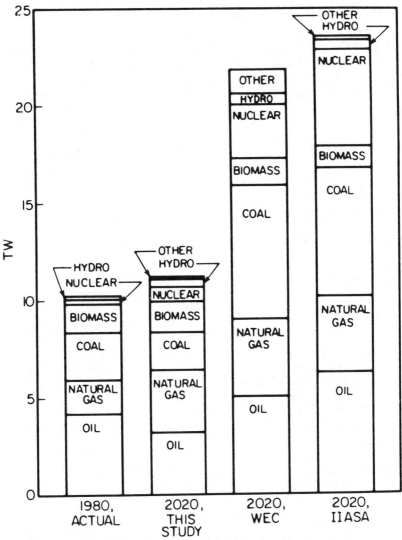

Figure 14 Primary energy use by energy source: actual use for 1980 and alternative projections for 2020. The IIASA (20) and WEC (29) projections are averages of those studies' high and low scenarios.

the 1980 level of 10.3 TW (Table 9), but less than half the levels projected in the IIASA and WEC energy studies.

The construction of our energy supply mix begins with the assumption that the overall level of fossil fuel use in 2020 is the same as in 1980, with the mix of oil, natural gas, and coal adjusted to reflect considerations of the atmospheric carbon dioxide buildup from the burning of fossil fuels, global security and the world oil price, and the relative abundances of oil and gas resources. Biomass is considered the "swing fuel," providing solid, liquid, and gaseous fuels needed in excess of those provided by fossil fuels.

Electricity's share of global final energy demand would increase from 10% in 1980 to 18% in 2020, reflecting a continuation of the ongoing electrification of the global energy economy. Despite this emphasis, requirements for new electricity supply would grow far more slowly (1.6%

Table 9 Global primary energy supply scenario (in TW)

	1980	2020
Nuclear power[a]	0.22	0.75
Hydropower[a]	0.19	0.46
Wind and photovoltaic electricity[a]	—	0.09
Fossil fuels		
Coal	2.44	1.95
Oil	4.18	3.23
Natural gas	1.74	3.23
Subtotal	8.36	8.41
Biomass		
Organic wastes		0.74[b]
Plantations or woodfarms		0.75[c]
Subtotal	1.49	1.49
Totals	10.3	11.2

[a] Nuclear energy is counted as the thermal energy released in fission (assumed to be 2.5 times the produced electricity in 2020); hydropower, wind, and photovoltaic energy as the electricity produced.

[b] We estimate that in 1980 the global production of organic wastes (forest product industry wastes, crop residues, manure, urban refuse) amounted to 2.8 TW. We assume that the level will increase in proportion to the population, reaching 4.4 TW in 2020. Because of competition with other uses we assume that only one sixth of these wastes are available for energy purposes in 2020.

[c] For an average yield of 10 tonnes per hectare per year, some 130 million hectares of plantations would be required by 2020.

per year on average during 1980–2020) than in most government and industry forecasts—so that there would still be considerable flexibility in putting together an electricity supply mix. With proliferation-related constraints on nuclear power, an expansion of hydropower, and a contribution from cogeneration comparable to that from nuclear power, there would be no need to expand fossil-fuel-based central station power generation beyond the present level (Table 10).

In what follows we describe how considerations of various supply constraints motivated the supply mix we have chosen to illustrate the flexibility gained with relatively low energy demand levels.

ATMOSPHERIC CARBON DIOXIDE AND THE BURNING OF FOSSIL FUELS   The need to adjust to a dramatically changed global climate in less than a century could be greatly reduced by reducing dependence on fossil fuels. Such efforts must focus on coal. While using all the remaining ultimately recoverable oil and gas resources would lead to an atmospheric carbon dioxide level of 440 ppm, or nearly 1.5 times the preindustrial level, using half of the remaining 9 trillion tonnes of coal resources would increase the CO_2 level to four times the preindustrial level.

Table 10 Global electricity supply scenario (in TW)

	1980	2020
Hydropower	0.19	0.46
Wind and photovoltaic electricity	—	0.09[a]
Cogeneration		
Biomass	—	0.14
Fossil fuel	—	0.13
Central station		
Nuclear	0.08	0.30
Fossil fuel	0.66	0.66[b]
Totals	0.93	1.78[c]

[a] Owing to the large uncertainties in the future of photovoltaics technology we do not specify how the wind/photovoltaics mix might be disaggregated. In the event that photovoltaics technology is not commercialized, all of this electricity would be provided by wind. However, if the promise of photovoltaic technology is realized (100), its contribution could be considerable.

[b] Three fourths of central station power generation is assumed to be based on coal at 40% average conversion efficiency and one fourth on natural gas at 50% conversion efficiency [using gas turbines with injection into the combustors of steam produced in turbine exhaust heat recovery boilers (102)].

[c] Electricity production is equal to the demand level shown in Table 8 divided by 0.89, to account for transmission and distribution losses.

We have modeled future levels of fossil fuel use by assuming that: (a) half of the released carbon dioxide remains in the atmosphere; (b) all estimated ultimately recoverable oil and gas resources are eventually used; and (c) coal production falls exponentially over time. The one free parameter, then, is the rate of exponential decline, which depends on the ultimate CO_2 ceiling level. If the ceiling were as low as 1.5 times the preindustrial level, coal would need to be phased out very rapidly, falling to one half the present level before the turn of the century—not a practical target. We assume instead a ceiling of 1.7 times the preindustrial level, which implies that coal use would have to decline only 20% during 1980–2020 (Figure 15). Coal use in 2020 with this scenario would be only 0.2–0.4 times as large as the levels projected in the IIASA and WEC scenarios (Figure 14).

THE WORLD OIL PROBLEM   In light of the facts that there is about as much recoverable gas left in the world as oil, and that the present rate of gas use is only two fifths that of oil, we assume a shift to gas to the extent that by 2020 gas and oil production rates become equal (Figure 14). Global oil use would be reduced thereby from 59 to 46 million barrels per day during 1980–2020. At this demand level there would probably be adequate oil supplies available outside the Middle East/North Africa (ME/NAf) region at production costs less than $30 per barrel (1982 $) to sustain dependence on the Middle East/North Africa region at the 1983 glut level of 15 million barrels per day.[11] This scenario would provide a hopeful outlook for oil-related global security problems and perhaps stable oil prices for the entire period to 2020.

THE LINK BETWEEN NUCLEAR WEAPONS AND NUCLEAR POWER   If growth of nuclear power were slow enough that the economic incentive to pursue plutonium recycle remained low everywhere (by avoiding uranium scarcity), the risks of latent proliferation in nonnuclear weapons states and of merging weapons and civilian nuclear power programs in weapons states could be greatly reduced.

Avoiding reprocessing and plutonium recycle technologies, though politically challenging, should in principle be much easier to accomplish today than was thought possible just a few years ago, since the economics of

---

[11] If world oil demand fell from 4.2 TW in 1980 to 3.2 TW in 2020, as in the scenario presented here, some 148 TW-years of oil would be required for the period 1981–2020. If in the period 1983–2020 production in the ME/Naf region were maintained at the 1983 "world oil glut level" of 1.06 TW (15 million barrels per day), cumulative oil requirements from regions other than the ME/NAf region in this period would be 105 TW-years. For comparison, world oil resources remaining outside the ME/NAf region and estimated to be ultimately recoverable at a price less than $26 per barrel (1982 $) is some 132 TW-years [see (20), Table 17-6, p. 531].

CONSTRAINED ANNUAL COAL PRODUCTION

VARIATION WITH ULTIMATE ATMOSPHERIC CO2 LEVEL

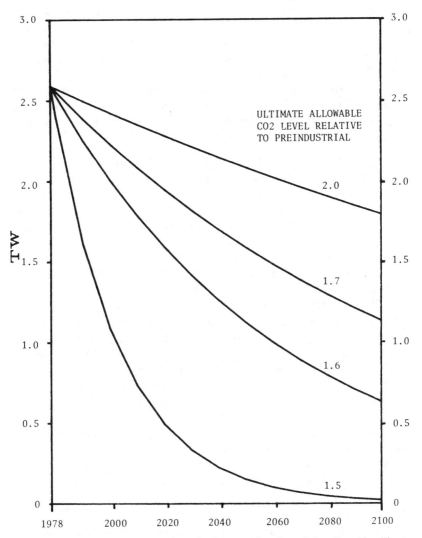

Figure 15 Constrained annual coal production as a function of the allowable ultimate atmospheric CO_2 level, assuming that a CO_2 constraint is reflected entirely as a constraint on the use of coal, along the lines discussed in the text.

reprocessing and recycle are not now favorable and would become only marginally favorable at very high uranium prices.[12]

Since even a ban on nuclear fuel reprocessing and plutonium recycle would not prevent proliferation via the clandestine recovery of plutonium from spent fuel—the risk of which increases with the extent of worldwide nuclear power development—it would seem desirable to go further and make nuclear power an energy technology of last resort, limited to situations in which viable alternative energy technologies are not available. With this perspective we assume that installed nuclear generating capacity increases from the 1980 level of 120 GW(e) to some 460 GW(e) (approximately the level generally expected for the year 2000) and then levels off (Figure 14). This implies that beyond the turn of the century the only nuclear power plants that would be built are those that would replace retired units.

RENEWABLE ENERGY RESOURCES   One way to cope with the global risks posed by overdependence on conventional supplies would be to shift to greater dependence on renewable resources. While the costs of such resources in large amounts are far more speculative at present than the costs of more efficient end-use technology, renewable energy resources can play at least a minor role in the period of interest. We have singled out hydropower, windpower, photovoltaic power, and bioenergy as promising options, which we assume provide for energy needs in excess of what is provided by fossil fuels and nuclear power.

Hydropower is especially promising in developing countries, where only 7% of economical reserves have been developed. We assume that the hydropower share of global electricity increases from 20% in 1980 to 25% in 2020, by which time the worldwide level of hydroelectric development would be about half the level projected for 2020 by the World Energy Conference in 1980 or about one fifth the total .technically usable hydroelectric potential (87). We limit wind and photovoltaic power combined to 5% of total power production.

We assume that biomass is provided by a 50–50 mix of organic wastes and biomass grown for fuel on energy plantations or farms, and that efforts are made to use biomass energy renewably. Such efforts would be facilitated by the fact that the need for biomass, as the swing fuel, would be only slightly higher in 2020 than in 1980 (Figure 14). But there would be an

---

[12] On the basis of economic calculations presented in (86), the uranium price would have to rise to $100 per pound of U_3O_8 (triple the present price) before the cost of plutonium recycle would be able to compete with current once-through fuel cycles. And even if the price of uranium should increase to $150 per pound the cost advantage of recycle would amount to less than 2% of the bus-bar cost of power generation.

enormous increase in the useful energy services obtained from this biomass, because we have emphasized efficient biomass conversion and end-use technologies.

SUPPLY FLEXIBILITY   We do not claim to have presented here an energy supply mix that is optimal in an economic sense or otherwise. Rather we have chosen a mix, not obviously constrained by costs or other factors, that shows that at or near present global energy demand levels plausible energy supply mixes can readily be identified that would not lead to the troublesome problems that appear unavoidable at much higher demand levels.

IMPLEMENTATION

We have argued that it is possible to identify long-run energy strategies consistent with and even supportive of the achievement of a sustainable world society. While such strategies differ radically from conventional supply-oriented strategies, our analysis leads us to conclude that they are both technically and economically feasible. But it is unlikely that market forces can be relied on to bring about the kind of energy future described here. New public policy initiatives are needed both to make markets work better than they do today and to correct inherent market shortcomings (38).

In developing countries, energy policies need to be closely coordinated with development policies generally, ensuring in particular that adequate energy is allocated to basic human needs programs and to rural needs, especially cooking, agriculture, and rural industry.

For industrialized and developing countries alike, important policies needed at the national level include:

1. Elimination of subsidies for energy supply
2. Pricing policies that sensitize consumers to long-run marginal costs for new energy supplies
3. Stabilization of the world oil price seen by consumers by a variable oil tax that is adjusted with changes in the world oil price and inflation
4. Better information to consumers about cost-effective opportunities for investments in energy efficiency
5. Redirection of capital toward consumer investments in energy efficiency
6. Regulation of energy performance for activities, such as automotive fuel economy, in which energy performance is readily measurable and easily understood
7. Changes in the incentive structures for energy utilities so that they will provide energy services to their customers (by investments in energy conservation as well as supply), and not simply energy supplies

8. Subsidies to the poor for energy services associated with the satisfaction of basic human needs in developing countries, and for investments in energy efficiency in industrialized countries
9. Support for research and development on end-use energy.

At the global level policies are needed both to support national policies in developing countries of the type just outlined (by providing capital support for institution-building, capital, etc), and to deal with global externalities associated with oil, the nuclear weapons–nuclear power connection, and the carbon dioxide problem.

Since each oil importer would gain from the lower world oil price that would result from the oil import reduction efforts of others, and from some coordination of internal oil prices, it may be desirable to coordinate national oil import reduction efforts with some kind of oil import reduction agreement (88). While such an agreement would lead to lower revenues for oil exporters, exporters would benefit in the long run from the resulting reduced uncertainties about future world oil demand.

To deal with the nuclear power–nuclear weapons link, a denuclearization treaty may be useful. Such a treaty would involve both the avoidance of nuclear fuel reprocessing technologies (on the parts of nuclear weapons states as well as nonweapons states) and real commitments by the superpowers to move away from dependence on nuclear weaponry (14).

The gradual coal phase-out described here is one of the most significant policy challenges implicit in the energy strategy outlined here. The CO_2 ceiling we have assumed would require limiting coal use in the long run to about one fourth of the coal available at prices less than half the world oil price in 1982. However, coal is a dirty fuel requiring much more capital investment than either oil or gas to use in environmentally acceptable ways. The prospect of stabilizing the world oil price (and thereby natural gas prices as well) by pursuing energy-efficiency improvements worldwide would make coal a much less desirable fuel than in energy futures with growing energy demand. Still, a carbon dioxide control treaty, involving perhaps agreed-upon coal taxes or other control mechanisms, may be needed. The fact that nearly 90% of the world's estimated geological coal resources lie in just three countries (the Soviet Union, the United States, and China) suggests that an agreement among these countries may lead to a solution of the problem.

There remain many unanswered questions about how best to implement end-use energy strategies. But the kinds of policies most likely needed involve the coordinated use of familiar policy instruments. We do not believe that a precondition for bringing about what may be construed as a radical energy future is the creation of a radical new world order.

CONCLUSIONS

The global energy outlook for the long term would be far more hopeful than what is implied by conventional energy projections, if energy planning shifted in emphasis from supply expansion to improvements at the point of energy use.

Such end-use energy strategies would provide energy services more economically than supply-oriented energy strategies, freeing up economic resources for other purposes. North–South tensions would be eased because of the resulting more equitable use of global resources. It appears that with such strategies energy supply would not be a constraint on the development of developing countries, basic human needs could be met, and living standards could be increased considerably beyond the satisfaction of basic human needs in the next several decades. Finally, end-use energy strategies would provide considerable flexibility in the choice of energy supplies, allowing us to mitigate greatly the global problems posed by expanded use of oil, fossil fuels generally, and nuclear power. In short there are hopeful prospects for evolving a global energy future compatible with the achievement of a sustainable world society.

There are reasons to be optimistic that end-use energy strategies can be implemented on a wide scale. First, a good track record has already been established in decoupling energy and economic growth; between 1973 and 1982 there was no net increase in energy use in OECD countries, even though GDP increased 20% (Figure 9). Second, much more can be accomplished with existing technology (and, in fact, the global scenario presented here does not depend on technological breakthroughs). Third, the pace of technological change is swift; indeed, most of the technical opportunities on which the major findings of the present analysis are based emerged only in the last few years, as part of a worldwide "quiet revolution" of innovation in end-use technology. Finally, while market forces acting alone will not likely lead to the kind of energy future we have described here, we do not believe that radical or unprecedented institutional changes are needed.

Our analysis stops short of identifying a truly sustainable long-run energy future, as we have not looked beyond the year 2020. However, we are much less worried about the future beyond that date than about getting just that far. Before then much new technology will emerge that will make long-range problems even more manageable. Energy planners should strive not to solve the energy problem for all time, but rather to pursue an evolutionary energy strategy consistent with and supportive of the achievement of a sustainable world.

We have not shown that the global energy balances we have arrived at are consistent with what can be achieved on a regional or country basis; that exercise remains to be carried out. A comprehensive global perspective on the energy problem must evolve as an integration of perspectives on individual countries and regions. But our analysis does suggest that there are no obvious significant global constraints to an energy future consistent with the solutions to other important global problems, and it thus provides a motivation for pursuing detailed country and regional analyses along these lines.

The major impediment to an end-use energy strategy is the present inadequate industrial infrastructure for marketing energy services. Since our institutions for managing energy supply have evolved over many decades, while the energy crisis that first focused attention on energy demand occurred but a decade ago, it is premature to make judgments as to precisely what institutional changes would be best to help build up this infrastructure. Planners should be prepared to try new approaches, learn from mistakes, and take pride in successes.

ACKNOWLEDGMENTS

The authors acknowledge support from the Energy Research Commission of Sweden, the International Development Research Centre of Canada, the Max and Anna Levinson Foundation, the Rockefeller Brothers Fund, the Swedish International Development Authority, and the World Resources Institute for this research.

Literature Cited

1. Forestry Department, UN Food & Agric. Organ. 1981. A global reconnaissance survey of the fuelwood supply/requirement situation. *Rep. of the Technical Panel on Fuelwood and Charcoal on its Second Session of the Preparatory Committee for the United Nations Conference on New and Renewable Energy, Rome, Italy.* A/Conf. 100/PC/34. Rome, Italy: United Nations General Assembly
2. The World Bank. 1984. *World Dev. Rep. 1984,* Oxford Univ. Press
3. Williams, R. H. 1985. *A Low Energy Future for the US.* Cent. Energy and Environ. Studies, Rep. No. 186. Princeton, NJ: Princeton Univ.
4. The World Bank. 1983. *The Energy Transition in Dev. Countries,* Washington, DC
5. Office Policy Planning Anal., US Dept. Energy. 1983. *Energy Projections to the Year 2010; a Technical Report in Support of the National Energy Policy Plan.* Rep. DOE/PE-0029/2. Washington, DC
6. Johansson, T. B., Williams, R. H. 1985. Reducing energy demand in industrialized countries, in *Proc. Symp. Greenhouse Problem Policy Options,* ed. D. Abrahamson, P. Ciborowski. Minneapolis, Minn: Hubert H. Humphrey Inst. of Public Affairs
7. Geller, H. S. 1985. *The Potential for Electricity Conservation in Brazil.* São Paulo, Brazil: Companhia Energetica de São Paulo
8. Ghai, D. P., Khan, A. R., Lee, E. L. H., Alfthan, T. 1977. *The Basic-needs Approach to Development.* Geneva: ILO
9. Streeten, P., Burki, S. J., Haq, M. U., Hicks, N., Stewart, F. 1981. *First Things First: Meeting Basic Needs in Develop-*

ing Countries. New York: Oxford Univ. Press

10. Quibria, M. Q. 1982. An analytical defense of basic needs: The optimal savings perspective. *World Dev.* 10: 285–91

11. UN Food & Agric. Organ. 1981. *Agriculture: Toward 2000.* Rome

12. Blechman, B. M., Hart, D. M. The political utility of nuclear weapons. *Int. Security* 7: 132–56

13. Feiveson, H. A. 1978. Proliferation resistant nuclear fuel cycles. *Ann. Rev. Energy* 3: 357–94

14. Feiveson, H. A., Goldemberg, J. 1980. Denuclearization. *Econ. and Polit. Weekly* XV: 1546–48

15. Carbon Dioxide Assessment Committee. 1983. *Changing Climate.* Rep. of the Board on Atmospheric Sci. and Climate, Natl. Res. Council. Washington, DC: Natl. Acad. Sci.

16. Kellogg, W. W., Schware, R. 1981. *Climate Change and Society: Consequences of Increasing Atmospheric Carbon Dioxide.* Boulder, Colo: Westview

17. Bolin, B., Degens, E. T., Kempe, S., Ketner, P. 1979. *The Global Carbon Cycle.* Rep. of the Sci. Committee on Problems of the Environ. and the Int. Council of Sci. Unions. Chichester: Wiley

18. Biswas, M. R. 1978. *United Nations Conference on Desertification in Retrospect.* Laxenburg, Austria: Int. Inst. Applied Systems Anal.

19. Pimentel, D., Dazhong, W., Eigenbrode, S., Lang, H., Emerson, D. et al. 1985. *Deforestation: Interdependency of Fuelwood and Agriculture* Draft. Ithaca, NY: College of Agric. and Life Sci., Cornell Univ.

20. Energy Systems Program Group of the Int. Inst. Applied Systems Anal. 1981. *Energy in a Finite World—A Global Systems Analysis.* Cambridge, Mass: Ballinger, 834 pp.

21. Lovins, A. B., Lovins, L. H., Krause, F., Bach, W. 1981. *Energy Strategy for Low Climatic Risk.* Rep. for the German Fed. Environ. Agency

22. Chateau, B., Lapillone, B. 1977. *La prevision a long terme de la demande d'energie: Propositions metholologiques.* Paris: Editions du CNRS (In French)

23. Lapillone, B. 1978. *MEDEE-2: A Model for Long Term Energy Demand Evaluation.* IIASA Rep. RR-78-17. Laxenburg, Austria: Int. Inst. Applied Systems Anal.

24. Häfele, W., Rogner, H.-H. 1984. A technical appraisal of the energy scenarios? A rebuttal. *Policy Sci.* 17: 341–65

25. Keepin, B. 1984. A technical appraisal of the IIASA energy scenarios. *Policy Sci.* 17: 199–275

26. Khan, A. M., Hoelzl, A. 1982. *Evolution of Future Energy Demands till 2030 in Different World Regions: an Assessment Made for the Two IIASA Scenarios.* IIASA Rep. RR-82-14, 118 pp. Laxenburg, Austria: Int. Inst. Applied Systems Anal.

27. Williams, R. H., Dutt, G. S., Geller, H. S. 1983. Future energy savings in US housing. *Ann. Rev. Energy* 8: 269–332

28. Bloodworth, I. J., Bossanyi, E., Bowers, D. S., Crouch, E. A. C., Eden, R. J. et al. 1978. *World Energy Demand—The Full Report to the Conservation Commission of the World Energy Conference.* Guildford, United Kingdom and New York: IPC Sci. and Technol.

29. World Energy Conf. 1983. *Energy 2000–2020: World Prospects and Regional Stresses,* ed. J.-R. Frisch. London: Graham & Trotman. 259 pp.

30. Edmonds, J., Reilly, J. 1983. A longterm global energy-economic model of carbon dioxide release from fossil fuel use. *Energy Econ.* 5: 74–88

31. Edmonds, J., Reilly, J. 1983. Global energy production and use to the year 2050. *Energy-Int. J.* 8: 419–32

32. Edmonds, J. A., Reilly, J., Trabalka, J. R., Reichle, D. E. 1984. *An Analysis of Possible Future Atmospheric Retention of Fossil Fuel CO_2,* DOE Rep. DOE/OR/21400-1. Washington, DC: US Dept. Energy

33. Rose, D. J., Miller, M. M., Agnew, C. 1983. *Global Energy Futures and CO_2-Induced Climate Change.* Rep. MITEL 83-015. Cambridge, Mass: MIT

34. Energy Inf. Admin., US Dept. Energy. 1982. *Projected Costs of Electricity from Nuclear and Coal-Fired Power Plants, Vol. 1.* Rep. DOE/EIA-0356/1. Washington, DC. 145 pp.

35. Nordhaus, W. D., Yohe, G. W. 1983. Future paths of energy and carbon dioxide emissions. In *Changing Climate,* Rep. of the Carbon Dioxide Assessment Committee. Washington: Natl. Acad. Press. 496 pp.

36. Colombo, U., Bernardini, O. 1979. *A Low Energy Growth Scenario and the Perspectives for Western Europe.* Rep. prepared for the Commission on the European Communities. Panel on Low Energy Growth

37. Seidel, S., Keyes, S. 1983. *Can We Delay a Greenhouse Warming? The Effective-*

ness and Feasibility of Options to Slow a Build-Up of Carbon Dioxide in the Atmosphere. Rep. of the Strategic Studies Staff of the Office of Policy and Resources Management. Washington, DC: US Environ. Protection Agency

38. Goldemberg, J., Johansson, T. B., Reddy, A. K. N., Williams, R. H., *Energy for a Sustainable World*. Unpublished

39. von Hippel, F., Levi, B. 1983. Automotive fuel efficiency: the opportunity and the weakness of existing market incentives. *Resources and Conservation* 10:103–24

40. Hausman, J. 1976. Individual discount rates and the purchase and utilization of energy-using durables. *Bell J. Econ.* 10:33–54

41. Meier, A., Whittier, J. 1983. Consumer discount rates implied by purchases of energy-efficient refrigerators. *Energy-Int. J.* 8:957–62

42. Ruderman, H., Levine, M. D., McMahon, J. E. 1984. In *Doing Better: Setting an Agenda for the Second Decade*. Proc. A.C.E.E.E. 1984. Summer Study on Energy Efficiency in Buildings. Santa Cruz, pp. F208–18. Washington, DC: Am. Council for an Energy-Efficient Economy

43. Berndt, E. R. 1978. Aggregate energy, efficiency, and productivity measurement. *Ann. Rev. Energy*, 3:225–73

44. Steen, P., Johansson, T. B., Fredricksson, R., Bogren, E. 1981. *Energy—for what and how much*? Stockholm: Liber Forlag (In Swedish) Summarized in Johansson, T. B., Steen, P., Bogren, E., Fredricksson, R. 1983. Sweden beyond oil: the efficient use of energy. *Science* 219:355–61

45. Larson, E. D., Williams, R. H., Bienkowski, A. 1984. *Material Consumption Patterns and Industrial Energy Demand in Industrialized Countries*. Cent. Energy Environ. Studies, Rep. No. 174. Princeton, NJ: Princeton Univ.

46. Ross, M., Larson, E. D., Williams, R. H. 1985. *Energy Demand and Materials Flows in the Economy*. Cent. Energy Environ. Studies, Rep. No. 193. Princeton, NJ: Princeton Univ.

47. Dutt, G. S., Lavine, M., Levi, B., Socolow, R. 1982. *The Modular Retrofit Experiment: Exploring the House Doctor Concept*. Cent. Energy Environ. Studies, Rep. No. 130. Princeton, NJ: Princeton Univ.

48. Sinden, F. W. 1978. A two-thirds reduction in the space heat requirement of a Twin Rivers townhouse. *Energy and Buildings* 1:243–60

49. Johansson, T. B., Steen, P. 1985. *Perspectives on Energy—On Possibilities and Uncertainties in the Energy Transition*. Stockholm: Liber Forlag (In Swedish)

50. Wall, L. W., Flaherty, J. 1983. In *What Works: Documenting Energy Conservation in Buildings*, ed. J. Harris, C. Blumstein. Proc. 2nd Summer Study on Energy Efficient Buildings, pp. 257–75. Washington, DC: Am. Council for an Energy-Efficient Economy

51. Energy Inf. Admin., US Dept. Energy. 1983. *Non-Residential Buildings Energy Consumption Survey: 1979 Consumption and Expenditures; Part 2: Steam, Fuel Oil, LPG, and All Fuels*. Washington, DC: USGPO

52. Solar Energy Res. Inst. 1981. *A New Prosperity: Building a Sustainable Energy Future*. Andover, Mass: Brick House. 454 pp.

53. Int. Energy Agency. 1984. *Energy Balances of OECD Countries, 1970/1982*. Paris: Organ. Econ. Coop. & Dev.

54. Volvo. 1984. *Volvo LCP2000—Light Component Project*. Gothenburg, Sweden: Volvo Personvagnar AB

55. Horton, E. J., Compton, W. D. 1984. Technological trends in automobiles. *Science* 225:587–93

56. Sekar, R. R., Kamo, R., Wood, J. C. 1984. Advanced adiabatic diesel engines for passenger cars. In *Adiabatic Engines: Worldwide Review*, pp. 79–87. SAE Publication SP-571. Warrendale, Pa: Society of Automotive Engineers

57. Office of Mobile Source Air Pollution Control, US Environ. Protection Agency. 1982. *Preliminary Perspective on Pure Methanol Fuel for Transportation*. Ann Arbor, Mich: US Environ. Protection Agency

58. Neitz, A., Chmela, F. 1980. Results of MAN-FM diesel engines operating on straight alcohol fuels, in *Proc. Int. Symp. on Alcohol Fuels Technol*. Gueruja, São Paulo, Brazil

59. von Hippel, F. 1981. *US Transportation Energy Demand*. Cent. Energy Environ. Studies. Rep. No. 111. Princeton, NJ: Princeton Univ.

60. Ross, M. H. 1984. Industrial energy conservation. *Nat. Resour. J.* 24:369–404

61. Berg, C. A. 1979. *Energy Conservation in Industry: the Present Approach, the Future Opportunities*. Rep. prepared for the President's Council on Environ. Quality. Washington, DC

62. Solow, R. M. 1957. Technical change and the aggregate production function. *Rev. Econ. and Stat.* 34:312–20

63. Gyftopoulos, E. P., Lazaridis, L. J., Widmer, T. F. 1974. *Potential Fuel Effectiveness in Industry.* Rep. to the Energy Policy Project of the Ford Foundation, Cambridge, Mass: Ballinger

64. Hane, G. J., Hauser, S. G., Blahnik, D. E., Eakin, D. E., Gurwell, W. E. et al. 1983. *A Preliminary Overview of Innovative Industrial Materials Processes.* Rep. PNL-4505, UC-95F prepared for the US Dept. Energy, Richland, Wash: Pacific Northwest Lab.

65. Birchall, J. D., Kelly, A. 1983. New inorganic materials. *Sci. Am.* 248:88–95

66. Williams, R. H. 1978. Industrial cogeneration. *Ann. Rev. Energy* 3:313–56

67. Ladomatos, N., Lucas, N. J. D., Murgatroyd, W. 1978. Industrial energy use—I: power losses in electrically driven machinery. *Int. J. Energy Res.* 2:179–96

68. Ben-Daniel, D. J., David, E. E. Jr. 1979. Semiconductor alternating-current motor drives and energy conservation. *Science* 206:773–76

68a. Geller, H. S. 1985. Ethanol fuel from sugar cane in Brazil. *Ann. Rev. Energy* 10:135–64

69. Williams, R. H. 1985. *Potential Roles for Bio-Energy in an Energy-Efficient World.* Cent. Energy Environ. Studies. Rep. No. 183. Princeton, NJ: Princeton Univ.

70. Foley, G., Moss, P. 1983. *Improved Cooking Stoves in Developing Countries.* Earthscan Technical Rep. No. 2. London: Int. Inst. for Environ. & Dev.

71. Lokras, S. S., Sudhakar Babu, D. S., Bhogale, S., Jagadish, K. S., Kumar, R. 1983. *Studies on Cookstoves Part I: Development of an Improved Three-Pan Cookstove.* Cent. for the Application of Sci. & Tech. to Rural Areas (ASTRA). Bangalore, India: Indian Inst. Sci.

72. Baldwin, S., Dutt, G. S., Geller, H. S., Ravindranath, N. H. 1985. *Improved Fuelwood Stove: Glimmerings of Success.* Cent. Energy Environ. Studies, Rep. No. 184. Princeton, NJ: Princeton Univ.

73. Shukla, K. C., Hurley, J. R. 1983. *Development of an Efficient, Low NO_x Domestic Gas Range Cook Top.* Rep. GRI-81/0201. Chicago, Ill: Gas Res. Inst.

74. Reddy, A. K. N. 1981. A strategy for resolving India's oil crisis. *Current Sci.* 50:50–53

75. Leach, G., Lewis, C., Romig, F., van Buren, A., Foley, G. 1979. *A Low Energy Strategy for the United Kingdom.* The Int. Inst. for Environ. & Dev.

London: Sci. Rev.

76. Krause, F., Bossel, H., Muller-Reissman, K.-F. 1980. *Energie-Wende, Wachstum und Wohlstand ohne Erdol und Uran.* Frankfurt: S. Fisher (In German)

77. Norgard, J. S. 1979. *Husholdninger og Energi.* Copenhagen: Polyteknisk Forlag. (In Danish)

78. Olivier, D., Miall, H., Nectoux, F., Opperman, M. 1983. *Energy-Efficient Futures: Opening the Solar Option.* Earth Resources Res. Ltd. London: Blackrose

79. Brooks, D. B., Robinson, J. B., Torrie, R. D. 1983. *2025: Soft Energy Futures for Canada.* Prepared by Friends of the Earth for the Dept. Energy, Mines & Resources and Environment. Canada: Dept. Energy, Mines & Resources

80. Goss, R. M., Strang, D. J., Partridge, C. L., Ball, J. V., Snoswell, P. D. et al. 1983. *BP Statistical Review of World Energy 1982.* London: The British Pet. Co.

80a. Goldemberg, J., Williams, R. H. 1985. *The Economics of Energy Conservation in Developing Countries: the Consumer Versus the Societal Perspective. A Case Study of the Electrical Sector in Brazil.* Cent. Energy Environ. Studies, Rep. No. 189. Princeton, NJ: Princeton Univ.

81. Dept. Int. Econ. Soc. Affairs. 1981. *World Population Prospects As Assessed in 1980.* New York: United Nations

82. Nordhaus, W. D. 1977. The demand for energy: an international perspective. In *International Studies of the Demand for Energy,* ed. W. D. Nordhaus, p. 273. Amsterdam: North-Holland

83. Energy Modeling Forum. 1980. *Aggregate Elasticity of Energy Demand.* Vol. 1. EMF Rep. 4. Stanford, Calif: Stanford Univ. 50 pp.

84. Doblin, C. P. 1982. *The Growth of Energy Consumption and Prices in the USA, FRG, France, and the UK, 1950–1980.* Rep. RR-82-18. Laxenburg, Austria: Int. Inst. Applied Systems Anal.

85. Energy Inf. Admin. 1984. *State Energy Price and Expenditure Report 1970–1981.* Rep. DOE/EIA-0376(81). Washington, DC: US Dept. Energy

86. Sandberg, R. O., Braun, G. 1984. *Economics of Reprocessing—US Context.* Paper presented at the Am. Nuclear Society's Topical Meet. on Financial and Econ. Bases for Nuclear Power. Washington, DC

87. Fed. Inst. Geosci. and Natural Resources. 1980. *Survey of Energy Re-*

sources 1980. Prepared for the 11th World Energy Conference, Munich, 8–12 Sept. 1980. London: World Energy Conference
88. Chao, H., Peck, S. 1982. Coordination of OECD oil import policies: a gaming approach. *Energy—Int. J.* 7:213–20
89. United Nations. 1983. *1981 Yearbook of World Energy Statistics.* New York: United Nations
90. Hall, D. O., Barnard, G. W., Moss, P. A. 1982. *Biomass for Energy in Developing Countries.* Oxford: Pergamon
91. Am. Paper Inst. 1981. *Statistics of Paper, Paperboard, and Woodpulp.* New York
92. Energy Inf. Admin. US Dept. Energy. 1982. *Housing Characteristics, 1980,* a report contributing to the Residential Energy Consumption Survey
93. Int. Energy Agency. 1982. *Energy Balances of OECD Countries, 1976/1980.* Paris
94. Ribot, J. C., Rosenfeld, A. H., Flouquet, F., Luhrsen, W. 1983. In *What Works: Documenting Energy Conservation in Buildings.* ed. J. Harris, C. Blumstein. Proc. 2nd Summer Study on Energy Efficient Buildings, pp. 242–56. Washington, DC: Am. Council for an Energy Efficient Economy
95. Schipper, L. 1982. *Residential Energy Use and Conservation in Sweden.* Rep.

LBL-14147. Berkeley, Cal: Lawrence Berkeley Lab.
96. Norford, L. K. 1984. *An Analysis of Energy Use in Office Buildings: the Case of ENERPLEX.* PhD thesis. Dept. Aerospace and Mechanical Engineering, Princeton, NJ: Princeton Univ.
97. Welmer, K. 1981. *A Method to Make Use of a Building's Heat Storage Capacity in a Controlled Manner to Save Energy.* BFR Rep. R104: 1981. Stockholm: Swedish Building Energy Res. Council (In Swedish)
98. Volkswagen. 1981. '*Car 2000' Research Vehicle of Volkswagenwerk AG*
99. Fawcett, S. L., Swain, J. C. 1983. *Prospectus for a Consumer Demonstration of a 100 MPG Car.* Columbus, Ohio: Battelle Memorial Inst.
100. Demeo, E. A., Taylor, R. W. 1984. Solar photovoltaic power systems: an electric utility R & D perspective. *Science* 224: 245–51
101. Energy Inf. Admin. US Dept. Energy. 1984. *Annual Energy Outlook 1983, with Projections to 1995.* Rep. DOE/EIA-0383(83). Washington, DC: USGPO
102. Marine and Industrial Projects Dept. (General Electric Co.), Engineering Dept. (Pacific Gas & Electric Co.). 1984. *Scoping Study: LM5000 Steam-Injected Gas Turbine.* San Francisco: Pacific Gas & Electric Co.

SUBJECT INDEX

A

Abu Dhabi
  natural gas exports of, 474
  natural gas trade partnerships
    of, 467-68
Acetylene
  amorphous silicon alloy
    formation and, 11
Addis Ababa, Ethiopia
  energy policy in, 130
Advanced Boiling Water Reactor, 440-41
Advanced Pressurized Water
    Reactor, 437-39
Afghanistan
  natural gas exports of, 474
  natural gas production in, 465-66
  natural gas production/
    consumption in, 466-67
  natural gas reserves in, 464-65
  natural gas trade partnerships
    of, 467-68
Africa
  biomass energy resources in,
    408
  cooking fuels used in, 115
  deforestation in, 407, 422
  forest resources in, 409
  fuelwood consumption in, 411
  reforestation in, 424
  urbanization in, 109
African Rift Valley
  methane in, 69
Aggregate demand
  price elasticity of, 335-38
Agricultural chemicals
  production of
    fuel consumption in, 187
Agriculture
  commercial energy and, 217
  industrial energy and, 641
  mobile engines in, 173
Air conditioning
  international penetration of,
    396
  residential energy use and,
    367
Alcohol fuels
  see Ethanol fuels
Aldehyde pollution
  ethanol fuels and, 156-57
Aluminum
  electrolysis of
    efficiency improvements in,
      176

Aluminum industry
  electricity consumption in,
    176
American Council for an Ener-
    gy-Efficient Economy, 311
American Society of Heating,
    Air Conditioning, and
    Refrigeration Engineers,
    578-79
American Society of Mechanical
    Engineers Pressure Vessel
    Code, 443
Ammonia
  amorphous silicon alloy
    formation and, 10
  production of
    fuel consumption in, 184-85
Amorphous semiconductors
  doping of, 3
  forbidden energy gap of, 7
Amorphous silicon
  carrier transport mechanism
    of, 8
  deposition of, 8-12
  device structures, 12-17
  hydrogenated, 2-3
  optical absorption coefficient
    of, 5
  substitutional doping of, 8
Amorphous silicon films
  photoconductivity of, 2
Amorphous silicon hydride
  optical properties of, 3
Amorphous silicon : hydrogen
    absorption curve
  factors influencing, 6
Amorphous silicon : hydrogen so-
    lar cells
  performance of, 16
Amorphous silicon solar cells, 1-31
  collection efficiencies for, 23
  current-voltage characteristics
    of, 20-21
  device characteristics of, 17-26
  heterojunction, 14-15
  optical absorption properties
    of, 3-6
  photogenerated carrier trans-
    port in, 6-8
  p-i-n type, 14-15
  spectral response of, 21-24
  stacked junction, 15-17
  voltage-dependent current col-
    lection in, 24-26

Amorphous silicon solar cell
    technology
  advantages of, 2
  economic factors in, 26-28
  United States policy and, 29-30
  worldwide efforts in, 28-29
Anadarko Basin
  deep domains of
    gas abundance in, 71
Andaman Islands
  oil fields of, 69
Angola
  petroleum exploration in, 220-22
    contractual agreements gov-
      erning, 230
    political risk and, 236
  petroleum resource base of,
    223
Animal dung
  use as fuel, 420
Appliances
  see Electric appliances
Arab oil embargo, 520, 558,
    565-67, 592
ARCO Solar 6.5 MW Carrisa
    Plains project
  estimated cost of, 27-28
ARCO Solar thin-film silicon so-
    lar cell
  current-voltage characteristics
    of, 21
  quantum efficiency of, 22,
    25
  spectral response of, 25
Argentina
  petroleum exploration in, 220-22
    contractual agreements gov-
      erning, 230
  petroleum resource base of,
    222-24
Argon
  presence in gas fields, 70
Argonne National Laboratory,
    457
Arizona
  oil and gas fields of
    helium occurrence in, 70
Arizona Public Service Com-
    pany, 439
ASDEX-U tokamak, 86, 89
ASEA/ATOM, 445-50
Asia
  biomass energy resources in,
    408

CUMULATIVE INDEXES

CONTRIBUTING AUTHORS, VOLUMES 1–10

A

Ahearne, J. F., 8:355–84
Alpert, S. B., 1:87–99
Alterman, J., 3:201–24
Anderson, B. N., 3:57–100
Aperjis, D. G., 9:179–98
Auer, P. L., 1:685–713

B

Baldwin, S., 10:407–30
Barnes, R., 8:193–245
Beaujean, J. M., 2:153–70
Beckerman, S., 3:1–28
Berg, C. A., 1:519–34
Berg, D., 1:345–68
Berndt, E. R., 3:225–73
Birk, J. R., 5:61–88
Block, H., 2:455–97
Bodle, W. W., 1:65–85
Bohi, D. R., 7:37–60; 9:105–54
Bolin, B., 2:197–226
Bolton, J. R., 4:353–401
Boshier, J. F., 9:51–79
Boyer, L. L., 7:201–19
Bradley, S., 7:61–85
Broadman, H. G., 10:217–50
Brondel, G., 2:343–64
Brooks, H., 4:1–70
Brown, H., 1:1–36; 5:173–240
Brown, K. C., 8:509–30
Brown, N. L., 5:389–413
Brownell, W. A., 9:229–62
Budnitz, R. J., 1:553–80
Bullard, C. W., 6:199–232

C

Carver, J. A. Jr., 1:727–41
Charpentier, J. P., 2:153–70
Chen, R. S., 5:107–40
Cicchetti, C. J., 4:231–58
Colglazier, E. W. Jr., 8:415–49
Colombo, U., 9:31–49
Comar, C. L., 1:581–600
Constable, R. W., 1:369–89
Corey, G. R., 6:417–43
Costello, D., 5:335–56
Craig, P. P., 1:535–51; 10:557–88
Crews, R., 5:141–72

Cuevas, J., 8:95–112
Curtis, H. B., 2:227–38

D

Darmstadter, J., 1:535–51; 3:201–24; 7:261–92
Deese, D. A., 8:415–49
Denning, R. S., 10:35–52
Dickler, R. A., 7:221–59
Doctor, R. D., 7:329–69
Douglas, D. L., 5:61–88
Doub, W. O., 1:715–25
Dunkerley, J., 3:201–24
Dutt, G. S., 8:269–332

E

Eaton, D., 1:183–212
Eckbo, P. L., 4:71–98
Eibenschutz, J., 8:95–112
El Mallakh, R., 2:399–415

F

Fallows, S., 3:275–311
Farhar, B. C., 5:141–72
Fay, J. A., 5:89–105
Feiveson, H. A., 3:357–94
Fesharaki, F., 6:267–308; 10:463–94
Fickett, A. P., 1:345–68
Fisher, A. C., 7:1–35
Flaim, S., 6:89–121
Friedman, K. M., 4:123–45

G

Geller, H. S., 8:269–332; 10:135–64
Giraud, A. L., 8:165–91
Glassey, C. R., 6:445–82
Godbole, S. P., 9:427–45
Gold, T., 10:53–78
Goldemberg, J., 7:139–74; 10:613–88
Goldschmidt, V. W., 9:447–72
Golueke, C. G., 1:257–77
Gould, R. R., 9:529–59
Gray, J. E., 10:589–612
Greenberger, M., 4:467–500
Greene, D., 8:193–245

Gregory, D. P., 1:279–310
Grenon, M., 2:67–94
Griffin, K., 10:285–316
Gururaja, J., 2:365–86
Gyftopoulos, E. P., 7:293–327

H

Häfele, W., 2:1–30
Hall, C. A. S., 3:395–475
Hall, D. O., 4:353–401
Hambraeus, G., 2:417–53
Hamrin, J., 7:329–69
Harris, D. P., 4:403–32
Harte, J., 3:101–46
Hartman, R. S., 4:433–66
Helliwell, J. F., 4:175–229
Hendrie, J. M., 1:663–83
Heron, J. J., 8:137–63
Hertzmark, D., 6:89–121
Hewett, C. E., 6:139–70
Hibbard, W. R. Jr., 4:147–74
High, C. J., 6:139–70
Hill, G. R., 1:37–63
Hirst, E., 8:193–245
Hoffman, K. C., 1:423–53
Holder, G. D., 9:427–45
Holdren, J. P., 1:553–80; 5:241–91
Hollander, J. M., 4:1–70
Howarth, R., 3:395–475
Hubbard, R. G., 10:515–56
Huntington, H. G., 10:317–40

J

Jacoby, H. D., 4:259–311
Jassby, A., 3:101–46
Johansson, T. B., 10:613–88
Jones, C. O., 4:99–121
Just, J., 4:501–36

K

Kahane, A., 10:341–406
Kahn, E., 4:313–52
Kalhammer, F. R., 1:311–43
Kalt, J. P., 5:1–32
Kamath, V. A., 9:427–45
Ketoff, A., 10:341–406
Kouts, H., 8:385–413
Kratzer, M. B., 10:589–612

705

CHAPTER TITLES VOLUMES 1–10

Annual Reviews Inc. ORDER FORM

A NONPROFIT SCIENTIFIC PUBLISHER

4139 El Camino Way, Palo Alto, CA 94306-9981, USA • (415) 493-4400

Annual Reviews Inc. publications are available directly from our office by mail or telephone (paid by credit card or purchase order), through booksellers and subscription agents, worldwide, and through participating professional societies. Prices subject to change without notice.

- **Individuals:** Prepayment required on new accounts by check or money order (in U.S. dollars, check drawn on U.S. bank) or charge to credit card — American Express, VISA, MasterCard.
- **Institutional buyers:** Please include purchase order number.
- **Students:** $10.00 discount from retail price, per volume. Prepayment required. Proof of student status must be provided (photocopy of student I.D. or signature of department secretary is acceptable). Students must send orders direct to Annual Reviews. Orders received through bookstores and institutions requesting student rates will be returned.
- **Professional Society Members:** Members of professional societies that have a contractual arrangement with Annual Reviews may order books through their society at a reduced rate. Check with your society for information.

Regular orders: Please list the volumes you wish to order by volume number.
Standing orders: New volume in the series will be sent to you automatically each year upon publication. Cancellation may be made at any time. Please indicate volume number to begin standing order.
Prepublication orders: Volumes not yet published will be shipped in month and year indicated.
California orders: Add applicable sales tax.
Postage paid (4th class bookrate/surface mail) **by Annual Reviews Inc.** Airmail postage extra.

ANNUAL REVIEWS SERIES		Prices Postpaid per volume USA/elsewhere	Regular Order Please send:	Standing Order Begin with:
			Vol. number	Vol. number
Annual Review of ANTHROPOLOGY (Prices of Volumes in brackets effective until 12/31/85)				
[Vols. 1-10	(1972-1981)	$20.00/$21.00]		
[Vol. 11	(1982)	$22.00/$25.00]		
[Vols. 12-14	(1983-1985)	$27.00/$30.00]		
Vols. 1-14	(1972-1985)	$27.00/$30.00		
Vol. 15	(avail. Oct. 1986)	$31.00/$34.00	Vol(s). _____	Vol. _____
Annual Review of ASTRONOMY AND ASTROPHYSICS (Prices of Volumes in brackets effective until 12/31/85)				
[Vols. 1-2, 4-19	(1963-1964; 1966-1981)	$20.00/$21.00]		
[Vol. 20	(1982)	$22.00/$25.00]		
[Vols. 21-23	(1983-1985)	$44.00/$47.00]		
Vols. 1-2, 4-20	(1963-1964; 1966-1982)	$27.00/$30.00		
Vols. 21-23	(1983-1985)	$44.00/$47.00		
Vol. 24	(avail. Sept. 1986)	$44.00/$47.00	Vol(s). _____	Vol. _____
Annual Review of BIOCHEMISTRY (Prices of Volumes in brackets effective until 12/31/85)				
[Vols. 30-34, 36-50	(1961-1965; 1967-1981)	$21.00/$22.00]		
[Vol. 51	(1982)	$23.00/$26.00]		
[Vols. 52-54	(1983-1985)	$29.00/$32.00]		
Vols. 30-34, 36-54	(1961-1965; 1967-1985)	$29.00/$32.00		
Vol. 55	(avail. July 1986)	$33.00/$36.00	Vol(s). _____	Vol. _____
Annual Review of BIOPHYSICS AND BIOPHYSICAL CHEMISTRY (Prices of Vols. in brackets effective until 12/31/85) *(Formerly* Annual Review of Biophysics and Bioengineering)				
[Vols. 1-10	(1972-1981)	$20.00/$21.00]		
[Vol. 11	(1982)	$22.00/$25.00]		
[Vols. 12-14	(1983-1985)	$47.00/$50.00]		
Vols. 1-11	(1972-1982)	$27.00/$30.00		
Vols. 12-14	(1983-1985)	$47.00/$50.00		
Vol. 15	(avail. June 1986)	$47.00/$50.00	Vol(s). _____	Vol. _____
Annual Review of CELL BIOLOGY				
Vol. 1	(1985)	$27.00/$30.00		
Vol. 2	(avail. Nov. 1986)	$31.00/$34.00	Vol(s). _____	Vol. _____
Annual Review of COMPUTER SCIENCE				
Vol. 1	(avail. late 1986)	**Price not yet established**	Vol. _____	Vol. _____
Annual Review of EARTH AND PLANETARY SCIENCES (Prices of Volumes in brackets effective until 12/31/85)				
[Vols. 1-9	(1973-1981)	$20.00/$21.00]		
[Vol. 10	(1982)	$22.00/$25.00]		
[Vols. 11-13	(1983-1985)	$44.00/$47.00]		
Vols. 1-10	(1973-1982)	$27.00/$30.00		
Vols. 11-13	(1983-1985)	$44.00/$47.00		
Vol. 14	(avail. May 1986)	$44.00/$47.00	Vol(s). _____	Vol. _____

ANNUAL REVIEWS SERIES		Prices Postpaid per volume USA/elsewhere	Regular Order Please send:	Standing Order Begin with:

Annual Review of **ECOLOGY AND SYSTEMATICS** (Prices of Volumes in brackets effective until 12/31/85)

[Vols. 1-12	(1970-1981)	**$20.00/$21.00**]		
[Vol. 13	(1982)	**$22.00/$25.00**]		
[Vols. 14-16	(1983-1985)	**$27.00/$30.00**]		
Vols. 1-16	(1970-1985)	**$27.00/$30.00**		
Vol. 17	(avail. Nov. 1986)	**$31.00/$34.00**	Vol(s). _____	Vol. _____

Annual Review of **ENERGY** (Prices of Volumes in brackets effective until 12/31/85)

[Vols. 1-6	(1976-1981)	**$20.00/$21.00**]		
[Vol. 7	(1982)	**$22.00/$25.00**]		
[Vols. 8-10	(1983-1985)	**$56.00/$59.00**]		
Vols. 1-7	(1976-1982)	**$27.00/$30.00**		
Vols. 8-10	(1983-1985)	**$56.00/$59.00**		
Vol. 11	(avail. Oct. 1986)	**$56.00/$59.00**	Vol(s). _____	Vol. _____

Annual Review of **ENTOMOLOGY** (Prices of Volumes in brackets effective until 12/31/85)

[Vols. 9-16, 18-26	(1964-1971; 1973-1981)	**$20.00/$21.00**]		
[Vol. 27	(1982)	**$22.00/$25.00**]		
[Vols. 28-30	(1983-1985)	**$27.00/$30.00**]		
Vols. 9-16, 18-30	(1964-1971; 1973-1985)	**$27.00/$30.00**		
Vol. 31	(avail. Jan. 1986)	**$31.00/$34.00**	Vol(s). _____	Vol. _____

Annual Review of **FLUID MECHANICS** (Prices of Volumes in brackets effective until 12/31/85)

[Vols. 1-5, 7-13	(1969-1973; 1975-1981)	**$20.00/$21.00**]		
[Vol. 14	(1982)	**$22.00/$25.00**]		
[Vols. 15-17	(1983-1985)	**$28.00/$31.00**]		
Vols. 1-5, 7-17	(1969-1973; 1975-1985)	**$28.00/$31.00**		
Vol. 18	(avail. Jan. 1986)	**$32.00/$35.00**	Vol(s). _____	Vol. _____

Annual Review of **GENETICS** (Prices of Volumes in brackets effective until 12/31/85)

[Vols. 1-15	(1967-1981)	**$20.00/$21.00**]		
[Vol. 16	(1982)	**$22.00/$25.00**]		
[Vols. 17-19	(1983-1985)	**$27.00/$30.00**]		
Vols. 1-19	(1967-1985)	**$27.00/$30.00**		
Vol. 20	(avail. Dec. 1986)	**$31.00/$34.00**	Vol(s). _____	Vol. _____

Annual Review of **IMMUNOLOGY**

Vols. 1-3	(1983-1985)	**$27.00/$30.00**		
Vol. 4	(avail. April 1986)	**$31.00/$34.00**	Vol(s). _____	Vol. _____

Annual Review of **MATERIALS SCIENCE** (Prices of Volumes in brackets effective until 12/31/85)

[Vols. 1-11	(1971-1981)	**$20.00/$21.00**]		
[Vol. 12	(1982)	**$22.00/$25.00**]		
[Vols. 13-15	(1983-1985)	**$64.00/$67.00**]		
Vols. 1-12	(1971-1982)	**$27.00/$30.00**		
Vols. 13-15	(1983-1985)	**$64.00/$67.00**		
Vol. 16	(avail. August 1986)	**$64.00/$67.00**	Vol(s). _____	Vol. _____

Annual Review of **MEDICINE** (Prices of Volumes in brackets effective until 12/31/85)

[Vols. 1-3, 5-15, 17-32	(1950-52; 1954-64; 1966-81)	**$20.00/$21.00**]		
[Vol. 33	(1982)	**$22.00/$25.00**]		
[Vols. 34-36	(1983-1985)	**$27.00/$30.00**]		
Vols. 1-3, 5-15, 17-36	(1950-52; 1954-64; 1966-85)	**$27.00/$30.00**		
Vol. 37	(avail. April 1986)	**$31.00/$34.00**	Vol(s). _____	Vol. _____

Annual Review of **MICROBIOLOGY** (Prices of Volumes in brackets effective until 12/31/85)

[Vols. 18-35	(1964-1981)	**$20.00/$21.00**]		
[Vol. 36	(1982)	**$22.00/$25.00**]		
[Vols. 37-39	(1983-1985)	**$27.00/$30.00**]		
Vols. 18-39	(1964-1985)	**$27.00/$30.00**		
Vol. 40	(avail. Oct. 1986)	**$31.00/$34.00**	Vol(s). _____	Vol. _____

Annual Review of **NEUROSCIENCE** (Prices of Volumes in brackets effective until 12/31/85)

[Vols. 1-4	(1978-1981)	**$20.00/$21.00**]		
[Vol. 5	(1982)	**$22.00/$25.00**]		
[Vols. 6-8	(1983-1985)	**$27.00/$30.00**]		
Vols. 1-8	(1978-1985)	**$27.00/$30.00**		
Vol. 9	(avail. March 1986)	**$31.00/$34.00**	Vol(s). _____	Vol. _____

DATE DUE

GOSHEN COLLEGE - GOOD LIBRARY

3 9310 01076940 2